THE PREHISTORY OF METALLURGY IN THE BRITISH ISLES

THE PREHISTORY OF METALLURGY IN THE BRITISH ISLES

R. F. Tylecote

THE INSTITUTE OF METALS
LONDON

Book 506
Published by The Institute of Metals
1 Carlton House Terrace London SWlY 5DB

and
The Institute of Metals
North American Publications Center
Old Post Road, Brookfield VT 05036
U S A

First published in hardback 1986

Reprinted in paperback 1990

British Library Cataloguing in Publication Data

Tylecote, R. F.

The prehistory of metallurgy in the British Isles.
1. Metallurgy—Great Britain—History
2. Mineral industries, Prehistoric—Great Britain
I. Title
669'.009361 TN57

ISBN 0-901462-96-9

Typeset by Fakenham Photosetting Ltd,
Fakenham, Norfolk

Printed and bound by Antony Rowe Ltd,
Eastbourne

Contents

List of Figures

Chapter 5

Chapter 6

Chapter 7

Preface

This book is basically a revised version of my 1962 book, *Metallurgy in Archaeology*, published by Edward Arnold, London. It treats the subject matter in much the same way (i.e. more or less chronologically), but the epilogue has gone for the reason that much of the work suggested in it has already been carried out. It has been replaced by a chapter on fuels and fuel ashes, which seemed to be an aspect that did not receive sufficient attention in the original version.

The object of the first book was to gather together the wealth of information on metallurgical aspects of archaeology contained mainly in the appendixes to excavation reports. These appendixes had been contributed by metallurgists, chemists and other specialists in techniques. Today, efforts are being made to integrate the scientific work and to build it more into the body of the archaeological report so that its general relevance may be discussed and conclusions made.

A large proportion of sites excavated produce industrial material such as slags, furnaces and other metallurgical debris. However, up to the present time few excavations have been carried out with the prime object of elucidating metallurgical techniques. It is to be hoped that in the future many more excavations of this type will be made.

Since this is a subject of interest to metallurgists as well as archaeologists, the work has not been treated as a textbook on metallurgy for the archaeologist but is addressed to both archaeologists and metallurgists. In doing this I was only too well aware that I was likely to satisfy neither group, making the subject too technical for the archaeologist and not sufficiently fundamental for the metallurgist. I have tried to meet both these objections as far as possible, and I hope I shall be forgiven for any shortcomings. To assist the reader a glossary of technical terms will be found in the appendixes.

It is, of course, obvious that a book such as this cannot be written without the active help of friends and colleagues, and it is my happy duty to acknowledge this. First, the help of my wife who has done an immense amount of the research required, visited museums, and checked the references and tables. Without her this book would never have been written.

Others that I would especially like to thank for helpful discussion are: Prof. L. Alcock, Dr H. J. Axon, Prof. Dr H-G. Bachmann, Dr. H. Battey, Justine Bayley, Neil Beagrie, Leo Biek, W. Britnell, Dr Dennis Britton, Colin Burgess, D. S. Butler, Jim Cherry, Dr H. F. Cleere, R. E. Clough, Dr P. T. Craddock, D. W. Crossley, C. M. Daniels, M. Davies-Shiel, B. Earl, L. N. W. Flanagan, Dr Noel Gale, Brian Gilmour, Dr T. A. P. Greeves, H. W. M. Hodges, Margaret and Tom Jones, Dr H. McKerrell, Alan McWhirr, Donald Maxwell, James Money, Dr J. P. Northover, Dr W. A. Oddy, Dr George Parker, Prof. Dr J. Piaskowski, Dr R. Pleiner, Prof. Bruce Proudfoot, Drs B. and J. Raftery, Prof. Beno Rothenberg, Dr C. J. Salter, Dr Brian Scott, Dr Colin Shell, Prof. Cyril S. Smith, R. B. K. Stevenson, C. F. Tebbutt, W. N. Terry, Ing. E. Tholander, Dr A. E. Wraith, and Prof. Dr U. Zwicker.

In the course of my writing I have received generous help from the following correspondents all of whom I sincerely thank: Joan and Roy Day, Prof. Barri Jones, G. Stainthorpe, John Steane, L. L. Ketteringham, Martin Biddle, Colin Brewer, Stafford Linsley, Roger Hetherington, Dr W. H. Manning, H. J. Case, L. V. Grinsell, K. S. Painter, J. S. Wacher, and Dr Dennis Britton who supplied the analyses of the Todmorden Bronzes. These were carried out at the Research Laboratory for the History of Art and Archaeology at Oxford with the aid of a grant from the British Academy.

I would also like to thank the members of the Historical Metallurgy Society, who have given me so much assistance in all sorts of ways and allowed me to use their drawings as a basis for many of the Figures included in the book. Many of the drawings have been taken from the Journal of the Historical Metallurgy Society.

I am exceedingly grateful to the Cumberland and Westmorland Antiquarian and Archaeological Society for permission to reproduce Figs. 38 and 148; the Chemical Society for Fig. 1; the Joint Museum of Antiquities of the Society of Antiquaries of Newcastle upon Tyne and the University of Newcastle upon Tyne for Figs. 67 and 105; the Society of Antiquaries of Newcastle upon Tyne for Fig. 36; the Newcomen Society for Fig. 23; Messrs Garraway Ltd for Fig. 48; Metal Treatment and Drop Forging for Fig. 71; the National Museum of Ireland for Figs. 2, 41, 54, 57, 61 and 63; the Royal Society of Antiquaries of Ireland for Fig. 41; the Society of Antiquaries of London for Figs. 28, 39 and 53; *Norfolk Archaeology* for Fig. 100; *Ulster Journal of Archaeology* for Fig. 45; the British Museum

(Natural History) for Fig. 6; The Metals Society for Figs. 72, 79, 94, 127, 128 and 154; the Public Record Office for permitting the reproduction of Figs. 24 and 25 (Crown Copyright reserved), Dr A. Raistrick for Fig. 23; Messrs Johnston and Bacon Ltd for the outline maps used in Figs. 7 and 22; L. L. Ketteringham for her drawings used in Figs. 130, 131, 132 (Fig. 130 was based on an MS in the Bodleian Library, Oxford and Fig. 131 on an MS in the British Museum (Sloane 3983)); the Society of Medieval Archaeology for permission to reproduce Figs. 119, 121, 123 and 124; the Council for British Archaeology for permission to reproduce Figs. 89, 130, 131 and 142; the Society for Post-Medieval Archaeology for Fig. 158; Messrs Routledge and Kegan Paul for Fig. 144; Cambridge University Press for Fig. 55 and Fig. 145 (Crown Copyright reserved); David Crossley and the Royal Archaeological Institute for Figs. 141, 152 and 159; the National Museum of Antiquities of Scotland for Fig. 87; the Trustees of the British Museum for Figs. 6, 9, 14, 27, 33, 49, 51 and 53; the Ulster Museum for Fig. 65; Norfolk Museums Service for Fig. 81; the National Museum of Wales and G. C. Boon for Fig. 10; *Technology and Culture* for Fig. 143; the *British Welding Journal* for Fig. 111; South Shields Corporation for Figs. 112 and 114; the Surrey Archaeological Society for Fig. 132; Colin Shell and Dr Joan Taylor for Figs. 16 and 55.

Many of the analyses have been specially made for this book. For this work I would like to thank the United Steel Companies (through the good offices of Dr H. F. Cleere); Prof. Dr H-G. Bachmann of Degussa, Hanau, West Germany, and R. E. Clough of the Institute of Archaeology, London, using equipment under the care of Ian Young in the Dept. of Geology, University College; and the XRF unit in the Dept. of Geology, Bedford College, under the direction of Dr G. Mariner.

Finally, I would like to thank George Finch of the Library, University of Newcastle upon Tyne, and Peter Dorrell and Stuart Laidlaw of the Institute of Archaeology, London for their admirable work on some of the photographs.

R. F. Tylecote

Institute of Archaeology
London
August 1985

Introduction

The arrangement of this book arises from the fact that in the early periods only non-ferrous metals were worked, and it is therefore natural to consider each metal in turn and to discuss its use throughout the whole period. This discussion finally culminates in a chapter on metal fabrication. With the coming of iron, it is convenient to consider the iron ore resources of the country, and the principles of extraction. After this the subject is broken up into periods and the main changes are not those of technique but of scale and organization. The penultimate chapter introduces the blast furnace. Since the period of history now begins, this process is discussed only briefly and with special reference to its products.

It will be noted that there is a marked absence of references to historical information that has been used and discussed by other writers on this subject. This is mainly because there is very little historical material applicable to the British Isles, and what little there is has been written by observers who were not specialists in the techniques they witnessed; therefore their descriptions are often difficult to interpret in terms of technical detail. I have felt that it is better to rely upon the factual evidence of the material itself rather than the statements of observers.

The prehistorical period of British metallurgy starts shortly after 2500 BC and can be said to end with the appearance of Agricola's *De Re Metallica* – the first real treatise on metallurgy – in 1556. Although this was not published in Britain, it was based on the experience of the foremost workers in Europe on this subject, and since they were invited to Britain to redevelop the metalliferous mines, this work can be said to represent the state of metallurgical knowledge in Britain also. Only one other earlier book has any contribution to make: Theophilus' *Diversarum Artium Schedula*, which probably dates from the 11th century. This contains a useful account of some aspects of metallurgy, but the descriptions of furnaces are not sufficiently complete to make a generally accepted reconstruction easy.

Since *Metallurgy in Archaeology* was published in 1962 there has been a great deal of work done by archaeologists on the typology of early British metalwork. Owing to the techniques of ^{14}C and thermoluminescence dating it has been possible to date some of the bronzes by their associated material, and so it is now much easier to assign a date to a non-associated metal object. In the case of iron, which normally contains very little carbon, it is becoming possible with more refined ^{14}C dating techniques to envisage the possibility of direct dating. Our knowledge of early copper-base artifacts has been increased enormously by the detailed examination of founder's hoards. Much early material in museums has now been examined and reassessed as to its date and country of origin, but the detailed chronology of the early periods is still under active discussion by archaeologists.

In complete contrast to our knowledge of artifacts is the situation regarding metal extraction. Very little is known about where these processes took place and the techniques used. While nothing is known about early metal mining in Great Britain itself, there are ^{14}C dates for early mine timbers in Ireland going back to the mid-second millennium BC. Although there are some well known early mining and smelting sites in England, such as Alderley Edge in Cheshire, very little work has been done on mining and no work has been done on smelting. In order to try to present a complete picture I have had to draw on furnaces found outside the British Isles.

Even the medieval tin-smelting sites on Dartmoor have not yet been excavated, so there is still much work to be done. For these reasons the chapters that follow on copper and tin have had to be confined to the study of artifacts and their composition and production. The chapters on iron show a much more complete picture of smelting and smithing.

1 *The native metals*

For the purpose of this study the term native is applied to metals that are found in the metallic state and which are not produced from ores by smelting. Gold, and some copper occur in this form, and meteoric iron is also in this sense a native metal. These three metals were once lying on the earth's crust waiting for man to make use of them. Silver also occurs in native form, but is thought to have existed in such a manner only in deep veins, and was therefore not available to early man.[1] Apart from these, the only other native metal of use to early man was arsenic.

One can assume that Palaeolithic man came across the native metals in his struggle for existence. Gold he must surely have noticed and used, although there seems to be no evidence of this. Copper would be less obvious, since it would be coated with a film which would disguise its lustre, but its weight would be conspicuous. Somewhere, though, he must have picked up a small metallic meteorite – the British example from Rowton, Salop, weighs only 3·6 kg – and noticed that it was heavier than stone.

Ross found the Cape York Eskimos using meteoric iron in 1818.[2] It is probable that primitive implements were made from this material before the Eskimos came into contact with iron-using communities elsewhere. One piece of evidence for the early use of meteoric iron lies in the presence of a copper chisel in a crack in the large meteorite at Descubridora in Mexico.[2] It is not known whether the chisel is made of native or smelted copper, and there is no means of dating these early uses of iron or copper.

Native copper was used by the pre-Columbian peoples of North America; deposits are still relatively abundant there. Among the more advanced peoples of Eurasia, however, this source of metal has been exploited on a large scale for a much longer period; small amounts are still coming from most copper mines being worked at the present time.

Egyptian tombs have yielded meteoric iron dated to about 3500 BC. Iron from later tombs such as those of Tutankhamun (1340 BC) and Princess Aashait at Deir-el-Bahari (2050 BC) has also been found to be meteoric.[3]

One interesting example is a Shang Dynasty bronze axe in which a piece of meteoric iron has been inserted as the cutting edge.[4]

Very little evidence remains of man's earliest use of native metals, but obviously when man learnt how to shape metals and finally smelt ores, a new world was open to him. He had available a range of materials capable of producing shapes, colours, and mechanical properties unobtainable in stone. In the succeeding centuries he has learnt to use these to the full, fashioning the ornaments, weapons, and utensils of modern civilization.

GOLD

Gold mostly occurs in the native state and since it has such an obvious lustre one can be fairly sure that it was the first metal known to man. Being so ductile and having a low work-hardening capacity, it can be hammered without annealing (i.e., softening with heat).

As native metals occur mainly on the surface one cannot be sure how many deposits of gold there have been. Gold is recorded as having been found in significant quantities in Cornwall,[5] Scotland (Sutherland [now Highland],[6] Leadhills, and Wanlockhead[7]) and in Wales at Dolaucothi[5] and Dolgellau.[8] In Ireland it has been found in the Wicklow Mountains[9] and the north[10] and the south-west.[8, 10]

The main areas of gold deposits in Wicklow are in the valley of the Dodder about 16 km south of Dublin and in the streams flowing from Croghan Kinshelagh in the south of the county. Here a nugget weighing 0·58 kg was discovered in the 18th century.[11] Small nuggets have been found in Cornwall but there seems to have been no systematic working here at any time.[5]

In south Scotland,[6] and Wicklow,[11] no deep mining of gold appears to have been carried out in early times, and all the gold has come from the surface, mainly in the vicinity of streams. Medieval miners at Leadhills do not seem to have found gold below the surface, but bronze picks are recorded from surface workings which suggest that gold working may have been quite early in this area.

In the case of Dolaucothi, some deep mining was carried out in Roman times. Smyth[12] in 1846 recorded the fact that Roman pottery and ornaments had been found. A watercourse could be traced at intervals for 13–16 km along the hillsides, terminating in a pond or tank near the mines. Two adits run into the hill a distance of 50 m with a slight uphill gradient. These tunnels are 2–2·3 m high by 1·6–2·0 m wide and seem to be unusually spacious and quite unlike the later medieval workings. Considering their experience in Spain and other parts of the Empire, there is no doubt that the Romans were capable of such work. Davies[13] has no doubt as to the Roman dating, drawing attention to the graffiti on the gallery walls and the use of a heavy iron hammer and millstones of Roman type. Whittick[14] dates the excavated finds to the late 1st, and 2nd centuries. The existence of milled auriferous quartz in pits in the bed of the River Cothi suggests ore-dressing before the construction of the aqueduct, but probably still within the period of Roman exploitation. Recent work by Lewis and Jones[15] has confirmed this and added to the overall picture.

The method used for mining appears to have been fire-setting; in this method the rock faces were heated by lighting fires against them, and then cracked by throwing water on them.[14] Stoping (excavating a vein) had been carried out 27 m below the adit level,[13] which indicates that some form of drainage machine must have been used. In one stope were found parts of a water wheel and part of a shovel-shaped cradle, both of which may be seen in the National Museum of Wales, Cardiff.

There is no doubt that Dolaucothi was one of the few deep Roman mines known in this country and certainly the most technically advanced. Many of the workings are of the open-cast type and the spoil heaps contain some 500 000 t of material showing that work must have been carried out by breaking the quartz rock in which the gold occurs as fine particles, and washing the mixture of fine sand and gold with running water, so moving the lighter particles of quartz and allowing the heavier particles of gold to sink. The gold-dust would then be put into crucibles and melted in a charcoal fire with the aid of bellows; a temperature of at least 1063°C would be necessary for the process.
the process.

The gold mining area in south Scotland stretched from Linlithgow in the north to Wanlockhead and Leadhills in the south, with a width of about 16 km.[7] Apart from the bronze pick already referred to, there is no evidence of early working. By AD 1239 there is a reference to monastic working of lead at Glengonnar Water, but it is not until the minority of Queen Mary (*c.* 1550) that there is a reference to gold working. By this time the working of lead was at least temporarily abandoned, since lead was sent from elsewhere to refine the gold. In 1562, in a period of 30 days, 3·6 kg of gold were sent to Edinbugh. One nugget, found mixed with sulphur (? pyrites), weighed 0·35 kg. By this time, gold was being worked in both Linlithgowshire and Leadhills, and supplied the metal necessary for Mary's gold coinage and probably for her crown.[6, 8, 16]

By about 1575, it would seem that the surface deposits of natural gold were becoming exhausted, since it is recorded that the King granted to a Fleming, A. Petierson, the right to extract gold and silver from copper ores.[7] It should be said here that the types of ore to be found in the Wanlockhead–Leadhills area are among the most numerous in the country, although it is not uncommon for several ores to be found in the same area and often in the same vein. Both lead and copper ores contain precious metals, and by the later medieval period the majority of the precious metal output of the British Isles (mostly silver) was obtained from such ores by the process of cupellation (*see* Chapter 4).

Alloys containing 80%, or even more, of gold can be found native, but substantial amounts of silver and small amounts of base metals do often occur in native gold. Whereas the separation of base metal impurities from this metal can be carried out by the cupellation process, which was certainly well known before Roman times, the removal of silver requires an even more complex process, described by Agatharchides in about 100 BC and recorded by Diodorus Siculus.[17] This description introduces the principle of refining gold-dust with the aid of salt. With this knowledge the process could easily be extended to cover the separation of silver from solid gold. This would involve the granulation of the gold alloy (which could be done by melting and pouring into water), smelting with salt, and leaching out the silver chloride thus formed with acid, ammonia, or brine. It is unlikely that many people went to the trouble of separating silver and gold before Roman times, and this may account for the comparatively large amount of 'electrum' (a whitish gold–silver alloy, containing more than 40% silver when white) found in the Mediterranean world. Later, much of the electrum would be made by the intentional alloying of gold with silver, particularly when it was discovered how to make electrum look like pure gold by superficially dissolving out the silver in acid.

The traditional way of testing the purity of gold or silver is by means of the touchstone. This consists of a piece of fairly fine-textured but abrasive stone on which the gold object is rubbed. The streak or mark obtained can be compared with that made by objects or trial plates of known purity, and a very accurate estimate obtained of the purity of the unknown. In the case of gold alloys containing silver only, estimates may be made of the gold content to an accuracy of 1%,[18] but with complex silver–copper-containing alloys, it is not so reliable and the technique has to be supplemented with acid leaching.

In view of the considerable evidence for the working of noble metals at Hengistbury Head,[19] suggesting a pre-Roman mint, it is not surprising to find that this site yielded a touchstone. A small piece of of siliceous shale was found which had been used for this purpose.

The composition of natural gold

While most of the gold produced today comes from the finely disseminated gold dust found in quartz rocks and from refining other non-ferrous metals, from early time until the 19th century a large amount must have come from the nuggets which have received so much publicity. There is no doubt that, apart from its silver content, natural gold is relatively pure, which provides us with one way of recognizing it.

Liversidge[20] reports the analyses of nuggets from Australia, shown in Table 1. From these we see that Australian gold contains 1–11·5% silver, and very often traces of non-metallic constituents, such as ferric oxide, alumina, and quartz. There is no reason to believe that gold nuggets from other sources are very different. Analyses of British gold are given in Table 2,[21] and these show that it has much the same composition.

Table 1 Composition of natural gold nuggets of Australian origin (after Liversidge[20])

Provenance	Weight, g	Au, %	Ag, %	Others
Orange (NSW)	1·08	93·45	6·0	SiO_2 + $Fe(OH)_3$
Adelong (NSW)	0·40	92·75	6·5	$Fe(OH)_3$
Queensland	0·70	87·95	11·5	$Fe(OH)_3$
Wellington (NSW)	0·99	93·35	6·0	$Fe(OH)_3$
Peak Hill (NSW)	0·92	97·90	1·0	—
Hill End (NSW)	Au; containing a cavity lined with quartz crystals			
Parkes (NSW)	Au + $Fe(OH)_3$ or Fe_2O_3			

In the Witwatersrand,[22] gold particles can be irregular and formed around inert mineral inclusions. Their characteristics indicate a transport distance of only 5–30 km. In this short distance and under the right conditions of chemical attack the silver can be leached out of the gold thus increasing the fineness with distance of transport. In the Witwatersrand, however, the conditions are such that the fineness stays constant. On the other hand, gold particles can be dissolved by biochemical processes and the silver content can be enriched by a factor of two.

1 **Structure of a gold nugget from West Australia; ×2**

The crystal structure of nuggets (Fig. 1), shows that they have been subject to cold deformation and have been annealed. This structure possibly results from the welding-up of fine gold particles released from the quartz rocks by weathering since gold is so easily

Table 2 Analyses of British natural gold

Provenance	Weight, kg	Au, %	Ag, %	Other elements, %
Irish nugget[11] (Wicklow)	0·58	92·0	5·1	—
Wanlockhead[16] (*c.* 16th cent.)	—	86·6	12·4	Fe 0·35
Gold dust[10] (Sperrin)	—	90·0	9·8	Cu <0·1 Sn <0·01

(After Taylor[21])

Provenance	Ag, %	Cu, %	Pt, %	Pb, %	Sn	Other trace elements, %
Ogofau Mine, Carmarthen	~15	~0·010	—	0·12	—	Hg 0·3
Gwyn Mine, Dolgellau	~15	~0·010	—	—	—	Hg 0·07
Arenig Mine, Cwm Prysor	~10	~0·010	—	—	—	Hg ~1·0 Zn 0·5; As 0·05
St David's Mine, Bontddu, Dolgellau	~ 5	~0·14	—	—	—	Hg ~1·0
Clogau, Gwynedd	~ 4	~0·010	—	tr.	—	Hg 0·02 Bi tr.
Clogau, Gwynedd	~ 3	0·13	—	0·80	—	Hg 0·1 Bi 0·6 Sb tr.
'Drift', Highland	~ 7	<0·010	—	—	~0·005	Hg <0·01
'Drift', Highland	~ 7	~0·015	—	—	~0·004	Hg 0·02
Suisgill Burn, Highland	~ 8	~0·010	—	—	—	Hg tr.
Suisgill Burn, Highland	15–20	<0·010	—	—	—	Hg tr.
Longcleugh Burn, Leadhills, Strathclyde	~ 5	<0·010	—	—	—	Hg 0·03
Allt-an-T'Fhionaraibh River, Highland	~ 5–10	<0·010	—	—	—	Hg <0·01
Windgate Burn, Leadhills, Strathclyde	~ 5	~0·015	—	—	—	Hg tr.
Martin Gold Nugget, Leadhills, Strathclyde	~ 5	~0·010	—	0·04	—	Hg <0·01
Kildonan, Highland	~ 5	~0·017	—	—	—	Hg <0·01

Table 3 Analyses of gold objects, %

Object and provenance	Date	Au	Ag	Cu	Sn	Fe	Others	Remarks	Reference
Torc terminal from Cairnmuir, Peebles	1st cent. AD	85	9	3·5	tr.	2·5	—	Probably alloyed	29
Torc terminal from North Creake, Norfolk	1st cent. AD	37	pres.	pres.	1·0	—	—	Alloyed	30
Tubular Torc, Hoard A Snettisham	1st cent. BC	69·6	27·0	3·1	—	0·2	rem. 0·2	Alloyed	28
Loop-terminal torc, Snettisham	1st cent. BC	28·3	43·6	27·0	tr.	—	Pb tr.	Alloyed	28
Ring, Snettisham	1st cent. BC	29·2	44·4	25·3	—	—	—	Alloyed	28
Ring-terminal torc, Hoard E, Snettisham	1st cent. BC	58·1	38·3	2·9	—	tr.	Zn tr.	Alloyed	28
Bracelet, Hoard E, Snettisham	1st cent. BC	95·6	3·6	0·6	—	—	Zn tr.	Probably natural	28
Buffer-terminal torc, Hoard E, Snettisham	1st cent. BC	79·6	17·4	2·6	tr.	—	—	Alloyed (?)	28
Scottish Crown: (*a*) arches	16th cent. (?)	87·4	10·0	2·6	—	—	—	Probably alloyed	16
(*b*) fillet	16th cent. (?)	88·1	10·0	1·9	—	—	—		
Ardagh Chalice Ireland	Early Christian	73·5	23·5	3·0	—	—	—	Alloyed	31
Lunula (Irish)	EBA	88·64	11·05	0·12	—	—	0·19	Natural	25
Hook of ribbon torc (Irish)	MBA	71·01	24·09	4·67	—	—	0·23	Alloyed	25
Ribbon torc (Irish)	MBA	71·54	23·67	4·62	—	—	Pb 0·17	Alloyed	25
Ribbon torc (Irish)	MBA	79·48	18·01	2·48	—	—	0·03	Alloyed	25
Ring (Irish)	MBA	86·72	12·14	1·16	—	—	tr.	Probably alloyed	25
Wire ring (Irish)	MBA	85·62	12·79	1·47	—	—	0·12	Probably alloyed	25
Torc (Moulsford	12th cent. BC	74·3	12·00	13·70	0·2–0·5	—	Pb 0·2–0·5	Alloyed	26
Torc 1 (Ipswich)	1st cent. BC	80·2	12·4	4·40	—	—	—	Alloyed	27
Torc 2 (Ipswich)	1st cent. BC	78·1	10·1	11·8	—	—	—	Alloyed	27
Torc 3 (Ipswich)	1st cent. BC	85·5	10·7	1·3	—	—	—	Alloyed (?)	27
Torc 4 (Ipswich)	1st cent. BC	88·7	11·3	1·8	—	—	—	Alloyed	27
Torc 5 (Ipswich)	1st cent. BC	82·0	16·9	1·6	—	—	—	Alloyed	27
Torc 6 (Ipswich)	1st cent. BC	84·0	15·0	0·3	—	—	—	Natural	27

pressure welded.[23] They could have been welded by the action of water upon the rocks which would drive the gold particles together. The trapping of clay, which can be seen in some of the Australian nuggets, strongly suggests this. In the course of the centuries the gold has recrystallized, giving the annealed structure found in the nuggets of today.

Gold nuggets as large as 72 kg in weight have been found,[24] and there is little doubt that most of the gold recovered before Roman times was found in the form of nuggets. Only more recently has mineral dressing progressed to the stage where it is possible to recover gold-dust so fine as to be invisible to the naked eye.

There is now quite a wealth of analyses of British and Irish gold objects. Some of these are given in Table 3; Hartmann has included some in his ambitious programme on European gold,[32] and Taylor[21] has published many more.

Maryon[33] reports that out of 22 Irish gold objects spectrographically analysed all contained some silver, all but one contained some copper, most some lead, all but two some tin, two contained some bismuth, and many contained titanium, vanadium, and zinc.

A spectrographic analysis[34] of a tin-filled penannular ring (Late Bronze Age, *c.* 500 BC) from Limerick, showed some silver and a trace of copper to be present in the gold. Evans[35] reports that some of the gold objects from Broighter (Londonderry) such as the boat (Fig. 2) were in fact made of electrum; but the torcs and the collar were 'pure gold', while some chains appeared to be 'dull', which probably means that they contained copper.

More recent gold finds, such as those from Snettisham[28] and Ipswich[27] have been more fully analysed, with the results given in Table 3. At least one of the Early Iron Age torcs shown in this table appears to be unalloyed native gold, but there is plenty of evidence to show that by the Middle Bronze Age (12th century BC) gold was being alloyed with both copper and silver. The gold in the Scottish crown of 1540, however, seems to consist of metal alloyed with a small amount of copper to offset the whiteness caused by the silver. It probably came from the Leadhills–Wanlockhead area.[16]

We can safely say that all gold, whether obtained from nuggets or dust, is likely to contain silver exceeding about 1%. Since comparatively complex processes are required to separate copper and silver from gold, metal containing more than 99% metallic gold is likely to belong to the last two millennia, and with more than 1% copper, after the Middle Bronze Age.

The non-metallic impurities such as quartz and ferric hydroxide would be removed by melting, and therefore gold artifacts containing these materials are probably made from beaten natural gold as distinct from remelted gold. Furthermore, melted metal containing an appreciable amount of silver will show a characteristic structure after casting. However, this structure would be destroyed by subsequent working processes, which would give a metal indistinguishable

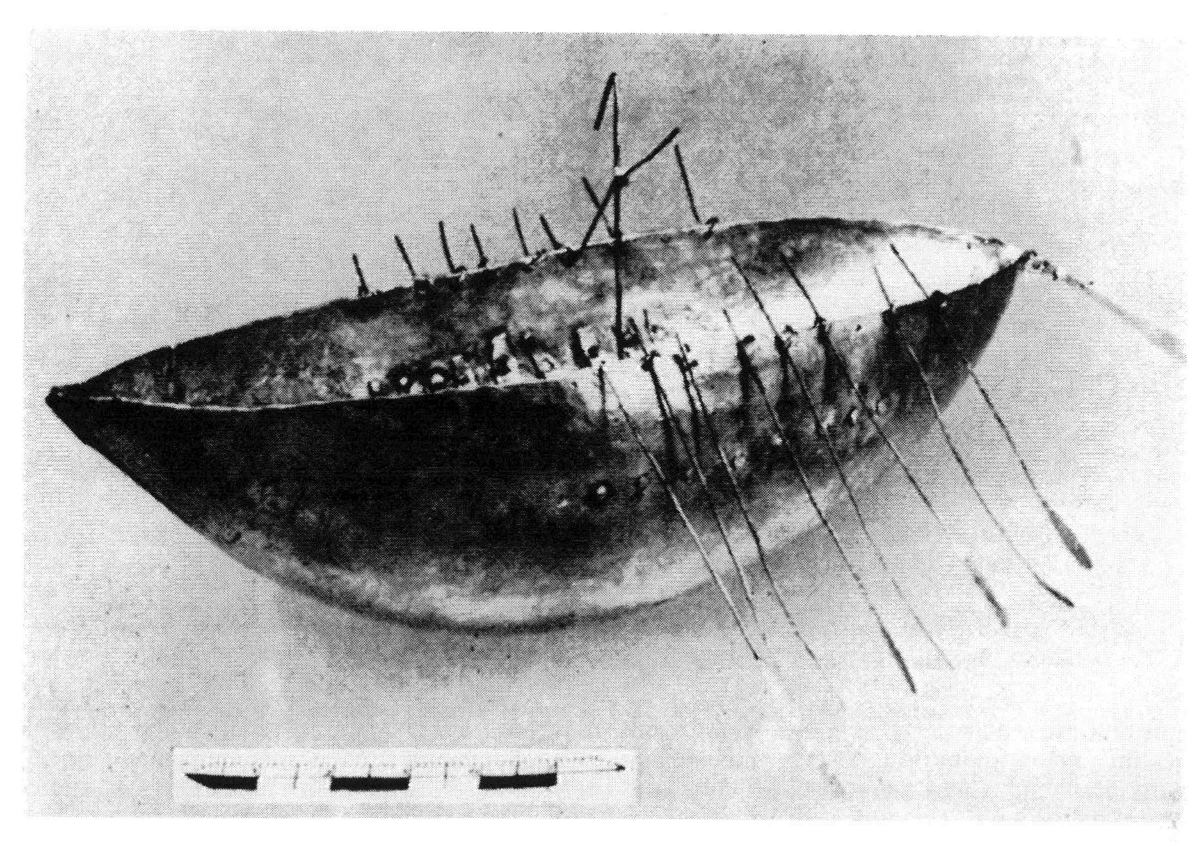

2 The Broighter Boat (1st century AD)

from unmelted natural gold except for the inclusions in the latter.

NATIVE COPPER

Copper, being less noble than gold, is more likely to be attacked by weathering agencies. For this reason there is some controversy about the manner of the formation of native copper.[36] It is almost certainly formed by the precipitation from solutions resulting from reactions occurring in the oxidation zone of a copper deposit.

Native copper has been found in Cornwall, Scotland, and Ireland.[37] It can be of extremely high purity as shown in Table 4, but in an increasing number of cases very high arsenic contents have been found (arsenic is probably present in native form) and it is not possible to say that because a copper object is of very

Table 4 Spectrographic analyses of native coppers, and an Irish axe probably made from native copper, wt-% (after Coghlan[37])

Element	Approx. limit of detection	Native copper: Cornwall, Lizard	Native copper: Cornwall, Gwennap	Native copper: Ireland, Arran Moor	Native copper: Ireland, Waterford	Native copper: Scotland, Renfrew	Irish axe Pitt-Rivers Museum
Sb	0·0002	n.d.	n.d.	n.d.	n.d.	n.d.	0·05
As	0·002	n.d.	n.d.	0·002	n.d.	n.d.	n.d.
Bi	0·0001	n.d.	0·0001	0·0003	0·0003	n.d.	0·005
Co	0·0001	n.d.	n.d.	n.d.	n.d.	n.d.	n.d.
Fe	0·0001	0·0002	0·0002–0·003	0·05–0·2	0·05–0·2	0·001	0·01
Pb	0·0001	n.d.	0·0004–0·006	0·0005	0·0005	0·0003	0·0004
P	0·005	n.d.	n.d.	n.d.	n.d.	n.d.	n.d.
Mn	0·001	n.d.	n.d.–0·005	n.d.	n.d.	n.d.	0·001
Ni	0·0001	0·01–0·06	n.d.	0·0003	n.d.	0·0003	0·0003
Ag	0·001	0·002–0·03	0·003	0·005	0·005	0·002	0·005
Sn	0·0005	0·0005	0·001–0·005	n.d.	n.d.	0·0005	n.d.
Zn	0·001	n.d.	n.d.	n.d.	n.d.	n.d.	0·001
Te	0·001	0·005	n.d.	n.d.	n.d.	n.d.	n.d.
Au	—	—	n.d.	—	—	—	—
Mg	0·0001	tr.	—	~0·05	n.d.	n.d.	tr.
Si	—	0·008–0·1	0·1 (approx.)	0·05	tr. <0·005	0·08	0·07
Al	0·005	n.d.	tr.	~0·05	n.d.	tr.	tr.

n.d. = not detected (i.e., less than limit of detection, if any)

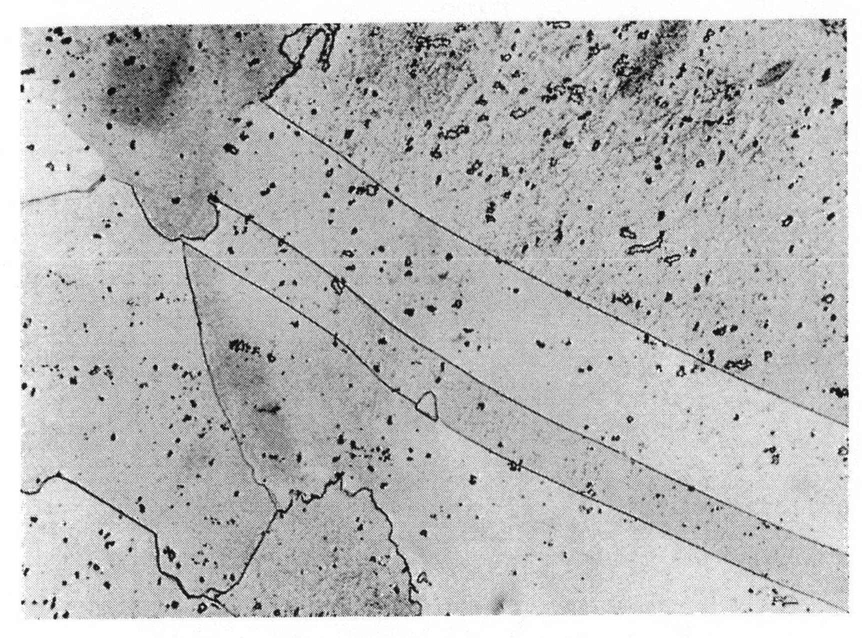

3 Structure of Canadian native copper; × 50

high purity it must have a native origin. Apart from arsenic, the major impurities are iron and silicon which are probably mainly present as $Fe(OH)_3$, and silica, as in native gold.

Only by electron-probe analysis and micro-examination can one distinguish smelted coppers from native coppers. The analysis of an Irish axe, shown together with the analyses of British and Irish native coppers in Table 4, indicates that it could well have been made from native copper. Silica would not be so readily eliminated from native copper as from gold, during melting, and some of the iron from the iron oxide would be dissolved in the copper if it were melted under reducing conditions. One major impurity in native copper according to Smith[38] is silver and this occurs in a segregated form giving areas of high-silver copper which should be detected by the electron microprobe.

Coghlan[39] thinks that native and smelted coppers are not distinguishable, except by micro-examination. But the author finds it difficult to believe that early smelted coppers can be as pure as those shown in Table 4. A pure ore from Limerick (*see* Table 6), contains about 0·5% Fe and 0·5% Pb, some of which would remain in the final copper, unless they had been carefully refined out.

Native copper[40] is found in many forms: as small grains or pellets, spongy or laminated masses, or in a more dense form. These masses have been found in sizes varying from 0·5 kg to large blocks weighing several tonnes. They all appear to contain extensive non-metallic inclusions. Coghlan[40] experimented with the malleability of this material, and attempted to forge a copper chisel from a massive native copper block. During cold-forging it soon became hard, brittle, and cracked, and had to be annealed twice to obtain a chisel edge.

A metallurgical examination of a piece of native copper from the Lake Superior region of Canada was made by the author. It was found to consist of large grains about 3 mm across (Fig. 3) and some small but very hard inclusions which probably consisted of silica and ferric oxide. Such material has been worked by natural agencies such as rock movements or impact with water-borne boulders, and afterwards annealed, either by heating for a short time above 200°C or merely by leaving for a very long period at ordinary temperatures, i.e., in the course of geological time. The original hardness was 77 HV which showed that the metal had not been softened in geological time. After heating for 1 h at 500°C the hardness was found to be 38 HV which is the sort of figure to be expected from fully softened pure copper.

Native coppers differ considerably in their capacity for cold work. Coppers with few insoluble impurities such as $CaCO_3$, Fe_2O_3, and SiO_2 can be cold worked to a considerable degree without cracking. High purity Michigan copper can be hardened to 140 HV by a reduction in thickness of 75% (*see* Fig. 4), and readily annealed to restore the hardness to less than 60 HV.[41]

These tests show that primitive man could fashion some native copper into small artifacts by cold work without knowing how to soften it, but to make larger implements he would have to anneal it at intervals. This was done in the case of the Old Copper Culture of North America in about 3000 BC.[41]

Thompson[42] says that this process might have been discovered as a result of the frustrations experienced in early attempts to work this metal, a piece might have been thrown into the fire in disgust and found to be soft and ductile when recovered because of this unintentional annealing.

We can safely conclude that the process of annealing would quickly follow the use of native copper, but the process of melting would be appreciably more complicated and would not follow directly, since it requires a temperature of 1083°C which cannot be obtained in a domestic fire without a forced draught.

However, the principle of melting metals may already have been discovered by experience with lead, which can be readily smelted and melted in the domestic fire. Once the melting process was applied to lead, the melting of native copper, which requires a forced draught, would sooner or later follow.

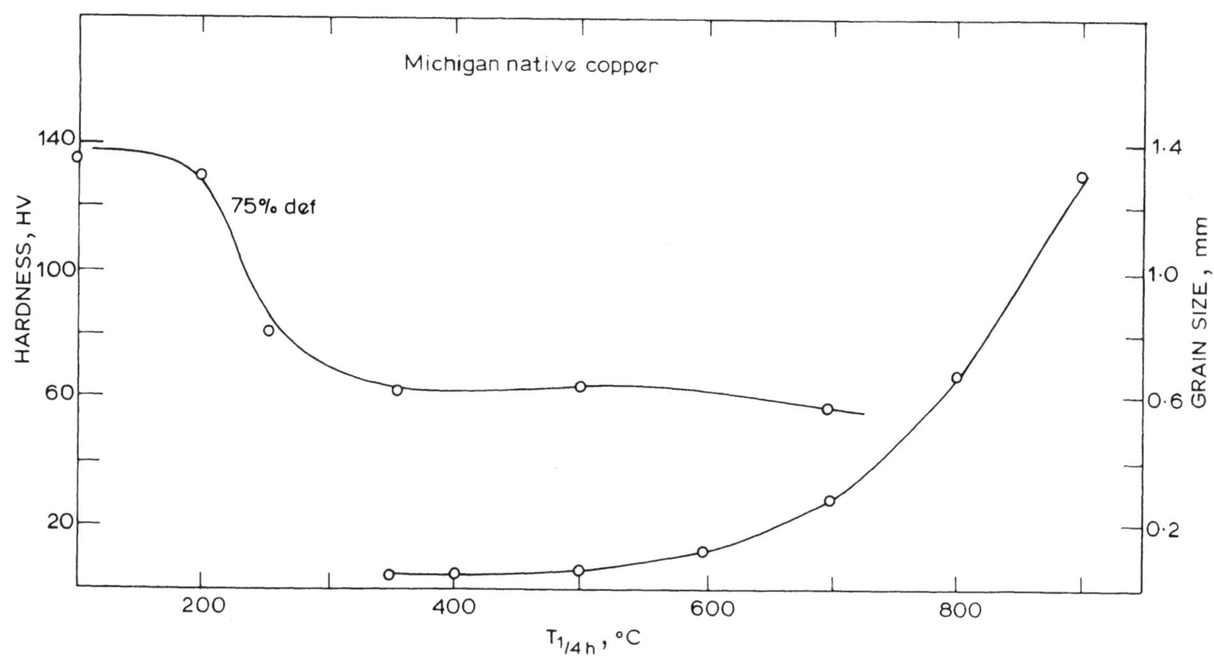

4 Hardness of Michigan native copper (after Schroeder and Ruhl[41])

NATIVE IRON

Native iron comes from two sources, from outer space in the form of metallic meteorites, or terrestrial iron in basaltic rocks. It occurs in the basalt from the Giant's Causeway in Ireland and no doubt in other similar rocks in the British Isles, and in placer deposits in New Zealand and, nearer home, in Piedmont.[43] But it occurs as extremely fine grains, which would require a very complex mineral dressing process for their recovery. All these deposits appear to have been formed by the interaction of iron oxides with organic minerals. Larger pieces are known to occur on Disko Island, off the coast of Greenland. These may be as large as 0·5 cm and contain up to 2% Ni. Such pieces may be inserted into organic hilts to make saw-like knives.[44]

Meteoric iron, on the other hand, has been found in early implements and must have been of great importance. This iron is characterized by a high nickel content, and usually by the presence of an appreciable amount of phosphorus and some other less important elements. In 1916, Zimmer[2] listed the metallic meteorites which were known up to that date, and found that there was at least 250 t of meteoric material extant. Of this, 99·4% was claimed to be malleable and therefore capable of being turned into useful implements.

Since 1916, some meteorites have been re-analysed, and it seems that many of the earlier analyses for nickel were slightly low. In particular, some of the meteorites which had less than 4% Ni have been found on re-analysis to be in the more normal region of 5–9%. Furthermore, many of the earlier analysts did not distinguish between nickel and cobalt. These metals are very similar chemically and most nickel ores contain cobalt. It is now clear that the cobalt content of meteorites is in the region of 10% of the nickel. About 350 metallic meteorites have been analysed,[45] and Fig. 5 shows that most of the meteorites fall into the range 5–10% Ni. In no case did a specimen contain less than 4% Ni, and the maximum so far is about 60% Ni.

As Zimmer shows,[2] a large proportion of the total number of meteorites known has been found in the New World. As there is no known scientific reason for this, it must be assumed that their distribution was originally even, and that most of those that fell in the Old World have been discovered by the natives and used.

Only one metallic meteorite is known from these Islands, and this fell at Rowton, Shropshire in 1876 and is now preserved in the British Museum (Natural History). It contains 8·58% nickel, which therefore makes it representative of the vast majority, and it weighs about 3·5 kg. This particular example has a thin coating of oxide and an outer zone about 6 mm thick in which the structure has been altered by the heat produced during atmospheric flight (Fig. 6).

We can assume, therefore, that the natives of this country have received from time to time a supply of metal from outer space. We shall never know how much was available for use. The main questions are: how easy was it to use, and how is the archaeologist or metallurgist to recognize it and distinguish it among a large quantity of prehistoric iron-work, which is chiefly made of smelted metal?

The first question has, to a large extent, been

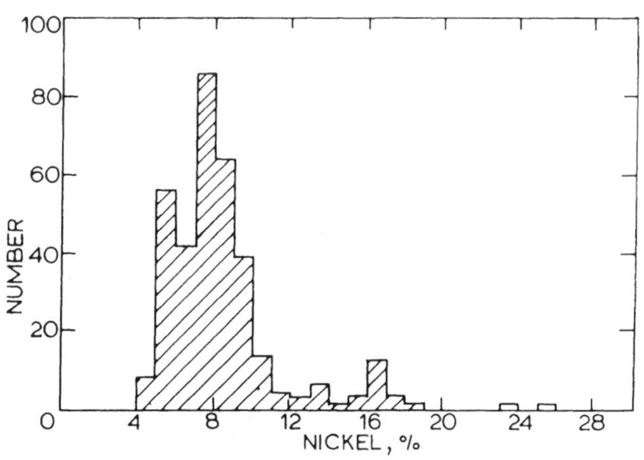

5 Nickel contents of metallic meteorites

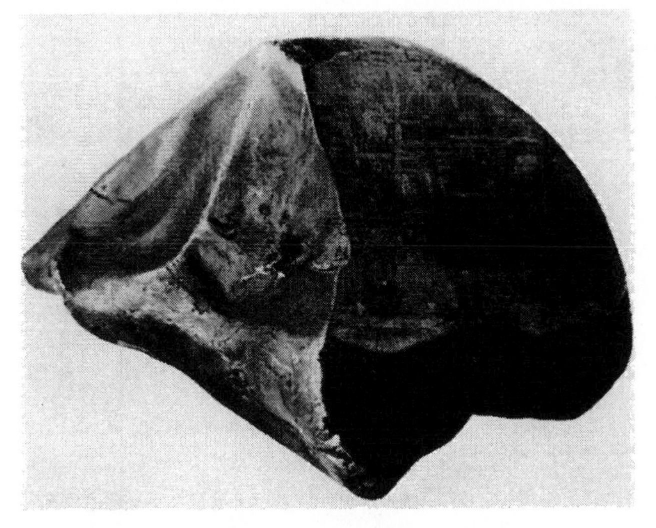

6 Meteorite from Rowton, Salop (19 cm across)

Table 5 Analysis of meteorites

Element, %	San Angelo[46]	Canyon Diablo, Arizona[48]
Fe	91·96	92·6
Ni	7·86	7·0
Si	0·011	—
P	0·099	0·25
S	0·032	—
Co	—	0·1
Cu	0·04	0·005
Ge	—	0·01
Ca	—	0·002

answered by Zimmer[2] and Richardson.[46] Although most of the iron meteorites are malleable in the cold state, the amount of forging that can be carried out in this condition is very limited. Pure iron–nickel alloys work-harden like iron itself, and the degree of cold working is strictly limited by the high degree of hardness obtained. Meteorites behave similarly but the brittle constituents present in some reduce the amount of deformation that may be applied without cracking. While large pieces of meteorite crack and splinter on hammering, according to Zimmer, some of the small pieces so formed can be hot-hammered into arrowheads. This difficulty is due to the coarse crystalline structure of some meteorites, particularly those exceeding about 6% nickel.

Richardson tried hot-forging small pieces, weighing about 40 g, of the San Angelo meteorite, which contained 7·86% Ni and had a hardness of 250 HV. One piece was hot-forged at about 800°C and yielded under the hammer, but tended to open up along certain crystal planes. He also had trouble welding these pieces at a higher temperature (1200°C) for the same reason and therefore reached the conclusion that meteorites with nickel contents greater than about 6% would not be easily worked by primitive people. Yet Zimmer gives sufficient evidence to show that meteorites with higher nickel contents than this are hot-forgeable and many with lower nickel contents cold-forgeable by primitive people.

I. M. Allen[47] has successfully cold-forged a small knife from a piece of meteoric iron of unspecified composition.

The Cape York (West Greenland) meteorite containing 7·95% Ni has been used by natives to make knives and scrapers by cold-hammering.[2] Zimmer cites many other examples of the use of meteoric iron for implements, and there is little doubt that small pieces of the most numerous groups of meteorites (5–9% Ni) can be cold-hammered.

One can conclude therefore that the low-nickel meteorites were readily forged by primitive people but that many of the meteorites with higher nickel contents are not so easily cold- or hot-forged. Clearly the easiest way would be to cold-forge and anneal alternately until the required form was obtained, but there is little doubt that most meteorites could be hot-forged by a competent smith.

As far as recognition is concerned, chemical analysis is the best guide but is not infallible. Table 5 lists the analysis of two meteorites, which should be examined together with Tables 77 and 78 (Chapter 6). The nickel content is quite unlike most early smelted irons; some modern steels approach these nickel levels, but these are made by alloying, i.e., the intentional introduction of nickel and can easily be distinguished. Iron ores, with few exceptions, never contain more than 0·2% Ni. But there are a few high nickel ores such as the garnierites that might produce metal of higher nickel content.

Therefore, if the nickel content is over 4% there is a good chance of meteoric origin. Most of the meteorites with less than about 6% Ni are known as hexahedrites and show little structure and therefore may be difficult to distinguish microscopically from pure irons.[49] Meteorites with a nickel content between 7 and 13% are mainly octahedrites which have a very characteristic Widmanstätten structure consisting of a network of bands of a low nickel constituent known as kamacite, bordered by a higher nickel constituent taenite. These bands intersect in three or four directions. Filling the angular interstices are fields containing plessite, a mixture of the above two constituents. There is a third group which is almost structureless known as ataxites. These fall into two compositional groups: up to 7% and between about 12 and 20%.

About 75% of the total are octahedrites with nickel contents exceeding 6%. Hexahedrites and nickel-poor ataxites comprise about 15%, and nickel-rich ataxites about 10%. This means that 75% have a very characteristic structure, provided that they have not been heated to high temperatures after landing. Simcoe[48] carried out tests to determine the extent to which the natural characteristic structure of meteoric iron is obliterated by heating. He estimated that the structure could not be significantly altered by short-time heating below 650°C. But the alteration undergone in the outer zones of some meteorites due to heating during the 20 s or so that it took to penetrate the earth's atmosphere shows that substantial changes may be brought about by heating during smithing where the temperature may reach 1200°C. These changes may well be sufficient to obliterate much of the characteristic structure of the octahedrites comprising the larger proportion of the meteorites. Desch, in his discussion on Zimmer's paper[2] has also made this point.

However, not all occurrences of nickel in iron are believed to be meteoric. For example a socketed axe from Poland[50] with about 4%Ni and another similar axe from Sweden[51] with an average of 0·58%Ni have been recently examined and claimed to be non-meteoric. It seems, therefore that there are or were supplies of Fe–Ni ores such as the garnierites which could be responsible for such compositions. Nickel would certainly be preferentially reduced during the reduction of iron oxides to iron.

There is one structural feature which is common to both early man-made iron and some meteoric irons. This is the presence of Neumann bands, which are formed in wrought iron as a result of heavy impact in forging at low temperatures. In meteorites, they have been formed by the deceleration on entering the atmosphere or by shock on landing. Their presence need cause no confusion, when the nickel content and structural details are taken into account.

Li Chung[4] has considered the problems of recognition of meteoric iron in corroded artifacts and shows that nickel distribution in the two-phase structure of the octahedrites is impossible to remove by corrosion and can still be seen by electron-probe analysis of the rust. Also, the slag inclusions that are so common in man-made iron are absent in meteoric iron.

REFERENCES

1 W. GOWLAND: *Archaeologia*, 1917–18, **69**, 121–60.
2 G. F. ZIMMER: *J. Iron Steel Inst.*, 1916, **94**, 306–49.
3 J. K. BJORKMAN: 'Meteors and meteorites in the ancient Near East', Pub. No. 12; 1973 Center for Meteorite Studies, Arizona State Univ.
4 LI CHUNG: *Ars Orient.*, 1979, **11**, 259–89.
5 W. GOWLAND: *J. R. Anthropol. Inst.*, 1912, **42**, 235–87.
6 L. SCOTT: *Proc. Prehist. Soc.*, 1951, **17**, 16–82.
7 J. R. S. HUNTER: *Trans. Geol. Soc. Glasgow*, 1884, **7**, 373–92.
8 MINERAL RESOURCES CONSULT. COMMITTEE, INST GEOL. SCI.: 'Gold. Mineral Dossier No. 14'; 1975, London, HMSO.
9 J. S. JACKSON: in 'Proceedings of 5th Atlantic Colloq.' (ed. M. Ryan); 1978, Dublin.
10 S. BRIGGS, J. BRENNAN, and G. FREEBURN: *J. Hist. Metall. Soc.*, 1973, **7**, (2), 18–26.
11 G. A. J. COLE: 'Memoir and map of localities of minerals of economic importance and metalliferous mines in Ireland'; 1922, Dublin, The Stationery Office.
12 H. WARINGTON SMYTH: *Mem. Geol. Surv. GB*, 1846, **1**, 480.
13 O. DAVIES: *Archaeol. Cambrensis*, 1936, **91**, 51–7.
14 G. C. WHITTICK: *Trans. Newcomen Soc.*, 1931–2, **12**, 57–84.
15 G. D. B. JONES, L. J. BLAKEY, and E. C. F. MACPHERSON: *Bull. Board Celt. Stud.*, 1960, **19**, (1), 71–84; P. R. LEWIS and G. D. B. JONES: *Carmarthen Antiq.*, 1970, 6, 9 pp.
16 A. J. S. BROOK: *Proc. Soc. Antiq. Scotl.*, 1889–90, **12**, 89–92.
17 DIODORUS SICULUS: Book III, 14, (Trans. C. H. Oldfather), Vol. II, 121; 1935, London, Loeb Edition.
18 W. GOWLAND: *J. Inst. Met.*, 1910, **4**, 4–41.
19 J. P. BUSHE-FOX: 'Excavations at Hengistbury Head, Hampshire, in 1911–12', Report No. 3, Soc. Antiq., London, 1915.
20 A. LIVERSIDGE: *J. Chem. Soc.*, 1897, **71**, 1125–31.
21 J. J. TAYLOR: 'Bronze Age goldwork in the British Isles'; 1980, Cambridge University Press.
22 K. HALLBAUER: *Gold Bull.*, 1978, **11**, (1), 18–23.
23 R. F. TYLECOTE: 'The solid phase welding of metals', 191; 1968, London, Edward Arnold Ltd.
24 G. CHAD NORRIS: 'Address delivered to the Chemical Society, Univ. of Gold Coast, at Achimota, 1949'; (undated), London, West African Gold Co. Ltd.
25 C. F. C. HAWKES and R. R. CLARKE: in 'Culture and environment. Essays in honour of Sir Cyril Fox', (ed. I. L. L. Foster and L. Alcock); 1963, London.
26 E. T. HALL and G. ROBERTS: *Archaeometry*, 1962, **5**, 28–31.
27 J. BRAILSFORD and J. E. STAPLEY: *Proc. Prehist. Soc.*, 1972, **38**, 219–34.
28 R. R. CLARKE: *Proc. Prehist. Soc.*, 1954, **20**, 27–86.
29 R. R. CLARKE: *Archaeol. J.*, 1951, **106**, 59–61.
30 E. VOCE: *Man*, 1951, **51**, 139–41.
31 M. STOKES: 'Early Christian art in England'; 1888, London.
32 A. HARTMANN: *Celticum (suppl. to Ogam)*, 1965, **12**, (98), 27–44.
33 H. MARYON: *Proc. R. Ir. Acad. (C)*, 1937–8, **44**, 181–228.
34 A. G. G. LEONARD and P. F. WHELAN: *Sci. Proc. R. Dublin Soc.*, 1929, **19**, 55.
35 A. EVANS: *Archaelogia*, 1897, **55**, 391–408.
36 P. E. SPIELMAN: *Nature*, 1926, **118**, 411.
37 H. H. COGHLAN: *Man*, 1955, **55**, 6–8; also 117.
38 C. S. SMITH: in 'Actes XI Congr. Int. d'Hist. Sciences', Vol. VI, 237–52; 1965, Warsaw–Krakow.
39 H. H. COGHLAN: *Archaeol. J.*, 1942, **22**, 22–38.
40 H. H. COGHLAN: *Man*, 1951, **51**, 90–3.
41 D. L. SCHROEDER and K. C. RUHL: *Am. Antiq.*, 1968, **33**, 162–9.
42 F. C. THOMPSON: *Man*, 1958, **58**, 1–7.
43 H. H. READ: 'Rutley's elements of mineralogy', 24 ed., 486; 1957, London.
44 V. F. BUCHWALD: Lecture to the Historical Metallurgy Society at Newcastle, on 19 Sep. 1980. *See also* J. M. BIRD and M. S. WEATHERS: *J. Geol. Soc.*, 1977, **85**, 359–71. (I am indebted to Dr C. J. Salter for this information.)
45 G. T. PRIOR: 'Catalogue of meteorites', (revised M. H. Hey); 1953, London. This work is being continually revised by the British Museum (Natural History). A new edition is expected.
46 H. C. RICHARDSON: *Am. J. Archaeol.*, 1934, **38**, 555.
47 H. H. COGHLAN: 'Notes on prehistoric and early iron in the Old World', (ed T. K. Penniman and B. M. Blackwood), 179; 1956, Oxford, Pitt Rivers Museum.
48 C. R. SIMCOE: *Met. Progr.*, 1956, **70**, 72.
49 S. H. PERRY: 'The metallography of meteoric iron', United States National Museum, Bulletin 184, Smithsonian Inst., Washington, 1944.
50 J. PIASKOWSKI: *J. Iron Steel Inst.*, 1960, **194**, 336–40.
51 E. HERMELIN, E. THOLANDER, *et al.*: *J. Hist. Metall. Soc.*, 1969, **13**, (2), 69–94.

2 *Copper and copper alloys*

The spread of copper metallurgy to the British Isles has long been thought of as being via a Mediterranean or Atlantic route to the west of Ireland.[1] Unfortunately there is not much evidence for this. If one accepts ^{14}C dates from metal-using Near Eastern sites such as Cayönü, Ali Kosh etc., as proof of the origins of metallurgy, i.e. with the early civilizations of the Near East, then one can either postulate diffusion from them or independent development elsewhere. It has often been suggested that independent origins are more common than they were previously considered to be.[2] The sources of native copper are widespread and it can be argued that it would not be long before the successive early stages of physical metallurgical development were reached, leading finally to the smelting of copper ores. How this very important step was taken has not been gone into in great detail but it usually occurred after the introduction of pottery and therefore in societies which had reached a sufficient pyro-technological level.

The social aspect of technology is an important one as may be seen from the late introduction of copper metallurgy in South America. It is unlikely that this was the result of a late diffusion current, as most copper-using societies in the rest of the world had got well beyond the Chalcolithic stage when South American metallurgy started. It had to go through the various stages such as pure copper and arsenical copper, to tin bronze: the same series as in Europe and the Near East.

When the Eurasian countries were in the Chalcolithic phase, the South American civilizations had not reached a sufficiently high stage of social development to take advantage of any diffusion current existing. Therefore, social development and technological knowledge are necessary. Independent development is unlikely; so much so that when the social climate is right an existing diffusion current from a more highly developed civilization will probably supply the necessary technology.

However, independent development is a possibility, but terms must be carefully defined. One cannot doubt that the independent use of native copper was as widespread as native copper itself. Unfortunately it is almost impossible to distinguish native copper from smelted copper once it has been melted. While small artifacts such as beads and rings could be made out of native copper by hammering small pieces, large artifacts such as the Danubian axes would require melting and casting. This is a quite advanced technique and, although it is very probable that the idea of melting such metals as lead in a crucible was well known, it took an appreciable time to develop the process of melting copper in a crucible at 1 100°C. At the moment it is not known whether this came first or whether the community was reached by the diffusion of the technique.

In southern and south-eastern Europe there are many early mining sites where copper ores were mined, as distinct from native copper. Even without smelting evidence one can assume the use of copper minerals for copper production. One such site is Ai Bunar in Bulgaria, another is Rudna Glava in Yugoslavia, and a third is probably the Iberian metalliferous region in the south-west.

Approximate calibrated ^{14}C dates for the mining of copper ores are as follows:

Ai Bunar	4000 BC
Rudna Glava	3800 BC
Iberia	3000 BC
Mitterberg	2500 BC

These dates suggest an east to west movement of people seeking copper ores from the 5th millennium BC. It is unlikely that they were seeking copper ores merely for pigments since there were traces of smelting in the first and third areas.

These sites tend to confirm the old theory of Mediterranean diffusion of technology. We have as yet no evidence for the spread of copper mining across Europe from south-east to north-west, apart from the Austrian (Mitterberg) evidence which is somewhat later (2500 BC). It is therefore possible to envisage diffusion of copper smelting from the Near East taking place along the Mediterranean by the use of the minerals on the northern shores in the years around 4000–3000 BC and reaching Ireland by about 2500 BC.

Meanwhile there was another, trans-European, current from south-east to north-west which arrived in mainland Britain a little later, about 2200–2000 BC, and which can be identified with Beaker traditions of burial. By 2000 BC these two currents had merged and spread in many directions.

One interesting feature of early British metallurgy is the rapid introduction of the use of tin bronze. Whereas in the Near East it took from 6000 BC to 3000

BC to introduce bronze, in Britain it was being used in 2000 BC only 400 years after the introduction of a copper-using technology. It has been suggested that the tin was panned from the stream beds of Co. Wicklow and no doubt other rivers of Ireland in sufficient quantities to make about half the tin needed (200 kg) for the bronze made in EBA Ireland.[3]

It is certain that in Britain, Cornish deposits were exploited and tin bronze gradually ousted arsenical copper throughout the British Isles in the years 2000–1800 BC. The earliest date for the production of metallic tin in Britain is given by the piece of tin smelting slag found in a ritual burial at Caerloggas, Cornwall, dated to about 1800 BC.[4] Before this, bronze could have been made either by adding smelted tin metal or by adding high-grade tinstone pebbles under a reducing cover to a copper charge in a crucible.

It is very probable that in the early phases of the Bronze Age the metallurgists were a class apart, and were not entitled to ceremonial burial.[1] Thus detribalized, they were liberated from the bonds of local custom and they therefore had the freedom to travel, and perhaps settle where they could find markets for their products, even abroad. Since copper and, more particularly, tin ores are restricted to certain well defined areas of the British Isles, the early metallurgist had to have his own personal trade contacts. It is perhaps possible that he mined and smelted his own material and then carried the finished products round certain limited areas, rather like the Breton onion seller of today. Alternatively, the itinerant caster with his few stone moulds travelled the area calling at intervals on the smelters in permanent settlements in order to replenish his supplies.

The Late Bronze Age altered much of this. The use of clay moulds, instead of a few stone moulds, widened the technique; bronze workers seem to have been more common and processes were capable of greater and more rapid output. Much of the metal of this period occurs in the form of hoards. These are mainly of two types. Firstly, those which seem to have been deposited as part of some ritual requirement, often in marshy ground as part of some water cult; these are often bent or mutilated. Secondly, those which are clearly founder's hoards, containing scrap metal and pieces of copper ingot. The pieces are often broken or bent to make them more suitable for remelting.

Many of these hoards can be dated to the later phases of the LBA and it is possible that the smith never returned to collect his hoard because the demand for this metal had ceased. Hoards of this period contain tens, or even hundreds of items. Secondary (scrap) metal seems to have constituted a large proportion of the total amount of metal used. The bronze founder no longer needed to seek out a piece of suitable stone and carve it, but could now make a clay mould or cast a bronze mould whenever it was necessary. Most of the fragments of clay moulds found are from two-piece moulds and are not single-piece 'investments', which tends to show that a straightforward moulding process was used in which the two pieces of moulding material were pressed around a wooden or metal pattern to obtain the impression. The pattern was removed by stripping off one of the half-moulds from the pattern, whereupon the pattern could be removed from the other half-mould and re-used. After drying, the two half-moulds could be reassembled like the stone moulds but could be held together by an outer wrapping of mould material.

The alternative technique is the 'lost-wax' process in which a wax model is invested with clay, then dried, and the wax melted out. This needs a new pattern every time or some technique for the rapid reproduction of wax models. There is no evidence for the use of this process in the British Isles before the Early Iron Age.

By the Late Bronze Age there was a good deal of international trade and an enormous increase in output. It seems that bronze workers travelled considerable distances on pack-horses, so large are the hoards now found. The number of Late Bronze Age objects surviving is large; the hoards exceed those of all previous phases put together by ten times, and as a rule contain a larger number of castings. Previously, copper or bronze was reserved mainly for weapons and smith's tools, but now the community could afford to use cauldrons, buckets, carpenter's tools, and ceremonial objects, after the reduction in price ensuing from increased output. Naturally, the smelting branch had to expand its activities, and the supply of metal from Ireland was supplemented by the exploitation of sources in Wales, N. England, Scotland, and from the Continent.

LOCATION OF COPPER ORES

A glance at the map (Fig. 7) at once shows that the sources of copper ores are confined to Ireland, Scotland, and the western seaboard of England and Wales, with one or two exceptions. The distribution of copper ores is in many ways similar to that of

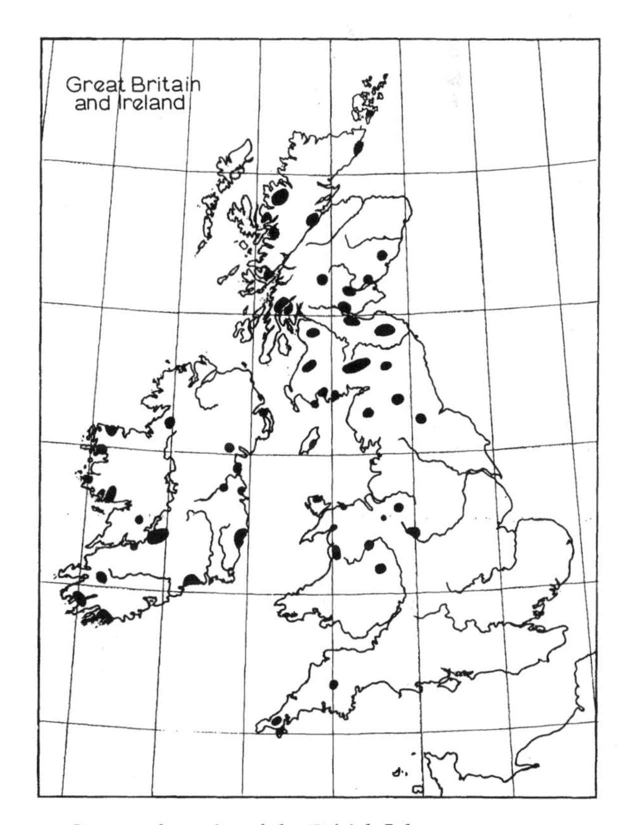

7 Copper deposits of the British Isles

Table 6 Copper deposits in the British Isles

Ireland	Scotland		Wales	England
Avoca (Wicklow)	Crinan (Strathclyde)	Moffat (Dumfries & Galloway)	Great Ormes Head (Gwynedd)	Alderley Edge (Cheshire)
Dungarven (Waterford)	Islay (Strathclyde)	Girvan (Strathclyde)	Plynlimon (Dyfed)	Keswick (Cumbria)
Tramore (Waterford)	Mull (Strathclyde)	Garnoch Valley (Strathclyde)	Amlwch (Anglesey)	Richmond (Yorkshire)
Skibereen (Cork)	Morvern (Highland)	Burray (Orkney)		Alston Moor (Cumbria)
Kenmare (Co. Kerry)	Loch Duich (Highland)	Rousay (Orkney)		Ecton (Derbyshire)
Killarney (Co. Kerry)	Russel (Loch Kishorn, Highland)	Wick (Highland)		Cheadle (Staffordshire)
Lough Derg (Tipperary)	Loch Garbhaig (Kinlochewe, Highland)	Loch Ness		Wenlock (Salop)
Lough Corrib (Galway)	Loch Fyne (Strathclyde)	Loch Tay		Mullion and Callington (Cornwall)
Connemara	Luce Bay (Dumfries & Galloway)	Strathearn (Tayside)		Bere Alston (Devon)
N. Mayo	Cree Valley (Dumfries & Galloway)	Glenesk (nr Montrose, Tayside)		Coniston (Cumbria)
Pollboy (Leitrim)	Fleet Valley (Dumfries & Galloway)	Ochil Hills (Tayside)		Llanymynech (Salop)
Lough Shinny (Dublin)	Dee–Ken Valley (Dumfries & Galloway)	Calton Hill (Edinburgh)		Peckforton Hills (Cheshire)
Beauparc (Meath)	Urr Estuary (Dumfries & Galloway)	Lammermuir Hills (Lothian)		
Brownstown (Meath)		Leadhills (Strathclyde)		
Dungannon (Tyrone)		Sandwick (Shetland)		

Based on references 5–11, with additions

lead, and in marked contrast to iron, the ores of which are more widely distributed, with a bias towards the central and eastern regions of England. The places where copper ores are known to occur are listed in Table 6.

It was probably not long before peoples entering the country by way of the Irish Sea found plentiful supplies of material exposed in outcrops, whereas those entering the country from the east would be disappointed, and would take time to develop and exploit the ores of the western seaboard and Ireland.

It is probable that south Ireland was the main source of copper in the west; native copper and copper ores outcrop in the cliffs of Waterford near Bunmàhon. Outcrops of native copper have also been reported from Cornwall (Mullion),[12] and ores outcrop in south Scotland in Wigtownshire [now Dumfries and Galloway].

Primitive tools such as stone hammers and wooden shovels have been reported from many workings. In the absence of other more reliable dating evidence these finds cannot be accepted as proof of prehistoric working, for primitive tools were used in mining until comparatively recent times. But the Mount Gabriel mine has yielded timber giving a date in the 14th century BC.[13]

Most non-ferrous ores are complex. The copper ores often contain substantial quantities of lead and zinc and, occasionally, tin. The slag from Hayle (*see* Table 12) shows this, and many coppers contain lead. In an area such as Cornwall it would be possible for the smelting of a tin-bearing copper ore to give a bronze.

Copper from Mullion is not found in contact with tin ore, but copper and tin ores occur together in the Camborne–St Just area.[12] Jope[14] has discussed the joint occurrence of copper and lead ores. Such ores are associated near Conlig (Co. Down) and Alderley Edge, and probably in Wigtownshire where a plano-convex ingot was found which also contained tin.[15] Most of the non-ferrous metals occur in heavily mineralized areas where a large number of minerals occur together and it would be very easy to introduce galena (PbS) which, owing to its weight, would not separate from copper on washing. The Leadhills area of south Scotland is an example of this type of deposit, and cerussite, galena, malachite, and chalcopyrite have been reported from this district.[16]

Figure 7 shows the deposits of copper ores that are known to exist and which have been exploited at some time or other. Those that outcrop today are not necessarily the only ores to have been exploited by early man. Many of the ores which are now exploited by deep mining methods may have had outcrops which were used in early times, traces of which have now disappeared. The lack of evidence of early mining in a later developed deposit is no evidence for its neglect by primitive peoples. Later methods of mining are notorious for covering up traces of mining by earlier miners, particularly when the earlier methods were open cast.

Some mining sites have been examined intensively with a view to determining the date of their first exploitation. Alderley Edge in Cheshire is an example of this.[7,8] In historical times this site was worked for copper from about 1717 to about 1879 and again in the 20th century. In 1876 it yielded 15 000 t of ore, from which 301 t of copper were extracted. A large number of stone hammers were found in the old workings in 1874 and Boyd Dawkins made an examination of these and the workings.[8] The hammers were found in open cast surface workings – about 2·4–3·6 m deep. A hundred stone tools are said to have been discovered together with an oak shovel. Lead is also found at Alderley Edge and both lead and copper were worked in the 18th and 19th centuries. The lead occurs as galena and the copper as malachite.

Some circular pits sunk near the surface into superficial sandstone beds were found to be rich in greenish nodules of chrysocolla ($CuSiO_3.2H_2O$), and stone hammers were discovered nearby. More hammers, which can be seen in the Manchester Museum, were found in lead and copper beds near Mottram St Andrew to the north-east.[7] Pits 3·6 m deep had been cut into the rock near the Engine Vein, Alderley, and some holes had been made at right angles to these. In the debris filling these pits, slag from copper smelting was found, and thin layers of charcoal and calcined ore were found at Dickens Wood mine.

A section made by the Geological Survey in 1866 is shown below and indicates the complex mineralization of this deposit. It should also be mentioned that iron bloomery slag was found to the south-west near Welsh Row.[8]

		Thickness
Top	Sandstone	1·40 m
	Clay with copper at bottom	0·78 m
	Ferruginous sandstone and lead	1·90 m
	Cobalt beds	1·40 m
	White sandstone with cerussite ($PbCO_3$)	1·53 m
Bottom	Ferruginous sandstone with Mn, Co, and iron ores	3·70 m

This site is typical of many in the British Isles. The mode of occurrence of the ores suggests an early exploitation, but the remains are not closely datable; in the absence of closely datable artifacts such as pottery, they could belong to any period between the Late Bronze Age and the Early Medieval period. The simultaneous occurrence of copper and lead together with manganese and cobalt should be noted, although these last two elements should be confined to the slags.

Another site extensively investigated is that at Great Ormes Head, Gwynedd.[17] The mine dumps were found to contain stone hammers and pestles which were probably thrown out of depths cleared and mined in the 19th century. As this evidence was inconclusive for dating the earliest use of these ores, Davies[17] examined the base of the hill near the Gogarth Hotel, where he found sherds of 2nd century Romano-British ware. Stone hammers were also found, similar to those found on the dumps above. No ore or slag was found and very little charcoal; smelting could therefore not have been carried out in the vicinity, probably because of the absence of wood.

Gardner[9] examined these remains in connection with the possible significance of a hoard of Roman coins found near Little Ormes Head in 1907. Roman occupation, according to Gardner, was proved by the coin hoards and sections of 50 cm diameter plano-convex ingots found in 1921, north-east of Dinarth at Llandrillo-yn-Rhos. The mines were last worked between 1835 and 1847, when they are said to have produced £240 000 worth of copper.

The Roman period has produced about 16 plano-convex ingots varying in weight from 12 to 22 kg. Some of these have been inscribed or stamped (*see* Fig. 10), suggesting a period of well-organized exploitation of the British mines as is certainly the case for the lead mines of the Roman period. The great majority have been found in North Wales and Anglesey, and the fact that most of them were found in the latter suggests quite clearly that the well known Parys and Mona mines just south of Amlwch were the sources of the ore. If so, all trace of such working was destroyed in the intense exploitation of the 19th century. The ore won in the 19th century was hard to work and had a low copper content (1·5–2%),[18] but it is almost certain that the surface ores mined by earlier workers were softer and richer. The mines on Great Ormes Head could account for some of the ingots found in other parts of North Wales.

In England, the mines at Keswick were worked by AD 1250, the Devon mines in Edward III's reign and the Cornish by Henry VI's reign.[19] Percy[18] records that ores near Richmond (Yorks.) were smelted in the 18th century at Middleton Tyas and recent work has been done on this site by Raistrick[20] and Hornshaw.[21] The smelting of ores from Ecton, Derbyshire, was carried on at Ellastone near Ashbourne in about 1686. In 1750, ores from Ecton were smelted at Whiston, and in 1780, Staffordshire ores were being smelted at Cheadle.

In Ireland, smelting of the Avoca, Wicklow, ores was being carried on at Ballymurtagh in 1780–90,[18] and mining has recently been resumed in this area,[22] and in other areas of Ireland.

COMPOSITION OF COPPER ORES AND ITS SIGNIFICANCE

Copper ores such as those shown in Table 7 may be divided into two types:

1 Oxidized ores which contain copper in the form of malachite ($CuCO_3.Cu(OH)_2$), chrysocolla ($CuSiO_3.2H_2O$), olivenite ($Cu_2(AsO_4)OH$), or cuprite (Cu_2O). Such ores are probably formed by the natural oxidation or weathering of sulphide ores in successive stages: $Cu_2S \rightarrow CuSO_4 \rightarrow CuCO_3$, often with the aid of surface waters. Such oxidized ores are likely to occur on top of the ore deposit, deeper mining producing the less oxidized ores and finally, the sulphides.

2 Sulphide ores such as chalcocite (Cu_2S) or chalcopyrite ($CuFeS_2$) and the so-called grey ore – *fahlerz* – containing both tetrahedrite (Cu_3SbS_3) and tennantite (Cu_3AsS_3).

The majority of copper sulphide ores contain substantial amounts of iron sulphide, i.e. iron pyrites, which would be converted to oxide on weathering and would be present to a greater or lesser extent in the ore body. The majority of oxide ores seem to be relatively free of iron, and the iron content has often been made the criterion for deciding whether an oxide or a sulphide ore has been smelted, but the need for a ferruginous flux in smelting a siliceous ore negates this argument.

In most cases the copper is dissolved from the surface ores, and this gives rise to a zone between the weathered surface ores and the deeper sulphide ores, where a process of secondary enrichment has occurred. In this zone the primary sulphides are enriched by the deposition of copper and other minerals leached out of oxidized zones above. This is the zone in which the grey *fahlerz* ores occur and these, because of their high concentration of copper, are supposed to have been one of the most sought-after sources of ore for the earliest smelters. In this area, above the water table, are also found deposits of enriched oxide minerals.

Table 7 Analyses of copper ores, %

	Cu	Sn	Pb	Ag	Ni	Co	As	Sb	Bi	Fe	Zn	Mn	P	Al	Mg	Si	Ref.
Lower limit of detection		0·005	0·0001	0·001	0·001	0·001	0·02	0·001	0·0001	—	0·055	0·00005	0·01	0·005			23
Cuprite Cornwall		n.d.	0·01	0·001	n.d.	n.d.	0·25	0·05	n.d.	0·02	n.d.	n.d.	n.d.	0·05	0·0001	0·5	23
Malachite Somerset		n.d.	0·0001	n.d.	n.d.	n.d.	0·1	0·1	n.d.	0·03	n.d.	0·03	0·02	0·08	0·0005	0·01	23
Malachite Camborne, Cornwall		0·02	0·001	n.d.	n.d.	n.d.	0·25	0·02	n.d.	0·02	n.d.	0·0001	0·01	0·07	0·0001	0·1	23
Malachite Limerick		n.d.	0·1	n.d.	0·003	n.d.	n.d.	0·08	n.d.	0·1	0·1	0·1	n.d.	0·15	0·0005	0·08	23
Chalcocite St. Just, Cornwall		0·02	0·0001	0·002	n.d.	n.d.	0.5	n.d.	0·0001	0·5	n.d.	0·0001	n.d.	0·03	0·0005	0·02	23
Grey ore Tetrahedrite Ballycummisk		n.d.	n.d.	0·01	n.d.	0·001	~ 5·0	~10·0	0·01	0·25	0·05	n.d.	n.d.	0·02	0·001	~5·0	23
Chalcopyrite Ballycummisk		0·005	n.d.	0·001	0·001	0·0003	n.d.	n.d.	0·005	~30	0·005	0·0005	n.d.	0·05	0·0001	0·01	23
Grey ore Kenmare, Kerry	40·3	—	—	0·15	—	—	16·7	3·7	—	4·54	3·18	S = 25·3		HgS = 0·56		5·29	5
Chalcopyrite Nantyreira, Wales	28·7	—	1·16	n.d.	n.d.	n.d.	—	0·36	n.d.	0·30	n.d.						24
Chalcopyrite Nantyrarian, Wales	13·1	—	0·60	n.d.	>tr.	n.d.	—	0·04	n.d.	14·8	18·8						24
Chalcopyrite Nantyricket, Wales	20·9	—	0·51	tr.	>tr.	n.d.	—	>tr.	>tr.	24·3	>tr.						24
Chalcopyrite Schull, Co. Cork	>30	n.d.	0·002	0·001	n.d.	n.d.	n.d.	0·005	n.d.	~30	n.d.	n.d.	n.d.	0·01	0·001	0·003	25
Bornite Levant, Cornwall	>30	0·1	0·002	0·002	n.d.	n.d.	0·05	0·01	0·03	>20	n.d.	0·02	n.d.	0·2	0·001	0·2	25
Olivenite Gwennap, Cornwall	>30	n.d.	0·005	n.d.	n.d.	n.d.	~20	n.d.	n.d.	0·1	0·01	0·001	1–2	0·02	0·002	0·1	25
Chalcocite Levant	>30	~5	0·01	0·01	—	—	0·3	0·005	0·003	>20	~ 3	0·001	n.d.	0·3	0·002	0·05	25
Tetrahedrite Loch Tay	>30	0·005	0·003	~1	0·02	n.d.	~ 5	~20	0·001	~ 3	~ 3	0·01	n.d.	0·01	0·03	0·4	25

Tetrahedrite St. Blazey, Cornwall	>30	0·01	~0·3	~2	n.d.	n.d.	0·5	~20	n.d.	~ 3	~ 3	n.d.	n.d.	0·001	0·001	0·03	25
Tetrahedrite Allihies, Co. Cork	>30	n.d.	0·002	0·02	0·01	—	~ 1	~20	0·02	~ 5	~ 1	n.d.	n.d.	0·02	0·002	0·03	25
Tetrahedrite Ballycumisk, Co. Cork	>30	n.d.	0·002	0·03	0·01	—	0·5	~20	0·01	~ 5	~ 1	0·05	n.d.	0·4	0·3	~6	25
Tennantite W. Wheal Jewel, Cornwall	>30	0·5	0·1	0·03	0·02	—	~ 5	< 0·01	0·005	~ 2	n.d.	0·05	n.d.	0·02	0·001	~5	25
Tennantite Wheal Jewel	>30	n.d.	0·002	0·005	n.d.	—	~ 7	< 0·01	n.d.	~ 5	n.d.	n.d.	n.d.	0·001	0·0001	0·02	25
Tetrahedrite Lanreath, Cornwall	>30	n.d.	0·05	~2	0·01	n.d.	0·5	~20	0·003	~ 3	~ 1	n.d.	n.d.	0·001	0·001	0·02	25
Tetrahedrite Par, Cornwall	>30	0·03	0·03	~2	n.d.	n.d.	0·5	>20	0·001	~ 5	~ 3	n.d.	n.d.	0·02	0·001	0·1	25
Tetrahedrite Cornwall	>30	n.d.	0·01	~2	n.d.	n.d.	~ 2	>20	0·03	~ 3	~ 3	n.d.	n.d.	0·003	0·001	0·02	25
Limit of detection*		0·005	0·002	0·0005	0·01	0·01	0·05	0·005	0·0005	0·001	0·01	0·0001	0·3	0·001	0·0001	0·002	25

n.d. = not detected.
* Note that limits of detection of results for Ref. 25 are different from others.

Also sought but not found, with approximate limits of detection, %:

Li	0·1	Ba	0·1	Tl	0·01?	Pt	0·01	Nb	0·1?
Na	0·1	W	0·05	Ge†	0·01	Y	0·1	Ta	0·1?
K	2	Sc	0·1	Cd	0·01	Zr	0·01	Cr	0·02
Be	0·01	Ce	1	Hg	0·1?	Hf	0·01	Mo†	0·02
Sr	0·17	Ga	0·01	Au	0·01?	Th	0·1	La	0·1

† Applies only to results from Ref. 25

Most copper ores are relatively inconspicuous, but the weathered surface with its high iron content (i.e. the *gossan* left behind by dissolution of the copper-bearing minerals), gives an indication of what lies below. The grey ores are, of course, sulphides, and contain appreciable quantities of arsenic, antimony, and silver, but unlike the deeper deposits are relatively low in iron, which has remained in the oxidized zone above.

As iron oxide is the main component of copper smelting slag, most of it should be removed from the metal into the slags during smelting or refining. An oxidizing roast of the ore before smelting will tend to oxidize iron rather than copper, and will assist the iron to find its way into the slag. Iron is therefore not considered to be an element which will give a useful indication of the type of ore used and its provenance. The same can usually be said for cobalt and manganese.

The compositions of various types of ore from England, Ireland, and Wales are given in Table 7 and further analyses of Irish ores have been given by Butler.[26] The two specimens of grey ore from Ireland contain appreciable arsenic and antimony, and relatively small amounts of iron, compared with the 30% present in the sulphide, chalcopyrite.

Obviously, the best way of deciding which type of ore has been used to make an artifact is to consider the relative proportions of elements that are as reducible or as oxidizable as copper itself. These will appear in the smelted copper in the same relative proportions as in the ore, and will be little affected by slight differences in smelting technique. Certain elements, such as silver, gold, and nickel will be concentrated during smelting and therefore will occur in a greater quantity in the smelted product than in the ore. Using rich ores, the concentration factor is not likely to be much greater than about three.

Arsenic and antimony behave very similarly under smelting conditions and are therefore likely to be present in the smelted copper in the same ratio as in the ore. For example, from Table B (Appendix), we see that for Irish artifacts the ratio of arsenic to antimony is generally 3:1. This would suggest an ore deposit with these elements present in much the same ratio. The ratio in the deposit at Kenmare, Co. Kerry (Table 7) is approximately 4:1. Furthermore, the silver content in this deposit is 0·15% which, due to concentration during smelting would give rise to a content in the metal of about 0·3%. This is precisely the average silver content of the group shown in Table B (Appendix). The Ballycummisk deposit has an As/Sb ratio of about 1:2, but such a ratio does not seem to be present in any early copper artifact, although it is found in Later Early Bronze Age artifacts (Table C, Appendix). There must have been many other deposits of this type which have now been worked out.

The ore from Kenmare, and one of the Welsh ores contain a substantial amount of zinc but this element is very rare in pre-Roman metal work. Northover's analyses[27] for Wales and Ireland rarely show more than 0·16% Zn in BA bronzes and one must assume either that these ores were not used or that zinc has been almost completely eliminated in the smelting process owing to its relative oxidizability and volatility.

Whereas many of the plano-convex ingots are probably the product of smelting, one from Carleton, Dumfries and Galloway must be the result of re-melting alloyed material because of its lead and tin content.[15] The occurrence of lead and copper ores together has already been discussed and there is a definite possibility of lead being in the ore before smelting. The presence of 1·85% tin is more puzzling. No appreciable amount of tin is present in the British ores shown in Tables 7 or 8. Coffey[30] has investigated this question and found that tin occurred in considerable quantities in some of the Cornish copper ores arriving in South Wales at the end of the 19th century. It can also occur in copper ores from non-tin districts, especially in the grey ores. A black (impure) copper smelted at Swansea in 1900 from Cornish ores contained 0·7% tin. One of the ores contained 0·94% tin and 12·3% copper which would give a ratio of copper to tin of 93:7, assuming equal losses on smelting. A piece of a plano-convex ingot from the Aylesbury hoard[31] contained 0·76% tin, and an ingot of LBA type from Plymouth harbour[32] had 0·4% (approx.) tin (*see* Table 9).

Although tin is not normally thought of as being a constituent of Irish copper ores (*see* Table 7), in the Goldmines River in the copper district of Avoca, 1·6 kg of tin ore was found in 68 kg of sand.[30] Tin also occurs in a lode at Dalkey, Co. Dublin, together with lead and zinc.

Jackson[11] estimates that the Goldmines River, and other rivers in the area would have been adequate to supply all the tin required for the modest amount of tin-bronze made in Ireland during the EBA.

THE INTRODUCTION OF SMELTING

Before the introduction of smelting, two high-temperature techniques were available to early man: the ability to melt native metals and the ability to make pottery. The melting of gold would require a temperature of 1 063°C, and the melting and pouring of pure copper at least 1 150°C. A temperature of the order of 800°C is needed for firing a usable pot.

Since the open camp-fire is not capable of giving a temperature of more than 700°C, some other method of producing a higher temperature must have been used at an early date. Most pottery kilns make use of an induced draught, i.e. a chimney, to attain higher temperatures, but it is difficult to use this type of furnace to smelt or melt copper. A better method would be the insertion of a pair of bellows in such a position in the ordinary camp-fire as to raise at least a portion of it to a higher temperature.

Coghlan[37] carried out some experiments on copper smelting with open fires. He suggests that an open hot wood fire would give a temperature of 600–700°C, while pure copper carbonate (malachite) requires a temperature of 700–800°C for reduction. He found that a charcoal fire as shown in Fig. 8*a* exposed to a natural wind in March reached a temperature high enough to reduce malachite. But the atmosphere within was not sufficiently reducing, i.e. the CO/CO_2 ratio was not high enough, so the malachite was only calcined to CuO and no metal was produced. With just sufficient air, charcoal will burn to carbon monoxide:

$$2C + O_2 = 2CO$$

Table 8 Composition of some British and Irish copper ores, %

	Grey ore (*fahlerz*)*		Enriched ore† malachite + covellite	Chalcopyrite‡		Malachite§	
	Cornwall	Devon	Middleston Tyas	Wales (Parys Mt.)	Partly roasted flotation conc. Ireland (Avoca)	Alderley Edge, Cheshire	
Cu	15·5	12·5	—			1·58	7·21
Cu_2O			28·35	—	—	—	—
S	23·7	15·6	—	—	—	—	0·18
CuS	—	—	54·8	—	—	—	—
Cu_2S	—	—	—	10·0	18·2	—	—
Fe	41·7	15·0	—	—	—	1·02	2·49
FeO	—	—	—	22·44	23·4	—	—
Fe_2O_3	—	—	1·0	—	—	—	—
FeS	—	—	—	8·80	39·0	—	—
Sb	5·6	4·1	—	—	0·026	—	0·08
As	3·1	0·8	—	—	0·02	2·06	4·23
Ni	—	—	—	—	—	—	—
PbO	—	—	—		0·48	—	0·19
ZnO	—	—	—	0·037	3·76	—	0·09
SiO_2	8·5	47·3	1·63	49·5	14·5	84·70	75·96
MnO_2	—	—	—	—	0·064	0·50	—
CaO	—	—	—	—	—	—	—
CO_2	—	—	9·23	—	—	—	—
H_2O	—	—	4·30	—	—	—	—
$BaSO_4$	—	—	—	—	—	4·50	—
Total	98·1	95·3	99·31	90·8	99·4	99·12	90·43

— = not estimated
* After Napier, Ref. 28
† After Hornshaw, Ref. 21
‡ Ref. 29
§ Ref. 7

and malachite will be reduced:

$$CO + CuCO_3 = 2CO_2 + Cu.$$

However, if too much air is present a good deal of the carbon monoxide will be burnt to carbon dioxide:

$$2CO + O_2 = 2CO_2$$

leaving insufficient carbon monoxide to reduce the malachite.

In a second experiment, Coghlan put the malachite into a flat pottery dish, with an upturned porous pot over it, so making a sort of kiln or covered crucible. Otherwise it was as shown in Fig. 8*a*. After several hours at a red heat, a well reduced copper sponge was obtained which appeared to have been melted. This was repeated, but the malachite was first ground, and the result this time was a bead of copper. This result is rather unexpected since no reducing agent was put inside the pot. It can only be concluded that by placing the pot within the fire, the oxidizing effect of the external wind was much reduced, so allowing the carbon monoxide to penetrate the porous pot before it was oxidized externally to carbon dioxide. As a result of these experiments Coghlan believes that copper was first accidentally smelted in a pottery kiln. Possibly a copper ore had been used as a pigment and the conditions were so reducing that the pigment was converted to metallic copper and ran down the pot. However, there must have been a further step away from the kiln towards the hole in the ground or 'bowl hearth' and the use of bellows.

Bellows are seen in an Egyptian drawing of a bronze casting or smithing fire dated 1450 BC.[39] They must have been a very early introduction, since suitable skins from which they could be made were so readily obtainable. With the use of bellows the bowl furnace became a practical proposition and the crucible or kiln would become obsolete as a method of smelting.

Before we consider the smelting operations carried out on copper ores it would be as well to discuss the principles of the process. As we have seen, copper ores may be oxidized, i.e. sulphates, carbonates, or pure oxides; or sulphides, i.e. chalcocite (Cu_2S). The oxidized ores are found at the top or surface of an ore deposit and are the result of atmospheric oxidation or weathering of the sulphides. Naturally, early copper smelters would discover the oxidized ores and smelt them with relatively little trouble, and later, start to work on the sulphide ores. But we cannot give a date to this transition, and it is extremely probable that the technique of smelting sulphide ores was known very early. In the 15th century BC Mt Gabriel copper mine in Co. Cork, Ireland,[3] were found stone tools and low-grade chalcocite, chalcopyrite, and tetrahedrite – all sulphide minerals. However, it is very likely that the sulphide minerals were being used well before this. Many authorities have attempted to decide whether oxide or sulphide ores have been smelted by looking for traces of sulphur in the product. Unfortu-

Table 9 Composition of lumps of copper from founder's hoards and other Bronze Age sites

Provenance	Date	Description/ weight, kg	Composition, % Cu	Sn	Pb	S	Fe	Ag	As	Sb	Ni	Bi	Zn	Co	O_2	Reference
Irish	LBA?	Piece run from furnace	rem.	0·001	0·21	~1·0	0·01	~0·05	0·4–0·5	>0·02	0·03	0·000 1	0·05	tr.	~1·0	34
Gillan, Cornwall	LBA	Many pieces	98·74	0·008	0·001	0·74	0·027	—	0·002	0·005	0·004	0·000 2	<0·002	0·000 4	0·5	33
Weston Wood, Surrey	LBA	Many pieces	99·8	n.d.	0·13	0·039	0·037	0·018	n.d.	n.d.	n.d.	<0·01	n.d.	—	0·08	34
			99·7	n.d.	0·092	—	0·02	0·084	n.d.	n.d.	0·073	<0·01	n.d.	—	—	
Plymouth, Devon.	LBA type	2·56	94·27	~0·4	0·05	low	0·01	0·005	0·10	0·025	0·02	—	0·04	—	>0·04	34, 32
Aylesbury, Bucks.	LBA	19 lumps (8 kg)	97·3	0·76	nil	—	0·032	—	—	—	—	—	0·032	—	—	31
Hertford Heath	LBA		97·5	—	0·09	—	0·1	0·36	0·95	1·3	0·03	0·025	0·060	—	—	35
Gray's Thurrock, Essex	LBA	1 lump	92·56	0·003	0·02	+	0·005	0·02	0·009	0·004	0·05	0·000 4	0·001	—	+	35
Watford, Herts.	LBA	4 pieces	99·9	n.d.	n.d.	—	n.d.	0·022	n.d.	n.d.	n.d.	0·042	n.d.	—	—	31
Stuntney, Camb.	LBA		99·9	<0·1	<0·02	—	<0·006									
			99·9	<0·1	<0·02	—	<0·006									36

— = not sought; n.d. = not detected; + = present; tr. = trace

(a)

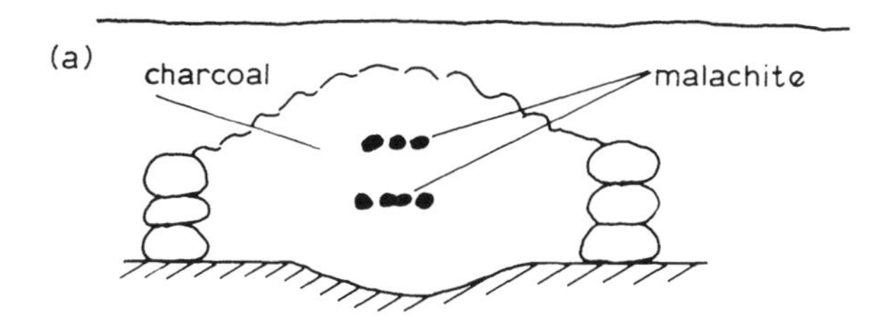

(b)

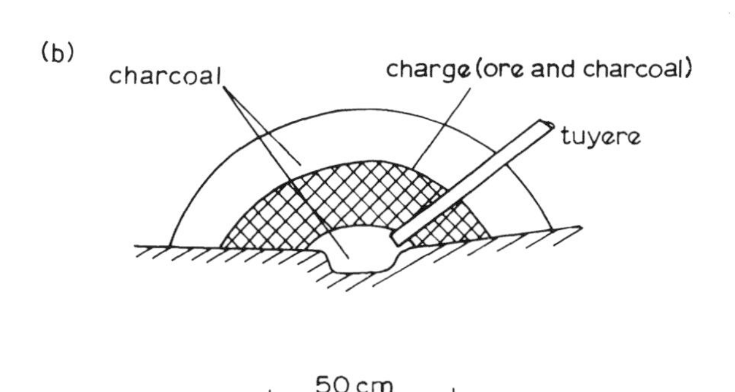

(c)

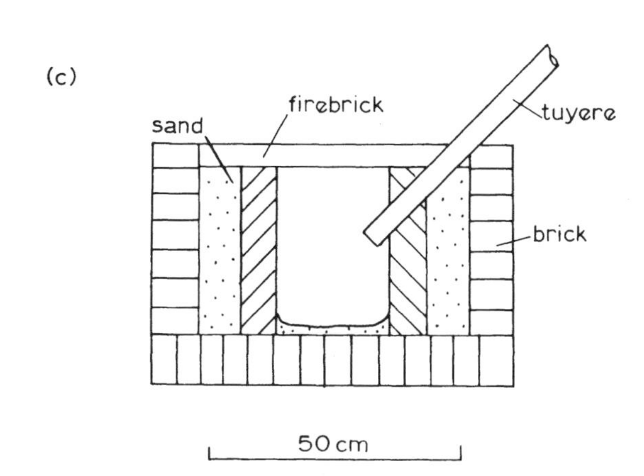

(d)

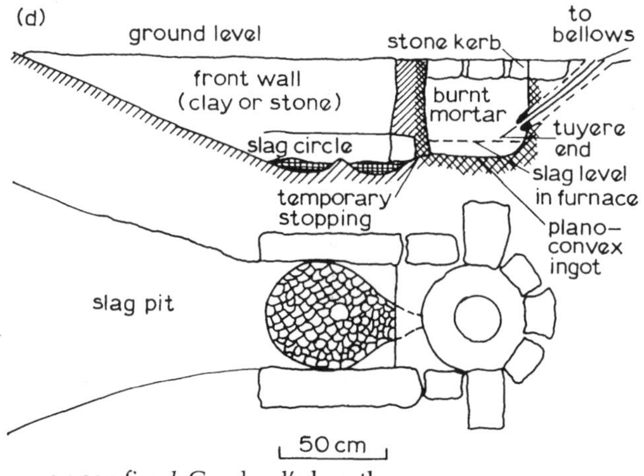

a open fire; *b* Gowland's hearth;
c experimental bowl furnace;
d furnace from Timna, Israel[38]

8 Copper smelting hearths and furnaces

nately, copper has a remarkable affinity for sulphur and traces are present in most coppers, and could have come from sulphates present in oxidized copper ores.

Because of the high melting point of pure copper (1 083°C), copper was usually smelted into prills or masses formed within the smelting furnace. The liquid copper could not be run out far, as its temperature was very little above the melting point. Early smelting furnaces undoubtedly consisted of a small hollow in a depression in the ground above which was placed the fire containing charcoal and oxidized ore, assisted with bellows. The copper was reduced from the ore, and dripped into the hollow below (Fig. 8*b*).

Metallurgically, no difficulty would be experienced with this operation, copper being readily reduced according to the reaction

$$CuCO_3 + CO = Cu + 2CO_2$$

Sulphide ores were either roasted to oxide, or first smelted to give sulphide and slag, and the resulting sulphide oxidized to copper. Percy[40] gives an account of primitive copper smelting by such methods in Sikkim. In one case, the copper sulphide ore with its silica gangue (unwanted mineral) was powdered, mixed with dung, and made into sausage-shaped lumps. These were then roasted on a hearth, so converting the sulphide to oxide according to:

$$Cu_2S + 2O_2 = 2CuO + SO_2$$

The copper oxide (CuO) was then put with charcoal into a 75 cm high shaft furnace in which the hearth was about 30–38 cm in diameter and 5–7·5 cm deep. The cupric oxide would be reduced according to the equation:

$$CuO + CO \rightarrow Cu + CO_2$$

and the silica gangue would be combined with some oxidized iron either added, or formed from the original ore in the roasting process to give a fayalite slag according to:

$$\underset{}{2FeO} + \underset{\text{(gangue)}}{SiO_2} \rightarrow \underset{\text{(slag)}}{2FeO.SiO_2}$$

This is similar in composition and properties to the primitive iron smelting slags. The 75 cm high shaft consisted of three sections which could be removed one by one when it was necessary to dismantle the furnace to remove the product. Three pairs of bellows were used, operating through three tuyeres, and the ingots produced weighed about 3 kg. (The size of the hearths was much the same as that of the Roman plano-convex ingots, listed in Table 10, and it is probable that these were made in similar furnaces.) The ingots were removed when nearly solid by inserting an iron rod into them and pulling them out.

The second method of smelting sulphide ores resembles present-day techniques. The sulphide ore is washed, and smelted in a furnace to separate the gangue and the copper sulphide and some iron sulphide. The copper sulphide undergoes no change in this operation and is left as a cake of 'matte' in the bottom of the hearth. A slag, similar in composition to the primitive smelting slag mentioned above but containing some copper sulphide, is run off and the cake of matte allowed to cool. This is then broken up, made into balls, and roasted to oxide. The oxide is then re-smelted in the original furnace with charcoal but this time is reduced to cakes of copper metal and some slag which solidifies on the surface of the ingot.

To show how easily oxide copper ores could be smelted, and how a tin–bronze could be produced by direct smelting of mixed minerals, Gowland constructed a primitive furnace. This consisted of a shallow sloping hearth with a central cavity 8 cm deep and 23 cm in diameter. A 2·5 cm diameter tuyere was

Table 10 Roman copper plano-convex ingots

No.	Provenance; date found; Grid Reference	Weight, kg	Size, cm Dia.	Size, cm Thick.	Present location	Inscription and remarks	Reference
1	Castellor, Llechylched	22	33	6·3	Cardiff	None	Cardiff Museum/41
2	Llanbeulan Parish, Anglesey (SH 335762)	15	28	5	—		41
3	Llanbeulan Parish, Anglesey (SH 335762)	Destroyed			—		41
4	Aberffraw, Anglesey, 1640 (SH 3568)	19	28	7·1	Mostyn Hall (cast at Cardiff)	SOCIO ROMAE – NATSOL	41 (*see* Fig. 10)
5	Bryndu, Rhosgoch, Anglesey, 1871 (SH 403902)	13	30	5	Cardiff	IVLS	41, 42
6	Bryndu, Rhosgoch, Anglesey, 1871 (SH 403902)	16	29	3·8	British Museum	IVLS (or IVFS) (CIL VII 1199)	41, 42
7	Bryndu, Rhosgoch, Anglesey 1871 (SH 403902)	12	29	—	British Museum	None	41, 42
8	Glanllynau Farm, between Criccieth and Afonwen, Gwynedd, 1908 (SH 457372)	20	33	6·1	?	IVLI IVLI and LCVSER	43
9	Glanllynau Farm, between Criccieth and Afonwen, Gwynedd, 1908 (SH 457372)	16	30	6·1	?	IVLI IVLI and LCVSER	43
10	Caernarvonshire (Gwynedd)	—	—	—	—	DOCN	Livens[44]
11	Carleton, Dumfries & Galloway, 1880	16	23	7·6	Edinburgh	Contains Pb and Sn	Whittick and Smythe[15]
12	4·83 km from Capel Curig, Gwynedd, near Carnedd Llewelyn, 1906 (SH 683644)	16	28	—	British Museum	OF-I XXXII SACV? (EE. IX 1259)	Whittick and Smythe[15]
13	Cerrig Dewi, Llangwyllog, Anglesey, *c.*1850 (SH 440779)	16	33	5	Unknown (Stanley 1873)	None	41
14	Cefn, Llanbenter, Aberffraw, Anglesey, 1827 (SH 3568)	14	31	5	Caernarvon 1873	None	41
15	Penmorfa, 1·61 km west of Tremadoc, Gwynedd, 19th century (SH 532424)	19	32	6·3	Cardiff	LENP Flat bottomed	45
16	Carnedd Llewelyn (S. Slope, Gwynedd) (SH 683644)	15	38	—	British Museum (purch. 1906)	ME (or MF) (EE. IX. 1258)	Whittick and Smythe[15]
17	Penmaenmawr, Gwynedd (SH 701754)	Part	*c.*30	*c.*5	Cardiff	1% Pb (*see* Table 11)	46

18	Dinorben, Clwyd, 1876 (SH 967757)	Part	*c*.30	*c*.5	(*see* Table 11)	46
19	Llanbedr-goch, Anglesey (SH 503804)	15	28	6·3		47
20	Edinshall, Borders	20·3				48
21	Tŷ n Dryfol, Aberffraw (SH 396725)					44
22	Parys Mt., Amlwch (SH 445900)				IVFS or IVLS	44
23	Parys Mt., Amlwch (SH 445900)				—	44
24	Tregele, Cemaes (SH 355926)	Part				49, 44
25	Bodfellion, Llandrygarn (SH 388783)					44
26	Tŷ Croes, Llanfaelog (SH 3474)					44
27	Llanfaethlu (SH 3187)					44
28	Olgra (SH 5181)					44
29	Llanfair-yng-Hornwy (SH 3190)					44
30	Dinarth, Llandrillo yn Rhos (SH 835800)	Part				44
31	Dinarth, Llandrillo yn Rhos (SH 835800)	Part				44

directed towards the cavity at an angle of 30° (Fig. 8*b*). A charge consisting of 7 kg of green carbonate ore containing 30% copper, 4·5 kg of 'cassiterite' containing 20% tin, 3·4 kg of limestone, and 4·5 kg of charcoal was heaped over the cavity. The charcoal and limestone were coarsely ground and mixed with the ores, and additional charcoal was placed in the cavity and ignited. The result was a bronze containing 22% tin.

Gowland's description is not very complete. What was the remaining 80% of the tin ore? And, assuming that the copper component in the copper ore was only malachite (50% Cu), what was the remaining 50% of the copper ore? Together with the added limestone we have about 1·7 + 3·5 + 3·6 = 9 kg of potential slag-forming constituents. If we assume that the gangue material of the copper ore was mainly silica and that of the cassiterite iron oxides, this material would form a fairly good slag through which the reduced ore would settle.

This calculation shows the essential difference between smelting a specimen of rich ore found in museums and geological collections such as those shown in Table 7 which in the main contain well over 30% of copper, and a real ore which today may contain only 1% Cu. In order to clarify this point and obtain data on the distribution of the minor and trace elements between the copper and the slag which, as we see above, forms the main product of the process, we carried out a number of experiments using a simple bowl furnace (Fig. 8*c*) and various ores.[24] Finding the ores available not sufficiently 'impure' for our purpose we also made up a number of artificial ores containing more of the elements Pb, As, Sb, Ni, and Zn so we could find out how much of these was recovered in the copper produced. Both oxide and sulphide ores were tested and it was necessary to add iron oxides to flux the siliceous gangue in the case of the former, and sand to flux the ferruginous gangue of the latter.

As long as the correct fayalite slag was formed and kept molten at a temperature of about 1 200°C, the reduced copper globules from the copper mineral would fall through it forming quite large lumps.

In the case of the sulphide ores we preferred to roast them first to avoid the two-stage 'matting' process. As long as the temperature did not exceed 850°C we found no difficulty achieving a complete or 'dead' roast although this part of the process proved rather smelly. Two of the sulphide ores came from the British Isles; one from Avoca, Co. Wicklow contained 18·2% Cu_2S, while that from Parys Mountain, Anglesey, Wales, ontained 10% Cu_2S (Table 8).

The artificial ores, oxidized and sulphidic, contained additions of about 5% of Mn, As, Zn, Ni, Pb, Sb, and Bi in the form of oxides or metals. The results showed that in the case of the direct reduction of oxide ores, all of the Mn was lost in the slag, all of the Ni and most of the As and Sb went into the metal, about half the zinc and the lead were lost, and most of the Bi was lost in the fume from the system.

Pre-roasting of the Avoca sulphide ore caused losses of As, Sb, and Pb but only a slight loss in the Zn, which was unexpected. No doubt, prolonging the roasting process could be made to reduce this element further. But smelting the roasted ore caused the majority of the remaining zinc to go into the slag.

Matte smelting of sulphidic ores causes much greater losses of the impurities. As the first, matting, stage is essentially an oxidizing process most of the easily oxidizable elements go into the slag or the fume. In fact, all but Ni, are markedly reduced. The second, roasting, stage continues this process so that the matting process, which could have been used on Irish ores from the 15th century BC onwards, is capable of giving a metal of much higher purity.

The main conclusion is that all the Ni, Ag, and Au will be recovered while varying percentages of the others will be lost depending on the process. Sulphur, oxygen, and iron will be the main impurities in the raw copper which, with care, can be removed in a crucible refining process.

SMELTING, MELTING, AND REFINING TECHNIQUES

Although no traces of early copper smelting furnaces have yet been found in Britain, a number of pieces of plano-convex ingots (Fig. 9) and other shaped pieces of pure copper have been discovered in Bronze Age founder's hoards and on other sites (Table 9). The sizes of these lumps vary from 0·1 to 7 kg, and it is very probable that they represent the raw material which the itinerant founder received from the copper smelter. The purity of these analysed is high, and it is clear that they are unalloyed. Unfortunately, complete analyses of many of the larger lumps of Late Bronze Age date have not been carried out, but most are comparatively low in lead and tin, and contain traces of arsenic and antimony. It would seem, therefore, that pre-Roman furnaces could produce pieces weighing up to 7 kg. Some of the pieces of LBA ingot material that have been fully analysed are remarkably pure and do not accord with the low purity and high iron content of the experimentally smelted material.[29] Thus, it appears that they are the product of a refining and re-casting operation of the sort that must have been used to produce the ox-hide shaped ingots of similar date in the Eastern Mediterranean.

Current experiments have shown that it is not easy to produce a well shaped ingot in the smelting furnace and that it is easier to remelt the prills and pieces of raw metal and at the same time refine them.

The main problem is to determine the type of furnace used exclusively for smelting in the later periods. We cannot attempt to answer this question for periods before the appearance of the Roman plano-convex ingots (sizes given in Table 10). These could have been formed in a furnace of the type shown in Fig. 8*d* with a crucible diameter of 23–38 cm. A mixture of an oxide or roasted sulphide ore flux and charcoal would be charged, and slag tapped from the surface at intervals. The ingot could have been left in the furnace and allowed to cool while residual fuel could have been removed from the surface to enable the ingots to be stamped. There are certain objections to this arrangement. For example, if the ingot was to be stamped *in situ*, the height of the shaft would have to be restricted.

Alternatively, and more probably, when the amount of copper smelted reached 14–23 kg, the furnace was tapped and the whole contents allowed to escape into a mould. In the case of copper with its high melting point it would not be possible to allow it to escape slowly from the furnace as it was being

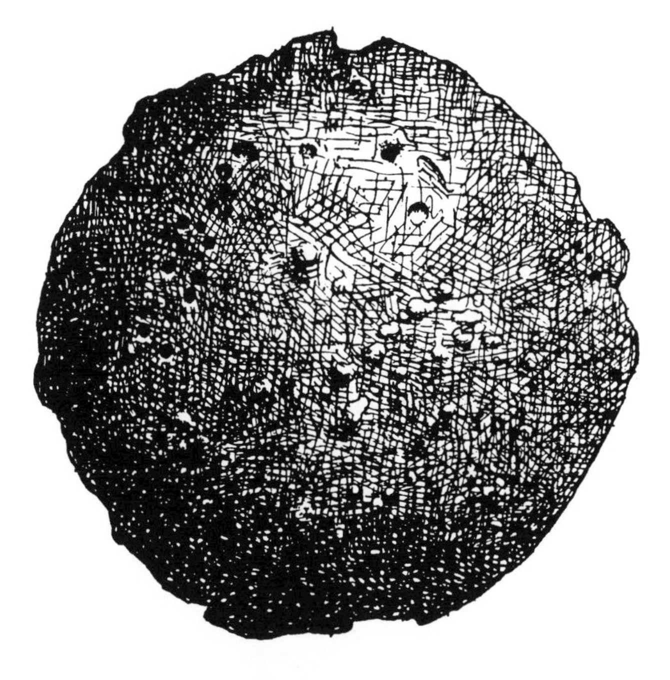

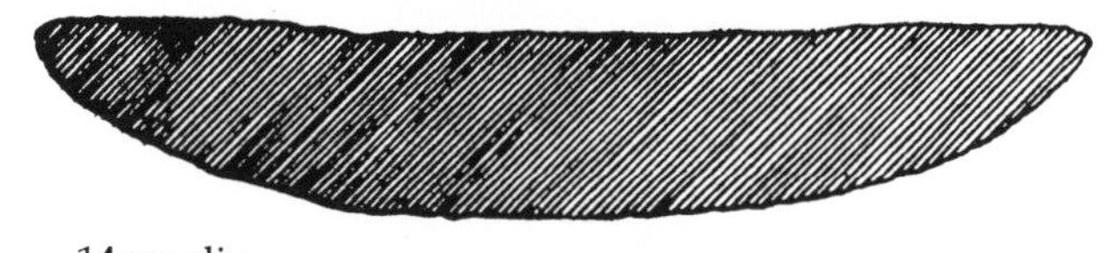

14 cm dia.

9 LBA plano-convex ingot from Worthing (by permission of the Trustees of the British Museum)

smelted, since it would be very little above its melting point and would tend to freeze before reaching the mould. In the case of lead, with a considerably lower melting point, this would be possible, and was clearly the method used in the Roman period.

The surfaces of the Roman copper ingots suggest that they were cooled in an oxidizing atmosphere, but that many of the stamps had been impressed before the ingots were completely cold, and while the copper was still soft.

Examination of those ingots in the National Museum of Wales suggests that they must have quite a high oxygen content, since they have a level surface and have not 'gassed'. Whittick and Smythe[15] analysed five of these ingots (Table 11), and concluded that mainly because of their low sulphur content, the Welsh specimens were the result of skilful smelting of pure oxidized ores, but, of course, the low sulphur content could be due to refining.

The two ingots from Amlwch probably have an oxygen content of about 0·5%; this would be responsible for the non-gassy appearance of the ingots, since oxygen in copper appears as the solid, cuprous oxide, which would produce sound metal. The ingot from Capel Curig is very similar to modern fire-refined copper and must have about the same oxygen content (0·04%).

Smythe was of the opinion that the ingot from Carleton, Wigtownshire [now Dumfries and Galloway], was a secondary remelted and alloyed product on account of its lead and tin contents. Its similarity to the Welsh ingots in size and shape strongly suggests that it was produced in much the same way, but the tin content suggests that it is more likely to be a Cornish than a Wigtownshire product as the copper ores in that area are not stanniferous.

One of the ingots from Bryndu (Table 10) had had additional metal poured on it, or tapped onto it, into which a stamp had been impressed. Another stamp had been put on the original surface of the same ingot while it was still hot enough to take it. The ingot from Aberffraw with SOCIO ROMAE NATSOL on it has

Table 11 Analyses of Romano-British plano-convex copper ingots, %

	A	B	C	D	E	F	G	H
	Amlwch (Anglesey)	Amlwch (Anglesey)	Carnedd Llewelyn (Gwynedd)	Capel Curig (Gwynedd)	Carleton (Dumfries & Galloway)	Penmaenmawr	Dinorben	Edinshall (Borders)
	Weight, kg							
	16·3	12·1	15·1	16·5	16·4	—	—	20·3
Cu	98·90	99·08	99·00	99·08	93·04		96·72	97·2
Ag	tr.	nil	0·03	0·03	0·14		nil	~ 0·1
Sn	tr.	nil	nil	nil	1·85		tr.	
As	0·03	tr.	0·06	tr.	tr.		nil	
Fe	0·11	0·04	0·09	0·10	0·13	tr.	nil	≤ 0·1
Ni	0·02	0·01	0·02	0·02	0·10	tr.	nil	
Zn	tr.	tr.	tr.	tr.	nil		nil	
S	—	0·04	—	—	0·02		—	
Insol.	0·35	0·07	0·18	0·13	—		—	
$CaCO_3$	—	0·39	0·73	0·41	—		—	
Pb	—	—	—	—	4·87	1·0	tr.	2·6
Sb	—	—	—	—	tr.		nil	≤ 0·1
Totals	99·41	99·63	100·11	99·77	100·15			

A, B, C, and D are in the British Museum
A to E analysed by Smyth, Ref. 15
F and G, Ref. 41
H National Museum of Antiquities of Scotland, Ref. 48

28 cm dia.

10 Roman copper ingot from Aberffraw, Anglesey (courtesy of the National Museum of Wales and G. C. Boon)

Table 12 Composition of copper smelting slags, %

	Cyprus (Apliki)[51]		Britain		
	LBA	Roman	Hayle[52] (18th cent.)	Redbrook[53] (19th cent.)	Modern[51]
SiO_2	21·3	28·8	—	55·04*	33·0
FeO	61·9	39·4	—	12·8	46·5
CaO	1·14	4·24	—	7·12	6·75
MgO	tr.	3·53	—	1·81	—
Al_2O_3	7·8	3·95	—	12·70	8·10
TiO_2	—	—	—	3·13	—
Mn	tr.	8·60	0·65	1·20	—
Cu	0·91	0·70	1·10	0·44	0·35
Sn	—	—	7·28	1·65	—
Pb	—	—	11·00	—	—
Zn	—	—	14·4	—	—
S	1·14	1·28	—	0·26	0·45
Alkali	—	—	—	2·23	—
Total	94·19	90·5	—	98·38	95·15

*Not all in solution: some present as free quartz
— = not analysed
tr. = trace

either had this inscription impressed when very hot, causing a lot of metal to be displaced round it, or has had additional metal run-on for the purpose (Fig. 10).

The Roman ingots are not as pure as some of the LBA ingots shown in Table 9, and they could be the direct product of smelting. However, the surfaces suggest that they have been cooled in an oxidizing atmosphere, i.e. outside the furnace, or are the result of re-melting and casting previously smelted metal.

So far, no large deposits of early copper-smelting slags have been found in the British Isles, but substantial deposits are recorded from Cyprus dated to the Late Bronze Age and Roman times (Table 12). The only distinctive feature is the copper and sulphur content, for in all other respects they are similar to iron-smelting slags of the bloomery period (approximately 700 BC to AD 1500). This does show, however, that when small quantities of ferruginous slags are found in areas where copper deposits are known to exist, they should be checked for copper. Large quantities (mounds) of slag in the British Isles are not likely to be the product of copper smelting, since the process was carried out on a much smaller scale.

The copper content of the slags shown in Table 12 is low: much lower than that of the crucible slags from melting. This is probably one way in which smelting and crucible slags can be distinguished. Smelting slags are low in copper mainly because of their fluidity and the high iron and manganese oxide contents. Melting (crucible) slags are mainly a product of the reaction between copper and the crucible and, as Tables 57 and 59 (Chapter 5) show, contain 20–28% of copper and little iron. The highest copper content of any early smelting slag so far reported is 5·8% for Enkomi,[51] although it can segregate to give higher values.

At Merthyr Mawr Warren a number of crucibles containing prills from copper smelting or casting with 3·6% Zn, rest copper, were found in a mound which was dated not earlier than 400 BC.[54] The pattern of the crucibles was very similar to those found by Bulleid and Gray at Glastonbury.[55]

At Glastonbury[55] a well preserved hearth was found, composed of three large stone slabs embedded in a raised circular area of clay 2 m in diameter. The largest stone was 0·86 m long. In the same area, a piece of a tuyere and fragments of crucibles were found.

One of the most interesting melting furnaces was found at Wilderspool.[56] Since, in most cases, melting entails alloying of the smelted copper with metals such as lead and tin, this operation is invariably done in crucibles. The small-scale melting furnaces would take only one crucible at a time but one of the Wilderspool furnaces seems to have been designed for taking at least seven crucibles. It measured 1·3 m by 1·0 m outside and was 0·84 m long inside. There were holes in the floor which supported the crucibles, and a crucible was actually found in one of them.

Since the crucibles were at floor level, the furnace could not have been like a pottery kiln in which the ware is supported above the fire; it must have been more like a baking oven, but with the fuel being burnt at the mouth of the oven while the crucibles were being heated. A suggested reconstruction is shown in Fig. 11. A similar type of furnace, recently found at Alésia in France, was used for the firing of moulds.[57] The temperature required (about 1 000°C), would be obtained by using an adequate chimney or by applying bellows to the stoke hole. This would in fact be rather

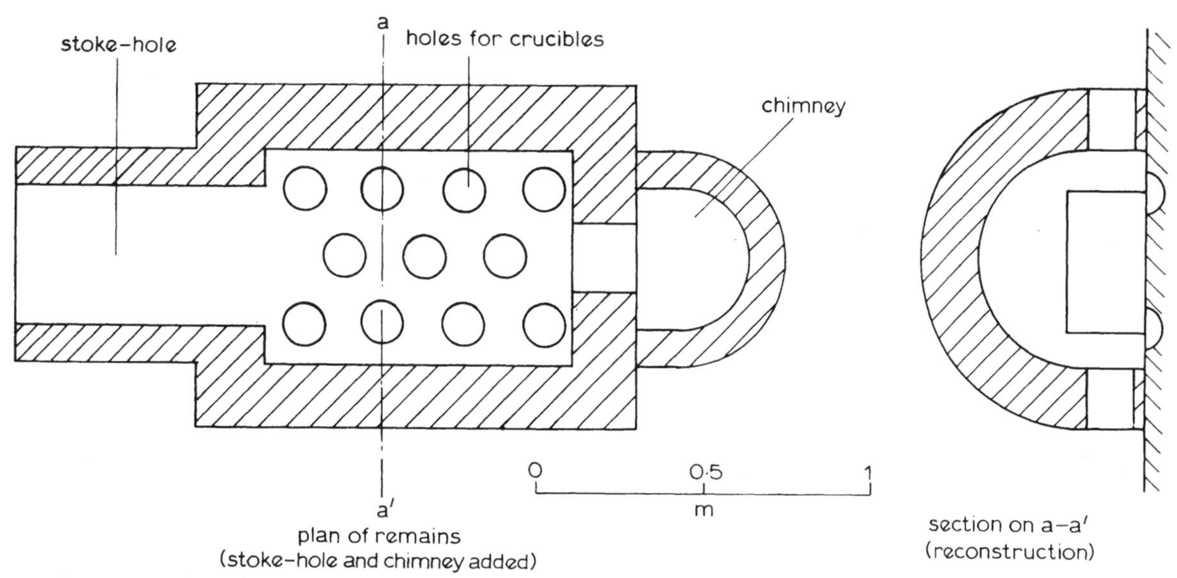

11 Reconstruction of Roman reverberatory crucible furnace found at Wilderspool, Lancs.

like a modern reverberatory furnace, but using crucibles instead of a hearth.

At Bac Mhic Connain in North Uist,[58] two furnaces were found in a wheel house not earlier than AD 200. One of these seems to be nothing more than a small melting or domestic hearth, but the other is an interesting rectangular stone-built furnace which might have been used for smelting. The furnace and hearth were connected by a floor of stone slabs, and the furnace foundations were 43 cm below the level of the smaller hearth. This is a shaft furnace of a type which has been recently investigated to see what temperatures can be obtained when using charcoal as a fuel.[59] It was found that provided that a sufficient chimney height was available, with a bed depth of 60 cm, a temperature of 1 100°C sufficient to melt copper could be obtained.

At Bac Mhic Connain the shaft stood 0·91 m above ground level, with an internal section of 35 × 27 cm. The west wall contained two openings, nearly the full internal width of the furnace in size, and separated by a stone 37·6 cm thick (Fig. 12). This furnace contained bronze slag and triangular pattern crucibles, but no ashes. Clay moulds were also found.

Another Hebridean site, on Barra[60] has also yielded the remains of bronze melting in the form of part of a crucible and a bronze residue containing very little tin. A hearth was found which seemed to have been connected with iron smelting, but was also probably used for non-ferrous metals.

A wheel house at Jarlshof, Shetland,[61] although producing no furnace structure, has provided evidence of a sand-filled casting pit in which moulds were probably supported while being filled. This pit was 0·28 m deep and would have supported long clay moulds, such as are used for swords at a slanting angle which would have assisted the proper filling of such a mould.[62] At a distance of 1·10 m away, a burnt area was found in which the paving stones were much calcined and the soil below burnt bright red over an area 0·5 m in diameter and 0·38 m deep. It would seem that there had been a succession of hearths in this area since it would not be possible to burn to a depth of 0·38 m by means of a single melting hearth. A bronze knife and fragments of clay moulds, but no crucibles, were found in this dwelling. In addition, fragments of clay moulds and bronze 'slag' were found in a midden. The bronze 'slag' contained in one case 22% Cu and

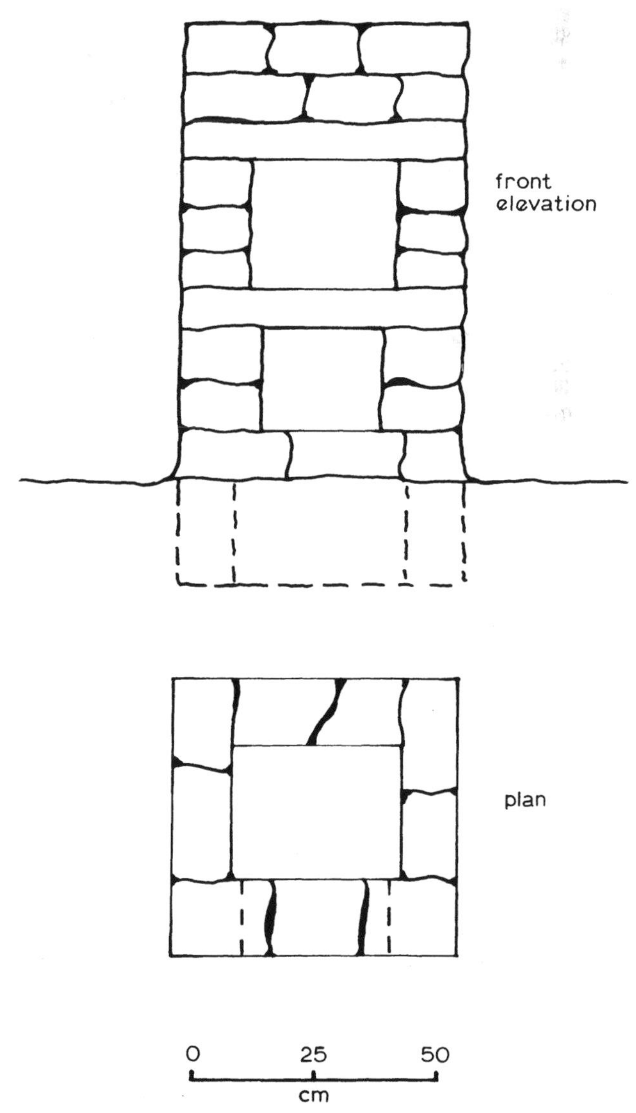

12 Stone-built furnace from Bac Mhic Connain, N. Uist, Outer Hebrides

2·88%Pb, and in another 10·6%Cu and 1·87%Pb, which again suggests the use of a Cu–Pb alloy. Fragments of a crucible which had been about 5 cm deep and 9·5 cm wide were found in another part of the same site.[63]

An excavation at the Roman site at Heronbridge[64] near Chester revealed a bronze-working hearth surrounded by a flagged floor. This hearth was about 0·76 m in diameter and was cut 15 cm into natural clay. At the bottom was a mass of siliceous bronze slag, accumulated over a long period. Several kilograms of slag were collected and found to consist of copper oxides (malachite and azurite) and lead silicate. This contained 33·2% Cu and 8·3% Pb and is almost certainly the product of melting copper–lead alloys. Fragments of clay moulds and crucibles were also found. This site was first occupied in the late 1st century AD.

Another Roman site which provided evidence of bronze working and crucibles is Wroxeter.[65] Furnaces were found here but since no smelting slag has been recorded it must be assumed that these were used for melting. A number of crucibles were found in the 1914 excavation and these, without exception, contained only traces of silver. It is extremely probable that the silver had been made by cupellation of lead on the site, since the remains of cupellation hearths were also found (*see* Chapter 4). Other evidence for bronze working in the Roman period has been found on many sites including Caistor by Norwich,[66] and Limbury.[67, 68]

Apart from crucibles, Irish sites have yielded little positive evidence of bronze working, but the crucible residues show clearly that melting had been carried out, in many cases in hearths used for smelting. Crucibles are reported from Garranes[69] (6th century AD) but there was no evidence of bronze smelting. Lough Faughan Crannog on the other hand yielded slags that have been qualitatively analysed.[70] Some were melting slags containing appreciable amounts of copper and tin, and traces of lead. Only in one case was the presence of tin questionable, and this appeared to be slag considerably diluted with occupation debris. In general the remains are those of crucible melting with strong circumstantial evidence of smelting. This site is dated to roughly the same period as Garranes, 7th to 10th centuries AD.

The above examples are only a selection of the sites on which industrial remains have been found. With the increase in the number of excavations in this country over the last 20 years, especially of settlement sites, a large number of industrial areas have been exposed. These range from Early Iron Age settlements such as Gussage All Saints in Dorset where a workshop for horse harness fittings was found,[71] to a bronze-working shop at Tintern Abbey dated to the 14th century AD.[72] Within this span, innumerable Roman sites such as Alcester, Manchester, and Usk have produced hemispherical melting hearths for non-ferrous metals, such as that at Heronbridge.

Considering the wide extent of bronze working, it is disappointing to have so little to report on the smelting side. But the above are all habitation sites and not mining sites and we have no examples of permanent smelting sites. It is possible that in the early periods smelting was carried out on the settlement sites but if so it has left little evidence.

By the Roman period, smelting was carried on near the mining areas as may be seen from the concentration of plano-convex ingots in North Wales. Medieval or later operations in these areas would be responsible for the disappearance of most of the evidence. Although most of the English and Welsh mining sites have suffered badly from recent exploitation, it would seem that some Irish mining sites might still have remains worth excavating.

COMPOSITION OF COPPER-BASE ARTIFACTS

It is now proposed to examine the artifacts of various periods with a view to detecting changes in smelting techniques and alloy technology. In considering the significance of any particular element or group of elements we must first separate the prepared alloys from the smelted products. This is not easy, but we will start with the general assumption that intentional alloying, as distinct from the presence of impurities, was not introduced until the coming of the simple tin bronzes in the Early Bronze Age.

The smelted metals or alloys can be divided into two groups:

(1) Those that show the state of metallurgical technique as deduced from the contents of arsenic or antimony or both.

(2) Those that point to special ore deposits, such as those containing traces of nickel or appreciable (>0·2%) amounts of silver.

The next division can be made on the basis of alloying. Here there are three groups:

(i) almost pure coppers containing less than 1·0% of any other element, i.e. impure coppers

(ii) tin bronzes containing more than 1% tin

(iii) other alloys containing more than 1·0% of any element.

In the present state of knowledge it is difficult to say whether alloys in group three are intentional or not. We know from the work of Brown and Blin-Stoyle,[36] for example, that during the Late Bronze Age lead was deliberately added to tin bronzes. Some Copper or Early Bronze Age artifacts contain as much as 7% arsenic;[73] this can have a very big effect on the mechanical properties and may have been intentional. However, the histogram for arsenic content given in Fig. 13 shows clearly that there are two groups: the 'pure' coppers in which arsenic is an unintentional impurity, and a group of arsenical coppers in which the arsenic content averages about 2%. The latter group almost certainly arises from a definite ore-body or mineral addition which has been selected either for the high strength of the metal made from it or because of its accessibility.

Copper and Early Bronze Age artifacts

Since the early analytical work of Coghlan and Case[23, 74] and Brown and Blin-Stoyle[36] a great deal of additional work has been done by Junghans, Sangmeister, and Schröder on Scottish and Irish material[75] and by the British Museum on British material.[76, 77] More recently, this has been supplemented by Northover and Burgess who have carried out work on the Welsh material for the Board of Celtic Studies[78, 79] and in doing so have studied the mode of distribution of

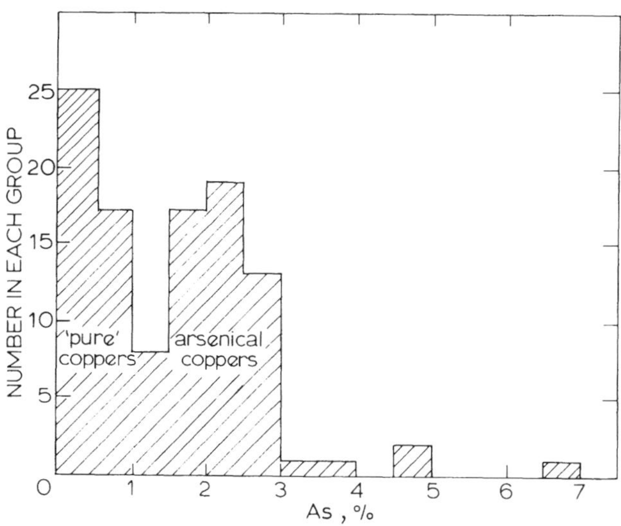

13 Arsenic contents of copper artifacts of the Copper and Early Bronze Ages (104 specimens)

the impurity and alloying elements in early copper-base alloys using the electron probe microanalyser (EPMA). This work has enabled Northover to reconcile alloying fashions with the archaeological phases of Burgess *et al.*, and this has both complicated the previous three-period structure of Early, Middle, and Late Bronze Age in Britain and thrown considerable light on the details.

In an earlier book,[80] I introduced the idea of a developing early Bronze Age and a full and later Bronze Age, leaving out the Middle Bronze Age. For Britain, Burgess and Northover[79] have divided the Copper–Bronze Age into 11 phases covering the period from the Early Beaker period to the Llynfawr–Hallstatt C period (Table 13).

The original three-period structure to which was added a fourth period – a Copper Age – has been made obsolete in overall Irish and British prehistoric metallurgy by the discovery of the fact that tin–bronzes were introduced very early into the metallurgical scene. As elsewhere, it is not surprising to find that calibrated ^{14}C dating has pushed back the beginning of the Copper–Bronze Age to hitherto undreamt of dates (2750 BC).

Strong regional differences are now apparent, with impure coppers from Ireland trying to hold their own on the British mainland against bronzes, probably introduced from the Continent.

Although it is possible that there was a pre-Beaker metallurgical tradition in Ireland we can certainly discount such a phase on the British mainland. Both Case[81] and Northover[79] group the earliest Irish type-sites such as Castletown Roche and Knocknague as being early Beaker, i.e. showing evidence of contact with people using Beaker burial traditions.

Case[81] believes that people with Beaker burial traditions were the first metallurgists in Ireland. He notices similarities in early thick-butted axe typology between Ireland and the Continent, but such examples are rare compared with Irish native types. He notes similarities with metal compositions which he equates with an Impact Phase in which continentally inspired (Middle Rhine) traditions have impinged upon Irish extractive metallurgy. He then considers the possibility of a putative pre-Beaker phase – perhaps the megalith builders – of people who have come from the east or the south (Iberia). However, no metal has been found in a primary context in an Irish

Table 13 Chronology of the Copper and Bronze Ages in Britain and Ireland (after Burgess and Northover[79])

Age	Date BC	Period	Industrial phases		Metals made: Irish	Metals made: British	Main artifacts made
EBA	2700–2100	Early Beaker	I	Castletown Roche	Cu (<2% As)	Cu–Sn	Flat axes
			II	Knocknague			Halberds
			III	Migdale			
	2100–1900	Late Beaker	IV	Willerby Wold	Cu–As	Cu–Sn	Knives
				Bush Barrow			Tanged spearheads
	1900–1600	Urn	V	Colleonard	Cu–Sn	Cu–Sn	
	1600–1500			Arreton			
MBA	1500–1400		VI	Acton Park I		Cu–Sn	Palstaves
	1400–1300			Acton Park II		(Cu–Sn–Pb)	Socketed spearheads
	1300–1200		VII	Taunton			Rapiers
	1200–1100	Early Urnfield	VIII	Penard I		Cu–Sn	
	1100–1000			Penard II			
LBA	1000–900	Late Urnfield	IX	Wilburton-Wallington		Cu–Sn + Cu–Sn–Pb	Socketed axes Palstaves
	900–700		X	Ewart Park			Leaf-shaped swords
	700–600	Hallstatt C	XI	Llynfawr		Cu–Sn–Pb Cu–Sn Fe	Hallstatt iron swords Bronzes

Table 14 Typical analyses from Early Bronze Age type-sites

Site	Artifact	Element, % Cu	Sn	Pb	Ag	Au	Ni	Co	As	Sb	B	Fe	Zn	Mn
Castletown Roche, Ireland[23]	Thick-butted axe	95·90	< 0·01	0·20	0·28	n.d.	<0·01	n.d.	2·00	0·93	<0·01	—	n.d.	n.d.
Knocknague (Kilbannen), Ireland[30]	Thick-butted axe 1874; 38	97·68	< 0·79*	n.d.	0·18	—	n.d.	n.d.	0·76	<0·79	—	n.d.	0·44	—
Migdale, Scotland[75]	Axe	rem.	>10·0	0	0·24	0	0	0	0·76	0·45	0	0	0	
Dorchester, Oxon.[82]	Knife	98·35	< 0·01	0·01	0·14	n.d.	0·2	n.d.	0·1	0·17	<0·01	—	n.d.	
	Rivet	90·00	6·60	0·21	0·085	n.d.	0·51	—	1·4	0·21	0·031	0·053	n.d.	
Bush Barrow[83, 84]	Dagger	90·0	9·6	—	0·01	—	<0·01	—	0·23	—	0·0042	0·006	—	—
	Dagger	95·0	—	<0·02	<0·005	—	<0·01	—	5·0	<0·05	0·0064	0·008	—	—
Colleonard, Scotland[75]	Axe	rem.	10·0	0	0·085	0	tr.	0	0·62	tr.	0	0	0	
Faversham[23]	Halberd	94·8	0·01	0·02	0·02	—	0·17	—	4·60	0·20	0·03	—	—	—
Arreton[83, 84]	Tanged spearhead	82	17·2	0·16	0·052		0·16		0·27	0·074	<0·005	<0·006	—	—

passage grave and the later wedge-shaped gallery graves have closer affinities with his Impact Phase.

The following is one example of alloy development. Irish halberds were invented in the copper-using period and are similar to the Portuguese Beaker type. They started as impure copper (<1% As), became Cu or Cu–As alloys, Cu–Sn alloys in Britain, and later copper–arsenic alloys with high arsenic (under international influence).

Northover[79] believes that the arsenic present in the early Irish-made artifacts was natural, while the later, high arsenical levels were due to added As minerals under Continental influence. But the easy availability of tin on the British mainland favoured the use of tin bronzes rather than the alien Cu–As tradition. About one-quarter of the arsenic and antimony in these alloys is present as inclusions of the type $Cu_2O \cdot 4Sb_2O_3 . 4As_2O_3$, the rest is presumably dissolved in the copper and therefore capable of hardening it. Although it is well known that high bismuth embrittles pure copper it was found that it segregates harmlessly when the copper is alloyed with As and Sn.

Intense As alloying (>3%) was only used for a brief period around the time of the introduction of tin bronze and was rapidly superseded by the more easily available and better tin bronzes.

The two traditions, the Irish impure copper and the British tin bronze, were carried on side by side until the Arreton Down phase (Wessex II). By about 1500 BC the British Isles were converted to tin bronzes throughout. The odd case of a Beaker arsenical copper knife from Dorchester (Oxon.) associated with a bronze rivet[82] is put down to re-use and re-riveting in a later period when the normal stock-in-trade of the smith was bronze. But the high nickel content of both rivet and knife suggests a continental origin (Table 14).

Coghlan and Case[23] have reported on the composition of a group of about a hundred Copper and Early Bronze Age artifacts. These comprise a large number of halberds and axes and a few knives. The results were divided into three groups.

In the first group, only 10% were tin bronzes. Of the 58 coppers represented all but three were of Irish provenance. The majority (38) were halberds. The main distinguishing characteristic of this group was a moderate arsenic and antimony, and a low nickel content (<0·03%).

Tin bronzes constituted 34% of the second group, and the group contained a much smaller proportion of halberds (11) to axes (13). It is mainly distinguishable from the previous group by a lower content of impurities, which implies an improved technique, although the difference tends to be marginal. The proportion of Irish artifacts is much smaller, suggesting that any improved technique coincides with the greater use of Scottish and English ores.

The third group has a high nickel content (0·1–0·35%) but in other respects is similar to group 1. Coghlan and Case[23] conclude that this represents a special ore deposit, probably Central European. Their group included only six specimens, one of which was a tin bronze. Only one specimen was of Irish provenance, the rest being English.

In group 1, arsenic contents reach 4·0%, and antimony contents 1·5%; the average ratio of As/Sb/Ag is of the order of 3:1:0·5. In group 3, although high arsenic contents of up to 7·0% are encountered, the silver and antimony contents are much lower and the three elements do not have the above ratio. Since arsenic and antimony tend to react similarly in smelting, the difference suggests a different class of ore rather than a better technique. The high nickel content reinforces this conclusion. It is noteworthy that in all three groups the lead content is low, reaching a maximum of 0·3%.

In order to determine those elements introduced by the addition of tin in the bronzes, Coghlan and Case[23] had a number of Bronze Age tin artifacts analysed. The results of this examination are given in Table 15 and it is clear that additions of tin contribute no impurities other than lead to the bronze examined. Considering the high purity of cassiterite (*see* Table 24, Chapter 3) one would not even expect contamination from this source and the tin used for the Lough Gara armlets must have a secondary origin.

Whereas the effect of arsenic on the strength of copper in the cast or annealed condition is relatively slight (*see* Table 16), as little as 1·04% As will raise the maximum strength of hammered copper from 440 to 585 MPa.[88] This is equivalent to increasing the hardness from 124 to 177 HV. Thus the enormous effect of arsenic on the hardness of a hammered cutting edge would not fail to have been noticed, and efforts would

Table 15 Composition of Irish tin artifacts of the Late Bronze Age (after Coghlan and Case[23])

Object and provenance	Element, % Cu	Pb	Ag	As	Sb	Sn
Armlet, Loch Gara, Sligo	0·010	0·12	<0·01	0·05	<0·01	98·0
Armlet, Loch Gara, Sligo	0·011	0·64	<0·01	0·05	<0·01	98·5
Armlet, Loch Gara, Sligo	0·039	0·30	<0·01	0·05	<0·01	99·6
Torc, Kilsallagh, Co. Longford	0·019	0·016	<0·01	0·10	<0·01	98·8

Table 16 Effect of arsenic and tin on the hardness of copper (after Refs. 85, 86, and 87)

Composition, %		Hardness, HV As cast	Worked and annealed	Work-hardened by 50% reduction
As	0	40	45	120
	0·25		68	122
	0·75		70	127
	2·0		85	140
	4·0		105	195
	6·0		110	210
	8·0	87	122	215
Sn	0	40	45	120
	2·0	50	50	140
	4·0	60	60	165
	6·0	70	65	185
	8·0	84	70	210
	10·0	100	80	230
	15·0	140	—	300

have been made to reproduce the conditions which gave this result. Antimony and silver will also contribute something to the hardness, and it must have been mortifying to early smelters when they found that by improving their technique, and so obtaining a purer copper, they had reduced its strength. One wonders how much early smelters knew about the hardening effect of impurities, and whether the arsenic and antimony contents were maintained at the level shown in Table B (Appendix) intentionally. Biek[89] concludes that where arsenic occurs in an otherwise pure metal, it was added deliberately perhaps as a high arsenic mineral. It has been noted that halberd rivets were often softer and had lower arsenic contents than the blades.[23]

This now seems to have a quite simple explanation.[90] When arsenical copper is heated to a hot forging temperature (800°C) some of the arsenic is lost as white fume owing to oxidation on the surface. Naturally, the smaller the artifact, the more arsenic is likely to be lost as the surface area/volume ratio increases and also the time required to make the smaller object. The volatilization of arsenic under these conditions shows itself as a white fume, and by observing the extent of this one can get some idea of the arsenic content and the amount being lost.

Coghlan and Case[23] attached importance to the presence of nickel in copper. Desch has discussed the possible occurrence of nickel in British ores and reports that nickel is not found in surface copper ores in Wales,[91] but does occur in the copper ores of Kirkcudbright and sparingly in Cornwall. In this connection it is worth noting that the plano-convex ingot of the Roman period found in south-west Scotland contained 0·10% Ni, which tends to support the possibility of south-west Scotland being the source of some nickeliferous copper ores. Isolated nickel contents greater than 1·0% have been reported by Brown and Blin-Stoyle,[36] but few as high as the 6·3% reported by Desch for the flat axe from Usk.[91] This is very probably an import.

Middle and Late Bronze Age artifacts

Brown and Blin-Stoyle[36] examined over 450 bronze artifacts of the Middle and Late Bronze Ages. These contained from 0·7 to 37% Sn, but the majority were near to the normal figure of 10%. Coghlan and Case's results on Early and Middle Bronze Age artifacts[23] presumably relate to material not re-used. By the Middle and Late Bronze Ages, an important component of the bronzes produced must have been re-used or secondary metal, but with rapidly increasing production the role of scrap metal may be relatively small and so the resulting metal should show changes in composition which might be correlated with changes of bronze types or with different mineral deposits.

The basic copper of the British bronzes shows a reduction in its overall impurity content, which was found to be statistically significant by Leese:[92]

Phase	Composition, %			
	As	Sb	Ni	Total
6. Acton Park	0·77	0·02	0·37	1·16
7. Cemmaes (Taunton)	0·58	0·10	0·47	1·15
8. Pennard	0·32	0·05	0·12	0·49

The reduction in Ni content suggests a different source of ore while the change in As and Sb could be due to differences in smelting technique. In Wales, the Acton Park phase saw the introduction of 7% or more lead to the bronzes, undoubtedly added as lead metal. This stage did not last and the lead content was progressively diluted, probably with lead-free scrap, until its revival in the south-east of England in the Wilburton phase (950–750 BC). Although about 2% Pb has a pronounced beneficial effect on the casting properties of a tin bronze, greater amounts than this are probably added as a diluent. The mechanical properties of leaded tin bronzes do not deteriorate progressively with added lead, and a lot depends on the rate of cooling.

It is probable that imported scrap metal was always a component of British Bronze Age trade. But this aspect has recently been highlighted by the discovery of underwater material off our coasts. One wreck off the coast near Dover[93] and another from Salcombe, Devon[94] produced early Urnfield metal probably from East France and Germany with Sb and As contents greater than the British average at the time: 0·4–0·5% As; 0·4–0·5% Sb; 0·25% Ni; 0·25% Ag. It is noteworthy that the Sb/As ratios are now of the order of 1:1 or sometimes higher.

In southern Britain the first phase of the Late Bronze Age is represented by the Wilburton hoard which, as it has high levels of As, Sb, Ni, and Ag, is probably heavily contaminated with scrap from the Continent. It also contains large amounts of additional lead. The Wallington material from northern Britain is free of lead,[95] but the ingot material often associated with this is of high purity (Table 9).

There were local sources of new metal such as that found at Breiddin, Powys, which had more Sb than As.[79] On the whole, by the end of the LBA the lead content had fallen to 2–3% as we see from the material from Llynfawr.[79]

The ingot material from this period is large and shows evidence of high purity, perhaps due to remelting and refining. It is unfortunate that we have so little ingot material from earlier periods. The 'mass' of metal from Ireland, the analysis of which is given in Table 9 may be an example of the earlier periods. Or, as Harbison has suggested,[96] some of the rough unfinished flat axes from Ireland may possibly be ingots.

Having established that lead was an intentional addition in the Middle and Late Bronze Ages, two things follow: the metal was added either to facilitate casting, i.e. to increase the fluidity of the bronze, or as a cheap diluent to an expensive metal. But it also establishes the important fact that lead was readily smelted by the early Middle Bronze Age, which, considering its apparent scarcity as a metal in its own right at this time in Britain, is of some importance. Furthermore, beaten bronze vessels had consistently less lead than the standard Late Bronze Age output. The beaten metal of the Heathery Burn Cave (Durham) bucket had 1·5% Pb, while a cast ring staple of the same bucket contained 7·6% Pb; a staple from Meldreth, Cambs., had 5·1% Pb; and a ring-handle from the Chrishall hoard had >15% Pb.[36, 97]

Classification by minor elements

We now have the results of several attempts to classify

Table 17 Impurity concentration arranged according to compositional groups, % (after Northover[79])

Early Bronze Age	
(A)	Principal impurities: As, Sb, Ag.
	Subgroups: A1 As>Sb>Ag
	A2 As<Sb>Ag
	A3 As≃Sb>Ag
	A* As>1·25.
	(A1* etc.: any material with As greater than 1·25)
(B)	Principal impurities: As, Ni.
	Subgroups: B1 As>0·75; Ni>0·06
	B3 As<0·75; Ni>0·06
	B4 As<0·25; Ni<0·06
	B1* As>1·25.
(C)	Principal impurities: nil, trace elements only.
(D)	Principal impurities: As, Sb, Ni, Ag. Subgroups based on (A): not common.
(E)	Principal impurities: As, Sb, Ni.
(F)	Principal impurities: As, Ag.
	Subgroups: F1 As>Ag
	F2 As≃Ag
	F3 As<Ag
	F* As>1·25.
(G)	Principal impurity: Ni only.
(H)	Principal impurities: Co, As, Ag, (Ni<Co).
Middle Bronze Age	
(M1)	Principal impurities: As 0·65–1·05; Ni 0·25–0·45; Co~0·05; Sb, tr.
(M2)	as (M1) but with As>1·05.
(N1)	As 0·35–0·70; Ni 0·25–0·5; Co~0·05; Sb 0·05–0·15.
(N2)	as (N1) but with Ni>0·50.
(O)	As>0·5; Ni<0·20; Ag 0·05–0·2; some Sb.
(P)	As 0·1–0·4; Ni 0·1–0·3; CO 0·05–0·15; Sb, tr.
(R)	Ni only.
Late Bronze Age	
(S)	Principal impurities: As, Sb, Ni, Ag, Co.
	As, Sb>0·4 + 0·25Ni + 0·25Ag.
	Subgroups: S1 Sb/As = 1
	S2 Sb/As = 2.
(T)	as for group (S) but with As, Sb<0·4.
	T3 Similar to (T) but with ~0·1As and Sb.

metal artifacts into compositional groups that are meaningful to the archaeologist, so as to enable him to relate the artifacts to ore sources and therefore determine their provenance. Unfortunately, while we know a great deal about the trace element composition of the metals we do not know so much about the ores.

Smelting experiments on copper minerals using simple apparatus have shown that out of a total of 49 elements analysed in the resulting copper by mass spectrometry only 12 were present in excess of 10 ppm (0·001%). These were: Pb, Sb, Sn, As, Ge, Zn, Ni, Fe, S, Si, Ag, and Mg. The possible sources for amounts less than this are so many (fluxes, furnace linings, fuel, etc.) that we may concentrate on these elements only.

The elements Si, Fe, and Mg are mainly present in the form of slag and therefore the quantity depends on the slag content which is normally between 1 and 3% of the metal. Sulphur is always associated with copper. The important elements appear to be: Ag, Ni, As, Sb, Zn, and Pb. The first two will be recovered *in toto* while the remainder will be affected by the smelting technique. Even so, Northover[79] finds them all significant and labels them 'principal impurities'. He would add cobalt to this list although it behaves more like Fe than Ni, and most of it tends to go into the slag.

Northover's conclusions are given in Table 17. Even when alloyed with significant amounts of Pb and Sn the impurity groups given in the table are clearly visible, demonstrating the fact that the additions of lead and tin were of high purity. Also, these groups are not vitiated by scrap metal conditions. When such additions are made they stand out and become significant in themselves.

In the case of As, amounts greater than 1·25% have a special significance and correspond with those shown in the histogram (Fig. 13) as 'arsenical coppers'. They either stem from specially selected ores or represent coppers to which additions of arsenical minerals have been made. It is unlikely that such additions would have been pure. The reduction in the As content with Archaeological Age is clear on moving down the table. It is only in the LBA that Sb equals As or exceeds it and such metals are almost certainly imports (see for example the ingot from the Hertford Heath hoard shown in Table 9).[35]

It should be noted that Zn does not figure in this scheme. It is volatile but no more so than As, but while the As like S has a strong affinity for the copper, the Zn tends to find its way into the slag.

Finally, it should be emphasized that this grouping scheme is only valid for the period for which it has been tested and therefore the objects must be dated by typological means or archaeological association before conclusions can be drawn. After all, the raw metal (black copper) of most of the world's copper smelters at the end of the 19th century contained 1·0% As and 0·6% Sb as well as substantial quantities of Fe and S.[98]

Metallographic examination of copper and bronze artifacts

Metallographic examination has now become standard practice in the examination of early metallic artifacts. It serves two purposes. One is to show how far the elements are segregated and therefore how meaningful is the elemental analysis of the artifact. The other is to indicate the manufacturing technique used and the state of the casting and smithing art. As far as bronzes are concerned the first objective has been assisted by the arrival of the electron-probe microanalyser (EPMA) which allows a scan of the whole section and can provide analyses of inclusions found thereby. This instrument is able to give a fairly complete analysis especially as far as the heavier elements are concerned. Unfortunately it is not so easy to determine the amount of oxygen present in copper.

It is also possible to use this technique for the analysis of slags, and slag inclusions in metal. Some examples of oxide and sulphide inclusions are shown in Table 18. The oxide inclusions are most likely to be the result of de-oxidation by the elements As and Sb. These would have been reduced along with the copper during smelting but would have been the first to be

Table 18 Composition of inclusions found in copper-base alloys, % (as found by Northover[79] and Sperl[99] using EPMA)

	(*a*) Raw copper from sulphide ore from Muhlbach, Austria[99]		(*b*) Raw copper from Kitzbühel[99]	
	Matrix	Inclusion	Matrix	Inclusion
Fe	1·59	6·99	0·07	0·72
Cu	98·68	71·70	94·17	79·82
As	0·08	0·0	0·23	0·0
Co	0·29	0·03	0·03	0·0
Ni	0·37	0·0	0·38	0·0
S	0·01	18·88	0·01	21·0

	(*c*) Halberd from Tonfannau[79]		(*d*) Flat axe from Merioneth[79]	
	Matrix	Inclusion	Matrix	Inclusion
Cu	98·3	—	98·04	—
Cu_2O	—	13·0	—	9·2
Sn	0·06	0·14	0·09	0·14
As	0·94	—	0·57	
As_2O_3	—	13·4	—	38·2
Sb	0·30	—	0·16	—
Sb_2O_3	—	73·6	—	52·6
Fe	0·03	0·02	—	0·03
Co	0·01	—	0·02	—
Ni	—	0·03	0·06	0·03
Zn	—	—	0·03	—
Ag	0·13	—	0·21	0·16
S	—	0·25	—	0·08

oxidized if the liquid metal was exposed to the air during casting.

The absence of sulphide inclusions is almost certain evidence for the use of oxide ores, but unfortunately their presence is not evidence of the use of sulphide ores. However, the overwhelming presence of sulphides and the absence of oxides may be taken as indicating the use of sulphides (*see* Table 18 (*a*) and (*b*)).

Since lead does not dissolve in copper-base alloys it is relatively easy to estimate its quantity by metallographic means. Also the degree of corrosion or lack of it is important in detecting fakes and in estimating the environmental conditions in which an object was found when its provenance is not known. In aggressive conditions intergranular corrosion can penetrate a long way into a specimen, particularly when assisted by poor casting technology.

A very large number of copper-base artifacts have now been examined metallographically[95, 100] and the number is increasing rapidly as the technique becomes a routine part of the overall examination. The shapes of the various Bronze Age implements referred to in this section are shown in Fig. 14.

The best way of obtaining some idea of the mechanical properties and therefore of the value of an implement as a cutting tool is to measure the hardness. This is normally done by an indentation test where a hardened steel ball is pressed into the prepared metal under a controlled load and the size of the indentation measured with a microscope (Brinell test). Alternatively a pyramidal diamond may be used, which makes a square-shaped impression (Vickers test). The result is normally expressed in kilograms per square millimetre and the two tests give almost the same figures for the range of copper-base alloys tested here. Differences would appear with hard materials such as steels owing to the deflection of the steel ball in the Brinell test but this test should not be used for hard materials. The results for the Brinell tests are given as 'HB' and for the Vickers test as 'HV' followed by the load used for the test in kilograms. Normally, loads between 1 and 30 kg give virtually the same readings unless the structure is very coarse, but loads less than 0·1 kg can give slightly high readings.

Some mechanical properties of the artifacts chemically analysed were reported by Coghlan and Case.[23] An Irish halberd, containing 2·61% As, 0·91% Sb, and 0·2% Ag, had a Brinell hardness in the cast portion of 79 HB, and in the cold-worked portion of 132 HB. This should be compared with the hardness of the cast tin–bronze of 120–160 HB, and pure copper of 40 HB. The high arsenic contents contribute appreciably to the hardness of these weapons.

A flat axe from Rhynie, Grampian, was examined by Voce.[101] This was one of a hoard of seven, three of which had been broken in half before being buried. The one examined consisted of an α bronze containing about 8% Sn and 0·5% As. It had not retained its original cast structure as it had been considerably worked. It would appear that some of this work had been done cold, during ritual fracturing along a line of porosity and inclusions, before being buried.

Penniman and Allen[102] have examined four Early Bronze Age halberds and a dagger. Three of these objects have been made from arsenical copper contain-

14 Copper and Bronze Age implements; not to the same scale (by permission of the Trustees of the British Museum)

1 Flat decorated axe from Glencar, Co. Sligo (Copper or Early Bronze Age, 20 cm long)
2 Flanged axe from Arreton Down, IOW (EBA, 20 cm long)
3 Palstave from Rougham, Norfolk (MBA, 14 cm long)
4 Palstave from the Thames at Whitehall (MBA, 16·5 cm long)
5 Looped palstave from Quy, Cambs. (LBA, 14 cm long)
6 Socketed axe from Walthamstow, Essex [now Greater London] (LBA, 12 cm long)
7 Winged axe from Whitstable, Kent (LBA, 15 cm long)
8 Socketed axe from Wingrave, Bucks (LBA, 10 cm long)
9 Method of hafting later flat axes and palstaves
10 Method of hafting socketed axes
11 Halberd blade from Maryport, Cumbria (Copper or Early Bronze Age, 30 cm long)
12 Rapier from the Thames at Surbiton, Surrey (MBA, 41 cm long)
13 Leaf-shaped sword from the lower River Lea (LBA, 70 cm long)
14 Leaf-shaped sword from Barrow, Suffolk (LBA, 68 cm long)
15 Basal-looped spearhead from Bottisham, Cambs. (MBA or LBA, 38 cm long)
16 Tanged spearhead from Arreton Down (EBA, 25 cm)
17 Spearhead with tang and ferrule from Snowshill, Glos. (EBA, 24 cm long)
18 Socketed spearhead from Arreton Down (EBA, 21 cm long)
19 Socketed axe from Worthing, Sussex (LBA, 10 cm long)

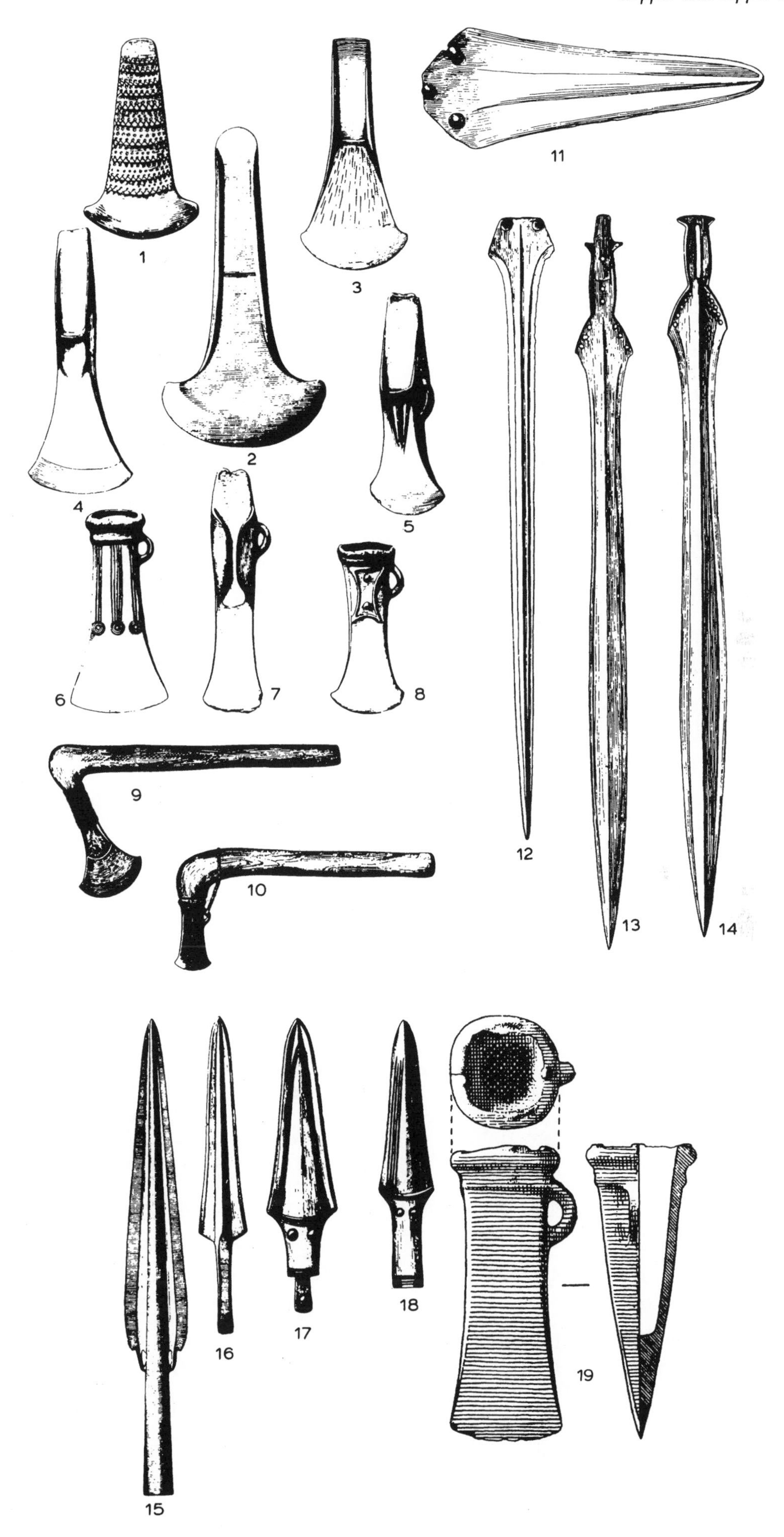
1
2
3
4
5
6
7
8
9
10
11
12
13
14
15
16
17
18
19

ing between 0·62 and 2·6% As and about 1% Sb. They have predominantly cast structures but have been forged to their final shape and cold-hammered at the edges. This has caused an increase in hardness from 73 to 153 HB. The artifacts appeared to have been cast from the pointed end in two-part moulds and slowly cooled. Some oxidation was apparent, as shown by the presence of oxides of arsenic and antimony. This seems to be very difficult to avoid, especially in the high-arsenic alloys (>2·5%) and may lead to brittleness. It is one of the reasons why high arsenic alloys are not used today.

Two halberds were made from tin bronzes containing 8·1 and 5·8% Sn. These had received much the same treatment as the arsenical coppers. This had had the effect of raising the hardness from about 65 to 133 HB. SnO_2 crystals were in evidence, showing that the material had been oxidized during production.

The rivets in this case had not been made of a more easily workable alloy than the blades themselves. A rivet in an arsenical halberd had the same structure and hardness as the blade, and had been made of the same material. A rivet in the bronze dagger contained 11·28% Sn, and had been finally forged cold after previous working and annealing. The hole in which it had been inserted was drilled from one side only.

Penniman and Allen[102] conclude that irrespective of material, the halberds had been made from blanks, cast in a two-part mould which had been made of stone or clay and pre-heated. They had then received a light forging to trim off the flash at the parting line. They had then been forged and annealed,[94] and finally the cutting edge had been cold-hammered and ground.

The above examples show that it was common practice to increase the strength of a cast alloy by working. However, this was not always done. Coghlan[103] reports on a cast copper, thick-butted, flat axe from Ireland, which contains about 1% each of arsenic and antimony and little else. This shows a cast, coarse dendritic, structure with porosity and some inclusions but no obvious cuprous oxide. It appears to have been slowly cooled after casting, possibly in a preheated closed stone or clay mould. This is not a good casting by modern standards; the porosity and inclusions may be due to the volatilization and oxidation of arsenic.

Voce[104] examined an ogival bronze dagger belonging to the Wessex culture of the Early Bronze Age. The cutting edge had been worked, but the hardness so obtained had been removed by subsequent heating. The fact that the cold-work resulting from decorating the surface with incised grooves still remained was proof that it had not been softened by heating in a funeral pyre. It appears, therefore, that either its maker was unfamiliar with the fact that cold-hammering hardens metal, or else had made a mistake.

An examination of an Irish flat axe made from a 1·5% As copper showed that it had been cooled fairly slowly in a clay or stone mould.[105] The melt was deoxidized and poured from a temperature of about 1100°C, i.e. just above the melting point of pure copper. After casting, the axe had been forged at the top (butt) and near the blade, but not in the centre.

For some reason the edge of this axe had also been left in the soft condition, and it was suggested it had been used as a pattern for clay moulding and had been annealed during the heating of the mould to harden the clay. While this is not impossible, it is rather surprising, since a two-part mould would be required and the pattern could be removed before firing the clay. It seems more likely that the axe has been annealed inadvertently either by faulty workmanship or, later, by falling into a fire. The softening temperature of 1·5% As copper would not be much above 300°C which could easily be reached by a domestic fire.

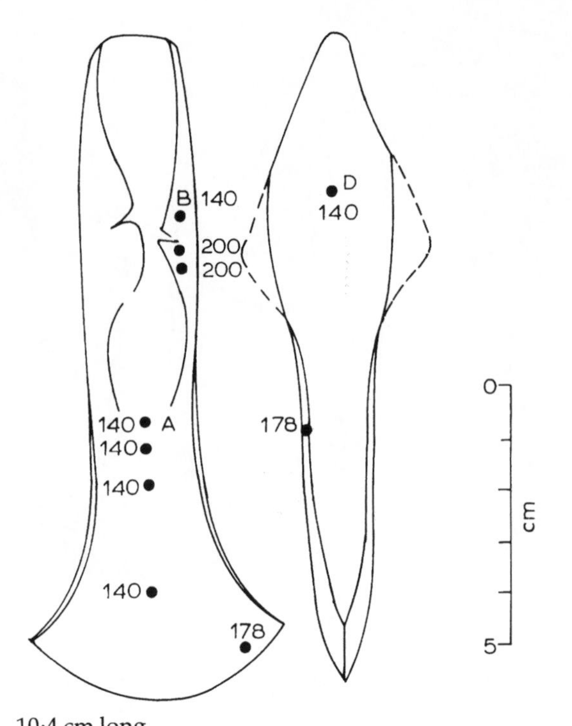

10·4 cm long

15 Hardness (HB) of MBA wing-flanged axe from Birtley, Tyne and Wear

A palstave from St Andrews, Fife, was found to contain 12–15% Sn. It had been cast in a two-part mould, and had been poured from a temperature of 1200°C. This is unnecessarily high, since a 12% tin bronze melts at about 950°C. Insufficient feeding had been employed so that the top was choked with dross. After casting, it had been reheated and the flanges lightly forged until the temperature dropped to below 550°C. The palstave was then reheated and a herringbone pattern engraved on the flanges. The patination on this palstave was very pronounced and consisted of a red layer next to the metal, followed by a black and green layer. A chemical analysis of the patina showed the presence of 43% sand, the remainder being mainly basic copper carbonate, together with some basic copper sulphate.

A Breton-type socketed axe containing 43% Pb and 1·46% Sn, said to have been found in the Vale of Mentieth was also examined.[105] No part of this axe had been forged at any stage in its manufacture. It contained no slag or dross, but about 1% oxygen. The lead was found to be evenly distributed. It had been cast in a two-part mould, the untrimmed flash still being visible along a plane through the loop of the axe. The core had been made of clay and had itself been made in a two-part mould as shown by the grooves on the inside of the socket, which are not quite down the centreline. Some clay and sand were found at the bottom of the socket, these being the remains of the core. Mitchell and Mitchell were convinced that the maker of this axe was well aware that it was not a usable implement, but the unfinished state seems to show that it was not a fraud as they suggested. It is now believed that these 'Breton' axes are some sort of currency or are, perhaps, ritualistic.

Raistrick and Smythe[106] examined a wing-flanged axe from Birtley, Tyne and Wear. This consisted of a high purity bronze containing 17·3%Sn. Here again the cast structure has entirely disappeared, and it was clear that a considerable amount of work, some of it cold, had been done on the original casting. After cold-working it had been annealed at a comparatively low temperature and again cold-worked in places such as the flanges. The hardness at different points is shown in Fig. 15 and we see that in the flanges the cold-working has increased the hardness to 200 HB and had caused some cracks. It is clear that the flange had been bent over by cold-hammering to make it grip the haft.

Experiments on pieces of metal of the same composition proved that the cast structure had been removed by annealing at 700°C followed by repeated hammering and annealing at a somewhat lower temperature. The actual results were as follows:

	Hardness, HB
Axe	
Thick parts	140
Edges and sides	178
Flange	200
Made-up alloy	
As cast	127
Worked and annealed	116–127
Annealed and cold-worked	182

Table 19 summarizes the hardnesses actually achieved by BA smiths. It may be seen from Table 16 that as-cast and annealed arsenical coppers have hardnesses of the order of 70 HV but can be work-hardened to give 150 HV. In fact, in BA practice, the hardnesses of pure and arsenical copper objects rarely exceed 100 HV (Table 19) but can go as high as 150 HV when the As and Sb exceed about 3·5%.

Cast tin bronzes should have hardnesses of the order of 100 HV and Table 19 shows that most of the bronze objects have been well forged and homogenized and start at about this figure. When the lead is low, edge hardnesses reach 170 HV owing to cold-working but drop to about 120 HV when appreciable lead is present. Clearly, BA smiths rarely achieved the figure of 224 HV which is known to be possible on 13% tin bronzes cast in stone moulds and hammer hardened.

Copper alloys and bronzes of the Iron Age

By the end of the LBA there must have been a lot of tin bronze in circulation. At the beginning of the Iron Age much of it was lost or merely dumped by smiths who either had little use for it or who went out of business altogether.

An increasing number of IA hillforts are now found to have had a BA occupation so one can assume that there was some continuity of metal working. It is possible that the realization that there was a connection between iron and copper, at least on the smelting side, had dawned on a large number of metal workers. For the smelters had been adding ferruginous fluxes such as hematite to siliceous copper ores for centuries and the fact that there was more than one metal in raw copper must have been noticed by many bronze workers. However, the idea that this 'impurity' was the new metal recently introduced by Hallstatt emigrants probably came as a surprise, and certainly iron working is an example of a diffused technique. Many smiths would be converted to the new metal. Others would continue to make copper-base alloys on a reduced scale, although it is possible that specialization did not occur at this time on settlement sites but only on the smelting sites.

One site recently investigated[65] is that of Gussage All Saints in Dorset, a hillfort which shows detailed evidence of an intensive industry for the manufacture of horse fittings. This is the first evidence in the UK for the use of the lost-wax process. Small bronze fittings were made, together with bronze-plated steel bridle-bits. This seems to be evidence for an integrated ferrous and non-ferrous industry. Unfortunately it is not known whether there were separate specialists doing the steel working and bronze working. Both aspects are represented on the site. The bronze was standard tin bronze with a low zinc content. It would seem that the steel was tinned before it was dipped in bronze to plate it.

Table 20 shows the composition of the few Early Iron Age bronze artifacts which the author has been able to find. The speculum coins (tin-money) are discussed later. The technique of adding lead which began in the Middle Bronze Age was continued. Substantial quantities of iron are now appearing. This may be the result of poor refining from the accumulation of iron caused by re-smelting many times and using iron stirring rods, or by the accidental incorporation of pieces of scrap iron. Isolated cases of high iron content (e.g. 2·78%) have been reported without comment by Brown and Blin-Stoyle[36] for Late Bronze Age artifacts and these undoubtedly come from ferruginous fluxes.

Although brasses and gunmetals were appearing on the Continent around this time[114] there is no evidence that they were finding their way into this country. The high silver content of the Hengistbury Bronze coin[110] can hardly be intentional and it would seem that some partially cupelled argentiferous copper had been used to make a bronze coin, or that it had been dipped into silver in order to plate it.

Voce has made a metallographic investigation of a Belgic (1st century BC–1st century AD) bowl from Welwyn.[112] This bowl appears to have been 'spun' from an alloy containing about 5%Sn, 1%Pb, 1%Fe, and about 0·25%Sb. It has been worked and annealed, and shows the presence of lead inclusions, lead being insoluble in bronze. It was finally left in the lightly cold-worked condition probably obtained by spinning and suffers from stress corrosion. This is a defect which is common in wrought materials which have been left in the cold-worked condition and not annealed. It results in cracking due to the combined effect of the internal stress left from the cold-working and corrosion from the atmosphere or soil. The subject of spinning is discussed in Chapter 5.

Composition of copper alloys of the Roman period

This period is marked by the widespread introduction

Table 19 Effect of working on hardness

Period	Object and provenance	Principal alloying elements, %	Hardness, HV			Reference and page
			As cast	Homogenized	Cold-worked	
EBA	Flat axe, Killarney	nil	—	94	102	Ref. 100, p. 27
	Flat axe, Ireland	nil	—	60	75	Ref. 100, p. 59
	Flat axe, Ireland	nil	—	57	75	Ref. 100, p. 66
	Flat axe, Ireland	1·78As	—	48	65	Ref. 100, p. 65
	Flat axe, Co. Mayo	1·3As	—	71	104	Ref. 100, p. 60
	Halberd, Irish	2·1As, 1·66Sb	—	75	153	Ref. 100, p. 107
	Halberd, Irish	2·6As, 0·9Sb	—	62	150	Ref. 100, p. 112
	Flat axe, Argyll	4·8As, 0·48Sb	62	—	95	Ref. 107
	Flanged axe, Slieve Gallion	12·9Sn		128	136	Ref. 100, p. 93
	Knife or dagger, Irish	5·8Sn	—	76	124	Ref. 100, p. 98
	Dagger, Irish	8·11Sn	—	65	133	Ref. 100, p. 102
	Flanged axe	12·1Sn	—	147	178	Ref. 100, p. 93
MBA	Palstave, Haydon Br.	10Sn	—	139	161	Ref. 95
	Palstave, Wallington	8–10Sn	—	70	124	Ref. 95
	Palstave, Norfolk	10·2Sn	—	126	147	Ref. 100, p. 127
	Palstave, Pusey, Oxon.	11Sn	—	113	170	Ref. 100, p. 128
	Palstave, Ireland	9·7Sn	—	113	141	Ref. 100, p. 138
	Haft flanged axe, Co. Mayo	11·6Sn	—	136	170	Ref. 100, p. 142
	Rapier, Corbridge	10Sn	—	127	164	Ref. 95
	Rapier, R. Thames	10·98Sn	—	57	104	Ref. 100, p. 150
	Knife, Maesbury	14Sn	—	134	262	Ref. 108
	Rapier, replica	13Sn	132	—	224	Ref. 62
LBA	Socketed axe, Irish	6Sn, 2·0Pb	—	88	121	Ref. 107
	Socketed axe, Ireland	11·7Sn 1·15Pb	—	83	104	Ref. 100, p. 196
	Socketed axe, Co. Tyrone	10·2Sn, 6·4Pb	—	109	119	Ref. 100, p. 199
	Socketed axe, Ireland	11·05Sn, 1·36Pb	—	88	124	Ref. 100, p. 200
	Sword	11·4Sn	—	83	112	Ref. 100, p. 226
	Spearhead, Ireland	10·2Sn	—	60	88	Ref. 100, p. 243
	Spearhead	8·4Sn	—	80	83	Ref. 100, p. 245

of zinc into copper alloys.

In the last section, zinc had appeared accidentally or intentionally in a few objects. The majority of analyses of Romano-British objects have been published by Craddock[76, 77] and are often from the collection in the British Museum. Others have been made by Smythe[115] and relate to North of England material principally emanating from the area of the Wall (Table 21). How

Table 20 Composition of Early Iron Age bronze objects

Object and provenance	Element, % Sn	Ag	Fe	Pb	S	Zn	As	Sb	Others	Reference
Lump, Hengistbury	nil	0·05	0·8	tr.	0·3					110
Ingate, Hengistbury	10·8	—	—	0·4	—					110
Lump, Hengistbury	9·8	0·04	—	tr.						110
Coin, Hengistbury	11·02	—	—	7·78	—					110
Coin, Hengistbury	17·41	—	—	0·37	—					110
Coin, Hengistbury	2·26	1·31	—	—	—					110
Coin (30 BC), Hengistbury	13·0	—	—	8·62	—					110
Coin, Hengistbury	11·7	—	—	9·6	—					110
Coin (100–75 BC), Sunbury	22·1	—	0·4	4·6	—					111
Droplets, Merthyr Mawr Warren	tr.	tr.	~1·2	2·3	—	3·6	tr.	tr.	Ni 0·3	54
Bowl, Welwyn, Herts.	~ 5·0	~0·01	>1·0	~1·0	—	n.d.	0·03	~0·2		112
Trumpet, Caprington, Strathclyde	9·61	—	—	nil	—					113 (p. 363)
Corroded strip, Glastonbury	<23·5	nil	tr.	nil	nil					55
Corroded strip, Glastonbury	15·7									55
Terminals, Glastonbury	13·5								O_2 8·3	55
Trumpet, R. Witham, Lincs.	12·0									113 (p. 363)

n.d. = not detected

Table 21 Analysis of some copper-base alloys of the Roman period

Object and provenance	Element, % Sn	Pb	Zn	Fe	Ni	Wrought or cast	Reference
Needle, Irish	2·36	—	14·7	—	—	—	50
Rosette, Silchester	nil	0·09	18·8	0·62	—	—	116
Stud, Silchester	nil	0·08	17·1	0·45	—	—	116
Needle, Southwark	1·03	1·07	13·0	—	—	—	117
Needle, Southwark	2·36	—	14·7	tr.	—	—	117
Cake, Housesteads	35·8	2·19	0·23	tr.	tr.	C	115
Ring, Housesteads	1·02	2·02	36·0	0·54	nil	C	115
Ring, Benwell	9·24	29·5	tr.	nil	tr.	C	115
Cake, Benwell	32·97	3·3	0·26	tr.	tr.	C	115
Plate, Corbridge	22·7	9·8	0·21	nil	0·03	C	115
Rod, Corbridge	12·2	6·9	2·7	tr.	0·09	C	115
Plummet, Benwell	7·9	9·0	7·1	tr.	0·05	C	115
Ornament, Corbridge	11·9	8·4	1·6	tr.	0·02	C	115
Key, Chesterholm	2·5	10·3	9·5	0·18	0·01	C	115
Ornament, Benwell	7·7	14·0	0·2	tr.	0·02	C	115
Ornament, Corbridge	6·0	5·9	6·7	tr.	0·03	C	115
Hairpin, Corbridge	5·2	3·6	9·7	0·2	0·03	W	115
Nail, Corbridge	5·2	nil	12·2	0·3	0·03	W	115
Bowl, (Cast) Benwell	9·6	6·8	1·6	0·1	tr.	C	115
Pin, Housesteads	0·4	0·4	12·6	0·28	nil	W	115
Staple, Corbridge	4·6	nil	10·1	0·11	0·04	W	115
Spiral, Housesteads	8·9	0·3	4·2	nil	0·04	W	115
Hairpin, Corbridge	1·8	nil	12·9	0·37	0·04	W	115
Wire, Corbridge	8·6	tr.	2·9	tr.	tr.	W	115
Wire, Corbridge	4·7	0·1	6·0	0·43	0·04	W	115
Tweezers, Housesteads	9·9	nil	1·5	tr.	0·05	W	115
Wire, Corbridge	8·1	nil	tr.	tr.	tr.	W	115
Ornament, Housesteads	10·0	—	—	—	—	W	115
Hairpin, Corbridge	3·4	1·3	0·54	0·22	tr.	W	115
Unidentified, Housesteads	6·0	—	—	—	—	W	115
Wire, Corbridge	4·7	nil	tr.	tr.	0·01	W	115
Spring, Corbridge	2·9	nil	—	—	—	W	115
Strip, Heronbridge	10·8	nil	—	—	—	W	64
Crucible contents, Wilderspool	9·73	—	1·53	—	—	C	56
Crucible spout, Springhead, Kent	19·9	11·3	23·8	—	—	C	118
Armlet, Aboyne, Grampian	6·76	4·4	1·44	—	—	—	117
Armlet, Aboyne, Grampian	3·64	—	9·13	—	—	—	117
Drummond Castle, Tayside (AD 50–150)	2·3	0·14	14·00	0·28	0·02	—	72
Drummond Castle, Tayside (AD 50–150)	3·2	0·14	9·40	0·16	0·02	—	72

far one can regard this material as being of British production and representing metallurgical practice in Britain is doubtful.

A plate from Colchester of 73/27 brass stamped V.H.E.T.B would seem to be a standard product sold to workshops for reworking.[119] However, there is little doubt that Roman techniques were rapidly disseminated in Britain, and that much metal was made in Britain in Romano-British workshops.

In early times, in fact before the 18th century in the West, zinc could not be made *per se*. Zinc melts at 420°C but boils at about 950°C, i.e. at a much lower temperature than any other common metal. In order to reduce it from its ores, it needs to be heated in contact with charcoal at 1 000°C. Unfortunately this is above its boiling point, so that it comes off as a vapour and is quickly converted back into zinc oxide, and therefore lost. One recent process uses zinc sulphide (zinc blende) as an ore, roasts it to zinc oxide, and then smelts it at 1 000–1 300°C in an almost closed vessel with coke to reduce it to metal vapour. This vapour is quickly condensed to metal at one end of the vessel which is kept just below the boiling point of zinc.

Up to the 18th century in Britain, brass was made by mixing the ore smithsonite ($ZnCO_3$), formerly called calamine, with copper under reducing conditions (i.e. charcoal). The ore was ground, and mixed with charcoal and granulated copper. This was heated in a crucible at about 950–1 000°C to reduce the zinc in the ore to zinc vapour which was absorbed by the solid copper granules. The temperature was then raised and the copper–zinc alloy melted. The addition of zinc lowers the melting point of copper from 1 083°C for pure copper to 1 000°C for a 20% zinc alloy. As far as can be seen from Table 21, the latter figure was rarely exceeded, and such a composition would be called 'gilding metal' today, not a brass. The commonest modern brasses contain about 30 and 40% zinc.

Before the Roman period the introduction of zinc into copper was almost certainly by smelting zinc and copper ores together. In the Roman period the technique of adding zinc as calamine was well understood, and was probably the only method used. It would not be easy to obtain such an accurate control over composition as with tin, which was added by this time in metallic form, so the variation of zinc contents shown in Table 21 from 1·5 to 36% is not altogether surprising. Usually, Roman copper-base alloys contain both tin and zinc, but there is a tendency for the wrought alloy to contain more zinc than tin, while the cast alloys usually contain more tin than zinc. In modern practice the brasses (Cu–Zn alloys) used for working contain 10–30% and those for casting usually 40% Zn. The bronzes now used for casting usually contain about 10% Sn and bronzes, with a few exceptions, are not now generally used for wrought products. The use of a brass containing 36% Zn in the ring from Housesteads is therefore similar to modern cast brass practice.

The other important point is that the addition of lead, started in the Late Bronze Age, is normal for cast products, such as statues but attempts have obviously been made to restrict the lead content of alloys required for working and gilding where it would be detrimental (Table 21). This has been most successful, and shows that the lead additions were intentional and not the result of contamination from secondary material. The average casting alloy is therefore what would now be termed a leaded gunmetal, whereas the wrought alloy containing about 5% each of tin and zinc does not approximate to any modern alloy.

Table 22 Mechanical properties of pure copper and some modern alloys approaching early alloys in composition (based mainly on Refs. 50 and 120)

Metal or alloy	UTS, MPa	Elongation, % (on 50 mm)	Hardness, HV
Pure copper (cast)	54	30–50	40
Pure copper (worked and annealed)	215	50	50
Pure copper (cold-worked only)	462	4	100
Brass (annealed) 70% Cu, 30% Zn	339	70	60
Gunmetal (cast) 85% Cu, 5% Sn, 5% Zn, 5% Pb	215	15	60
Gunmetal (cast) 88% Cu, 2% Zn, 10% Sn	292	25	65
Cast leaded bronzes			
85% Cu, 10·0% Sn, 5·0% Pb	215	15	70
80% Cu, 10% Sn, 10% Pb	185	—	65
76% Cu, 9% Sn, 15% Pb	169	—	60
73% Cu, 7% Sn, 20% Pb	154	—	55
70% Cu, 5% Sn, 25% Pb	139	—	50

Nickel is uniformly low which shows that high nickel–copper ores were not being exploited in significant amounts. Smythe[115] drew attention to the fact that the iron and zinc contents were related and it is probable that this is due to iron being an impurity in the zinc ore used.

The surprising fact is that pure copper was rarely used for artifacts in the Roman period, and one wonders whether this was due to the need to economize in copper as far as possible, or because the additional strength conferred by alloying additions was much sought after. A needle, such as one of those from Southwark,[117] made from an alloy containing 14% Zn would be substantially harder than pure copper when cold-worked, but additions of tin and lead would make a better casting alloy.

The ternary (three-metal) or quaternary (four-metal) alloys of copper, zinc, lead, and tin are so common during and after the Roman period, that one can consider that most of the so-called 'bronzes' are of this type. The zinc-containing ternary alloys do not correspond with any modern alloys, the nearest being the range of gunmetals, one of which contains 5% Sn, 5% Pb, and 5% Zn, and another of which contains 7% Sn and 2% Zn, both normally used in the cast condition. The strength of these alloys, compared with pure copper is shown in Table 22.

High-tin bronze (speculum-type) alloys were used for mirrors. The hardness and structure of mirrors from The Lunt, Coventry corresponds to a tin content of 24%.[143]

Silver–copper alloys are also found and appear to fall into two groups, the proportions being 80/20 and 50/50; most of them were found to contain gold. The silver is partly converted into chloride, and in some cases the copper has been added in the form of brass. Modern silver solder for soldering silver articles

Table 23 Composition of medieval copper alloys

Object and Provenance	Date (AD)	Element, %					Ref.
		Sn	Pb	Zn	Fe	Ni	
Crucible contents, Lough Rea, Galway	Early Christian	3·1	—	15·5	—	1·0	122
Crucible contents, Dunshaughlin, Meath	Early Christian	11·6	—	—	—	—	123
Crucible contents, Lagore Crannog, Meath	7–11th cent.	4·2	—	nil	1·3	2·2	122
Key, Ramsbury, Wilts.	9th cent.	5	nil	nil	—	—	124
Ring, Ramsbury, Wilts.	9th cent.	nil	nil	~30	—	—	124
Fragment, Ramsbury, Wilts.	9th cent.	10	—	nil	—	—	124
Foil, Kirkstall	1100–1500	6·07	1·17	4·89	—	—	125
Foil, Kirkstall	1100–1500	4·21	—	—	—	—	125
Foil, Kirkstall	1100–1500	4·82	0·51	6·44	—	—	125
Tap, Kirkstall	1100–1500	7·5	0·50	5·0	0·5	—	126
Pins, Wakefield	1350–1450	—	—	13–15	tr.	tr.	127
Pins, Wakefield	1500–1645	—	—	16	—	~1·0	127
Wire, Wakefield	1450–1500	—	—	19	tr.	tr.	127
Charity spout, Dymchurch	13–14th cent.	4·40	6·6	14·4	1·9	—	128
Memorial brass, Warwick	1453	3·6	1·2	8·2	2·6		129
Memorial brass	14th cent.	tr.	—	23·3	0·08		50
Memorial brass	1456	1·16	7·14	24·2			50
Memorial brass	1470	2·56	2·13	28·5			50
Memorial brass	1504	3·00	3·50	29·5	1·9		50
Bell metal, Thurgarton	12th cent.	22–25·5	3·5–4·5	1–2			130
Bell metal, Wharram Percy	1617	23·0	<1·0	—			
Bell metal, Chichester		24·0	2·70	0·03			
Bell metal, Gloucester		15·0	—	—			
Bell metal, Norton Priory		20	—	—			
Bell metal, Winchester	Saxon	18·8	4·35	0·1			131
Bell metal, Cheddar	12th cent.	20	nil	—			132
Mortars and measures, English	1308–1601	1–5	4–18	2			133

usually has a 70/30 composition, and one wonders whether the 80/20 wires were intended for such a purpose. No evidence of silver soldering has, however, been observed in Britain before the medieval period. It is possible that the objects made from the 50/50 alloy had been brightened to look like silver by leaching, i.e. dissolving out the copper from the surface.

Medieval copper alloys

There is an almost complete absence of analytical information regarding pre-Conquest medieval material. This is mainly because very little non-ferrous metallurgy was carried out on excavated English sites, and the Irish sites have yielded little more than crucible residues. A crucible residue from Lough Rea[122] showed that ternary alloys containing tin and zinc were in use (Table 23). Scandinavian evidence also shows the use of zinc-containing metal at this time,[134] and one is not surprised to find a brass from the Saxon site at Ramsbury[125] (Table 23). It is very probable that the brass circulating in this period was old Roman coinage. Simple tin bronzes were being produced on some Irish sites.[122, 123] Two of these three specimens contained high nickel and this supports still further the contention that nickel-containing copper alloys were indigenous.

From the post-Conquest period there are some pieces of foil and a tap from Kirkstall,[125] and a charity spout from Dymchurch.[128] The foil carried on the Roman tradition of low tin and zinc alloys for wrought material, and the spout shows the use of a higher lead content, together with a high zinc content for castings. But not a single piece of copper-base alloy from Winchester – apart from small pins – showed the presence of zinc in amounts exceeding 0·1%.[131] It would seem that the source of zinc for the making of gunmetal was beginning to come to an end by the 14th century and there was no new British zinc industry to take its place. However, by the late 14th century enough zinc or brass was being imported to make pins and memorial brasses (Table 23). The British brass industry did not get under way until the 16th century and even then had considerable difficulty in producing metal of adequate quality.

One of the largest consumers of copper-base metal was the bell founder, who was also responsible for the making of cauldrons, skillets etc., for wealthy households. Both were made from Cu–Sn alloys, although there was a great difference in the tin content. Mortars and measures tended to be high in lead and low in tin.[133]

CORROSION OF COPPER AND COPPER-BASE ALLOYS

Unlike tin and iron, copper and most copper-base alloys are very resistant to corrosion in both soils and water. The mere existence of enormous quantities of bronze objects dating from the Late Bronze Age and earlier is witness to their capacity for survival in the ground and coastal conditions obtaining in the British Isles.

In pure air the stable form of copper is the oxide tenorite (CuO). More acid conditions can produce malachite. If corrosion is not complete and some copper metal is left, then the red cuprous oxide (cuprite – Cu_2O) is normally present. A sulphurous urban atmosphere will produce brochantite ($CuSO_4.3Cu(OH)_2$), but this hardly ever forms in soils.

In the presence of both aeration and chlorides and in acid solutions the stable form of copper is the chloride paratacamite ($CuCl_2.3Cu(OH)_2$). On the whole the rate of corrosion is sensitive to acidity (pH), and aerated acid soils can give quite high corrosion rates.

As is well known, peaty moorland waters can be very benign to copper-base alloys. This is probably because of the build-up of a protective film but it may be due to the modification of the crystal form of the film and its growth-rate by an organic inhibitor. The result is that peats which are often thought of as being aggressive owing to their sulphur content can be positively benign. The state of the metal in a spear-head from Edington Burtle in the Somerset peat is remarkable. Although good workmanship has produced the high hardness of 250 HV, the benign conditions of the peat have suppressed the expected corrosion along the slip bands and destannification.

Destannification is a common modern problem with brasses although it is not so with bronzes. But with burial periods over 2 000 years it is not surprising to see it in the form of redeposited copper in the cavities of the bronzes. However, most early bronzes contain arsenic and it is well known that small amounts of As will inhibit this form of attack.

Stress corrosion cracking is another form of damage, which results from the combined effect of stress and corrosion. The stress may be externally or internally applied. The latter often arises as a result of cold-work which leaves uneven but balanced stresses in the worked component. Together, stress and corrosion may cause a metal to fail by cracking and the crack may be propagated by a continuing corrosion process which dissolves metal away at the root of the crack. This process is very sensitive to the corroding media and hence the environment. Some metals such as copper have a very narrow range of conditions under which such failure is possible. This type of corrosion has been detected on some bronzes, for example in an axe from Hasanlu in Anatolia, dated to the 3rd millennium BC, which was corroding in a very aggressive medium. Another example occurred in a 5% Sn bronze sheet from Jordan in a settlement of the 8th–7th century BC, and here the aggressive agent was probably ammonia. On the other hand coins of the 4th century BC to the 4th century AD were free of stress corrosion in spite of the fact that intercrystalline cracks caused in striking had filled with corrosion product which might have been expected to cause severe internal stresses. Clearly, bronzes have a high resistance to stress corrosion.[135]

SUMMARY

The early phases, which are mainly Irish show a predominance of copper and arsenical coppers and it is not until about 2000 BC that bronzes appear in England and Scotland. These are not common in Ireland and even on the mainland arsenical copper continues to be used, perhaps as imports from the Continent. Both metals have had their properties enhanced by heat treatment and work-hardening. By the MBA (1500 BC) tin bronzes are more or less universal and the tin content is relatively high (12–15%). In this period some artifacts contain what is believed to be additional lead. These high tin contents, together with homogenization of the cast structure and work-hardening of the edges give rise to some of the most metallurgically satisfying artifacts.

By the LBA (1000 BC) lead becomes a common but not a universal addition to the now standard 10% tin bronze. More intricate casting techniques have been introduced giving rise to thinner-walled castings, no doubt made possible by the lead addition. But the addition of lead is usually omitted for metal destined for wrought metal work. The quantity of artifacts made increases enormously but the quality seems to decrease, perhaps owing to a developing shortage of tin and the impending arrival of a new and more readily available metal: iron.

During the Early Iron Age there is a continuance of the differentiation between wrought and casting alloys, with low-tin, low-lead bronzes for the former and the higher tin and lead bronzes of the Late Bronze Age for the latter. The arsenic content is now low, but zinc had now made its appearance in at least one case.

In the Roman period zinc becomes more common either as brass or together with small quantities of tin in the wrought alloys; zinc also appears in smaller amounts in the casting alloys together with large tin and lead contents. Straight tin bronzes still occur.

In the pre-Conquest medieval period, zinc is occasionally found in some of the copper-base alloys, but the zinc was a scrap metal addition and brass was not to be made in Britain until the 16th century. Meanwhile bronze was still the main copper-base alloy, no doubt assisted by supplies of tin from Cornwall and Devon. Imported brass was used for pins and memorial brasses (Table 23).

REFERENCES

1 V. G. CHILDE: 'Prehistoric communities of the British Isles'; 1940, London.

2 C. RENFREW (ed.): 'British prehistory; a new outline'; 1974, London, Duckworth.

3 J. S. JACKSON: in 'Proc. 5th Atlantic Colloq.', (ed. M. Ryan), 107–125; 1978, Dublin.

4 H. MILES: *Corn. Archaeol.*, 1975, (14), 5–81.

5 W. B. WRIGHT and G. A. J. COLE: 'The geology of Killarney and Kenmare', 76; 1927, Memoirs of the Geological Survey of Ireland.

6 L. SCOTT: *Proc. Prehist. Soc.*, 1951, **17**, 16–82.

7 C. J. CARLON: 'The Alderley Edge Mines'; 1979, Altrincham.

8 C. ROEDER: *Trans. LCAS*, 1901, **19**, 77–118; 1905, **23**, 17–29.
9 W. GARDNER: *Archaeol. Cambrensis*, 1958, **107**, 64–71.
10 J. R. C. HAMILTON: 'Excavations at Jarlshof, Shetland'; 1956, Edinburgh.
11 G. A. J. COLE: 'Memoir and map of localities of minerals in Ireland'; 1922, Dublin, Stationery Office.
12 C. REID: *Man*, 1918, **18**, 9–11.
13 J. S. JACKSON: *Archaeol. Austriaca*, 1968, **43**, 92–114.
14 E. M. JOPE: *Ulster J. Archaeol.*, 1953, **16**, 37–40.
15 G. C. WHITTICK and J. A. SMYTHE: *Proc. Univ. Durham Philos. Soc.*, 1937, **9**, (2), 99–104.
16 J. R. S. HUNTER: *Trans. Geol. Soc. Glasgow*, 1884, **7**, 16–82.
17 O. DAVIES: *Archaeol. Cambrensis*, 1948, **100**, 61–66.
18 J. PERCY: 'Metallurgy; fuel; etc.'; 1861, London.
19 R. HUNT: *Mem. Geol. Survey GB*, 1846, **1**, 510.
20 A. RAISTRICK: *Naturalist*, 1936, 111.
21 T. E. HORNSHAW: 'Copper mining in Middleton Tyas', Publ. No. 6; 1975, North Yorks. Co. Record Office.
22 *Met. Bull.*, 1960, (4476), 10–11.
23 H. H. COGHLAN and H. J. CASE: *Proc. Prehist. Soc.*, 1957, **23**, 91–123.
24 'Report of the committee appointed to investigate early mining sites in Wales', 303; 1937, British Association.
25 H. H. COGHLAN and R. F. TYLECOTE: *Bull. Hist. Metall. Group*, 1970, **4**, (2), 79–80.
26 J. R. BUTLER: in 'Ores and metals', (ed. H. H. Coghlan), R. Anthropol. Inst., Occas. Paper No. 17, 34–53; 1963, London.
27 J. P. NORTHOVER: in 'Aspects of early metallurgy', (ed. A. Oddy), 63–70; 1977, Br. Mus. and Hist. Metall. Soc.
28 J. NAPIER: *Philos. Mag.*, 1852, **4**, 45–59; 192–201; 262–271; 345–355; 453–465. 1853, **5**, 30–39; 173–184; 345–354; 486–493.
29 R. F. TYLECOTE, H. A. GHAZNAVI, and P. J. BOYDELL: *J. Archaeol. Sci.*, 1977, **4**, (4), 305–355.
30 G. COFFEY: *J. R. Anthropol. Inst.*, 1901, **31**, 265–279.
31 C. BURGESS and D. COOMBS (eds.): 'Bronze Age hoards; some finds old and new', Br. Archael. Reps., British Series 67; 1979, Oxford.
32 R. F. TYLECOTE: *Int. J. Naut. Archaeol.*, 1980, **9**, (1), 67–68.
33 R. F. TYLECOTE: *Corn. Archaeol.*, 1967, **6**, 110–111.
34 R. F. TYLECOTE: in 'Festschrift für Richard Pittioni', Vol. 2, 157–172; Wien, Deutike.
35 P. T. CRADDOCK: in 'Proc. 5th Atlantic Colloq.', (ed. M. Ryan), 369–385; 1978, Dublin.
36 M. A. BROWN and A. E. BLIN-STOYLE: *Proc. Prehist. Soc.*, 1959. **25**, 188–208. (Analyses in *Archaeometry*, **2**, Supp.)
37 H. H. COGHLAN: *Trans. Newcomen Soc.*, 1939–40, **20**, 49–65.
38 R. F. TYLECOTE, A. LUPU, and B. ROTHENBERG: *J. Inst. Met.*, 1965, **95**, 235–243.
39 H. H. COGHLAN: 'Notes on the prehistoric metallurgy of copper and bronze in the Old World', 68; 1951, Oxford, Pitt Rivers Museum.
40 J. PERCY: 'Metallurgy; fuel; etc.', 389; 1861, London.
41 W. O. STANLEY: *Archaeol. J.*, 1873, **30**, 59–62.
42 T. F. EVANS: *Archaeol. J.*, 1873, **30**, 63–74.
43 C. E. BREESE: *Archaeol. Cambrensis*, 1908, **8**, 118–229.
44 R. G. LIVENS: *Trans. AASFC*, 1971, 247–253.
45 R. P. WRIGHT: *J. Rom. Stud.*, 1954, **44**, 102.
46 'W. J. H.': *Archaeol. Cambrensis*, 1939, **94**, 92.
47 W. W. WILLIAMS: *Archaeol. Cambrensis*, 1877, (4th series), **8**, 206–211.
48 J. CLOSE-BROOKS and J. TATE: private communication, 1980.
49 F. LYNCH: *Trans. AASFC*, 1966, 110–111.
50 W. GOWLAND: *J. Inst. Met.*, 1912, **7**, 23–49.
51 J. DU PLAT TAYLOR: *Archaeol. J.*, 1952, **32**, 133–167.
52 R. F. TYLECOTE: *Corn. Archaeol.*, 1966, **5**, 31–33.
53 Copper slag removed from site of smelter at Lower Redbrook on the Wye, 1967. Analysed by Murex Welding Processes Ltd, Waltham Cross, through the good offices of Dr W. Pumphrey.
54 C. FOX: *Archaeol. Cambrensis*, 1927, **84**, 44–66. Analysis of metal droplets reported by H. O'NEILL: *Foundry Trade J.*, 1949, (16 Jun.), 4.
55 A. BULLEID and H. ST. G. GRAY: 'Glastonbury Lake Village', Vol. 1, Vol. 2; 1911, 1917, Glastonbury Antiquarian Soc.
56 T. MAY: *Iron Coal Trades Rev.*, 1905, **71**, (11 Aug.), 427.
57 N. MANGIN and A. THOUVENIN: *Rév. Archéol. de l'Est et du Centre Est*, 1977, **27**, (3–4), 505–521.
58 J. G. CALLANDER: *Proc. Soc. Antiq. Scotl.*, 1931–32, **66**, 42–66.
59 A. E. WRAITH: *J. Hist. Metall. Soc.*, 1978, **12**, (1), 43–44.
60 A. YOUNG: *Proc. Soc. Antiq. Scotl.*, 1955–56, **89**, 290–328.
61 A. O. CURLE: *Proc. Soc. Antiq. Scotl.*, 1932–33, **67**, 82–136.
62 R. F. TYLECOTE: *Bull. Hist. Metall. Group*, 1973, **7**, (1), 1–5.
63 A. O. CURLE: *Proc. Soc. Antiq. Scotl.*, 1933–34, **68**, 224–319.
64 B. R. HARTLEY: *J. CNWAAHS*, 1954, **41**, 1–14.
65 J. P. BUSHE-FOX: 'Excavations at Wroxeter. 1912, 1913, 1914', Res. Comm. Reps. Nos. 1, 2, 4; Society of Antiquaries, London, 1913, 1914, 1916.
66 F. R. MANN: *J. Rom. Stud.*, 1939, **29**, 214.
67 R. F. TYLECOTE: *Bull. Hist. Metall. Group*, 1969, **3**, (2), 46–47.
68 *J. Rom. Stud.*, 1957, **47**, 214.
69 S. P. O'RIORDAIN: *Proc. R. Ir. Acad. (C)*, 1941, **47**, 77–150.
70 A. E. P. COLLINS: *Ulster J. Archaeol.*, 1955, **18**, 45–82.
71 R. F. TYLECOTE *et al.*: in Proc. 16th Int. Symp. on 'Archaeom. and archaeol. prospection', Edinburgh, 1976, 268–292.
72 P. COURTNEY: *J. Hist. Metall. Soc.*, 1982, **16**, (1), 22–23.
73 H. J. CASE: *Man*, 1954, **54**, 18–27.
74 H. H. COGHLAN: 'Studies of British and Irish celts', 1st series. Appendix by M. Cook and F. C. Thompson. *Man*, 1953, **53**, 91–101.
75 S. JUNGHANS, E. SANGMEISTER, and M. SCHRÖDER:

'Kupfer und Bronze in der Frühen Metallzeit Europas (SAM 2)', 3 Pts; 1968, Berlin, Verlag Gebr. Mann.

76 P. T. CRADDOCK: PhD thesis, Inst. Archaeol., University of London, 1976.

77 P. T. CRADDOCK: private communication, 1981.

78 J. P. NORTHOVER: in 'Aspects of early metallurgy', (ed. A. Oddy), 63–70; 1977, Br. Mus. and Hist. Metall. Soc.

79 C. BURGESS and P. NORTHOVER: 'The Welsh Bronze Age and its position in British Bronze Age metallurgy', in press.

80 R. F. TYLECOTE: 'A history of metallurgy'; 1976, London, The Metals Society.

81 H. J. CASE: *Palaeohistoria*, 1966, **12**, 141–177.

82 H. J. CASE: *Antiquity*, 1965, **39**, 219–222.

83 D. BRITTON: *Proc. Prehist. Soc.*, 1963, **29**, 258–325.

84 D. BRITTON: *Archaeometry*, 1961, **4**, 39–52.

85 J. R. MARÉCHAL: *Métaux*, 1958, (397), 377–383.

86 G. D. BENGOUGH and B. P. HILL: *J. Inst. Met.*, 1910, **3**, 34–97.

87 'Metals handbook'; 1939, American Society for Metals.

88 D. HANSON and C. B. MARRYAT: *J. Inst. Met.*, 1927, **37**, 121–161.

89 L. BIEK: *Man*, 1957, **57**, 72–76.

90 H. McKERRELL and R. F. TYLECOTE: *Proc. Prehist. Soc.*, 1972, **38**, 209–218.

91 R. E. M. WHEELER: 'Prehistoric and Roman Wales'; 1925, Oxford.

92 M. LEESE: unpublished PhD thesis, Inst. Archaeol., University of London, 1980.

93 R. F. TYLECOTE: *Int. J. Naut. Archaeol.*, 1977, **6**, (4), 269–283.

94 K. MUCKELROY and P. BAKER: *Int. J. Naut. Archaeol.*, 1979, **8**, (3), 189–210.

95 C. BURGESS (with Appendix by R. F. Tylecote): 'Bronze Age metalwork in Northern England, 1000–700 BC'; 1968, Newcastle upon Tyne, Oriel Press.

96 P. HARBISON: *N. Munster Antiq. J.*, 1966, **10**, 3–11.

97 M. J. HUGHES: *Archaeometry*, 1979, **21**, (2), 195–202.

98 L. U. SALKIELD: 'A technical history of the Rio Tinto Mines', private communication.

99 G. SPERL: *Met. Technol.*, 1980, **7**, (5), 212–217.

100 I. M. ALLEN, D. BRITTON, and H. H. COGHLAN: 'Metallurgical reports on British and Irish Bronze Age implements and weapons in the Pitt Rivers Museum', Occas. Paper No. 10; 1970, Oxford.

101 E. VOCE: *Man*, 1951, **51**, 6.

102 T. K. PENNIMAN and I. M. ALLEN: *Man*, 1960, **60**, (120), 5 pp.

103 H. H. COGHLAN: 'Notes on the prehistoric metallurgy of copper and bronze in the Old World', 2 ed., 19; 1975, Oxford, Pitt Rivers Museum.

104 H. H. COGHLAN (Appendix by E. Voce): *Archaeol. Austriaca*, 1958, (3), 57–69.

105 S. S. MITCHELL and M. E. CRICHTON MITCHELL: *Proc. Soc. Antiq. Scotl.*, 1934–36, **69**, 424–430.

106 A. RAISTRICK and J. A. SMYTHE: *Proc. Univ. Durham Philos. Soc.*, 1931–37, **9**, (2), 47–54.

107 H. H. COGHLAN: in 'Prehistoric man in Wales and the West; essays in honour of Lily F. Chitty', (ed. F. Lynch and C. Burgess), 183–188; 1972, Bath.

108 G. PARKER: '83rd–84th Ann. Rep. Wells Nat. Hist. and Archaeol. Soc.', 7–10; 1972–3.

109 G. A. WAINWRIGHT, M. SPRATLING, *et al.*: 'The Iron Age settlement of Gussage All Saints, Dorset', Department of Environment Archaeol. Rep. No. 10. HMSO, London, 1979.

110 J. P. BUSHE-FOX (Appendix by W. Gowland): 'Excavations at Hengistbury Head Hampshire 1911–12'; 1915, Oxford.

111 R. P. MACK: 'The coinage of Ancient Britain'; 1953, London.

112 E. VOCE: *Man*, 1951, **51**, 37.

113 J. EVANS: 'Ancient bronze implements of Great Britain and Ireland'; 1881, London.

114 P. T. CRADDOCK: *J. Archaeol. Sci.*, 1978, **5**, 1–16.

115 J. A. SMYTHE: *Proc. Univ. Durham Philos. Soc.*, 1938, **9**, (6), 382–405.

116 G. E. FOX; *Archaeologia*, 1900, **57**, (2), 229–256.

117 A. H. CHURCH: *J. Chem. Soc.*, 1865, **18**, 215–216.

118 W. S. PENN: *Archaeol. Cantiana*, 1957, **71**, 53–105.

119 J. MUSTY: *Antiq. J.*, 1975, **55**, (2), 409–410.

120 'Copper and its alloys in engineering and technology'; 1948, London, Copper Development Association.

121 R. F. TYLECOTE: metal analyses in B. HOBLEY: *Trans. Proc. Birm. Archaeol. Soc.*, 1966–67, **83**, 65–159.

122 R. J. MOSS: *Proc. R. Ir. Acad. (C)*, 1924–27, **37**, 175–193.

123 H. HENCKEN: *Proc. R. Ir. Acad. (C)*, 1950, **53**, 1.

124 J. HASLAM, L. BIEK, and R. F. TYLECOTE: *Med. Archaeol.*, 1980, **24**, 1–68.

125 R. HAYNES: *J. Iron Steel Inst.*, 1956, **183**, 359–361; correspondence 1957, **185**, 531.

126 L. ALCOCK and D. E. OWEN: *Proc. Thoresby Soc.*, 1955, **43**, 51–66.

127 C. CAPLE: private communication, 1980.

128 L. R. A. GROVE: *Archaeol. Cantiana*, 1956, **70**, 268.

129 J. M. FRIEND and W. E. THORNEYCROFT: *J. Inst. Met.*, 1927, **37**, 71–72.

130 P. W. CATHERCOLE and B. WAITES: *Trans. Thoroton Soc.*, 1959, **63**, 24–56.

131 M. BIDDLE and S. KEENE (eds.): 'Winchester Studies 7(2) – Crafts and industries other than ceramics'; 1984–5, Oxford University Press.

132 R. F. TYLECOTE in P. Rahtz (ed.), Br. Archaeol. Reps., British Series 65, 214, 1979, Oxford.

133 R. BROWNSWORD and E. E. H. PITT: *Metall. Mater. Technol.*, 1981, **13**, (4), 184–185.

134 K. LAMM: *Bull. Hist. Metall. Group*, 1973, **7**, (2), 1–7.

135 R. F. TYLECOTE: *J. Archaeol. Sci.*, 1979, **6**, 345–368.

3 *The production and properties of tin and its alloys*

The major deposits of tin ores in Western Europe are very limited, and Devon and Cornwall are among the very few places where primitive peoples could obtain such ores in the British Isles, although small amounts of tin ore have been found in the Wicklow gold deposits.[1] Excavations at Chûn Castle[2] showed that smelting was carried out in the pre-Roman Iron age (3rd–2nd century BC), and a La Tène-type brooch was found associated with slag near St. Austell.[3]

Recent excavations in the area of Bronze Age barrows at Caerloggas near St. Austell have revealed seven small pieces of glassy tin-smelting slag.[4] These have been deposited intentionally perhaps as part of some ritual, and are dated to the EBA, about 1600 BC. The slag was broken before burial and tin globules are encapsulated in and totally protected by the slag which has withstood weathering in the acid soils.

Many analyses of Cornish tin ores have been made. Some of these have been part of a larger programme to try to distinguish objects made from Cornish tin from those made with tin from elsewhere.[5] Others have shown that deep-mined Cornish ore contains cobalt and germanium.[6, 7] Some results of analyses of Cornish cassiterite are given in Table 24.

Once located, tin ore can be readily separated since, like lead, tinstone or cassiterite is much heavier than the rock in which it occurs and is usually black and hard. Its chemical formula is SnO_2, and the tin obtained from it has a melting point of 232°C, but a temperature of 1 000°C is required to smelt it, and more rigorous conditions than those needed for copper.

Tin ore is often found in the form of water-concentrated deposits or 'stream tin'. The eroded rocks containing the ore have been washed down by surface waters into the beds of streams during glacial times, with deposition in the valley bottoms well down-stream from their source rocks. This was subsequently covered by fluviatile and marine overburden devoid of workable tin to depths of as much as 24 m.

The tinstreamer had to cut a shaft through the overburden and wash the material at the bottom. This material varied in size from that of a split-pea to a fist, and the pebbles shown in Fig. 16 are typical. Naturally gold would also be deposited in this way and would be recovered during the working of tin streams.

There is no doubt that the first deposits worked were of this kind and that the vein deposits were not worked until much later. It would be interesting to know how early people smelted this material and whether they concentrated on the coarser material and ignored the finer particles of cassiterite, but unfortunately the evidence for early smelting is very meagre. Leeds[3] excavated Chûn Castle and found a large structure 3 m long by 1·7 m wide, roughly constructed of stone blocks which he claims was a smelting furnace. He says this was fired by three 'fire holes' each of which had a 'flue' leading from it, and there was a fourth flue connecting the three fire holes. The fire holes and flues were filled with black soil containing a large amount of charcoal.

A careful examination of Leeds's report has not convinced the present author that his interpretation is correct. If the structure was metallurgical it is possible that it was a crucible melting furnace.

TIN DRESSING

In the earliest times most of the tin requirement would have been satisfied by the collection of pebbles of tinstone such as were found at Trevisker[8] (Fig. 16). By

1 cm scale

16 Tinstone pebbles from St. Eval, Trevisker, Cornwall (courtesy of Dr Colin Shell)

Table 24 Composition of Cornish tin ores, %

Element	St. Mawgan in Pydar[8]	Cornish (after Rapp[5]) 1	2	3	4 (Dolcoath)
Sn	78·7				
Si	0·72				
Pb	tr.				
Cu	tr.				
Ca	0·14				
Ni	nil	(380)	0	(180)	0
Zn	nil				
Au	nil	(0·07)	(0·09)	(0·05)	(0·06)
Ag	nil	(4)	(1)	(2)	(2)
Mn	nil				
Ti	0·03				
Mg	nil				
Sb	—	0	0	0	0
W	—	0·19	0·18	0·1	(980)
Hg	—	(3)	0	(13)	(10)

Figures in parentheses are in ppm

medieval times, however, lower-grade ores were being worked which required breaking and concentrating to give a grade containing at least 65% tin metal (i.e. 70% SnO_2). Recent excavations on Bodmin Moor have revealed an open-cast excavation and a nearby dressing mill which concentrated the ore.[9] A leat served a water wheel which crushed the ore using stamps which were raised by cams on the wheel shaft and which ground the ore by crushing it between the stamp head, which was probably reinforced with iron, and a mortar stone (Fig. 19*b*). A current of water was arranged to wash away the broken ore into a launder where it was buddled in the time-honoured way. This took account of the large difference in density between the country rock, mainly silica with a specific gravity of 2·5 and the cassiterite with a density of 5·4. The lighter material was washed away over the side of the container or rills in the launder, and the cassiterite was left behind.

In later times, as Agricola shows,[10] crazing mills like corn mills were also used to mill the material finer and reference to such mills may be found in the literature from the 16th century onwards. These gave way to improved stamp mills in the 18th–19th centuries.

TIN SMELTING

Unfortunately no early tin-smelting furnaces have been found in Britain: the only ones reported are from South Africa.[11] Even here the evidence is meagre and only the foundations of saucer-shaped furnaces were found near Rooiberg. On the basis of these finds and some in Nigeria, an experimental furnace was constructed, as shown in Fig. 17. This consisted of a shaft 60 cm high and 13 cm in diameter with one tuyere. Ore with a tin content of 6·3% was used, which is much below the quality used in Britain in the medieval and later periods. It was clear that tin oxide was reduced but that the gangue kept the tin particles apart so that no massive tin was found. A better grade of ore (30·88% Sn) gave tin prills and larger masses but no plano-convex ingots and it was clear that these could only be made by re-melting the prills in a crucible and pouring into a mould. The purity of the tin was in the

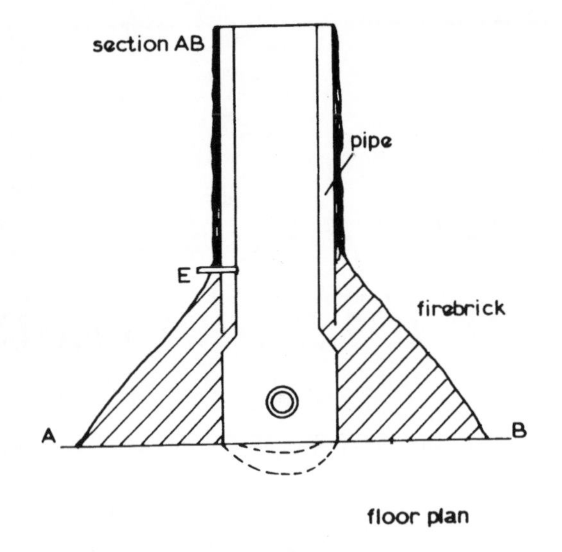

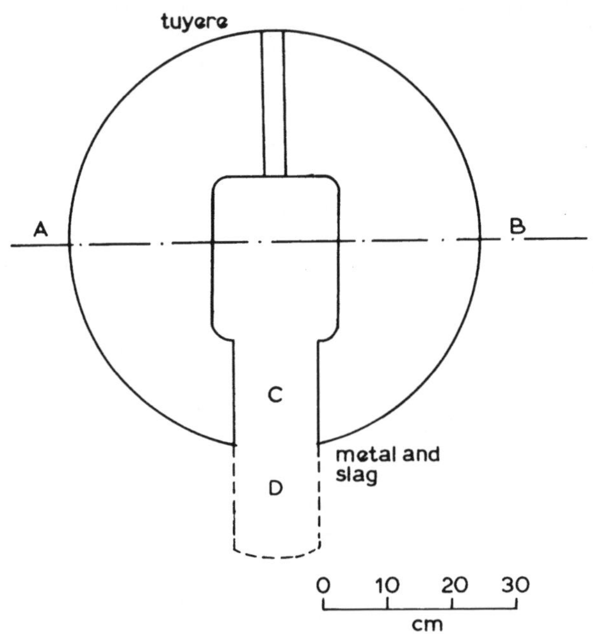

17 Rooiberg experimental tin-smelting furnace (after Friede and Steel[11])

range 98·8–99·15% and the slag composition is given in Table 26 (*below*).

In Britain, however, it was possible to obtain pebbles of pure cassiterite (Table 24) which contains 79% Sn, and the problems found in South Africa would not arise. Even so, temperatures as high as 1 000°C are needed and care must be taken to see that all the charcoal is burnt away at the end so that it does not keep the tin prills from coalescing.

By medieval times we begin to get detailed descriptions of the technique and the furnaces used. In 1778, Pryce[12] described a typical Cornish tin-smelting shaft furnace, known as 'the Castle' on account of its strength. It consisted of masonry bound with iron and had an internal cavity 60 cm square at the top and 36 cm square at the bottom, and was 2·18 m high. It was operated by two pairs of bellows at the back 2·4 m long and delivering air to a tuyere 25 cm from the bottom of the hearth. (Fig. 18.)

Tin ore and charcoal were laid in layers, the total charge being 400–600 kg of tin ore and 18–24 sixty gallon pecks of charcoal (about 1 120–1 500 kg). Smelt-

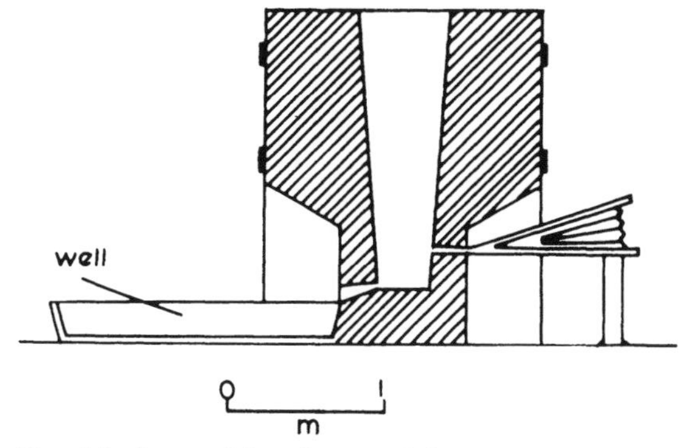

18 Cornish tin-smelting furnace (after Pryce[12])

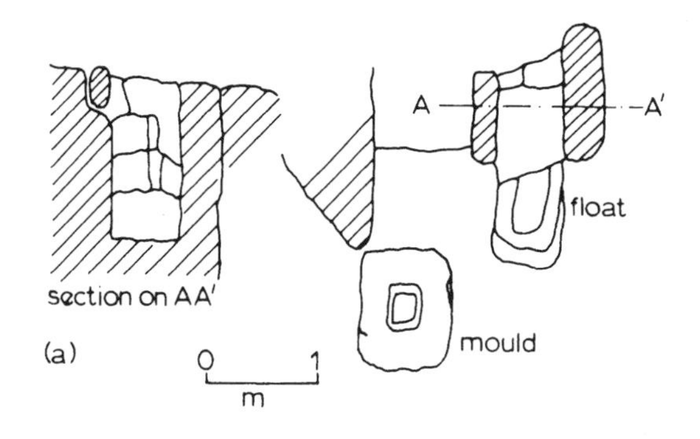

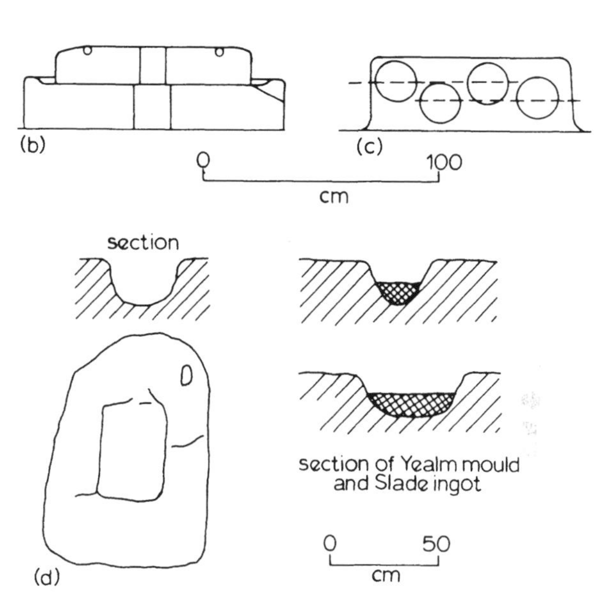

a details of furnace at Merrivale; *b* crazing millstones at Gobbet; *c* mortars or stamps at Little Horrabridge; *d* Colleytown mould

19 Remains of tin-processing equipment from Devon (after R. H. Worth[17])

ing took 12 h and the tin was tapped through a taphole 10 × 3·7 cm in section, into a stone well 2 m long and 0·3 m wide in front of the furnace and was ladled from the well into the moulds. This furnace closely resembles one shown by Agricola.[10]

At Trereife, near Penzance,[13] a furnace was found in a high bank of clay. It consisted of a cavity in the form of an inverted cone, 1 m diameter at the top, about 0·3 m diameter at the bottom, and 1 m deep. On one side was an opening which presumably admitted air, either induced or from bellows. The opening was low enough to discharge molten tin. Gowland[14] believes that the inside of the furnace was smaller than this because there must have been a lining which has since been removed.

On the top of the filled cavity was found a block of tin weighing 13·4 kg.[13] This measured 41 × 20 × 5 cm and had on it a complex inscription with the letters EIC in relief. It was found by Smythe[15] to contain 99·9% tin, like other Cornish tin ingots. Haverfield[16] regards the block as being of medieval date, and considering the similarity of the furnace to Pryce's furnace this is very probable.

While the remains of early working in Cornwall have largely been obliterated by more recent activity, this is not the case in Devon where a sharp decline set in at the beginning of the 17th century (Table 25). The remains of washing, grinding, and smelting installations still exist on Dartmoor and many have been recorded.[17, 18]

Table 25 Tin production in England

Date, AD	Production, t/a England (total)*	Devon only†
1200	200	
1300	200	
1400	300	
1450	500	
1500	600	
1550	800	
1600	1 200	
1650	1 800	
1800	2 522	
1824	5 000	
1853	5 763	
1872	9 560	
1400–1450		54
1450–1500		104
1500–1550		199
1550–1600		93
1600–1650		25

* Up to 1650 according to Majer;[56] after 1800 due to Barton[21]
† Worth[17]

After preliminary crushing in mortars or stamp mills (Fig. 19*b*) the ore was ground in 'crazing' mills which were normal vertically shafted 60 cm diameter millstones (Fig. 19); piles of sand can be seen outside some of the blowing houses. Some of the leats had a fall of about 4 m in a short distance; others were as much as 0·75 km long. Their water capacity was as much as 1·7 m^3 min^{-1} and the average diameter of the overshot wheels was about 3 m and the width 0·5 m; some of the wheels drove the crazing mills as well as the bellows. The bearings of the bellows wheels were of the open type and the shafts must have been placed below the bellows.

The blowing house at Outer Down near Chagford, Devon, seems a very good example of its type (*see* plan in Fig. 20). The wheel pit would house a wheel less than 3 m in diameter which would work bellows no more than 2 m long. The air would feed into a recess which would contain a furnace like the one shown in Fig. 19*a*.

The first furnaces were shaft furnaces and it is very doubtful whether the reverberatory furnace was introduced before 1700, by which time Dartmoor production had virtually ceased. Yet the Eylesbarrow

Table 26 Composition of tin slags, %

	Caerloggas, Cornwall[4] *c.* 1500 BC	Portugal[22] EIA	Rooiberg, S. Africa[11] IA.B6	Cornish[23]		Experimental slag from S. Africa[11] B5	Stable Hobba, Trereife, Cornwall* 18th cent.?
				Calenick 1702–1891	Carnelloe 1702–1891		
SiO_2	30	21·46	25·75	30·0		31·73	42·62
Al_2O_3	13	9·34	14·16	11·1		14·75	12·32
Fe_2O_3	—	64·43	34·64	3·9		8·21	18·22
FeO	4			32·6			18·22
MgO	—		2·97	—		3·52	0·31
CaO			3·30	1·6		<1·0	2·92
Na_2O			0·50			1·56	—
K_2O			2·51			1·60	0·05
TiO_2	1·5	0·96	0·76			0·86	2·04
P_2O_5		0·30	0·33	0·05		0·13	—
Cr_2O_3			0·12			<0·05	—
MnO		0·40	0·72	0·20	0·50	<0·05	0·47
ZrO_2			0·27				—
WO_3	3·0		—	—	—	—	—
Cu				1·48	0·7		0·11
Pb				0·02			—
Sb	(with Sn)			<0·01			—
As				<0·01			0·62
S	—	0·005	—	—	0·5	—	0·17
SnO	45	—	—	—	—	—	(21·0)
Sn	—	1·6	12·96	7·9	2·8	28·83	—

Slag from Outer Down Blowing House, Dartmoor had 13·4% Sn

* Analysed by R. E. Clough
() by difference

smelter had a 21 m long horizontal flue terminating in a vertical chimney. It is quite possible that this was an improved dust-catcher for the blast furnace since there is a wheel pit in a convenient position for driving bellows for a blast furnace.[19]

Some idea of the blast furnaces used in the 16th century can be obtained from Agricola.[10] These were 2·5–3 m high, and 60 × 30 cm internally. They were made of large sandstone blocks and were lined with clay. The tuyere was placed at the back opposite the taphole and directed downward. The hearth was a sloping sandstone slab and the taphole was always open. The tuyere was large so that the blast should not be too fierce. The metal and slag separated out in the large (0·3 m long) fore-hearth and the slag was able to overflow from this; a charcoal cover was maintained over the tin to avoid oxidation.

A furnace answering to this description was found at Merrivale in Devon (Fig. 19). The remains of the furnace were 1·5 m high and the internal dimensions were 50 × 60 cm. In front was a single piece of stone with a 33 cm wide depression in it which was probably the fore-hearth. The wheel must have been 3 m in diameter by 33 cm wide. This furnace had a granite mould nearby, capable of producing rectangular 'pigs' about 30 cm long by 20 cm wide with a rounded bottom section. A number of these mould stones have been found on Dartmoor, all of granite (Fig. 19). The metal, smelted with water power and local fuel, probably peat, was re-smelted (or refined) in the larger towns such as Exeter where it was re-cast into standard ingots weighing about 90 kg. The loss during this process was about 12%.[20]

Much less is known about the technical side of the Cornish tin industry. The blowing houses were huts built of stone and turf covered with thatch, that periodically caught fire or were intentionally burnt to recover the tin dust. The furnace was made of large stones held together by iron clamps and was blown with a water wheel. The metal was ladled out of the fore-hearth and refined by drossing in an iron 'pot'. heated like a lead-refining pot of the 19th century. After this it was ladled into granite moulds. The fuel seems to have been mainly charcoal.

By the end of the 17th century there were 26 smelting houses in Cornwall and two in Devon.[21] Those near the coast probably used imported coal although there is no evidence that coal was used for ore smelting before the introduction of the reverberatory furnace. However, tin slags were smelted in bloomery-type furnaces like lead slags and coal was probably used for this purpose.

Smelting in a primitive blast furnace with charcoal produced a very pure tin especially after a limited amount of drossing to reduce the iron content. The crust of a 17 kg plano-convex ingot from Tremethack Moor, Madron, contained only 1·45% of non-stanniferous material and most of this must have been picked up from the surrounding soil.

The early slags such as the EBA slag from Caerloggas (Table 26) were tin silicates with melting points of the order of 900°C, but later slags such as the Portuguese slag of the EIA were ferrous silicates showing that they had learnt to replace the tin with iron. The high-grade tin minerals available in the EBA would give rise to very little slag, so not much tin would be wasted. While the early smelters added iron ore to flux the siliceous gangue, modern smelters add appreciable amounts of lime. This causes a rise in the free-running temperature of the slag compared with the fayalite type of slag of the early iron age, medieval, and post-medieval times.

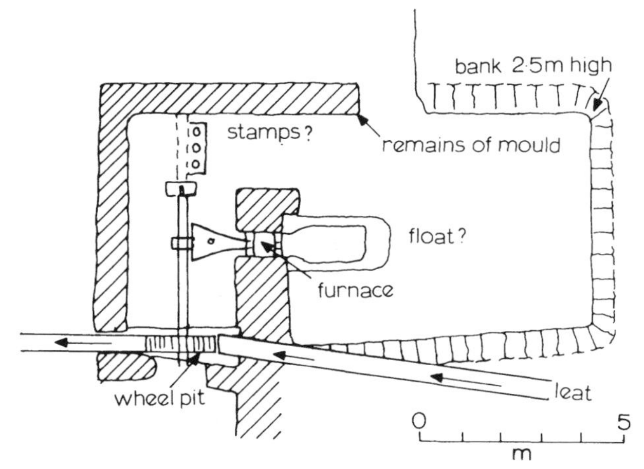

20 Reconstruction of Outer Down blowing house, Devon (after plan by R. H. Worth[18])

TIN INGOTS

About 25 tin ingots have now been found, the bulk of which consist of pure tin and are therefore primary produce (*see* Table 27). They consist mainly of three shapes: (*a*) plano-convex with some tendency to be oval; (*b*) elongated with round bottoms with a D-shaped section; (*c*) rectangular with a trapezoidal section. One ingot from St. Mawes, opposite Falmouth Harbour does not fit this classification and has often been referred to as 'astragoidal'. It could consist of two long rectangular ingots joined together at the centre.

Very few of these ingots have been found in a stratified context. The plano-convex ones are often referred to as 'Jew's House' tin which is clearly a reference to their supposed Phoenician origin. It would seem that these are the earliest type and may well date to the pre-Roman Iron Age. The rectangular ones are undoubtedly medieval and there are many granite moulds for this type which were being used well into post-medieval times.

The tapered ones (Fig. 20*f*, *g*) look as though they have been cast into a sloping mould which could have made ingots of the medieval lead type. If the 'EIC' on the Trereife ingot indicates the East India Co. then this type must be post-medieval. The Falmouth ingot is most probably of this date, although it has been suggested that its shape was designed to assist horse transport.

The Roman ingots of the Mediterranean emanating from the Iberian mines are purse-shaped, consisting of two rectangular ingots cast together with a handle. Borlase[25] reports the finding of two of this type on 'St. Austell Moor', 2·42 m down in a tin stream. They weighed 12·7 kg. In this and many other ways they are similar to those found off the coasts of the south of France and Sardinia and there is little doubt that they came here either as items of trade or were made here to conform with an internationally accepted form during the Roman period.

Many of the plano-convex type contain a very thick corrosion crust often exceeding 1 cm in thickness. This consists of SnO and SnO_2 and although the corrosion resistance of pure tin is not very high it is unlikely that these are less than 1 500 years old. The ingot from Chûn Castle is an example of this and like many others it has been so badly corroded that it has lost its original plano-convex shape, so much so that it now has the form of a thick disc (Fig. 20*c*).

TIN AND PEWTER ARTIFACTS

Metallic tin is a soft, ductile, silver-like metal with the low melting point of 232°C. The hardness[26] of pure metallic tin is 5 HV which is slightly greater than that of lead; the lead–tin alloys such as the solders and pewters may have hardnesses in the range 12–14 HV. Bars or strips of tin will emit a crackle or 'cry' when bent. Tin containing 10%Pb will still cry but less audibly. Zinc will destroy the cry even when as little as 1% is present.[27]

Small lumps of tin have been found in an early context such as those from the LBA hoard at Grays Thurrock,[28] in Essex. There is also a small piece of a high-tin segregate from Weston Wood with 60%Sn, 30%Pb, and 5–10%Sb.[29]

The few early artifacts found are the armlets from Lough Gara and the torc from Killsallagh, both in Ireland and listed in Table 28,[33] these probably belong to the EIA. The five fragments of torcs from Snettisham are now known to be made of silver.[40] There is also a damaged mount from the Llangwyllog hoard, which has been analysed and found to be pure tin (Table 28). So tin was being used as a decorative metal in this period but rather sparingly.

Table 28 lists the main tin and pewter products so far discovered, with their analysis where known. English pewter from the middle ages to the nineteenth century contained 20%Pb, while the standard for French pewter was 16·5–18%Pb. The lead content of pewter has been decreased considerably in recent years since it became known that lead may be dissolved from certain grades of pewter. The French standard was fixed in the belief that it was a safe figure for use with wine. Modern English pewter, now known as Britannia metal, does not contain lead but about 8%Sb and 2%Cu to harden the tin.[27]

It is difficult to decide the reason for the small amounts of Pb (2·7–4·6%) found in Roman tin, unless the small increase in hardness which results was appreciated.

In the Appleshaw finds,[37] the variation in lead content is very great, and the actual lead content of Roman pewter was probably left to the discretion of the maker. Lead contents in the range 23–35% confer considerable hardness, and since lead has always been cheaper than tin the high lead percentages would be much more economical. As Table 28 shows, the majority of the Roman tableware falls into the range 20–30%Pb, but pieces have been found containing over 50%Pb.

There is a further group of finds analysed by Gowland,[37] Smythe,[15] and others[39, 41] which are unclassified in the sense that they do not belong to any group of alloys used today apart from solders and lead–tin alloys for toy models. Since they contain more than 50%Pb they should really be considered as lead alloys, but three are items of tableware which were clearly intended to be pewter and they therefore merit consideration in this chapter. These are listed in Table 29. Many might be termed 'debased' pewters, and the fact that one contains 0·45%Cu suggests that pieces of bronze may have been partially dissolved in the alloy,

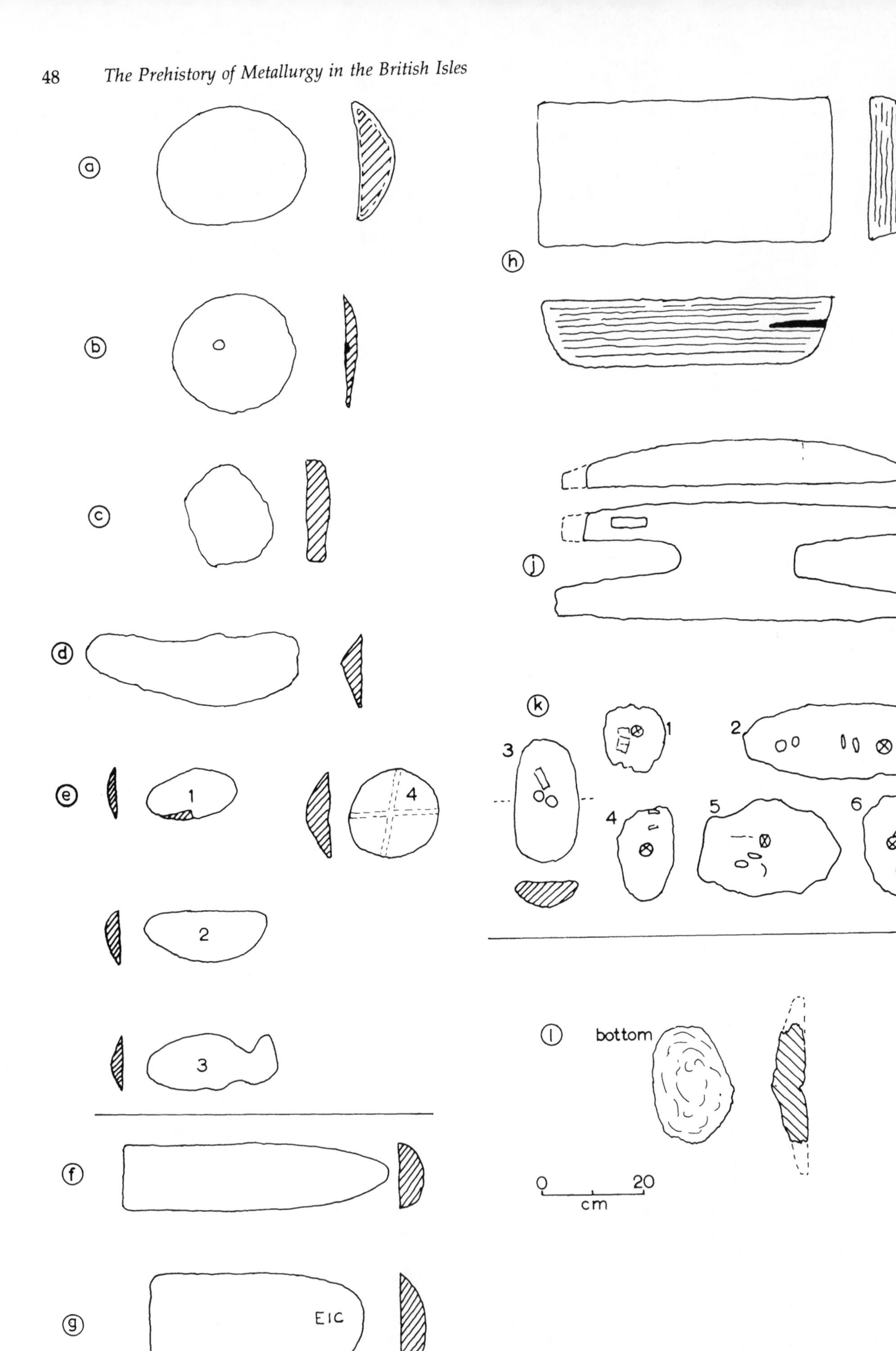

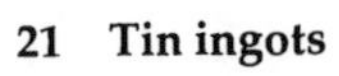
21 Tin ingots

Table 27 Tin ingots (analyses by Smythe[15])

1. Fowey (Truro Museum). Medieval. One of five, about 150 kg each. 63 × 30 cm. Various marks and holes (Fig. 21*h*).
2. Carnanton (Truro Museum). More like the Trereife ingot (No. 9) in shape. 13 × 46 cm. 17·8 kg weight. 99·9% Sn. At least three authorities claim to see an inscription DDNN on this ingot which dates it to the 3–4th century AD. (Haverfield[16] and Warner[24].) (Fig. 21*f*).
3. St. Mawes (Truro Museum). Much corroded. 68 kg. From Falmouth Harbour. 99·9% Sn (Fig. 21*j*).
4. Tremathack Moor, Madron. (Truro Museum). Plano-convex ingot of 'Jews' House' tin. Oval; 23 × 30 cm by 7–10 cm thick. 'Crust' is dense and 6 mm thick. Some of it is flaky. 17 kg (Fig. 21*a*).
5. Chûn (Truro Museum). Plano-convex ingot of tin, coated with corrosion product. Weight 5 kg. Oval, 20 × 15 cm. Thickness 38 mm. Also with it was a small piece of melted tin (Fig. 21*c*).
6. From Vellin Antron farm, Mabe. (Helston Museum). Plano-convex ingot. Found in 1913. One of 'half a dozen'. Tin content 96·88%. Oval; 10 × 13 cm by about 5 cm thick.
7. Near St. Austell. (Geological Museum, Penzance). Plano-convex ingot of 'Jews' House' tin found while building railway in 1858. Boat-shaped 46 × 30 cm by 3–7 cm thick.
8. Par Beach, St. Martin's, Scilly. (Penlee House Museum, Penzance). Plano-convex ingot. Thin and much corroded. Corrosion product is dense, but metal core is highly ductile and therefore pure. One of a number of finds found in a circular hut on the shore line. Dated to circa 300 AD; accompanied by a number of flaky pieces of tin corrosion product.
9. Trereife Ingot (Penlee House Museum, Penzance). 13·4 kg weight; 99·9% tin. 42 × 20 cm by 5 cm thick. (Contains letters EIC – East India Co?) (Fig. 21*g*).
10. Penwithick, St. Austell (Truro Museum, 45–1974). Elongated plano-convex ingot. Weight 7·73 kg. Length 37 cm; max. breadth 13 cm (Fig. 21*d*).
11. Prah Sands, Cornwall. Four ingots in good condition were found in a 'forest bed' amongst Quaternary trees. One of these (Fig. 21*e* no. 4) is plano-convex but is unusual in that it has 0·7 cm wide diametral lines engraved more or less at right angles on the bottom, resembling rope markings. There is a shrinkage cavity in the centre of the flat side. The ingot weighs 3·86 kg, is 24 cm in dia. and 5 cm thick, and is dated to about 660 AD. The others were more oval with plano-convex sections (*see* Fig. 21*e* nos. 1–3).
12. Pitton, near Porton Down, Wilts. A plano-convex ingot about 2 cm thick and 25 cm in dia. Weighs 10·1 kg; heavily mineralized with a substantial metal core. It also contains a central shrinkage cavity. Analysis showed it to be pure tin (Fig. 21*b*).
13. Battersea, River Thames. Six pewter ingots of plano-convex form found in the 19th century. Their composition varies from 94% to 50% tin. All but one is oval in shape and together they weigh 15·8 kg (Fig. 21*k* nos. 1–6).
14. Castle Dore, Cornwall. The highly oxidized remains of a small plano-convex ingot now weighing 705 g.
15. Lanlivery, Cornwall. A slightly oval plano-convex ingot found in the 19th century. This now weighs 21·02 kg and measures 26 × 23·5 cm by 10 cm thick. (Geological Museum, London; Inv. No. MI 12608.)
16. Trethurgy, Cornwall. Weight 7·03 kg to which must be added just over 5·5 kg of surviving corroded detached lumps. Originally it must have weighed at least 13 kg (Fig. 21*l*).

although this should not have been responsible for more than a trace of the tin content. These are probably all secondary alloys, in which the proportions of lead and tin were immaterial and probably accidental. While tin is a good deal lighter than lead, it would have looked like lead and therefore may have been incorporated accidentally. For use as a rivet the modern tinman's solder containing 66% Sn would be stronger than the 17% Sn alloy actually used, and sufficiently malleable. The coffin from Ilchester would be just as suitable made in lead, and there would appear to be no advantage in the alloy used. Attention must be drawn to the fact that some of the lead finds listed in Table 43, Chapter 4, contain a substantial amount of tin.

A very large proportion of the finds belong to the Roman period, which suggests that British tin deposits must have been exploited as intensively as those of lead, although there is far less evidence of this. The finding of stone moulds for casting pewter at Camerton and Lansdown in Somerset,[41–3] and Langton in E. Yorkshire,[44] suggests that Britain might have been an important manufacturing centre for Roman pewter based on supplies of tin from Cornwall and lead from the Mendips. It would appear that pewter manufacture was a feature of villa life in the Mendip area as we see at Gatcombe.[30]

The preferred alloy for organ pipes contains 70% Pb and therefore this subject will be discussed in the next section.[45] Pure tin organ pipes are known abroad but, so far, have not been reported in Britain. In the 12th century, Theophilus used soldered copper sheet,[46] but modern opinion believes that the best tonal properties are obtained with high-tin or high-lead alloys from the lead–tin system, and the latter are cheaper.

THE CORROSION OF TIN AND ITS ALLOYS

Metallic tin is only stable above 13°C.[47] At temperatures below this the form of tin which is stable is a non-metallic grey–brown material which usually exists as powder. While most metals undergo changes of some sort at various temperatures, tin is the only common metal which undergoes such a drastic change as an allotropic transformation from a metal (beta) to a non-metal (alpha) form at normal temperatures. However, the rate of change at temperatures just below 13°C (i.e. average English outside temperature), is extremely small, and tin would not be expected to change in an average lifetime. The first sign of a change is in the grain boundaries of the metal and leads to embrittlement. The change is inhibited to some extent by additions of other metals, and one wonders if the early use of pewter, an alloy of tin and lead, was due to the realization that additions of lead inhibit transformation, or to other factors such as increased strength or cheapness.

The majority of early pure tin products show some traces of tin corrosion. Richmond and Smythe[38] have described a Roman cup from High Rochester, Northumberland, which was in an advanced state of decay. Bannister[48] has discussed a scabbard-end of Roman or medieval date which was badly corroded. The Falmouth ingot shows numerous large blisters and

Table 28 Composition of tin and pewter finds

Object and provenance	Date	Element, % Sn	Pb	Cu	Ag	As	Sb	Remarks	Ref.
Rods, Auchtertyre (Grampian)	BA	78·66	21·34	—					57
Mount, Llangwyllog	LBA	99·0	0·7	0·1	—	—	0·045	Fe, 0·1; Ni, 0·01	31
Sceptre, Glastonbury	EIA	98·5	—	—					32
Armlet, Lough Gara	EIA?	98·00	0·12	0·01	<0·01	0·05	0·01		33
Armlet, Lough Gara	EIA?	98·50	0·64	0·011	<0·01	0·05	0·01		33
Armlet, Lough Gara	EIA?	99·60	0·30	0·039	<0·01	0·05	0·01		33
Torc, Killsallagh	EIA?	98·80	0·016	0·019	<0·01	0·10	0·01		33
Ingots, Battersea									
(0·82 kg)	Roman	94·0	4·59	0·115	0·0018		0·09		31
(5·07 kg)	Roman	68·4	31·5	0·064	0·0052		0·053		31
(3·09 kg)	Roman	67·6	30·9	0·067	0·0029		0·039		31
(1·23 kg)	Roman	67·4	31·1	0·052	0·0104		0·060		31
(3·50 kg)	Roman	54·0	43·3	0·051	0·0056		0·039		31
(2·10 kg)	Roman	50·4	43·9	0·075	0·0098		0·062		31
Tableware, Traprain Law (Lothian)	5th cent. AD	75·61	24·07					Found inside looted silverware	34
Button, Housesteads	Roman	99·43	0·82	nil					15
Lump, Benwell	Roman	97·70	2·73	nil					15
Cup, High Rochester	Roman	95·64	4·49	nil					15
Strip, Corbridge	Roman	94·50	4·50	1·23					15
Strip, Corbridge	Roman	94·00	4·63	1·40					15
Ring, Corbridge	Roman	66·79	33·53	—					15
0·45 kg ingot, Corbridge	Roman	94·78	5·37	—					36
Jug, Shapwick (Somerset)	Roman	88·00	12·22	—					15
Oval dish, Appleshaw (Hants.)	Roman	99·18	0·4	—					37
Bowl, Appleshaw	Roman	94·35	5·06	—					37
Circular dish, Appleshaw	Roman	90·55	8·31	—					37
Chalice-shaped cup, Appleshaw	Roman	76·41	23·04	—					37
Small dish, Appleshaw	Roman	72·36	26·09	—				Chi-Rho monogram	37
Flanged cup, Appleshaw	Roman	70·58	27·62	—					37
Circular dish, Appleshaw	Roman	64·75	34·66	—					37
Plate, Southwark	Roman	72·90	26·75	—				Inscribed: MARTINUS	37
Square dish, Icklingham (Suffolk)	Roman	71·80	27·32	—					37
Scabbard-end, Church Stretton (Salop.)	Roman or medieval	99·98	—	tr.					38
Tableware									
Abingdon Piggots	Roman	62·3	37·7		0·05				39
Abingdon Piggots	Roman	70·0	30·0	—					39
Sutton	Roman	62·2	37·8					Chi-Rho tazza	39
Sutton	Roman	67·8	32·2		0·05				39
Isleham Fen	Roman	72·5	27·5						39
Isleham Fen	Roman	76	24						39
Isleham Fen	Roman	69·2	30·8		0·05				39
Icklingham	Roman	79	21						39
Icklingham	Roman	79·5	20·5						39
West Row, Mildenhall	Roman	74·3	25·8						39
West Row, Mildenhall	Roman	73·4	26·7						39

Table 29 Lead–tin alloy finds (high lead)

Description	Date	Composition, % Pb	Sn	Ca	Fe	Ref.
Octagonal dish, Icklingham, Suffolk	Roman	53·54	45·75	tr.	tr.	37
Cup, Brislington, Somerset	Roman	54·80	45·38	nil	—	15
Coffins, Ilchester, Somerset	Roman	55·31	44·97	nil	—	15
		58·48	41·84			
Lump, Corbridge, Northumberland	Roman	56·30	43·94	nil	—	15
Spindle Whorl, Corbridge	Roman	61·60	37·94	0·45	—	15
Spindle Whorl, Corbridge	Roman	65·40	34·66	nil	—	15
Piece, Edlington, Yorks.	Roman	66·97	33·15	nil	—	15
Seal, Brough, Yorks.	Roman	69·97	29·83	nil	—	15
Seal, Brough	Roman	72·90	27·10	nil	—	15
Pottery rivet, Corbridge	Roman	82·23	17.55	nil	—	15
Dish, Whittlesey Mere or Burwell, Cambs.	Roman	57·0	43·0	—	tr.	37

Table 30 Analysis of corrosion products on tin,%

Compounds present	Plano-convex ingot, Tremathack Moor	Falmouth Harbour	Block, Trereife	Benwell	Corbridge	Scabbard-end
	Pre-Roman?	Pre-Roman?	Medieval?	Roman	Roman	Roman or medieval
	Original composition					
	Pure tin	Pure tin	Pure tin	2·73% Pb	5·37% Pb	Pure tin
	Reference					
	52	51	51	51	51	48
SnO_2	90·62	84·93		68·8	57·3	54·68
$SnO.H_2O$	0·43 (Sn)	2·61	44·7	27·9	34·1	43·45
$SnCl_2.2H_2O$	1·66 ($SnCl_2$)	5·03		—	4·4	
SiO_2	0·41	1·67		0·50	—	
H_2O	6·25	5·89		0·10	—	
$PbCO_3$	—			2·7	—	
$PbSO_4$	—				2·7	
$FeCO_3$	—				1·5	
CuO	—					0·81
SO_3	—					0·85
CaO	—					0·10
Fe_2O_3 etc.	1·04					0·12

several cracks and is certainly suffering from salt-water corrosion.[15]

In the restoration of corroded tin artifacts it is necessary to know something about the mechanism of corrosion of tin and how far transformation reactions are responsible for its deterioration. For there is no doubt that tin metal can revert to its original state more completely than many other metals, as may be seen from the state of many of the plano-convex ingots. Indeed, the fact that it can do this has led to some confusion over whether the occurrence of cassiterite in a furnace charge together with copper is proof of the use of the mineral to make bronze by mixing with molten copper under charcoal, or whether it is a corrosion product. Under the microscope it is easy to distinguish corroded tin from natural cassiterite.

The effect of temperature on the transformation reaction is very great[49] but under most circumstances it needs $-40°C$ to exceed a rate of 6 mm h^{-1}. Other factors such as deformation, vibration, pre-existing nuclei, purity, etc. are also relevant. It is not difficult to envisage a reaction which results in a volume increase of 25% producing its own stress and resulting deformation.

Whether or not transformation reactions are a cause of corrosion and deterioration is largely a matter of the ambient temperature and therefore what country the object is in. It appears that in Denmark, for example, it is assumed that transformation is a serious contributor to deterioration[50] while others believe that it is a rarity both in Europe and in the USA.[36]

Organ pipes at Nantes built in 1748 with 99·20–99·25% tin, while showing no signs of the alpha phase under X-ray diffraction, showed unmistakable metallographic signs that reversible transformations had occurred giving rise to fine 'powdery' crystals. Such pipes would have been submitted to vibration which the authors showed was a significant factor in favouring the beta–alpha transformation.[49]

Clearly, to avoid this preventable type of deterioration it is necessary to remove pre-existing nuclei by warming the object, to avoid stress, and to maintain a reasonable temperature, i.e. more than 13°C. Unfortunately, there are other causes of deterioration which are more common.

The 'crusted' plano-convex ingot from St. Mawgan in Pydar was found in a stratified level dated to mid-1st century AD, together with a pebble of polycrystalline cassiterite.[8] An X-ray diffraction examination of the crust, which amounted to the whole of the specimen, showed that it was 75% SnO and 25% SnO_2. It is interesting to note that alpha (grey) tin and also beta (white) metallic tin were both absent, i.e. less than 1%, but the possibility that grey tin has been instrumental in causing the accelerated corrosion of the metal cannot be entirely eliminated. Of course, chlorides may have similar effects and have been shown to be present by Smythe[51] on both sea and land finds (*see* Table 30). A 19th century chemical analysis of a similar ingot from Tremathack Moor is also given in Table 30.

In the St. Mawgan specimen, the whole section measuring 2·5 cm thick has been converted to oxide. While this is typical of tin on land it does seem to need some explanation, as the corrosion rate of this metal is far in excess of any other non-ferrous metal. A metallographic examination showed a very dense structure with radial cracks near the surface. Cracks had extended down to the centre but these had been filled by a darker phase than the more porous majority phase. The hardness of this dark phase was of the order of 1 000 HV (200 g). Clearly it must be the SnO_2. The majority (presumably the SnO) had a hardness of 300 HV (200 g), which is what one would expect from SnO. There were also minute crystals of a very light phase with a hardness of about 280 HV.

There is not a great amount of hydration. The cracking tendency on oxidation could be explained by the presence of alpha tin with its low density (SG = 5·75) compared with beta tin (SG = 7·28), but another possible explanation must also be put forward. Most oxides have larger volumes than the metal from which they form. In the case of SnO_2 the volume ratio is 1·32; with

Table 31 Composition of corroded pewter, wt-%

Element of compound	Camerton Somerset[41]	Llantwit Major, S. Glamorgan[55] A	B
Pb	30·9	33·5	3·5
Sn	40·6	41·0	~72·0
Fe_2O_3	30·93*	6	~6
Bi	—	tr.	tr.
O_2 } S	—	rem.	rem.

*Probably some of the iron is present as FeO.

Analyses by Metallurgy Department, University of Newcastle upon Tyne. Specimens kindly provided by G. C. Boon, National Museum of Wales.

SnO it is 1·28. In dry oxidation this volume increase can cause internal stresses which cause blistering and cracking. In corrosion, a good deal of the expansion is offset by leaching out the soluble components. In the case of tin these are slight and there is, therefore, a net expansion causing cracking.

The peculiarity of tin has recently been demonstrated in the case of a piece of a tin bronze dagger from the EBA barrow at Caerloggas.[4,53] The piece discovered was at first thought to be a replica made of glass or slag. It was found that it was in fact almost entirely tin oxide. Its coherence, in contrast to the pure tin crusts, was due to the leaching out of the more soluble copper salts during corrosion thus offsetting the volume increase of the 10–15% tin to tin oxides.

The SG of the dagger fragment was 2·36 implying a good deal of porosity. Analysis of the surface by X-ray fluorescence gave:

tin 50–80%
copper ~5%
iron ~5%
arsenic 10–20%.

The last figures imply that the original composition was strongly arsenical, which is in keeping with the EBA although it is recognized that As suffers extreme surface segregation.

Examination by X-ray diffraction gave an indexed structure that included varlamoffite (Sn,Fe) $(O,OH)_2$ which is quite common in corrosion products on bronze[35] and is probably the compound chemically analysed by Smythe[51] in Table 30 as $SnO.H_2O$. However two new compounds have recently been identified: romarchite (SnO) and hydroromarchite ($SnO.H_2O$), so it is quite possibly the latter.[54]

Table 31 gives the analyses of some corroded pewters. Analyses of corrosion products obtained by Smythe[51] and Bannister[48] are given in Table 30. The exact formulae of the tin oxides are still a matter of argument, but since they have little bearing on any deduction of archaeological value they will not be discussed here. What is sometimes important, however, is the evidence that can be drawn from a completely corroded specimen as to the composition of the original metal.

The figures for the tin–lead alloys given in Table 30 indicate that enrichment of tin in the corrosion products can be expected, mainly because the lead salts are more soluble in soils than the tin salts, although, as the Benwell example shows, this is not always the case. In the Corbridge ingot, the lead content falls from 5% in the metal to 2% in the corrosion product, while in an alloy that is predominately lead, the tin content of the corrosion product increases (Chapter 4). Thus the lead contents of the pewters represented in Table 31 were probably somewhat higher.

The processes of tinning are discussed in Chapter 5.

REFERENCES

1 J. S. JACKSON: in 'Proc. 5th Atlantic Colloq.', (ed. M. Ryan), 107–25; 1978, Dublin.
2 E. T. LEEDS: *Archaeologia*, 1927, **76**, 205–37.
3 J. D. G. CLARK: 'Prehistoric Europe – the economic basis'; 1952, London.
4 H. MILES: *Corn. Archaeol.*, 1975, (14), 5–8. The slag was later analysed by Dr C. J. Salter of the Department of Metallurgy, University of Oxford. I am very grateful for his permission to quote his results.
5 G. RAPP JR: in 'The search for ancient tin', (ed. J. S. Olin and T. A. Wertime), 59–63; 1979, Smithsonian Institution.
6 R. PEARCE: *J. R. Inst. Corn.*, 1871, **4**, 81–3.
7 'Trace elements in ores', Res. Rep. 1950–53, 26, Royal School of Mines, London, 1954.
8 L. M. THREIPLAND: *Archaeol. J.*, 1956, **113**, 33–81.
9 T. GREEVES: *J. Hist. Metall. Soc.*, 1980, **14**, (2), 104–5.
10 G. AGRICOLA: 'De re metallica', 294–415; 1950, Dover Publications Inc.
11 H. M. FRIEDE and R. H. STEEL: *J. S. Afr. Inst. Min.Metall.*, 1976, **76**, 451–70.
12 W. PRYCE: 'Mineralogia Cornubiensis: a treatise on minerals, mines and mining', 136, 282; 1778, London.
13 LE GRICE: *Trans. R. Geol. Soc. Corn.*, 1846, **6**, 43–5.
14 W. GOWLAND: *Archaeologia*, 1899, **56**, 267–322.
15 J. A. SMYTHE: *Trans. Newcomen Soc.*, 1937, **18**, 255–66.
16 F. HAVERFIELD: *Archaeol. J.*, 1892, **49**, 176.
17 R. HANSFORD WORTH: 'Dartmoor'; 1954, Plymouth.
18 R. HANSFORD WORTH: *Trans. Devon. Assoc.*, 1927, **69**, 343–5.
19 R. M. L. COOK, T. A. P. GREEVES and C. C. KILVINGTON: *Trans. Devon. Assoc.*, 1974, **106**, 161–214.
20 G. R. LEWIS: 'The Stannaries'; 1908, reprinted 1966.
21 D. B. BARTON: 'A history of tin mining and smelting'; 1967, Truro, D. Bradford Barton Ltd.
22 H. MAIA E COSTA and J. R. DOS SANTOS JR: *Trab. Soc. Portug. Antrop. Etnol.*, 1965–66, 173–80.
23 R. F. TYLECOTE: *J. Hist. Metall. Soc.*, 1980, **14**, (1), 1–16.
24 R. B. WARNER: *Corn. Archaeol.*, 1967, (6), 29–31.
25 W. BORLASE: 'The natural history of Cornwall', 163, (Pl. XX) 186; 1758, Oxford.
26 D. TABOR: 'The hardness of metals'; 1951, Oxford University Press.
27 H. J. L. J. MASSÉ: 'Pewter plate'; 1904, London, Bell.
28 C. H. BUTCHER: *Antiq. J.*, 1922, **2**, 105–8.
29 J. HARDING: private communication, 1964.
30 K. BRANIGAN: 'Gatcombe Roman Villa', Br. Archaeol. Reps., No. **44**, 1977, Oxford.
31 M. J. HUGHES: in 'Aspects of early metallurgy', (ed. W. A. Oddy), 41–50; 1977, Br. Mus. and Hist. Metall. Soc.

32 A. BULLEID and H. ST. G. GRAY: 'Glastonbury Lake Village', Vol. 1; 1911, Glastonbury.
33 H. H. COGHLAN and H. CASE: *Proc. Prehist. Soc.*, 1957, **23**, 91–123.
34 A. O. CURLE: 'The treasure of Traprain'; 1923, Glasgow.
35 H. S. CAMPBELL and D. J. MILLS: *Metall. Mater. Technol.*, 1977, **9**, (10), 551–7.
36 H. J. PLENDERLEITH and R. MARYON: *Conservation*, 1953, **1**, (2), 63–71.
37 W. GOWLAND: *Archaeologia*, 1896, **56**, (1), 13–20.
38 I. A. RICHMOND and J. A. SMYTHE: *Proc. Univ. Durham Philos. Soc.*, 1938, **10**, (1), 48–55.
39 J. LIVERSEDGE: *Proc. Cambs. Antiq. Soc.*, 1958, **53**, 6–10.
40 R. R. CLARKE: *Proc. Prehist. Soc.*, 1954, **20**, (1), 27–86.
41 W. J. WEDLAKE: 'Excavations at Camerton, Somerset'; 1958, Camerton.
42 T. S. BUSH: *Proc. Som. Archaeol. Soc.*, 1904–8, 119, 164 etc.
43 T. S. BUSH: *Proc. Soc. Antiq. Lond.*, 1908, **22**, 34–8.
44 J. H. GOODALL: *Yorks. Archaeol. J.*, 1972, **44**, 32–7.
45 W. LOTTERMOSER and J. MEYER: *Tin Uses*, 1963, (58), 1–6.
46 J. G. HAWTHORNE and C. S. SMITH (eds.): 'On divers arts: the Treatise of Theophilus'; 1963, The University of Chicago Press.
47 E. COHEN and K. D. DECKER: *J. Inst. Met.*, 1927, **38**, 383.
48 C. O. BANNISTER: *J. Inst. Met.*, 1926, **35**, 71–4.
49 C. BOUCHY and J. M. ROLAND: *Rev. Métall.*, 1969, **66**, 233–43.
50 K. HOLM: private communication, Lyngby, Denmark, 1979.
51 J. A. SMYTHE: *J. Inst. Met.*, 1940, **66**, 355–60.
52 J. H. COLLINS: *J. R. Inst. Corn.*, 1871, **4**, 83.
53 L. BIEK: in 'Proc. Bonn Archaeometry Conf. 1978', *Archaeo-physica*, 1979, **10**, 75–81.
54 R. M. ORGAN and J. A. MANDARINO: *Can. Mineral.*, 1971, **10**, 916.
55 V. E. NASH-WILLIAMS: *Archaeol. Cambrensis*, 1952–5, **102**, 89–163.
56 J. MAJER: 'Těžba cinu ve Slavkovském lese v. 16 Stoleti', 216, 1969, Prague National Technical Museum.
57 J. A. SMITH and S. MACADAM: *Proc. Soc. Antiq. Scotl.*, 1875, **9**, 435–43.

4 *Lead, silver, and antimony*

Lead and silver must be considered together, since about half of the silver produced today comes from lead ores. In antiquity this proportion must have been even higher. Antimony also is more likely to be associated with lead than with tin or copper although it is often added to pewters, but early pewters were usually leaded. Today a considerable amount of antimony is used to harden lead.

In Great Britain most of the lead ores are to be found in carboniferous limestone regions such as the Mendips, Derbyshire, and the borders of Durham and Northumberland. Lead has also been worked in Devon, Cumbria, and Ireland (*see* Fig. 22). In general the silver content of these British ores is today rather low (Louis[1] and Smythe[2] have listed it for ores mined in 1874 and 1923 and their figures are given in Table 32). With a few important exceptions this was also the case in early times; analysis of Roman lead suggests that the silver content was not normally high enough to warrant its recovery, and that unlike the Athenian lead–silver mines at Laurion, and some mines in Asia Minor which produced lead containing about 2 000 g of silver per tonne, the lead ores in this country were smelted mainly for their lead content.

The main lead ore is galena (lead sulphide, PbS) which is easily recognized by its great density and dark lustre. In appearance it is similar to coal, but is considerably heavier, with a specific gravity of 7·5 compared with 1·3 for coal. It tends to occur in the form of cubes and can be concentrated by breaking the ore-bearing rock and washing the lighter stone away in running water. In many deposits some of the galena has weathered to cerussite ($PbCO_3$). This was one of the minerals worked in the famous lead–silver mining areas at Laurion in Greece[4] and appears to have been a substantial component of the mineral worked in the Mendips.[5]

The lead content of the furnace charge naturally depends on the method of mining and concentrating the ore. Waldron[3] gives a figure of 50% lead for Mendip ores in 1875, but if required, it is possible to obtain furnace charges of almost 100% galena (87%Pb). In fact, some Derbyshire furnace charges of the Roman period must have contained nearly this amount.

Although silver, like gold, is considered to be one of the noble metals it rarely occurs in its native state. Gowland[6] believed that native silver played no part in man's culture. There are three main reasons for this. Firstly, when silver does occur in its native state it is in deep veins and not near the surface (the native silver produced until 1873 at Hilderston, near Linlithgow in Scotland, was deep-mined[7]). Secondly, when in its native state, silver is usually in the form of fine leaves and filaments, which could not be worked without melting, although a piece the size of a walnut was found at St. Just, Cornwall in 1753.[8] Thirdly, silver is easily converted to its chloride (cerargyrite, AgCl) by chloride-containing surface waters, a corrosion which is accelerated by small amounts of nitrates such as those present in dew.[6] The result is a whitish-grey mass probably overlooked by early excavators. Roman *denarii* were found in this state at Hengistbury Head.[9] The corrosion of silver probably accounts for the relative absence of this metal before the Early Iron Age, and certainly for the poor state of silver objects found in the last 2 000 years.

The silver content of lead ores is generally higher in the upper layers of a deposit, probably owing to concentration by dissolution of the lead. Gowland[6] considered that cerargyrite could have been produced by the dissolution of the lead content of oxidized galena or the corrosion of native silver, and readily smelted in the Bronze Age with a charcoal fire. Galena oxidizes to the carbonate, cerussite ($PbCO_3$), a small amount of which was found in the Roman villa at Llantwit Major, Glamorgan[10] and which contained 0·52% Ag (5 200 $g\,t^{-1}$). This makes it one of the richest silver minerals known in the UK. However, the nearest lead deposit in the area (Rudry–Machen) gave only 230 $g\,t^{-1}$ and it is probable that the cerussite was the product of a silver recovery process.

METHODS OF SMELTING

Galena is undoubtedly the mineral that primitive man could most easily smelt. Lead melts at 327°C and its oxide can be reduced below 800°C, in a domestic fire, burning charcoal or dry wood. Since galena is lead sulphide and not an oxide ore it must be first converted to oxide by roasting. For this a separate process is not required, since it can readily be carried out in the more oxidizing zones at the top of a furnace or domestic fire. Once part of the galena is roasted to its oxide (PbO) it reacts with the unroasted part according to the simple reaction:

$$2PbO + PbS \rightarrow 3Pb + SO_2$$

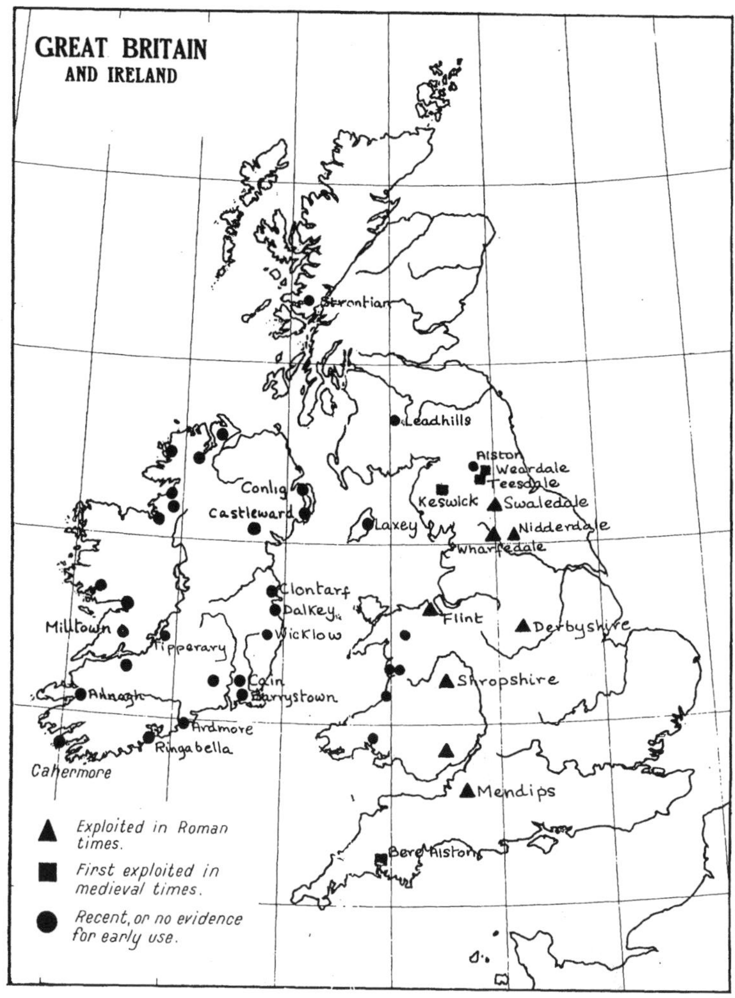

22 Occurrence of lead ores in the British Isles

Table 32 Silver content of lead from recently mined lead ores (ppm or g t^{-1})

	1847 Smythe[2]	1923 Louis[1]	More recent
Yorkshire	—	30–60	
Shropshire	nil	60–90	
Derbyshire	77–180	30–60	
Powys and Gwynedd	180	90–120	
Clwyd	210	90–180	143
Scotland	240	90–150	
Cumbria	280–370	180	
Ireland	300	—	
Durham and Northumberland	370	180	
Dyfed	450 (once 1 200–2 400)	90–120	
Rudry (Glamorgan)	750	—	230
Cornwall, Devon	1 200 (once 3 000–4 000)	1 200	
Isle of Man	—	900–1 200	
Mendips, Somerset	4 200 (Waldron[3] 1875)	—	77
Average for UK		120–150	

NB for comparison: 300 ppm = 0·03% = 10 oz/ton

producing lead, which falls to the bottom of the fire. This is known as a double decomposition reaction, since both the components, the sulphide and the oxide, are decomposed to give lead. In reality, the sulphide is acting as the reducing agent and the fuel takes no part in the reaction, it merely maintains a sufficiently high temperature. This should be compared with the equations for the production of copper from oxide ores (*see* p. 19).

An experiment was carried out by the author to show how easily galena could be smelted under relatively primitive conditions. A good fire of dried sticks was made in a brick-built brazier 60 cm high and 23 cm square internally. Since the major requirements are oxidizing conditions in the upper part of the fire and neutral conditions in the lower, large spaces were left between the upper bricks, while the lower bricks were closed together except for a single opening at the bottom.

After the fire had been burning for some time and the brickwork was very hot, 1·4 kg of undressed Weardale galena, in the form of 10–25 mm cubes, was sprinkled on the top. Some of the galena immediately fell to the bottom of the fire. However, the greater part was trapped near the top and subjected to roasting

Table 33 Composition of British lead and antimony slags, %

	Roman		Medieval						
	Pentre Ffwrndan[12]	Scarcliffe Park[13]	Arla Burn[13]	Wolsingham[13]	18th cent. Flints., Reverberatory[14]	Alston, ore-hearth[15]	Grassington, ore-hearth[15]	Matlock, slag-hearth[15]	Westerkirk, antimony slag*
	1	*2*	*3*	*4*	*5*	*6*	*7*	*8*	
SiO_2	58·2	15·50	37·15	23·50	12·52	28·5	—	13·0	28·7
PbO	32·3	—	5·70	0·54	48·87	3·0	34·0	1·0	nil
FeO	0·8	0·8	21·32	63·05	2·58	25·0	3·0	14·5	49·8
ZnO	—	0·25	1·52	0·05	7·52	10·6	—	2·5	nil
CaO	13·8	—	9·51	1·04	12·68	24·0	4·5	18·5	5·0
Al_2O_3	3·9	4·87	5·49	7·26	3·01	7·0	—	2·0	9·3
MgO	—	1·82	1·04	0·20	—	tr.	—	—	1·2
$PbSO_4$	—	7·26	—	—	9·85	—	23·5	7·0	—
BaO	—	—	—	—	—	—	33·5	30·0	—
CaF_2	—	—	—	—	—	—	1·5	13·4	—
$CaSO_4$	—	40·86	—	—	—	—	—	—	—
Na_2SO_4	—	11·96	8·38	11·90	—	—	—	—	—
K_2SO_4	—	4·30	2·54	1·76	—	—	—	—	1·5
Ag	(4)	(129)	(151)	(129)	—	—	—	—	—
Sb	—	—	—	—	—	—	—	—	nil
Total	109·0	87·62	92·65	100·1	97·03	108·10	100·00	101·9	95·5

Figures in parentheses are in ppm ($g\ t^{-1}$).

*Analysed at the Institute of Archaeology, University of London by R. E. Clough. (Slag matrix only; Sb, Pb present.)

conditions. The fire was allowed to burn itself out and the ashes were then searched for trickles of lead and slag. About 24 g of metallic lead was collected, the rest of the charge remaining as unaltered galena and yellow slag.

This slag was re-smelted in the same brick furnace, now filled with small coke and provided with an air blast of $100\ l\ min^{-1}$ through the hole at the bottom. The slag and galena from the first smelt were placed on top of the coke and a high temperature maintained until the fire had burnt itself out. After this about 0·25 kg of lead was collected at the bottom of the brazier, together with some black glassy slag weighing 1·14 kg, similar in appearance to known Roman and medieval slags and containing about 10% Pb.

This experiment shows that it would be possible to smelt lead in a hot fire surrounded by a ring of stone. The yield would be poor, but no knowledge of roasting or mineral dressing would be required. The main difficulty encountered was preventing the heavy ore from sinking to the bottom, before it was sufficiently oxidized. This could be overcome by a densely interlaced network of branches, which is essentially what was provided by the brushwood used in the Middle Ages.

Unfortunately, no detailed account of an early lead-smelting furnace has yet been published. Rahtz and Boon[11] found the remains of Roman open hearths at Herriot's Bridge in the Mendips, where partly smelted and unsmelted ores were found in the same levels as Samian ware. One of the most valuable sites excavated is that at Pentre Ffwrndan,[12] near Flint, in which the remains of a first century (*c.* AD 76) Roman smelting installation were found. Unfortunately, the remains of the furnaces were too fragmentary to make

Table 34 Analyses of some lead and antimony ores, %

	Scarcliffe Park[13]	Weardale[13]	Siphnos (Aegean)[17]		Laurion Greece[18]
	Galena	concentrate	Lead	Lead–antimony	Galena
	(Roman)	(Modern)	(Modern)	(Modern)	
PbS	90·97	75·11	96·8	29·5	79·0
Sb_2S_3	—	—	0·05	32·9	0·034 (Sb)
FeS_2	—	—	—	—	7·2
SiO_2	0·52	1·44	7·2	3·0	—
CaO	3·12	12·08	4·7	27·5	—
Al_2O_3	0·29	7·04	—	—	—
MgO	0·18	0·31	—	—	—
Na_2O	2·98	8·26	—	—	—
Zn	—	—	0·2	0·7	9·6 (ZnS)
Ag	0·006 1	0·005	—	—	0·035
FeO	—	4·94	0·7	1·1	—
Cu	—	—	—	—	0·82
Total	98·53	110·83		94·7	

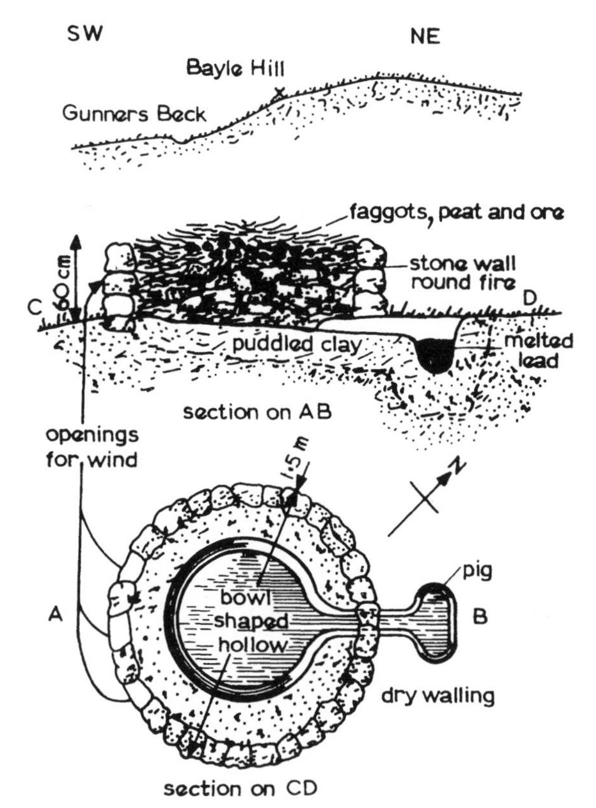

23 **Lead smelting hearth found at Gunnerside, Yorks. (after Raistrick[19])**

reconstruction possible, but a piece of ore was found and the galena separated from it had a silver content of 100 $g\,t^{-1}$Pb. Another piece of lead found on the site yielded 300 $g\,t^{-1}$, thus showing the order of variation that occurs in any lead ore deposit. Slag specimens were also analysed, with the results shown in Table 33.

The silver content of the slag is extraordinarily low (4 gt^{-1}Pb) and is not what would normally be expected from ore containing 100–180 $g\,t^{-1}$.

Primitive lead smelting tends to concentrate the silver into the first smelted lead. The silver content of the lead in the slag should therefore be somewhat lower than that in the lead produced and the difference will indicate the efficiency of the extraction process, the less efficient processes giving much lower values in the slag than the lead product. Experiments by the author gave the following results:

	g Ag/t Pb
silver in lead ore	112
silver in product of first smelt	254
silver in product of second smelt (reprocessing products of first smelt)	215
silver in slag left over from second smelt	185

Clearly the efficiency of the process carried out at Pentre Ffwrndan was very low.

We have more complete details of another Roman site, that at Scarcliffe Park on the Derbyshire–Nottinghamshire border.[16] Here the remains of the hearths consisted of shallow bowls scooped out of the clay, each approximately 1–2 m dia. The analyses of the slags and ores are given in Tables 33 and 34, and it is clear from these that they represent a very primitive process similar to the medieval 'bole hills'. The dark slag which contained a good deal of sulphate also contained galena, was non-magnetic, and showed no signs of having been molten. Even so, the lead content is relatively low and it appears that this slag is the product of smelting a high-grade ore from which the lead has liquated. Impure lead was also recovered containing appreciable galena trapped within it.

By the medieval period things had not changed very much. The bole hills appear to be nothing more than three-sided stalls in which coarse ore and brushwood were stacked and set alight. The slags produced were high in lead, a lot of it in the form of sulphate (*see* Table 33). Figure 23 shows an undated but probably medieval furnace found by Raistrick[19] near Gunnerside, Yorkshire.

The 16th century marks the introduction of water-powered smelting furnaces probably inspired by German experience. At this time the 'old order' was represented by foot-operated bellows-blown hearths (Fig. 24), very often erected on a platform so that they could be rotated to take the smoke away from the blowers.[20] These appear to have been only 50 cm deep with a tuyere placed near the top of the hearth. The 'new order' was represented by Burchard's furnace which was in fact a pair of blast furnaces very much like those shown by Agricola, each blown by a water-wheel and two bellows (Fig. 25). These shaft furnaces were about 1·5 m high and 1 m wide. It was claimed that the foot blast furnace which succeeded the natural draught boles had twice the fuel efficiency of the latter and made 3 t/week. Whatever the merits of the shaft furnace, the shorter furnace continued, but with water-powered bellows, and finally became the ore-hearth of the 18th century.

It would seem that the capacity of late medieval furnaces was much the same as that of Roman furnaces. Glanvil[21] describes the 17th century furnaces as having a well at the side of the hearth into which the lead ran. This held about 75 kg, about the same weight as a Roman pig.

In an open-hearth furnace the lead could be ladled out, after the removal of slag and residual charcoal. In larger furnaces it was tapped from a hole in the side, or from the well mentioned above. In both the Roman and later periods there is no doubt that pigs were tapped into moulds in a single operation, and not ladled out in small amounts.[22]

The slag formed during the smelting process consisted of lead oxide, together with sulphates and silicates. Those containing a large amount of lead often had the appearance of black glass. According to Gowland[23] primitive lead slags may contain 20–26% Pb. Spencer[24] gives the lead content of a Roman slag from Priddy as being 12%. Improvements in technique since the Middle Ages have gradually reduced this figure until, as a result of 19th century advancement, it is now 1·0%. The considerable lead content of ancient slags makes them readily distinguishable from those produced in modern times, and has made it profitable to re-smelt ancient slags to recover their residual lead.

According to Morgans[5] the yield of lead per ton of ore from the Mendips in Glanvil's time (1668) was about 55%. In about 1864, smelting furnaces were erected at St. Cuthbert's to process slags and debris in the vicinity containing about 5–10%Pb, which had already been worked since Roman times and, in fact,

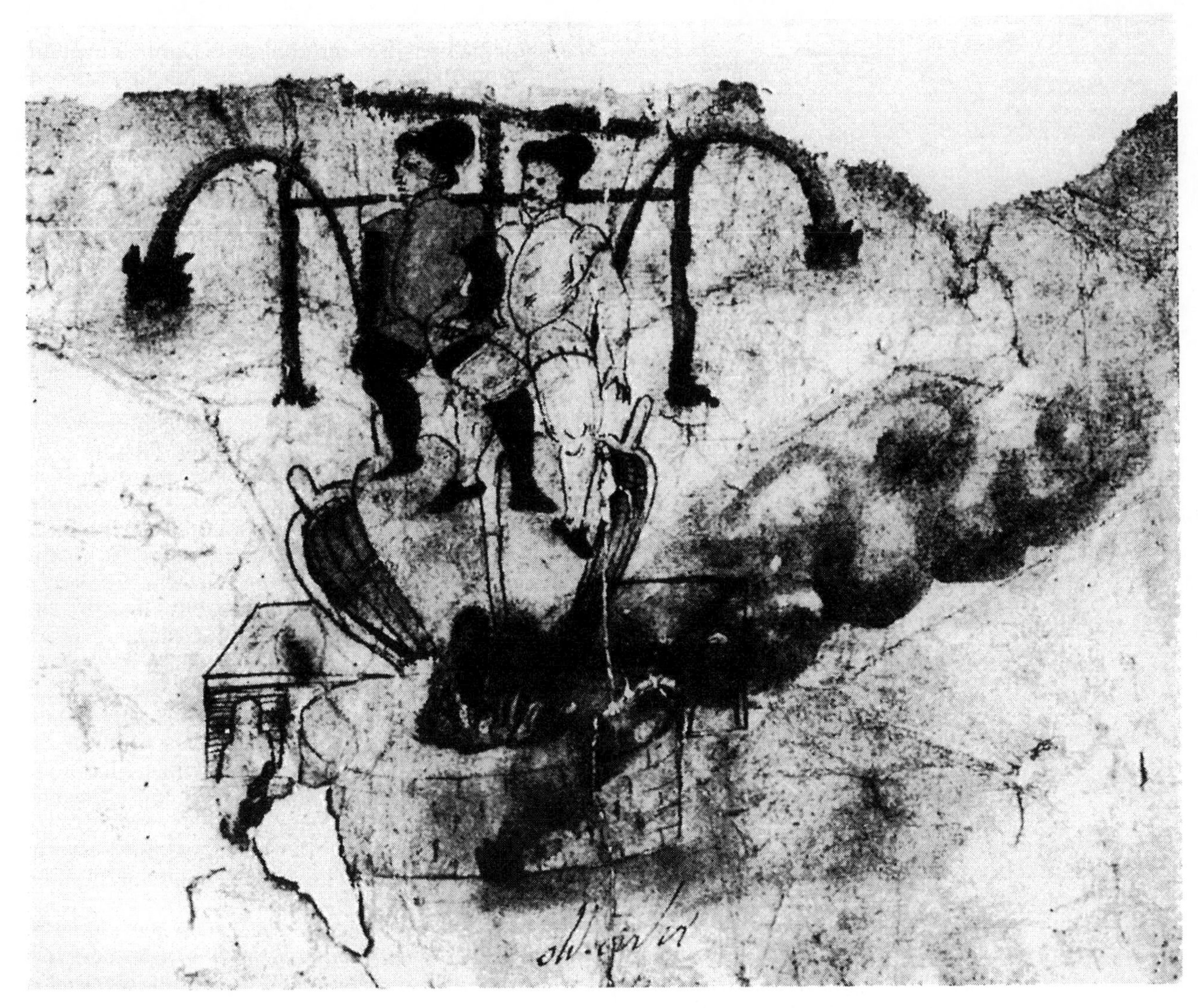

24 Foot-blown lead smelting hearth (PRO SP/12/122/63)

represented the deposits of medieval and later levels of operation. The resulting lead had a silver content of about 100 $g\,t^{-1}$.

At Bere Alston, in Devon[25] in the period 1297–1302, boles similar to those shown in Fig. 23 were used for smelting. They were fuelled with brush-wood. In addition shallower hearths were used to smelt slags containing 12–20% Pb, to recover lead from litharge (PbO) after cupellation, and to take the fines produced by breaking and washing which were too small for the taller boles. These were bellows-blown like the one shown in Fig. 24.

The hearths were built in the same way as iron bloom hearths and in 1582 were said to measure 30 × 45 cm in area.[26] They had to be worked at a very much higher temperature than the boles, requiring charcoal and a forced draught which was provided by water-driven bellows. Occasionally they were fuelled with peat or coal.

The use of this process, with two types of furnace, seems to have been universal in Derbyshire by 1584, when it was introduced into the more backward Mendips.[26] Here, the fixed bole had been used exclusively until about 1540, when it was replaced by the turn-hearth which was not dependent on wind direction.

Reference has been made to the large amount of lead left in the bole-hills slags, and retreatment of these slags in the ore-hearth or slag-hearth. In this process, one would expect the use of a flux to replace some of the lead in the slag, but it is not until 1701–2 that lime is mentioned in this capacity at the Grassington Mill, Yorkshire.[27] In 1729, the use of spar (limestone or fluorspar) in an ore-hearth in Derbyshire is mentioned.[28] Although some of the slags contain appreciable iron (Table 33) the only authority to recommend the addition of iron ore as a flux was Alonso Barba in 1640.[29]

THE RECOVERY OF SILVER FROM LEAD

Apart from the extraction of the silver (and gold if any was present), primitive lead was never refined. The simple nature of the smelting operation had the advantage of providing a relatively pure metal as far as base metals were concerned, and the lead content was often as high as 99·9% (*see below* Tables 38 and 43).

As already mentioned, most of the silver produced in early times, and a good deal of that produced today, comes from argentiferous lead ores such as galena. Today, a considerable amount of silver comes from other non-ferrous ores, in particular the copper ores. It is quite probable that silver was extracted from the copper ores of Devon and Cornwall in pre-Roman and

25 **Water-blown blast furnaces (PRO SP/12/122/63)**

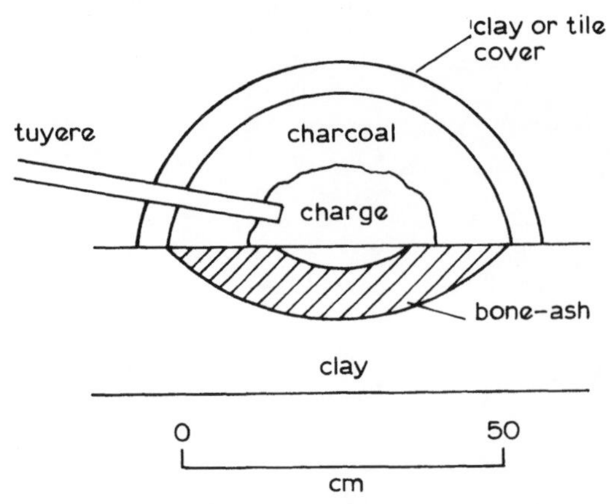

26 Reconstruction of Roman cupellation hearth at Silchester

Roman times. In 1812 copper ores from Wheal Brothers, near Callington, were yielding £3000 worth of silver per annum.[9]

The recovery of silver was carried out by cupellation, which has been practised for at least 3 000 years, and was the only process known in this country until 1833. It involves the oxidation of lead, to litharge, and of any other base metals which are dissolved in the litharge. The noble metals, silver and gold, are left unaltered while the litharge is absorbed by a bone-ash hearth, lost as fume, or skimmed off. The hearth must be shallow so that the maximum amount of molten lead is exposed to the oxidizing blast of air provided by bellows, and charcoal or wood must be used to maintain the temperature at about 1 000°–1 100°C.

Remains of Roman cupellation hearths have been found at Silchester, Wroxeter, and Hengistbury Head.[9,30,31] At Silchester the remains of two hearths were found in 1895 in what may have been a mint or other government establishment.[30] The lining of these hearths consisted of bone-ash (calcium phosphate) similar to that used today (*see* Fig. 26). This material is refractory and is made so as to be porous and absorbent. Gowland[30] analysed the hearths and from their composition (Table 35) came to the conclusion that one had been used for the cupellation of copper coins and the other for the cupellation of silver-containing

Table 35 Analyses of cupellation hearths from Silchester,[23] wt-%

	Hearth A		Hearth B	
	Hearth	Globules	Hearth	Metallic encrustation
Ag	0·13	2·98	1·14	32·61
PbO	54·3		50·47	
Cu_2O	20·86		26·54	
CaO	10·83		9·87	
P_2O_5	7·86		7·12	
SiO_2	0·42		0·41	
Fe_2O_3 } Al_2O_3 }	0·23		0·19	
Insol.	1·07		1·18	
H_2O } CO_2 }	2·57		1·18	
Au	—	Nil	—	0·39
Cu	—	78·13	—	56.64
Pb	—	16.14	—	8·81
Slag } O_2 }	—	2·75	—	1·55

Table 36 Composition of Hengistbury Head cupellation hearths,% (after Bushe-Fox and Gowland[9])

	A Large hearth 28 cm dia.	B Small hearth 23 cm dia.
Cu_2O	33·8	40·50
PbO	19·94	13·57
$Ca_3(PO_4)_2$	30·81	27·62
Ag	3·15	7·86
Au	0·12	0·52
$CO_2 + H_2O$	12·18	9·93

copper. This points to the refining of alloyed or debased coinage, or the extraction of silver from copper, rather than the recovery of silver from lead. In this process a base metal such as copper is oxidized, and the silver which is released from it dissolved in lead. The lead is then oxidized, leaving pure silver behind. Lead must be added at frequent intervals, to take up the silver from the copper. Gowland believed that lead was added too sparingly, and that a third refining operation would have been necessary, probably in another hearth. It is possible that the globules from Hearth A represent the original charge, which Gowland thinks could have been coins minted between AD 250 and AD 375, containing 1·5–17·4% Ag. The metallic encrustation on Hearth B represents a second refining in which the silver content has been raised to 32·6%.

An earlier excavation at Silchester, in 1893, produced a cupellation charge consisting of lead, copper, and galena, and oxides of copper and lead. Roberts-Austen[32] found it to contain 67·6% copper, 12·5% lead, and 0·18% silver, and it was probably used for the extraction of silver from coinage or copper–silver ores from Devon.

Excavations in 1911–12 at Hengistbury Head, Hampshire [now Dorset] yielded the remains of two more cupellation hearths and their contents.[9] The details are listed in Tables 36 and 37.

The smaller hearth, B, had been used in a later stage of the process, as is shown by the higher cuprous oxide (Cu_2O) content, and the higher silver content of its globules. A lump of partly cupelled material weighing 9 kg had apparently solidified in this hearth. Gowland[6] thought that the raw material used was argentiferous copper ore from the Callington area of Cornwall. This would first have had to be smelted to produce an impure copper, which could then be cupelled. It is probable that this was the site of a pre-Roman mint where silver was extracted from copper by cupellation.

Cupellation is incapable of separating gold from silver. Any gold will be left in the silver but can be

Table 37 Products of hearths at Hengistbury Head,% (after Bushe-Fox and Gowland[9])

	Globules from hearths		
	A	B	9 kg lump
Cu	84·03	43·2	50·45
Ag	15·9	56·8	46·43
Au	0·07	—	1·2

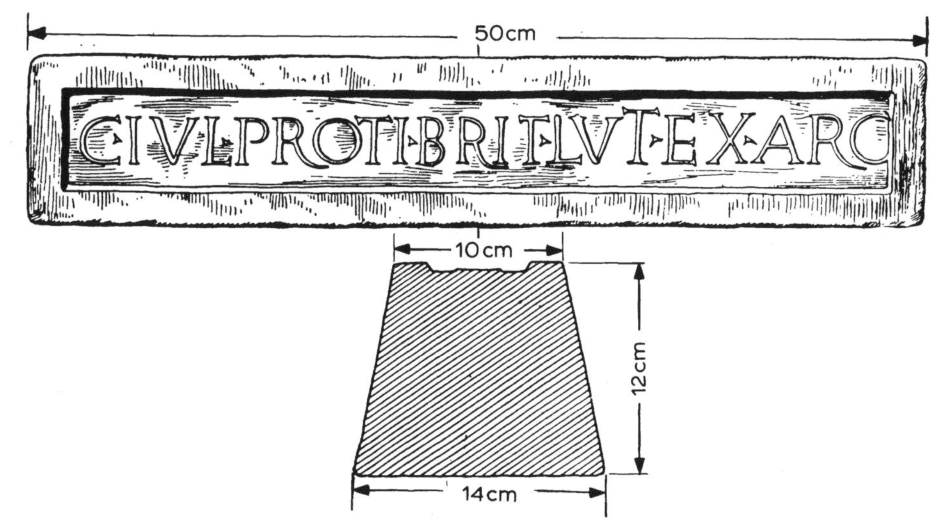

27 Dimensions of Hexgrave Park Pig (courtesy of Trustees of British Museum)

separated by granulating the impure silver in water, mixing it with common salt and clay and placing it in a furnace with charcoal at a red heat. The silver is converted to silver chloride, which is absorbed by the clay; the gold will be unchanged and can be washed out. The clay–silver chloride mixture may then be mixed with lead and cupelled.

Cupellation hardly ever goes to completion. It needs a good deal of lead to absorb the base metals in the silver and some of these as well as the lead get left behind. It was expected that cupelled silver might be distinguished from native silver by the lead content of cupelled material, as native silver seems to be devoid of lead. Cupellation experiments were therefore carried out in bone-ash cupels at 980°C in a muffle furnace with adequate air.[33] The starting material contained 5·80%Ag, 1·16%Bi, and 1·16%Cu, the rest being lead. After 2 h most of the lead had either been absorbed by the cupel or volatilized as PbO. The residual sessile drop or 'button' contained 97·8%Ag, 1·0%Pb, 0·3%Cu, and 0·9%Bi. The 1·0%Pb is very typical of most British silver and is probably proof of its production by the cupellation of lead. It is clear that the removal of large amounts of Bi and Cu by cupellation is a slow and difficult business as no doubt the workers at Silchester and Hengistbury Head were finding. Additions of lead have to be made at intervals to remove more of these elements, but the lead content is not likely to go much below 1%.

The recovery of the lead after cupellation is carried out by re-smelting the litharge in much the same way as the original lead ore, except that it is now a direct reduction process.

The extraction of the silver by cupellation is an expensive process amounting to about three times the original cost of producing the lead; the silver content of the ore must therefore justify this charge. In Roman times it appears to have been economic to desilverize lead containing more than 0·06%Ag (600 $g\,t^{-1}$).[23] By 1780, the economic figure was lowered to 0·027%Ag, while in 1923 it had become 0·009%. Today it is possible to extract silver down to 0·0003% if the original content warrants it.

The finding of a number of copper ingots containing 40–50%Pb and more than 1 000 $g\,t^{-1}$ silver at the Roman Villa of Lullingstone in Kent make one wonder whether another desilverizing process was being used in Roman times.[34] By melting silver-containing copper with lead it is possible to soak out the silver into the lead. Then the lower melting point lead can be liquated from the copper, carrying the silver with it. It is then easier to recover the silver from the lead by simple cupellation rather than by cupelling copper and adding lead. This process was not mentioned in the literature before the 16th century.

THE ROMAN LEAD INDUSTRY

The exploitation of the lead ores of Britain began very soon after the Romans arrived in AD 43. Two pigs of lead were found in the Mendips with inscriptions dating them to AD 49. By AD 74, inscribed pigs were being produced in the mines of Flintshire [Clwyd]. These pigs, which weighed on average 77·3 kg and were about 0·58 m long, 0·14 m wide, 0·13 m high (*see* Fig. 27), were inscribed in the early period with the names of the emperors in whose reigns they were made. They were probably mainly for export and were found in the vicinity of the mining and smelting areas, on roads leading to nearby ports, or at the ports themselves. One, found at St. Valéry-sur-Somme, weighed 75 kg. It was inscribed NERONIS AVG BRITAN L.II, and had probably come from either Flint or Shropshire.[23,35] Two more were found in Rome.

The latest dated pig to be found in Britain belongs to the joint reign of Antonius and Verus, AD 163–9. After this date it is believed that the ownership of the mines passed from imperial to private hands, for imperial stamps cease to be found. Eleven inscribed pigs which bear other stamps probably belong to the time when the mines were leased to private owners (e.g. C. Julius Protus, Fig. 27).

About eight inscribed lead pigs of Roman date have now been found in this country. Some of Gowland's engravings of these pigs are shown in Fig. 28. The lists given in Tables 38 and 39 have been assembled from various sources, and are believed to be correct. There has been some confusion, partly because several pigs have been lost since their discovery, and partly because others are known by more than one name. Whittick[40] has put forward the view that two from Shropshire in Gowland's list (Nos. 23 and 24) were in

Table 38 Roman lead pigs

No.	Gowland's no.	Ref.	Provenance and date found	Probable county of origin	Date made (AD)	Weight, kg	Analysis, % (ppm) Cu	Sb	As	Sn	Bi	Fe
1	4	23	Hexgrave Park, Notts.; 1848	Derbys.	2nd–3rd century	83·6	0·008	tr.	nil	nil	—	—
2	5	23	South Cave, Brough; 1890	Derbys.	2nd–3rd century	61·4	0·003 7	0·008 9	nil	nil	—	0·003 2
3	—	5	Brough-on-Humber; 1940	Derbys.	2nd–3rd century	86·6	0·006 2	0·010 0	—	nil	—	0·003 0
4	—	5	Brough-on-Humber; 1940	Derbys.	2nd–3rd century	86·9	—	—	—	—	—	—
5	—	5	Brough-on-Humber; 1940	Derbys.	2nd–3rd century	89·3	0·003 8	0·004 9	nil	nil	nil	0·001 5
6	—	5	Brough-on-Humber; 1940	Derbys.	2nd–3rd century	87·9	0·003 8	0·021 6	nil	nil	—	0·004 2
7	—	5	Brough-on-Humber; 1940	Derbys.	2nd–3rd century	36·4	0·002 5	0·002 4	nil	nil	—	—
8	6	23	Pulborough, Sussex; 1824	Derbys.	117–138	83·6	0·017 0·004 7	tr. 0·034 7	nil (<25)	nil (0·6)	—	—
9	7	23	Pulborough, Sussex; 1824	Derbys.	117–138							
10	8	23	Pulborough, Sussex; 1824	Derbys.	117–138							
11	9	23	Pulborough, Sussex; 1824	Derbys.	117–138							
12	3	23	Tansley Moor, Matlock; 1894	Derbys.		80·0	0·005 9 0·004 8	0·002 6 0·004 5	nil (5)	nil (0·4)	—	tr.
13	2	23	Matlock Bank; before 1783	Derbys.		37·7	0·022	0·008	nil	nil	—	—
14	1	23	Cromford Moor, Wirksworth; 1777	Derbys.	117–138	57·7	0·020 0·087	0·012 0·005 6	nil (<4)	nil (0·2)	—	—
15	13	23	Heyshaw Moor, Nidderdale; 1735	Yorks.	81	70·5	0·001 9					
16	12	23	Heyshaw Moor, Nidderdale; 1735	Yorks.	81	70·9	0·014 0·001 9	0·007 0·009 6	nil (<6)	nil (0·3)	—	—
17	14	23	Near Blagdon; 1853	Somerset	*c*.49	74·1	0·043 0·014 7	0·021 0·022	0·014 0·035	nil (0·8)	—	—
18	15	23	Bristol; 1865	Somerset	139–161	34·5	0·024 0·011 6	0·019 0·032	nil (<1)	nil (0·2)	—	—
19	16	23	Bristol; 1865	Somerset	139–161	40·5						
20	22	23	Snailbeach Mine; 1796	Salop.	117–138	87·7	0·038 0·026 9	0·004 0·003 8	nil (<2·5)	nil (0·4)		
21	27	23	Bossington, Stockbridge, Hants; 1793	Flints. or Salop.	60	70·9	0·034 0·013	0·006 0·016 4	nil (38)	nil (0·4)		
22	50	23	Hint's Common, Staffs.; 1772	Flints.	76	69·1	0·035 0·019 8	0·006 0·006 9	nil (<5)	nil (6)		
23	—	36	Cirencester (Manor Farm, Syde); 1962	Somerset	79	79·1	~0·005	n.d.	—	—	—	—
24	—	37	Rugby (Tripontium); 1966	Derbys.	2nd–3rd century	79·0	0·008 3	0·016	—	—	—	tr.

No.	Ni	Zn	Ag	Preserved at	Inscription	C.I.L. VII. Ref. no.
1	—	—	0·008 2	British Museum	C. IVL. PROTI. BRIT. LVT. EX ARG.	1216 Seen
2	0·000 2	tr.	0·008 2	Hull Museum	C. IVL. PROTI. BRIT. LVT. EX ARG (badly mutilated; pieces missing)	Seen
3	0·001 7	0·024 7	0·010 4	Hull Museum	C. IVL. PROTI. BRIT. LVT. EX ARG (piece missing from corner)	Seen
4	—	0·032 3	0·006 6	Hull Museum	C. IVL. PROTI. BRIT. LVT. EX ARG	Seen
5	0·000 9	0·009 0	0·005 6	Hull Museum	C. IVL. PROTI. BRIT. LYT. EX ARG	Seen
6	0·001 0	0·013 7	0·006 8	Hull Museum	SOC. LVT. BRIT. EX ARG (inscription not empanelled)	Seen
7	—	—	0·006·8	Hull Museum	(rough casting, no inscription)	
8	—	—	0·003 4 0·004 1	British Museum	(TI). CL. TR. LVT. BR. EX ARG	1215 Seen
9 10 11				Lost		
12	0·002 2	tr.	0·002 5 0·003 5	British Museum	P. RVBRI. ABASCANTI METALLI LVTVDARES	Seen
13	—	—	0·003 9	British Museum	L. ARVCONI. VERECVNDI METAL. LVTVD	1214 Seen
14	—	—	0·006 0 0·008 2	British Museum Skipton	IMP. CAES. HADRIANI AVG. MET. LVT.	1208 Seen
15	—	—	—		IMP. CAES. DOMITIANO AVG. COS. VII } (BRIG on side of both)	
16	—	—	0·006 6 0·008 4	British Museum	IMP. CAES. DOMITIANO AVG. COS. VII } (BRIG on side of both)	1207 Seen
17	—	—	0·024 5 0·042 3	British Museum	BRITANNIC AVG. F.I. (on side, V. ETP.)*	1202 Seen
18	—	—	0·003 4	British Museum	IMP. CAES. A(NTO)NINI AVG. PIIP. P. } probably from same mould	1210 Seen
19	—	—	0·002 4 0·002 7	Bristol Museum	IMP. CAES. A(NTO)NINI AVG. PIIP. P. } probably from same mould	
20			0·007 0 0·009 3	British Museum	IMP. HADRIANI. AVG.	1209c Seen
21			0·002 0 0·001 8	British Museum	NERONIS AVG EX KIAN IIII COS BRIT EX ARGENT† C N(I)PI ASCA(NI XXX IVL PM COS	1203 Seen
22			0·002 2 0·002 81	British Museum	IMP. VESP. VII. T. IMP. V. COS. (DECEA G on one side)‡	1205 Seen
23			0·003 24	Cirencester	IMP. VESP. AVG. VIII. BRIT. EX AR(G). SOC. NOVAE	Seen
24			0·006 3	Rugby	SOCIOR LVT. BR. EX. ARG	

*Whittick[38] reports the weight as being 73 kg and reads the inscription as: BRITANNIC(a) M(etalla) AVG(usti) L(egio) II, and the incuse inscription as V.ETPL.C and queries the date of AD 49.

†Whittick gives the weight as 75·3 kg and suggests the Mendips as the more likely source.

‡Whittick reads the incuse inscription on the side as DECEANGL and agrees the AD 76 date. He gives the weight as 68·3 kg. This is one of a pair cast in the same mould. The other was found at Tamworth nearby and weighed 68 kg. Its analysis was similar but not identical. However, it was well within the degree of reproducibility of the technique.

Table 39 Other Roman Lead Pigs (not fully analysed)

No.	Gowland's No.	Ref.	Provenance and date	Probable county of origin	Date made AD	Weight, kg	Preserved at	Inscription	EE or C.I.L. VII Ref. No.	Ag, g t⁻¹
25	10	23	Matlock Derbys.; 1777	Derbys.		78·5	Lost	TI. CL. TR. LVT. BR. EX ARG.	1215a	
26	11	23	Castleton, Derbys.; 1802	Derbys.			Lost	IMP. [– – –	1213	
27	17	23	Wookey Hole, Wells; 1544	Somerset	49		Lost	TI. CLAVD. CAESAR. AVG. P.M.TRIB. P. VII. IMP.XVI.DE. BRITAN. (Leland 1544)*	1201	
28	18	23, 39	Charterhouse; 1875	Somerset	*c.*70	78	The Priory, Roehampton	IMP.VESPASIAN. AVG. (on side: BRIT.EX ARG. VEB.)	EE. iii. 121a	
29	19	23, 39	Charterhouse; 1873	Somerset	139–161	101		IMP.CAES.ANTONINI.AVG.P(11) (palm branch, hammer, and circle)	EE. iii. 121d	27
30	20	23	Bruton, Somerset; 18th cent.	Somerset	164–169	~ 23	Lost	IMP.DVOR.AVG.ANTONINI/ET VERI ARMENIA-CORVM.	1211	
31	21	23	Sidney Place, Bath; 1822	Somerset	117–138	89	Roman Baths Mus., Bath	IMP.HADRIANI.AVG	1209d	20
32 } 33 }	23/24	23, 40	Aston Farm, Bishop's Castle, Salop.; 1767	Salop.	117–138	86 —	Netley Hall	IMP.HADRIANI.AVG (rim: MIN Brand branch and hammer)	1209a/b	
34	25	23, 41	Snead, Salop.; 1851	Salop.	117–138	84	Lost	IMP.HADRIANI AVG.LEG XX	—	
35	26	23	Minsterley, Salop.	Salop.	117–138	78·5	Liverpool	IMP.HADRIANI.AVG.	1209e	23
36	28	23	Broughton, Chester	Flints. [Clwyd]	74	81	Lost Chester (?)	IMP.HADRIANI.AVG. IMP.VESP.V.T.IMP.III.COS. (on side: DE CEANGI)	1209f 1204	 26
37	29	23, 41, 39	Roodee, Chester; 1886	Flints.	74	86·3	Chester (?)	IMP.VESP.AVG.V.T.IMP.III. (on side: DE CEANGI)	EE. vii. 1121	26 11
38–57	30–49	23	Runcorn	Flints.	84–96	Not known	All lost	IMP.DOMIT.AVG.GER.DECEANG IMP.VESP.VII.T.IMP.V.COS.	} these are the only two inscriptions known from this series	
58	51	23, 41	Common Hall St, Chester; 1849	Flints.		75	Chester (?)	CAESARI [S ? – –] NIVADON	1212	19
59	—	19	Hurst, Swaledale	Yorks.	117–138	77	Lost	(HADRIAN)		
60		26	Charterhouse; 1822	Somerset		45 }	Melted down; approx. weights only	(B.M.ADD.MSS. 33651.f103, Journals of Rev. J. Skinner in Br. Mus.)		
61		26	Charterhouse; 1822	Somerset		68 }				
62		26	Charterhouse; 1822	Somerset		90 }				
63			Charterhouse; 1822							
64		42, 43	Green Ore; 1956	Somerset	69–79	85	Wells	IMP.VESPASIAN.AVG.BRIT.EX. [ΛRG.VEB] LXV.TI.CL.TRIF.		Nil
65		42, 43	Green Ore; 1956	Somerset	69–79	86	Wells	IMP.VESPASIAN.AVG.BRIT.EX. ARG.ΛEB. LXXIIX LRΛD		tr.
66		42, 43	Green Ore; 1956	Somerset	69–79	90	Wells	IMP.VESPASIAN.AVG.LR.Λ[D]. BRIT.EX ARG. EB.TI.CL.TRIF. LXIIX.TRIF.		

67	42, 43	Green Ore; 1956	Somerset	69–79	85	Wells	IMP.VESPASIAN.AVG.BRIT. [EX A]RG [VEB] IMP.TI.C [L]. TRIF.TI.CL. TRIF.XTI.C.TI.C.		568
68	44	Clausentum, Bitterne; 1918	Somerset	68–79	81	Southampton	IMP.VESPASIAN AVG.BRIT.EX ARG.VEB. NOVEG SOC NO (one No. VI, the other VIII)		5
69					75·3				
70	45	Carsington, Derbys.	Derbys.		65·5	Carsington	<< X		
71	39	Cheshunt	Derbys.	117–138	84	Br. Mus.	IMP.CAES.HADRIANI AVG.		
72	19	Grassington, N. Yorks.	Yorks.	98–117	38·5	Lost	(TRAJAN)		
73	46	Ellerker, Humberside (alias Brantingham); 1957	Derbys.		80	Hull Mus.	SOC.LVT.BRIT.EX ARG. (same mould as No. 6 from Brough)		
74	47	Carmel, Clwyd; 1950	Flints.		61	Nat. Mus. of Wales	C.NIPI.ASCANI		37
75	41, 48	Bradwell, Derbys.; 1894	Derbys.		48	Sheffield			
76	41	Belby, Howden, Humberside; 1910	Yorks.		~ 51	Sold for scrap	SOCIORLFBR.EX ARG.		34
77	41	Oker Hill, Darley, Derbys; 1846	Derbys.		—	Lost			
78	41	Charterhouse; 1876	Somerset	69–79	82·7	Bristol Mus.	IMP.VESPASIANI.AVG	EE.iii.121b	
79	41	Charterhouse; 1874	Somerset	69–79		Lost	IMP.VESPASIA [---	EE. iii. 121c	
80	41	Wells; *c*.1530	Somerset	164–169		Lost	Same as No. 30		
81	41	Charterhouse; 1874	Somerset	164–169	2 fragments of what is probably the same pig	Taunton Mus.	---A]NTONINI [------] CORVM ---	EE. iii. 121e	
82	41	Charterhouse; 1874	Somerset	164–169		Taunton Mus.	AV]G[--- AR]MENIA[CORVM	EE. iv. p. 206	
83	41	Richborough; 1922–3	—	96–169		Richborough	IMP NERVAE CA [---		
84	41	Caerwent; 1947	—	—	16·8 (part only)	Cardiff (NMW)	L.]EG II AVG (across end)		
85	49	Brough; 18th cent.	Derbys.	—	Fragment	Lost	BREXARC	CIL 1217	
86	50	Yeaveley, Derbys.; 1975	Derbys.	1st cent.	59·42	Derby	SOCIORVM LVTVD (on side; BRIT/EX.ARG)		
87	50	Yeaveley, Derbys.; 1975	Derbys.	1st cent.	62·14	Derby	As above; from same mould		

*Whittick[38] believes that this piece was from a plaque and not an ingot, as DE BRITAN is unlikely on an ingot. This makes the precise dating of AD 49 unlikely.

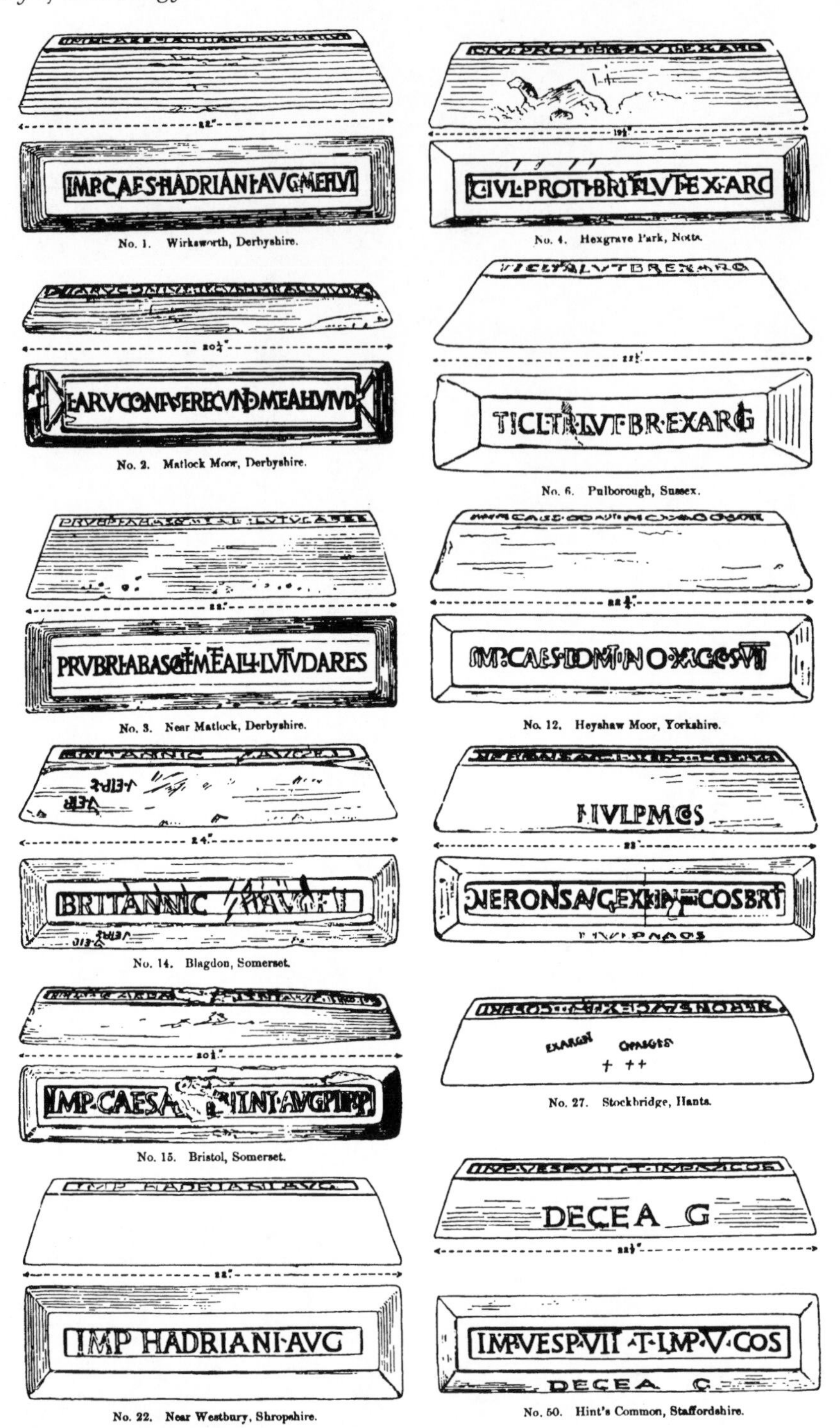

28 Some Roman lead pigs (after Gowland[23])

fact the same but as the inscriptions seem to differ this conclusion has not been accepted here.

The pigs from Bradwell[48] and Carsington,[45] both in Derbyshire, are uninscribed and may be medieval. Webster,[41] however, has classed the Bradwell pig as Roman, presumably because of its characteristic shape, and since the Carsington pig is similar in shape it has been included in Table 39. The few pigs so far found of definite medieval date, are all boat-shaped (*see* p. 71).

From their inscriptions and their provenances, it has been possible to identify most of the pigs with certain mining areas. The known mining areas and probable ports used for export are shown in Fig. 29. It may be seen that the mining areas, from south to north, are as follows: Somerset (Mendips), the Shropshire–Powys border area, Derbyshire (Wirksworth and Matlock), Flintshire [Clwyd] (Halkyn mountain), Yorkshire (Wharfedale, Nidderdale, and Swaledale). The most northerly pig found is from Hurst in Swaledale and is dated to AD 117–138.

One of the main ports was Clausentum (Bitterne, near Southampton), where two pigs were found in 1918.[44] A further pig, one of four found at Pulborough, Sussex, came from Derbyshire since it bears the same inscription as a pig from Matlock, Derbyshire,[51] of definite local origin. A pig found at Bossington, near Stockbridge, Hants., had the same composition as one

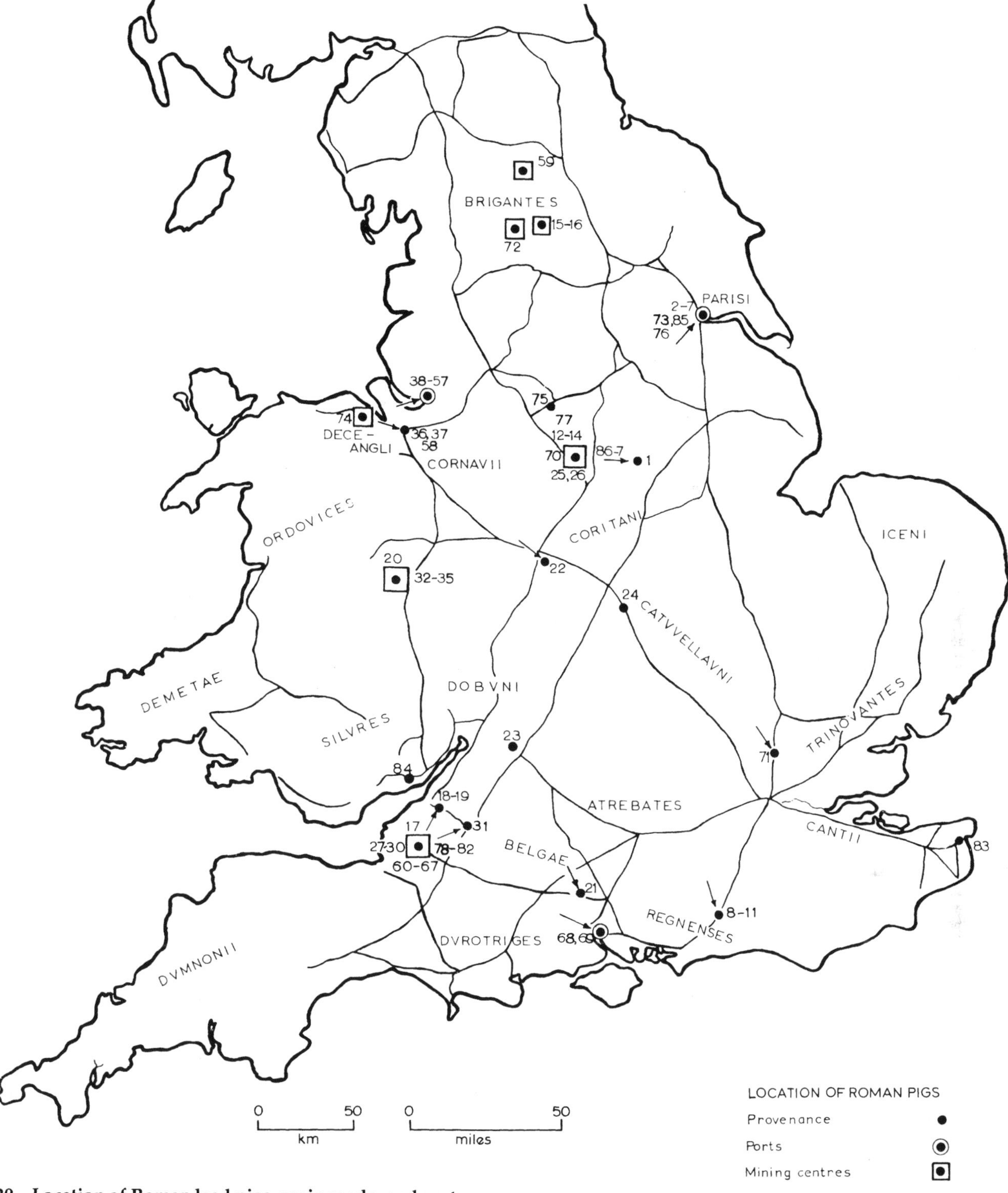

29 Location of Roman lead pigs, main roads, and ports

from Flintshire.[44] Although this does not guarantee an identical origin, it makes it a strong probability. It seems therefore, that Clausentum was used as a shipping port for pigs coming from several mining areas. Seven pigs were found at Brough-on-Humber,[52] or in the immediate vicinity[23] and it was undoubtedly one of the ports used by the Derbyshire and Yorkshire mining areas. Runcorn might have been one of the ports serving the Flint and Shropshire mining areas, for twenty pigs are reported to have been found in the river there.[23] Gough[26] thinks that Uphill, the terminus of a Roman road from the Mendips, may have been a shipping port for Mendip lead, but there is no direct proof of this. On the contrary, there is some evidence that some pigs went in an easterly direction.

All these facts point to a well organized industry in Roman times, the extent of which can be seen when one appreciates that over eighty pigs, weighing an average of 77·3 kg (about 6 t in all) are known to have been lost *en route*, and more must have been found and melted down in more recent times. Pigs of this weight are not easy to lose accidentally, or to pilfer, and can only represent a very small proportion of the total produced. Some of the Roman pigs from Somerset

30 Striations on model lead pigs and effect obtained by interrupting pouring (photo by J. A. Smythe)

and Derbyshire were found under stones as though they had been intentionally buried, the pilferer meaning to recover them at a later date.[43] It is interesting to note that at least one modern plant in the United States increased the weight of its lead ingots from the normal figure of about 50 kg to 4 t to reduce loss due to pilferage during transit.[53]

The pigs themselves have been cast in clay moulds, a wooden pattern being used to form the mould. The inscriptions were carved on panels attached to the wooden patterns so that they appear on the pig in relief (*see* Fig. 27). The inscriptions were mainly confined to the smaller bottom surface (as cast) of the pig, and usually appear in a sunken panel, although in some cases the letters are not empanelled but are merely raised proud of the surface. Some pigs have inscriptions on the sides and ends which are merely stamped, or are in relief and not in panels. Some of the pigs carry other markings. Three of these referred to by Whittick,[39] are a palm branch, a circle, and a hammer mark. These were always put on the end faces, and the latter appears to have been the mark of some Imperial passing officer at the mines. This mark should not be confused with a very similar mark made by pressing a lead pig in a sack against a hard surface.

The pigs seem always to have been cast into moulds

31 Continuity of crystal structure through the striations of a model lead pig (photo by J. A. Smythe)

of the right size for their intended weight. This is concluded from the fact that certain inscriptions are near the top of the mould and would only have appeared on the pig if the mould had been completely filled. But many pigs are of short weight because of accidents in casting or because of mutilation in antiquity or after discovery. Not too much importance, then, should be attached to 'low' weights. For example, No. 2 (Table 38) from South Cave, Brough, has had a large piece cut off it and another (No. 19) from Bristol, has been cast in a sloping mould.

Although none has been found, the moulds must have been substantially constructed. Smythe[52] concluded from an examination of the minor flaws visible in the surface of the pigs that no less than five had been cast in the same mould; one from Hexgrave Park, three from Brough, and the one from South Cave. It would appear, then, that however the pigs were removed from the moulds they were not levered out, for the resultant damage would have been considerable. The pigs must have been turned out of portable clay moulds, although a few do contain holes which could have been made by cast-in hooks. Baked clay moulds, from which an ingot of similar shape but smaller size could have been cast have been found near Carnac, France, but these were intended for salt.[54]

Palmer and Ashworth[43] thought that the horizontal striations visible along the sides of the pigs were the result of the moulds being filled by a small ladle, each ladle-full accounting for one striation. Elaborate experiments by Smythe[22] with model ingots weighing 1·14 kg show conclusively that this is incorrect; the interruption in the pouring of a pig produces a much more definite break in the structure than a striation (Fig. 30), and furthermore, interrupts the crystal structure, which the striations do not (Fig. 31).

Smythe concluded that the striations were due to surface-tension effects between metal and mould, and that the pigs were poured in one operation probably from a spout in a well at the side, or the bottom of the furnace itself. These striations are, in fact, a very common feature of metals slowly cast in permanent or semi-permanent moulds.

Not all the Roman lead pigs have been fully analysed but those that have are listed in Table 38. These are very pure and contain more than 99·9%Pb. They were first analysed by Gowland in the last few years of the 19th century.[23] Some have recently been re-analysed by Wyttenbach and Schubinger and there are some quite big disagreements on some ingots.[46] Where there are two analyses in Table 38 the first is Gowland's and the second Wyttenbach's. Generally Wyttenbach's neutron activation analyses give much lower figures for Cu, and similar figures for antimony and silver. Normally figures in this range cannot be considered as being to a better accuracy than ±25%, and copper in particular has a habit of segregating.

There remains the question of desilverization. A considerable number of the pigs bear the inscription EX ARG or EX ARGENT (Fig. 28) which was once thought to mean that the lead had been desilverized. But Gowland[23] noted that one of the pigs with this inscription (No. 1 in Table 38) had the second highest silver content of any he analysed and, moreover, a higher content than many which were uninscribed. Smythe accepted the earlier point of view in his book

Table 40 Silver contents of lead from galena and lead pigs from Brough-on-Humber (after Smythe[52])

	Silver content, g t^{-1} (ppm) Galena lead	Pig lead
Pig 3	23	104
Pig 3	26	104
Pig 5	26	57

of 1923,[2] but he finally settled the problem with the aid of specimens of galena and pig lead, from Brough-on-Humber.[52] Here he had the good fortune to be presented with two pigs (Nos. 3 and 5 in Table 38) which had fragments of galena embedded in the surface. These fragments were of the order of 25 mm³, and their presence at once suggested that the lead had not been desilverized in spite of the inscription EX ARG, since the galena must have been swept out of the smelting furnace with the molten lead, and would not have survived cupellation in this form.

The silver contents of lead made by Smythe from the galena, and the pigs in which the galena occurred, are given in Table 40. These figures show that the silver content of the galena was very low, but that that of the pigs was somewhat higher and not lower, as it would have been if they had been desilverized. Of course one would expect a higher silver content in the recovered lead than in the galena, as shown in the section on 'Methods of smelting'.

It is clear that these pigs had not been desilverized, and it is doubtful whether they would have been worth desilverizing. It seems that like many of us, the Romans were optimists and always referred to their lead mines as the 'silver–lead mines' or the 'silver mines' in spite of the fact that as far as this country was concerned very little silver would ever come out of them.

The highest recorded silver content of Romano-British lead is 0·056% (560 ppm), obtained on one of the pigs from Green Ore (No. 63 in Table 39).[43] This might have been worth desilverizing in Roman times, and contrasts with the silver contents of others from Green Ore which were very low. This difference would be explained if some pigs were lost before desilverization and others after, but it is difficult to see why steps should be taken to make an elaborate casting,

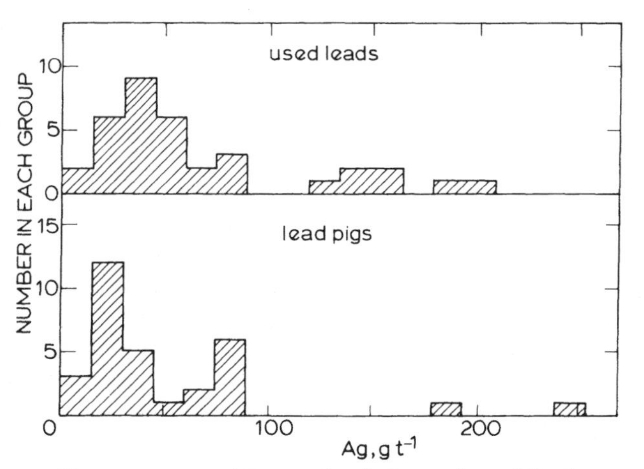

32 Silver contents of Roman lead pigs and used leads (after J. A. Smythe)

when it was to be destroyed later during desilverization.

The mean silver content of the Roman lead pigs from Clwyd is only 0·0025% (25 ppm). But, as already mentioned, a piece of lead found on the site at Pentre[12] contained 0·03% (300 ppm), and two analyses of the ores themselves (galena only) gave 0·01% (100 ppm) and 0·017% (170 ppm). It would appear, therefore, that if these are comparable the pigs produced from Clwyd ores might have been desilverized, although in view of the well known variability of the silver contents of lead ores one cannot be certain about this. (A recent sample of Halkyn ore assayed 122 ppm.)

The change in silver content of the lead ore from a single mine can best be shown by the following figures from Cwmsymlog, Dyfed.[55]

	Ag g/t (ppm)
AD 1604	734
1710	220
1811	107–122
1845	less than 80

In 1604 much of the lead was coming from a high-silver vein, but as time proceeded the percentage of the total from this fell, probably owing to the tendency for the upper levels to have higher silver contents than the lower. The comparatively high-silver ore from Pentre may represent a good vein, whereas the pigs themselves may be typical run-of-the mine material.

One can conclude, therefore, that in Roman times in spite of the inscription EX ARG, no attempt was made to desilverize the low-silver ores of Derbyshire, nor probably those of Yorkshire and Shropshire. The ores of the Mendips, Devon, and Cornwall and perhaps Clwyd, would be tempting material to work on, but the absence of definite remains of cupellation hearths at the mines makes this still an open question, although traces of the remains of cupellation hearths have now been found in the Mendips.[56] The fact that so many of the finds of Roman date contain such small amounts of silver (*see* Tables 39 and 43), suggests that lead was desilverized when worthwhile, but if so, it is difficult to believe that all the cupellation was concentrated at urban sites, such as Hengistbury Head, Silchester, and Wroxeter.

Smythe[57] has made a comparison of the silver content of Roman pig leads and used leads (i.e. objects) which is given in Fig. 32. This shows that the silver content of the lead from both sources is statistically similar, and that if any desilverization was carried out it was done, as might be expected, before the lead was cast into inscribed pigs. If this is so, it may be concluded that the pig from Green Ore containing 550 ppm silver was cast from lead that by some mischance had not been desilverized.

Although desilverization at Silchester and other town sites was carried out in calcium phosphate cupels, the absence of calcium phosphate on mining sites should not be accepted as evidence against desilverization, since it is possible to desilverize lead in shallow clay 'cupels' and to remove the litharge by skimming it off. Therefore, the presence of solid litharge can probably be accepted as evidence for desilverizing in the absence of definite remains of cupellation hearths. This appears to have been the case at Green Ore.[56]

33 Typical Roman silver ingots

Litharge has been found on many sites; in particular, Frenchgate, near Doncaster, Green Ore on the Mendips, and Caerleon-on-Usk. In most cases it is recognizable by its high density (SG 9·53). It is usually pinkish, consisting in the main of tetragonal (alpha) PbO with some copper and silver in it. But alpha PbO had been reported from the Nordeifel[58] and it is possible that in some cases the red coloration is due to cuprous oxide or cerussite ($PbCO_3$).

Several small silver ingots of a familiar Roman double-axe shape (Fig. 33) have now been found in the British Isles. Some of these are inscribed, and they have all been found in a 4th or 5th century AD context[59,60,61] (*see* Table 41). An analysis has been reported of one of these ingots (Table 46, below) and they are probably the result of the cupellation process. However, many of the objects found amongst the Mildenhall treasure for example, were manufactured outside Britain,[63] and it can be concluded, in view of the low average silver content of British ores and the high silver content of the lead ores in Greece and Asia Minor, that Britain was mainly a silver importer at all times.

THE LATER LEAD INDUSTRIES

So far, there is little evidence for lead smelting in the Migration period. The only record of Anglo-Saxon lead working relates to Derbyshire, where lead mines at Wirksworth were leased by the Abbess of Repton to an ealdorman Hunbert in 835.[64,65] The Domesday Survey of 1086 mentions the existence of at least seven lead works (*plumbaria*) in the Derbyshire area but very few elsewhere[66] (Table 42).

The early place-name evidence for lead mining or smelting is poor and the word 'bole' does not seem to come in before about 1500. Daniel[65] draws attention to the Old English element 'cost' which occurs in about 16 place names in the Peak District and which is probably synonymous with 'bole'.

There is no doubt whatever that by this period most of the silver was extracted from lead and copper ores by cupellation. In many of the coinages, silver gradually replaced gold from the 7th century onwards, and the greater part of this silver would have been derived from lead rather than copper. There is evidence of considerable international trade by the 10th century. For example, the total of 40 t of silver necessary to pay the English Danegeld on four separate occasions in the

Table 41 Silver ingots of the Roman Period

Provenance	Ref.	Weight, g	Date	Inscription
Richborough, Kent	59	343	Before AD 411	
Balline, Co. Cork	59	315	4th–5th century	EX OFFI I ATIS
	59	316	4th–5th century	EX OFC VILIS
Tower, London	59		AD 373	EX OF(FICINA) FL(AVII) HONORINI
Stanmore, Middlesex	59		AD 400–410	
Ballinrees, nr. Coleraine, Co. Londonderry	60	½ ingot	AD 410–420	CVRMISSI
	60	½ ingot	AD 410–420	EX OF(FICINA) PA/TRICI
Kent	60	340	Late Roman	EX OFF CVRMISSI
Din Lligwy, Anglesey	62	15·6	3rd–4th century	

1 Roman pound = 340 g; 1 oz (troy) = 31·17 g

Table 42 Derbyshire lead works mentioned in the Domesday Survey

Place	Number
Mestesforde	1
Wechesworde (Wirksworth)	3
Badequela (Bakewell)	1
Aisseford (Ashford?)	1
Crice (Crich)	1

10th and 11th centuries could represent the production of 400 000 t of lead with a mean silver content of 0·1%. It is not suggested that all this silver was from current English lead production; a lot of it was from coins that had been in circulation for some time and some of it was from other sources.[67] England must have been a very wealthy country and we know from the Sutton Hoo treasure that Anglian kings were able to amass quite a lot of personal wealth.

What was the reason for all this apparent wealth and what was the source of the silver? We know that Germany was paying for something obtained from England at this time by silver remitted to England. Rammelsberg in the Harz was the probable source of the metal but the Saxon mines no doubt made their contribution. The most likely import would have been wool. However, English lead production was high and most of the lead was being cupelled for silver although the mean silver content was low. During the two centuries after the conquest, English mineral exports increased by a factor of ten.[68] Lead ingots were exported for roofing purposes to France and a good deal of tin also went abroad. In addition, a large amount was used for plumbing in monasteries in England and abroad. In the 12th century 100 cartloads went from Newcastle to Rouen and 241 were exported from York. The latter probably came from the Yorkshire mines. Metal from the Derbyshire mines was exported via Boston and King's Lynn.

Mines yielding native silver are on record. The mine near Beinsdorf, Saxony, where 'pure silver and native copper were exposed to the daylight' and which was opened in 922 was unusual and can be compared with the silver mine at Hilderston, Scotland,[7] which produced some native silver up to 1873.

Although the present silver content of the Derbyshire ores is not exactly great (*see* Table 32), it is possible that the levels worked in this period were much richer than those worked later, for we know that a group of manors sent 40 lb (18 kg) of silver to the mint at Derby. In an account roll of 1322 there is no mention of silver from Wirksworth[69] in the same county. The mines at Alston, Cumberland were worked for silver and lead between 1100 and 1307 and the bishops were allowed by the Crown to retain the silver for their mint at Durham. By the 14th century, we read that great numbers were employed in the lead industry. Mention is made of 10 000 in the Mendip area in Edward IV's time. As the miners worked only part-time many could have been employed[70] but the smelters would have been few. However, a considerable number of women and children were employed as ore dressers.

The lead industry of medieval times was concentrated principally in the Derbyshire and Somerset areas where intensive competition existed. By the 15th century, lead production was expanding in the Northern Pennines and ore was being smelted on the bolehills at Wolsingham in Weardale. In 1427 a mine was opened at Harthope, near Wolsingham, which continued until 1458.[71]

In the 16th century Cardinal Wolsey (then Bishop) attempted to smelt lead with pit coal at a 'great' furnace at Gateshead, Co. Durham [Tyne and Wear].[1] However, in 1527 Wolsey leased the furnace and all the mines within his bishopric for a term of 30 years at a rent of £5 per annum, and nothing more is heard of any attempt to use coal exclusively as a fuel for over a hundred years.

The greatest development dates from the reign of Elizabeth I (1558–1603) when Germans were brought over to reorganize the mines of England and Wales, but not apparently the copper and tin mines of Cornwall. It was in this period (September 1582) that an Exchequer Special Commission (24 Elizabeth No. 1955) produced a report on patent infringement. William Humfrey, one of the chief English partners in the Mineral and Battery Works and assay master of the Royal Mint, claimed that he had patented an improved method of mineral dressing and of smelting lead ores which was being infringed by the Mendip and Derbyshire workers. It is difficult to be sure about the methods in question and it is possible that the Commission were also confused.[20,26]

One improvement was probably a jig or sieve for the improved washing of ores. The other was probably the foot-powered turn-hearth, which could be rotated so that the fume from the hearths did not upset the blowers (*see* Fig. 24).

One of the earliest British lead furnaces illustrated is Burchard's furnace of 1582[20] (Fig. 25). This is a shaft furnace similar to the smaller blast furnaces shown by Agricola in his book of 1556.[72] Whatever the merits of the shaft furnace, the shorter furnace continued but eventually with water-powered bellows and finally became the ore-hearth of the 18th century.

Boat-shaped lead pigs seem to have been the norm in this period. Two were found near Lisbunny Castle, Co. Tipperary,[73] one weighing 58·5 kg, and the other 65·6 kg, Four pigs were found at Rievaulx Abbey in Yorkshire,[74] weighing about 455 kg each. Another Tudor ingot, weighing 140 kg, was dredged from the sea off Whitstable.[75] These appeared to have been made by pouring into a trough cut into the ground, and may represent lead melted down at the time of the suppression in 1539. Pigs of this shape (Great Pigs) were being produced up to the end of the 17th century.[76]

TYPICAL LEAD AND SILVER FINDS

The principal use of lead in early times was for weights, such as net-sinkers and loom-weights, and as sheets for roofing, coffins, cisterns, and pipes.

Table 43 lists the principal finds of lead objects from this country and their analysis, where known. As may be seen, the vast majority of these date from the Roman period or later, and it can only be concluded that earlier lead has been melted down when found, and re-used.

The earliest recorded uses of lead in this country are probably the pommels for bronze swords from Tosson, Northumberland,[91] and patterns for making the

Table 43 Lead objects

Object	Provenance	Date	Weight, capacity, or dimensions	Analysis, %							Inscriptions	Reference
				Ag	Cu	Sb	Fe	Au	Sn	Others		
Net sinker	Meare, Somerset	250 BC–AD 50	20·8 g	0·007 7–0·009 3								77, 102
Piece	Bagendon, Glos.	AD 1–50	—	0·005 8								78
Water pipe	Chester	AD 79	102 kg	0·001 7							IMP. VESP. VIII T IMP. VII. COX CN IVLIO AGRICOLA LEG. AVG. PR. PR.	79
Sheet lead	Bath	AD 44–100	1·5–1·6 cm thick	0·002 7–0·004 8	0·037	0·024	tr.	tr.	nil			102, 23
Sheet lead	Caerleon	AD 75–400		0·002								102
Water pipe	Wroxeter	AD 78–380		0·005 6								102
Coffin	Sittingbourne, Kent	AD 250–400		0·009 8								102
Lump	Merlin's Cave, Wye Valley	AD 100–400	100 g	0·026 3								102
Piece	Hengistbury Head	EIA or Roman		0·005								9
Lump	Richborough, Kent	AD 43–388		0·007 8								80
Mass	Folkestone Villa	AD 78–388		0·007 2								80
Casting	Green Ore Somerset (1956)	AD 69–79		0·040								43
Piece of sheet	Heronbridge, Chester	1st or 2nd cent. AD		0·002 2								81
Piece	Caerhun, Conway	Roman		0·004 3								82
Coffin	Holborough, Kent	Roman (1st–2nd cent. AD)		0·003	0·015	tr.	—	—	0·69			83 84
Sheet	Usk	Roman		0·004 5	0·009 5	0·037 5	—	—	—	Zn 0·000 12		85
Rivet	Springhead, Kent	Roman		0·002	0·004	0·01	Present	—	0·08	Bi tr.		85
Casting	Springhead	Roman		0·006 7	0·047	0·004 8	0·000 94	—	0·66	Cd 0·012 Zn 0·022		85

Object	Site	Date	Size							Other	Ref.
Casting	Springhead	Roman		0·010 3	0·031	0·009 2	0·001 4	—	0·63	Cd 0·001 6 Zn 0·000 77	
Lump	Camerton, Somerset	Roman		—	9·75		0·64 (Fe_2O_3)	— —	0·12	Much corroded Pb 67%	86
Mass	Castledykes, Strathclyde	Roman		0·002 9	—						81
Ingot	Boxmoor, Herts.	Roman 3–4 cent.		0·01–0·02	0·06	~0·01	—	—	<0·1	As <0·001 Ni ~0·002	87
Lump	Verulamium	Roman		0·003	0·032	tr.	—	tr.	0·065		23
Bar	Silchester	Roman		0·002 3	0·026	tr.	—	tr.	0·354		23
Pump	Silchester	Roman		—	0·11	—	tr.	—	4·52		23
Pipe	Silchester	Roman		—	0·10	—	tr.	—	1·10		23
Lamp	Corbridge	Roman		—	—	—	—	—	a little		22
Vat	Ireby, Cumbria	Roman	46 l	—	—	—	—	—	1·3		98
Waste	Scarcliffe	Roman		0·003 4	0·014	0·220	—	—	—		13
Lump	Glastonbury Abbey	1130–84		0·033	—	—	—	—	—		98
Piece	Huttons Ambo, Yorks.	13th cent.		0·005	0·013						110
Sheet lead	Rievaulx, Yorks.	1131–1500		0·004							98
Lead pipe	Rievaulx	1131–1500		0·008 4							98
Cup	Roche Abbey	1147–1500		0·004 2							98
Lead pipe	Roche Abbey	1147–1500		0·003 4							98
Window lattice	Roche Abbey	1147–1500		0·007 3							98
Piece	Kirkstall Abbey	1200–1300		0·008 5	0·20–0·63	tr.	0·004–0·01	tr.			89
Pipe	Finchale Abbey, Durham	c.1600		0·012 3							22
Sheet lead	Finchale Abbey	c.1600		0·000 4							22
Bullet	Marston Moor	1664	17·8 g	0·007 3							
Sounding lead	HMS Association	1707	3·1 kg	0·006	0·01						98
						0·005	—	—	0·001		90
Bullet	HMS Association	1707	12 mm dia.	0·009	0·003	n.d.	—	—	n.d.		90
Bullet	HMS Association	1707	16 mm dia.	0·004	0·02	n.d.	—	—	n.d.		90

n.d. = not detected.

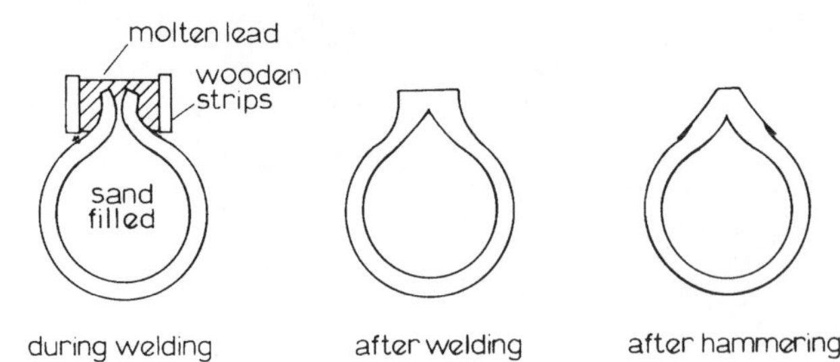

34 Method of manufacture of Roman lead pipe

36 Length of lead pipe from Red House, Corbridge, showing box joint (length of pipe 1 m)

moulds of socketed axes of the Late Bronze Age from Anwick, Lincs.[92] If lead was cast into a bronze mould the pattern so produced could have been used to make a clay mould. This could be made in two halves to allow removal of the pattern, or it could be made in one piece and heated so as to melt out the lead. The latter method would avoid the complication of the two-part mould and is in principle similar to the modern 'Mercast' process, in which mercury is used instead of lead. This question is discussed further in Chapter 5.

Small pieces of impure lead and lead–tin solder have been found in association with late Bronze Age objects.[93]

Sheet lead was made by casting lead onto flat sand beds with wooden sides lined with clay, and not by rolling, which is the method used today. Pipes were made by bending cast sheet round a wooden former, filling the inside with sand, and welding the edges together by pouring on additional metal, which was retained in place by means of strips of wood or metal acting as a mould (Fig. 34). The temperature and the quantity of additional metal had to be sufficient to heat the raised edges of the pipe to their melting point in order to form a bond between them and the newly applied metal, and this was rather wasteful. Figure 35 shows a section of a pipe from the Roman site at Wall near Lichfield, Staffs., which had been made in this way. The metal in the weld has a rather larger grain size than the wall of the pipe owing to the fact that the pipe has been cold-formed and has therefore recrystallized with a finer grain size, whereas most of the weld has not been worked. It appears, from the fine crack at the lower left-hand edge of the weld, that the weld had been hammered slightly to give it its final inverted V shape.

The joints between lengths of pipe were made by first cleaning the pipe ends, then slashing them diagonally and fitting one within the other. The gap at the top was covered with a piece of clean lead sheet. In the example shown in Figs. 36 and 37 from Red House, near Corbridge, Northumberland[65] this gap was 8 cm long, necessitating a piece of sheet about 11 cm long to cover it. Finally a three-piece mould was placed round the ends leaving a gap about 1 cm wide all the way round the pipe which was filled with molten lead.

Modern pipes are made by an extrusion process in

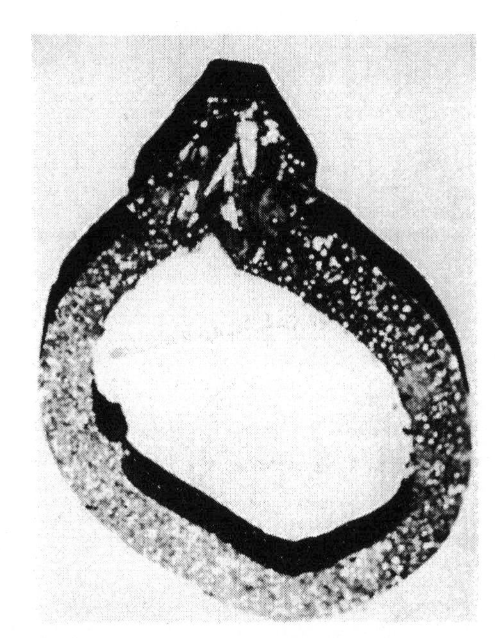

35 Section of lead pipe from Wall, Staffs. (photo by J. A. Smythe)

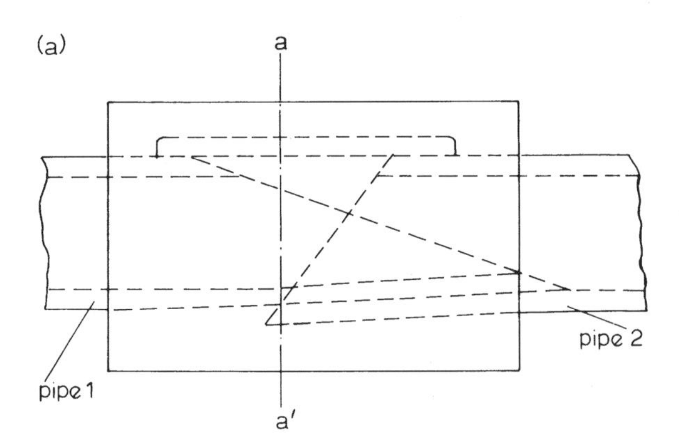

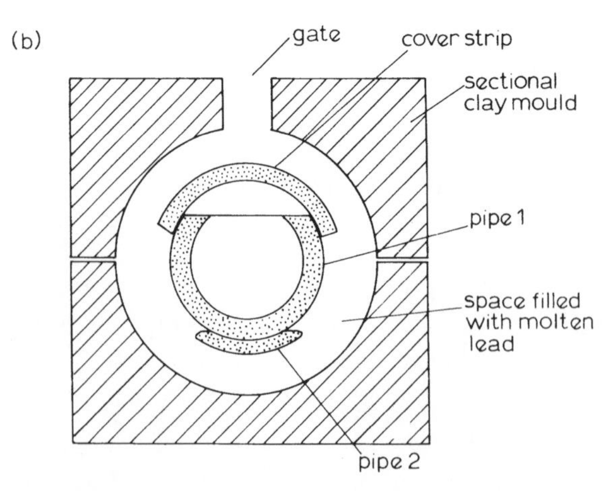

a joint as found; *b* section on a–a′ with mould in position ready for filling

37 Method of joining pipes by means of moulded box joints

Table 44 Dimensions of lead vessels (mainly after Curwen[97])

Provenance	Ref.	External dia., cm	Height, cm	Capacity, l	Remarks
Icklingham, Suffolk	95	81	33	173	Chi–rho and Alpha–Omega
	95	—	—	73	Found in early 18th century, now lost. Alpha?
Bourton-on-the-Water, Glos.	94	76–86	34–35·5	182	Found in a building near a well
	94	93–99	41	296	Found in a building near a well
R. Ouse, near Huntingdon	94	76	40	182	
Unknown (near Cambridge)	94	67	32	116	
Wiggonholt, Pulborough, Sussex	96	77	48	209	Chi–rho
Ireby, Cumbria	98	50–53	18	46	

which lead is squeezed through a die. The method of 'lead-burning', in which heat from a gas flame is used to melt together the edges of two sheets, is the modern equivalent of the primitive welding process.

The cylindrical cisterns found at various sites including Bourton,[94] Icklingham,[95] and Wiggonholt[96] are all of the same pattern (*see* Table 44). They were made by bending two lead sheets and welding them up into a cylinder, with a circular plate at the bottom which was 'dished' so that its circumferential weld was not on the edge but round the vertical sides. Decorations were cast in the sheet by imprinting a pattern in the sand-bed on which it was cast, as in the production of 17th century cast-iron firebacks.

The vat from Ireby, Cumberland[98] (Fig. 38) was made in the same way as the cisterns, but the manner of welding is slightly different. The edges were serrated as if with a cleaver, to provide a better hold for the molten metal which was then applied to the joint. This method gives a good mechanical joint without requiring a large amount of additional metal to provide enough heat to fuse the edges of the sheets. It also produces a neater join than that of the lead pipe, although one would not expect its liquid holding properties to be as good. The shape of the weld in this case was probably obtained by using a clay mould. The sheet and the weld were made of the same metal: lead containing 1·3% Sn. This amount of tin would not have lowered the melting point appreciably and was obviously an unintentional addition although it had the effect of raising the hardness from 3 to 6·7 HV. According to Richmond[98] the vessels from Icklingham and Wiggonholt, and those now at Cambridge in the Museum of Archaeology and Ethnology, had been joined in the same way. Rectangular containers such as coffins or lead liners to wooden coffins were made from a single sheet, sometimes welded at the corners (Fig. 39).

In some cases, lead–tin alloys or solders of rather lower melting point than lead were used in joints (Table 45). A pure lead pipe from Silchester had been 'burnt' with an alloy containing 4·52% Sn, which would be rather stronger than lead.[23] A force pump was soldered with a solder containing 25·3% Sn, which is very similar to modern plumber's solder.[23] The re-melting of lead objects soldered with lead–tin solders was often the reason for appreciable tin contents in the Roman leads, as in the case of the Ireby vat.

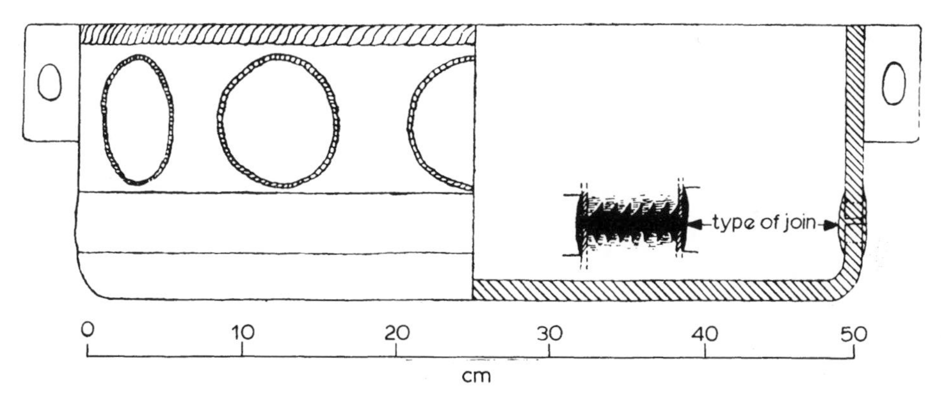

38 Roman vat of lead from Ireby, Cumbria, showing method of construction (courtesy of the Cumberland and Westmorland Archaeological Society)

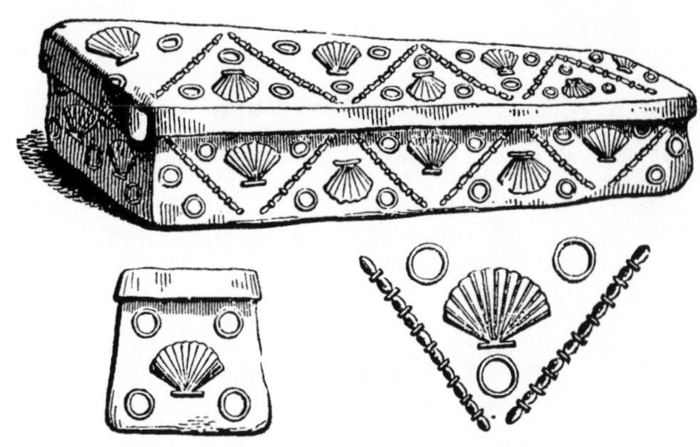

39 Lead coffin from Colchester (after Gowland;[23] courtesy of the Society of Antiquaries, London)

When high purity lead is cast, the grain structure tends to be large. This is reduced in size by hammering. In a lead bullet from Corbridge the casting pipe formed when the object was cast in a closed mould, was hammered in, and the bullet finally shaped by hammering.[22] In a sling-bolt, also from Corbridge, the cast structure is well developed but, although the edges have been hammered, the casting pipe has not been crushed.[22]

A number of other lead–tin alloy finds have been recorded, which are listed in Table 29 (Chapter 3). Those from Edlington and Corbridge could be lumps of solder with 33–44% Sn (plumber's solder). Although the addition of tin strengthens lead, the maximum effect is reached at 15% Sn, so that the choice of 17·6% for the Corbridge pottery rivet is almost ideal.

One of the most important groups of later lead–tin artifacts is organ pipes. Although Theophilus recommended the use of copper for the pipes of his 11th century organ,[100] sometime during the middle ages saw the introduction of tin and lead–tin alloys because they gave a better tone. Today the standard alloy is about 80% Pb with a hardness of about 10 HV (1 kg). It is usually cast to give 'spotty' metal. This grainy effect is obtained by pouring out a thin layer of molten metal on to a cloth-lined slab so that 60/40 eutectic colonies are nucleated and slowly grow to meet at the colony boundaries where the shrinkage is concentrated, so giving the grainy effect. This process must have been discovered by accident and exploited.

Soon after the introduction of movable-type printing to Europe in about 1440 the lead–tin–antimony alloys were used as type metal. These were usually tin-rich pewterer's metal in the early days, but later became lead-rich with about 8% Sb.[101]

Friend and Thorneycroft[102] and others, have examined the silver contents of many finds, with a view to determining the degree of desilverization that was possible at a given period. However, owing to the enormous difference in the silver content of the British ores and their low average silver contents it is a dangerous practice to base conclusions on individual finds. As already pointed out, the initial silver content determines whether desilverization is economically worthwhile, and unless it is known whether the lead was, in fact, desilverized, the final silver content is meaningless. As far as Table 43 is concerned, one can only note the relatively high silver content of two lumps from the Wye Valley and Glastonbury, and the moderate value for the net-sinker from Meare, which seems to be the earliest piece of British pure lead yet analysed.

The corrosion of silver in the presence of water containing even slight traces of salt is much more rapid than that of lead. It is probably this, and the fact that precious metals would be repeatedly re-melted, that accounts for the relative rarity of finds of silver.

As Gowland pointed out,[6] the absence of silver finds from the Early Bronze Age does not prove that it was unknown in Great Britain. No silver object has been found which is earlier than the Iron Age, and in Scotland none earlier than Roman times. The material found at Balline, Co. Limerick,[59] obviously represents a silversmith's or coiner's hoard, and is dated to the late 4th or early 5th century AD. Apart from the complete ingots (details given in Table 41), there were three pieces of silver plate, and parts of two other ingots.

Four hoards of silverware have now been found in the British Isles: at Traprain Law, East Lothian,[103] Mildenhall,[63,104] Water Newton,[105] and Coleraine, Ireland.[91] Various other pieces have been found, including the Mileham dish[106] and the Corbridge lanx;[107] the latter is made of cast silver sheet 2–3·8 mm thick, and was probably made at Ephesus.

The Mildenhall treasure comprises 34 pieces of silver tableware, including bowls, dishes, goblets, spoons, and ladles. The technique of manufacture is typical of late Roman silverware in general and is discussed in Chapter 5. The finer pieces were undoubtedly of Mediterranean workmanship, and were very probably made in Rome itself in the 4th century.

Table 45 Composition of lead–tin solders

	West Caister, Norfolk, LBA[93]	Wookey Hole Som., EIA[99]		Silchester, Hants., Roman[23] Force pump	Lump of solder
Pb	65·00	38·5	70·0	73·6	61·84
Sn	34·00	33·8	16·5	25·30	38·01
Cu	0·10	—	—	0·12	tr.
Fe	0·02	—	—	—	tr.
Ag	0·015	—	—	—	—
Sb	0·03	—	—	—	—
As	0·04	—	—	—	—
Zn	0·06	—		—	—
Loss		27·7	—	—	—
Impurities (unidentified)	—	—	13·5	—	—

Table 46 Composition of Roman silver from Britain

Object	Ref.	Composition, % Ag	Cu	Pb	Au	Sn
Mirror from Wroxeter	108	86·6	12·2	1·2	—	—
Pieces of bowls from Mildenhall	63	95·3	2·3	—	0·5	—
Fragments from Traprain Law	109	94·08 96·20 95·30	4·97			
Corbridge lanx	107	92·6				
Beaker from Great Horwood	110	93·6	4·4	1·0	1·0	—
Double axe ingot from Kent	61	95·2	4·1	1·22	0·81	—
Roman *patera*	111	rest	3·44	0·33	0·47	—
Bowls from Water Newton						
No. 8	105	97	2·4	—	—	0·5
No. 9	105	96	4·0	—	—	0·1

According to Brailsford,[63] it is not impossible, however, for the less accomplished pieces to have been made in Britain. Some pieces of the treasure have been analysed and the results are given in Table 46. A number of fragments from two fluted bowls were found to contain 95·3%Ag, 2·3%Cu, 0·5%Au, the rest being lead and silver chloride. This may be compared with an average silver content of 95·2% for three samples from the Traprain treasure. Modern sterling silver contains 7·5%Cu which is added intentionally to harden it and it is probable that this was also the case in Roman times. Roman silver could, alternatively, have been made from copper–silver ores, or from silver cupelled from debased coinage, such as may be the case in the remains from Silchester. The lead content is typical of silver from the cupellation process.

A silver mirror found at Wroxeter,[108] dated to the 3rd century, was found to contain 12·2%Cu, and is probably an import.

A fluted bowl from Mildenhall[63] has handles which at some time after burial have become detached. A similar bowl from Coleraine[60] also displays this feature, which suggests the preferential corrosion of solder, which, in the case of the silverware from Traprain Law[109] was found to contain 85%Sn and 15%Pb.

At least one piece of silver from the latter hoard had been joined by pressure welding. It appeared to be the longitudinal seam of a vessel or tankard. The edges had been lapped over and hammered.[112] As silver requires a temperature of 500°C for this operation and a heated anvil, the use of this technique implies a long standing tradition which perhaps had its origins in the cold welding of the gold boxes shown in Chapter 5. The hardness of the silver sheet adjacent to the weld was 92 HV and that along the weld line was 118 HV. This shows a slight increase in hardness due to welding.

In a moist environment, dissimilar metals behave as electrolytic cells and the baser of the two metals (in this case the solder) is preferentially corroded, while the nobler metal is not attacked. The latter, however, will still be subject to the normal corrosion processes, which may be very destructive.

The enormous Pictish chains which may be seen in the National Museum of Antiquities in Edinburgh are made of circular bar links 13 mm in diameter and are dated to AD 700. They contain 19·2%Cu and 4·2%Au, and it has been suggested that they were probably made from melting down looted, gilded, silver plate.[113]

Naturally, complicated objects such as the mazers (chalices) are made out of more than one piece and these can have different compositions. Scottish medieval cups and mazers were found to have copper contents of 4–20%, gold in the range 0·2–2%, zinc in the range 0·2–3·3% and the expected lead in the range 0·05–0·9%.[114]

CORROSION PRODUCTS ON LEAD AND SILVER

Lead, unlike silver, can survive long periods of interment without complete alteration. The chemical composition of the white corrosion patina on lead and its alloys has been determined by Smythe[115] and Haynes,[89] and the results are given in Table 47. The specimen analysed by Smythe came from Roman lead at Corbridge containing 2·0%Sn and 97·8%Pb. The specimen examined by Haynes from Kirkstall Abbey was dated to the 13th century. The lead contained about 0·37%Cu and 0·0085%Ag.

The large amount of silica and alumina in the Kirkstall sample shows that it contained a good deal of attached soil. One conclusion from these results is that the percentage of additional elements, such as copper and tin, has increased owing to dissolution of the lead compounds. The tin content has increased from one-fiftieth to about one-seventh, while the copper content of the Kirkstall specimen has increased from 0·5% to 1% (when allowance is made for the entrained soil).

The analysis of a completely corroded find would give some guidance to the nature of the alloy used, but allowance would have to be made for differential corrosion and dissolution rates.

More recent work has identified the compounds present. The Roman cemetery at Poundbury, Dorset has yielded lead coffins at various times since 1855, and the corrosion deposits usually consist of an adherent layer of alpha PbO overlain by a thinner and less adherent layer of $PbCO_3$ (cerussite). Roman scrap lead from Fishbourne was very similar.[116]

The corrosion deposit on lead from underwater objects has a rather different composition. The Roman anchor stock from Porth Felen, Wales, was found to be mainly $PbSO_4$ (anglesite) and $PbCl_2$ (cotunnite) with small amounts of $PbCO_3.PbCl_2$ (phosgenite) and

Table 47 Corrosion products on lead,%

	Roman (2·0%Sn) Corbridge[115]	13th century (0·4%Cu) Kirkstall[89]
SnO_2	10·63	—
PbO	74·90	49·25
SiO_2	0·30	33·19
Al_2O_3	—	6·19
FeO	—	1·87
MgO	—	0·9
CaO	0·15	0·40
CuO	—	0·52
S	nil	0·013
Loss (H_2O, CO_2 etc.)	13·26	7·19
Cl	0·12	—

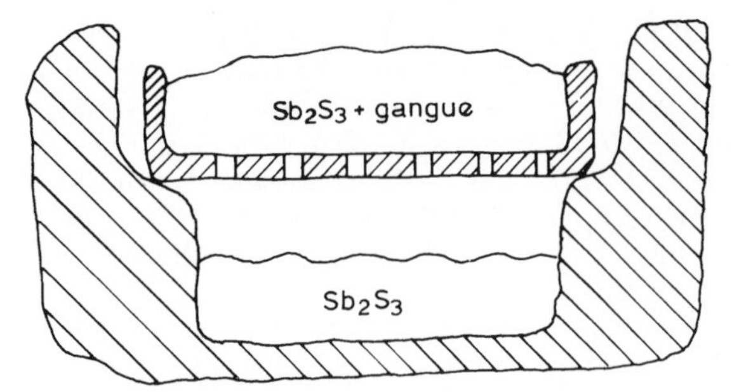

40 Detail of technique used for antimony sulphide extraction in the 18th century (dimensions not known)

cerussite also present. Under these conditions PbO seems to be absent.[117]

The silver coins (1st and 2nd century *denarii*) which were found in a completely corroded state at Hengistbury Head[9] consisted of: 73·7%Ag, 24·2%Cl and 0·44%Au, with traces of Pb, Cu, and Fe. Thus they were almost entirely converted to silver chloride, but considering the presence of no more than a trace of copper, must have originally consisted of high-purity silver.

ANTIMONY

Although antimony is one of the few metals that occur in the native state its occurrence in this form is extremely rare, and most of the world's antimony used to be extracted from the minerals stibnite (Sb_2S_3) or jamesonite ($Pb_4FeSb_6S_{14}$). In very early times stibnite was used as an eyepaint because of its intense blackness and as a colourant in glazes, beads, and faience, where in conjunction with lead it confers a yellow colour. Its use as a cosmetic is however much rarer than usually thought mainly owing to the fact that there were many other more common black pigments.

Its earliest appearance in this country is in copper-base alloys and these were probably imported. Its presence in lead is usually in trace element proportions as we have seen, but many European lead ores contain much of it (Siphnos for example[17]) and we find it in the slags from lead smelting. Clearly few efforts were made to recover it in its metallic state until the medieval period when we first hear of it in alchemy and medicine, and see it alloyed with tin and lead in pewter and typemetal. There is a great deal of confusion between *stibnite*, *regulus*, and the use of the word antimony to mean the sulphide. Agricola,[119] mentions the purification of the sulphide to free it of its gangue elements, but he does not tell us how to produce the metal. Biringuccio[120] used the word *antimony* to mean the sulphide, and *regulus* to mean the metal. It is doubtful if he knew how to reduce the metal from the sulphide with iron or copper, but it is clear that this was done somewhere in late medieval times and cakes of antimony metal were made available to pewterers and type-founders.

The first detailed account of the extraction of antimony in Britain comes from an 18th century Scottish source and relates to the mining and smelting of antimonial minerals at Glendinning House, near Westerkirk, Eskdale in Southern Scotland.[121]

Here the prepared antimony sulphide, which was probably jamesonite ($Pb_4FeSb_6S_{14}$), was heated in a furnace to about 600°C, in a perforated clay vessel in such a way that the Sb_2S_3 could liquate and be collected in a vessel below (Fig. 40). The solid gangue was left behind and would form a slag in the perforated container and be discarded.

In this way the antimony sulphide was concentrated. It was then heated in a crucible with metallic scrap iron and an alkali flux. This was placed in a hotter furnace (about 1 000°C), whereupon the reaction

$$3Fe + Sb_2S_3 \rightarrow 3FeS + 2Sb$$

took place. The iron sulphide, which was partly oxidized and slagged with the alkali flux, was discarded, and the metallic antimony (regulus), which melts at 630°C, was refined by mixing with antimony from a previous smelt and more flux to remove residual iron.

An examination of this site in 1962 by C. M. Daniels produced a number of slag blocks clearly related to this process. These were about 28 cm dia. by 10 cm thick and tapered slightly to give a bottom dia. of 25 cm. They weighed about 11 kg. Metallographic examination showed them to consist mainly of a silicate slag with randomly orientated lath-like crystals, containing prills of grey cast iron and stringers of impure antimony. The grey cast iron consisted of graphite and pearlite and had a hardness of 256 HV (1 kg). The antimony prills were quite brittle with another phase which was probably iron; they had a hardness of 108 HV (1 kg). The hardness of pure antimony is between 30 and 60 HV (1 kg).[122] The composition of the metal was 90–97·5%Sb, 2·5–4·3%As, and 0–5·5%Pb. The composition of the slag is given in Table 33. It would appear that the flux was wood ash and the iron had been well oxidized. The slag is a typical fayalite slag and resembles that from non-ferrous smelting generally, and the iron bloomery process.

These slag blocks were connected with the second stage of the process and they confirm that iron was used as the reducing agent. Today, antimony is smelted in this country by roasting the sulphide to the volatile oxide, Sb_2O_3, which is caught in filters, leaving the gangue behind to form a slag. The composition of the slag is: 40%SiO_2, 21%FeO, 18%CaO, 7%MgO, 7%Al_2O_3, and 2–3%Sb. Most of the product is used in the form of oxides but metal is made in a small hearth furnace by the reduction of oxides with charcoal under a flux. Its main uses are as a hardener for lead and tin alloys, and as a compound for colourants and matches.

REFERENCES

1 H. LOUIS: *Chem. Ind.*, 1923, **1**, (252), 286.
2 J. A. SMYTHE: 'Lead'; 1923, London.
3 C. WALDRON: *Trans. Cardiff Nat. Soc.*, 1875, **7**, 1–5.
4 C. E. CONOPHAGOS: 'Le Laurium Antique'; 1980, Athens.
5 T. MORGANS: *Trans. Inst. Min. Eng.*, 1902, **20**, 478–94.
6 W. GOWLAND: *Archaeologia*, 1917–18, **69**, 121–60.
7 H. AITKEN: *Trans. Fed. Inst. Min. Eng.*, 1893–94, **6**, 193.
8 J. H. COLLINS: *J. R. Inst. Corn.*, 1903–5, **16**, 103–19.
9 J. P. BUSHE-FOX (Appendix by W. GOWLAND): 'Excavations at Hengistbury Head, Hampshire in 1911–12', Rep. No. 3, Soc. Antiq., London, 1915.

10 V. E. NASH-WILLIAMS: *Archaeol. Cambrensis*, 1952–5, **102**, 89–163.
11 P. A. RAHTZ and G. BOON: *J. Rom. Stud.*, 1954, **44**, 99.
12 D. ATKINSON and M. V. TAYLOR: *Flints. Hist. Soc.*, 1924, **10**, (1), 22 pp.
13 R. J. HETHERINGTON: PhD thesis, University of Newcastle upon Tyne, 1979.
14 J. PERCY: 'Metallurgy of lead'; 1870.
15 P. A. DUFRÉNOY *et al.*: 'Voyage métallurgique en Angleterre'; 1837, Paris.
16 H. C. LANE: 'Field surveys and excavation of a Romano–British rural native settlement at Scarcliffe Park, East Derbyshire', Res. Rep. No. 1, Derwent Archaeol. Soc., 1973.
17 G. A. WAGNER, W. GENTNER, H. GROPENGIESSER, and N. H. GALE: in 'Scientific studies in early mining and extractive metallurgy', 66, O.P. No. 20, Br. Mus., 1980.
18 E. PERNICKA: *Erzmetall*, 1981, **34**, (7/8), 398.
19 A. RAISTRICK: *Trans. Newcomen Soc.*, 1927, **7**, 81–96.
20 M. B. DONALD: 'Elizabethan monopolies'; 1961, Edinburgh.
21 J. GLANVIL: *Philos. Trans.*, 1667, **3**, 525; 1668, **3**, 767.
22 J. SMYTHE: unpublished work.
23 W. GOWLAND: *Archaeologia*, 1901, **57**, (2), 359–422.
24 L. J. SPENCER: *Geol. Mag.*, 1889, **36**, 71.
25 L. F. SALZMAN: 'English industries of the Middle Ages'; 1923, Oxford.
26 J. W. GOUGH: 'The mines of Mendip', 2 ed. (revised); 1967, Newton Abbot, David & Charles.
27 A. RAISTRICK: *Proc. Univ. Durham Philos. Soc.*, 1950, **10**, 529–40.
28 J. MARTYN: *Philos. Trans.*, 1729, **36**, 22.
29 ALVARO ALONSO BARBA: 'El arte de los metales'; 1770, Madrid. (Facsimile 1925) (1 ed. 1640, Mexico).
30 W. GOWLAND: *Archaeologia*, 1900, **57**, (1), 113–24.
31 J. P. BUSHE-FOX: 'Excavations at Wroxeter, 1913', Rep. No. 2, Soc. Antiq., London, 1914.
32 G. E. FOX and W. H. ST. J. HOPE: *Archaeologia*, 1894, **54**, (1), 199–238.
33 H. MCKERRELL and R. B. K. STEVENSON: in 'Methods of chemical and metallurgical investigation of ancient coinage', (ed. E. T. Hall and D. M. Metcalf), Special Publication No. 8, 195–210; 1972, London, R. Numismatic Soc.
34 P. T. CRADDOCK and M. S. TITE: 'Report on the composition of five copper–lead ingots found at the Lullingstone Villa, Kent', Res. Lab. No. 4697, Br. Mus., 23 Aug. 1981.
35 M. BESNIER: *Rev. Archéol.*, 1920, **12**, 211–44; 1921, **13**, 36–76; 1922, **14**, 98–121.
36 G. C. WHITTICK: private communication, 25 May 1963.
37 J. A. REYNOLDS: 'A lead pig from Tripontium' (1st inter. rep. H. Cameron), *Trans. Proc. Birm. Archaeol. Soc.*, 1966–7, **83**, 175–6.
38 G. C. WHITTICK: *Britannia*, 1983, **13**, 113–23.
39 G. C. WHITTICK: *J. R. Soc.*, 1931, **21**, 256–64.
40 G. C. WHITTICK: *Trans. SANHS*, 1932, **46**, (2), 129–35.
41 G. WEBSTER: *Flints. Hist. Soc.*, 1952–3, **13**, 5–33.
42 R. P. WRIGHT: *J. Rom. Stud.*, 1957, **47**, 226–30.
43 L. S. PALMER and H. W. ASHWORTH: *Proc. SANHS*, 1958, **101**, 52–88.
44 W. DALE: *Proc. Soc. Antiq. Lond.*, 1918–19, **31**, 36–9.
45 R. W. P. COCKERTON: *Antiq. J.*, 1953, **73**, 110.
46 A. WYTTENBACH and P. A. SCHUBINGER: *Archaeometry*, 1972, **15**, (2), 199–210.
47 E. DAVIES: *Archaeol. Cambrensis*, 1950, **101**, (1), 83–4.
48 J. C. COX: *Archaeol. J.*, 1895, **52**, 25–44.
49 C.I.L. 1217.
50 R. G. HUGHES and J. DOOL: Two Roman pigs of lead from Derbyshire (SK 38503990); note received 17 Oct. 1975.
51 I. A. RICHMOND (Appendix by J. A. SMYTHE): *Derbys. Archaeol. J.*, 1938, **59**, 53–65.
52 J. A. SMYTHE: *Trans. Newcomen Soc.*, 1939–40, **20**, 139–45.
53 J. L. BRAY: 'Non-ferrous production metallurgy', 2 ed., 296; 1953, London.
54 M. U. JONES: private communication, 1963.
55 W. J. LEWIS: *Ceredigion*, 1952, **2**, 27 *et seq*.
56 H. W. ASHWORTH: *Mendip Nature Res. Comm. J.*, 1970, (Mar.), 17 pp., (duplicated).
57 J. A. SMYTHE: private communication (compiled from figures available in 1936).
58 H. G. BACHMANN: *Bonner Jahrbuch*, 1977, 617–22.
59 S. P. O'RIORDAIN: *Proc. R. Ir. Acad. (C)*, 1945, **51**, 35–82.
60 H. MATTINGLEY and J. W. E. PEARCE: *Antiquity*, 1937, **11**, 39–45.
61 K. S. PAINTER: *Antiq. J.*, 1972, **52**, 84–92.
62 E. N. BAYNES: *Archaeol. Cambrensis*, 1908, **8**, 183–210.
63 J. W. BRAILSFORD: 'The Mildenhall treasure', 2 ed.; 1955, London, Trustees of the Br. Mus.
64 W. DE G. BIRCH: 'Cartularium Saxonicum', Vol. 1, 579; 1885, London.
65 M. DANIEL: *Bull. PDMHS*, 1980, **7**, (6), 339–41.
66 H. ELLIS: 'A general introduction to Domesday Book'; 1833, London.
67 P. H. SAWYER: *Trans. R. Hist. Soc.*, 1965, **15**, 145–64.
68 M. CARUS-WILSON: in 'Medieval England', (ed. A. L. Poole), Vol. 1, 230; 1938, Oxford.
69 J. H. LANDER and C. H. VELLACOTT: V.C.H. Derbyshire, Vol. 2, 323.
70 I. BLANCHARD: *Econ. Hist. Rev.*, 1978, **31**, 1–24.
71 V.C.H. Durham, Vol. 2, 348.
72 G. AGRICOLA: 'De re metallica', 357.
73 J. RAFTERY: *J. R. Soc. Antiq. Irel.*, 1958, **88**, 115.
74 C. DUNNING: *Archaeol. J.*, 1952, **32**, 199–202.
75 I am indebted to Miss Louise Millard, Curator of the Royal Museum, Canterbury for this information.
76 A. G. MACCORMICK and L. WILLIES: *Bull. PDMHS*, 1976, **6**, (3), 144–5.
77 H. ST. G. GRAY and A. BULLEID: 'The Meare Lake village'; 1948, Taunton.
78 Excavation directed by Mrs E. M. Clifford. Analysis by Dr Ruddock.
79 *Proc. Soc. Antiq. Lond.*, 1899–1901, **18**, 98.
80 J. N. FRIEND and W. E. THORNEYCROFT: *J. Inst. Met.*, 1927, **37**, 73.
81 Information from the late Dr J. A. Smythe.

82 P. K. B. REYNOLDS: *Archaeol. Cambrensis*, 1936, **91**, (222), 210–45.
83 R. F. JESSUP: *Archaeol. Cantiana*, 1954, **68**, 1–61.
84 M. FARRELL: undergraduate dissertation, University of Wales, Cardiff, 1977.
85 W. S. PENN: *Archaeol. Cantiana*, 1957, **71**, 53–105.
86 W. J. WEDLAKE: Excavations at Camerton, Somerset, Camerton Excavation Club, 1958.
87 Personal communication from Dr J. Penney of Associated Lead Manufacturers Ltd to L. Biek, 10 Dec. 1974.
88 M. W. THOMPSON: *Archaeol. J.*, 1957, **114**, 69–91.
89 R. HAYNES: *J. Iron Steel Inst.*, 1956, **183**, 359–61.
90 H. S. CAMPBELL and D. J. MILLS: *Metall. Mater. Technol.*, 1977, **9**, (10), 551–6.
91 J. EVANS: 'Ancient bronze implements', 285; 1881, London.
92 C. G. READ: *Proc. Soc. Antiq. Lond.*, (2nd series), 1895–7, **16**, 327–34; *Inventaria Archaeol. GB*, 51.
93 P. T. CRADDOCK and M. S. TITE: 'Examination of a hoard of Bronze Age metal from West Caister, Yarmouth', Rep. No. 4713, Br. Mus., 1981.
94 H. E. DONOVAN: *Trans. BGAS*, 1933, **55**, 377–82; 1934, **56**, (116), 99–128.
95 *Archaeol. J.*, 1942, **22**, 219–20.
96 R. P. WRIGHT: *J. R. Soc.*, 1944, **34**, 89.
97 C. CURWEN: *Antiq. J.*, 1943, **23**, 155–7.
98 I. A. RICHMOND: *Trans. CWAAS*, 1946, **45**, 163–71 (N.S.).
99 H. E. BALCH: *Archaeologia*, 1911, **62**, (2), 565–92.
100 J. G. HAWTHORNE and C. S. SMITH (eds.): 'On divers arts: the treatise of Theophilus'; 1963, The University of Chicago Press.
101 H. J. L. J. MASSÉ: 'Pewter plate', 22; 1904, London, Bell.
102 J. N. FRIEND and W. E. THORNEYCROFT: *J. Inst. Met.*, 1929, **41**, 105–17.
103 E. BURLEY: *Proc. Soc. Antiq. Scotl.*, 1955–6, **89**, 118–26.
104 H. MARYON: *Man*, 1948, **48**, 25, 38.
105 K. S. PAINTER: 'The Water Newton Early Christian silver'; 1977, London, Br. Mus.
106 'Guide to the antiquities of Roman Britain', 38, Pl. VI; 1951, Br. Mus.
107 O. BRENDEL: *J. Rom. Stud.*, 1941, **31**, 100–27.
108 D. ATKINSON: 'Report on excavations at Wroxeter 1923–7'; 1942, Birmingham Archaeol. Soc.
109 A. O. CURLE; 'The treasure of Traprain'; 1923, Glasgow.
110 D. E. STRONG: 'Greek and Roman gold and silver', 216; 1966, Methuen, London.
111 W. GOWLAND: *J. R. Archaeol. Soc. Irel.*, 1912, **42**, 235–87.
112 R. F. TYLECOTE: *Brit. Weld. J.*, 1962, **9**, 512–22.
113 Information kindly supplied by the National Museum of Antiquities, Edinburgh.
114 H. MCKERRELL: *Proc. Soc. Antiq. Scotl.*, 1971–2, **104**, 309–15.
115 J. A. SMYTHE: *J. Inst. Met.*, 1940, **66**, 355–60.
116 R. F. TYLECOTE: *J. Archaeol. Sci.*, 1983, **10**, 397–409.
117 W. S. ROBINSON: MSc thesis, University of Wales, Cardiff, 1980.
118 A. LUCAS (revised J. R. HARRIS): 'Ancient Egyptian materials and industries'; 1962, London.
119 G. AGRICOLA: 'De re metallica' (translated from Latin edition of 1556 by H. C. Hoover and L. H. Hoover); 1950, New York, Dover Publications Inc.
120 V. BIRINGUCCIO: 'Pirotechnia'; 1959, New York, Basic Books.
121 J. SINCLAIR: 'The statistical account of Scotland', Vol. 2, 525–8; 1794, Edinburgh.
122 D. TABOR: 'The hardness of metals'; 1951, Oxford.
123 C. M. DANIELS: *Archaeol. Aeliana*, 1959, **37**, 85–176.

5 Methods of fabrication

The preceding chapters have dealt with the production of various non-ferrous metals. It is now intended to describe the methods that were available for turning these metals into useful objects. All the evidence goes to show that in the case of copper the smiths worked on plano-convex ingots of the type detailed in Fig. 9 (Chapter 2), which they received from the smelters. These were broken up into conveniently sized pieces for re-melting and alloying in crucibles. In most cases a good deal of scrap metal would be added. The result was either cast into the objects required or into ingots for hammering into sheet or rod.

CASTING

The technique of casting must have been invented long before the process of smelting. Massive pieces of native metal were probably never melted, but the fine particles of gold resulting from washing by panning would need to be melted before they could be used. How did early man imagine that these bright particles could be melted in a hot fire? The answer to this question is not known, but metallurgical crucibles go back at least to the beginning of the 3rd millennium BC.

The first cast components known in the British Isles appear to be the flat copper axes cast in a stone mould. Today copper is sometimes thought to be a difficult metal to cast in the pure state. Some alloys are a good deal easier, but as Coghlan[1] has shown it should have been quite within the experience of Copper Age craftsmen to make good castings in a closed mould. However, the majority of flat copper axes were probably made in open stone moulds (as shown in Fig. 41), in which case any defects such as blowholes and oxide would be confined to the upper surface, and could have been scraped off. The flat copper axe weighing more than 100 g is a relatively wasteful use of a metal which in early times was difficult to obtain. As casting technique developed from the open to the closed mould, it would not be long before the idea of the core presented itself as a method of making a hollow component and so saving metal. The socketed spear-head is one of the earliest and the axe one of the commonest cored castings.

In order to produce a casting, the metal must first be melted in a clay or stone crucible in a fire which will have to be raised to a sufficiently high temperature with the aid of bellows.

When metals are heated in solid fuel-fired or gas-fired furnaces, they absorb gas from the fuel. The gases present in the furnace may be carbon monoxide, carbon dioxide, sulphur dioxide, and hydrogen sulphide, but by far the most deleterious is water vapour which arises from the combustion of hydrogen in the fuel. The water vapour formed is reduced by the metal being melted according to the equation:

$$2Cu + H_2O = Cu_2O + 2H$$

and the hydrogen so formed readily enters the metal. It stays dissolved until solidification when it is rejected as gas bubbles, which may spoil the casting.

Properly carbonized charcoal should not contain hydrogen, but this process was probably never quite carried to completion, and some hydrogen would be derived from this source. The other source would be the water vapour in the atmosphere.

In modern practice, troubles due to hydrogen are avoided by melting the metal under oxidizing conditions, in which case all the hydrogen is at once converted to water vapour which is not dissolved. Cuprous oxide is formed, however, and this is rendered innocuous by the addition of an efficient deoxidizer such as phosphorus or lithium.

Oxygen is rarely analysed because of the large sample required, but Cope[2] has determined the oxygen contents of some coins. These were in the range 0·02–0·15%, which is well within the limits for modern fire-refined tough pitch copper. Naturally the alloying elements, such as lead and tin, help to deoxidize the copper, and good practice will ensure that the oxides are removed as slags and drosses. Raw copper such as the plano-convex ingots can have very much higher oxygen contents as may be seen from the 0·5% in the ingot from Gillan, Cornwall (Table 9, Chapter 2).

Most metals shrink on solidification, and continue to contract slightly as they cool down to the ambient temperature. This shrinkage may cause cracks and cavities in the casting. In modern practice this is countered by 'feeding', i.e. providing a large mass of metal above or near the casting which stays molten long enough to supply metal during the solidification of the casting. In this way a 'pipe' or hollow can be avoided. In a thin-section casting such as a sword or socketed axe, piping does not normally occur, as the shrinkage can be made good by the surfaces of the casting moving together slightly. It would seem that the majority

Table 48 Stone moulds of the Copper and Early Bronze Ages (mainly after Hodges[9, 10])

Provenance	Implement	Publication (Museum Register No.)	Present whereabouts
Borough Moor, Kintore, Grampian	Axe	*Proc. Soc. Antiq. Scotl.*, **38**, 492	Edinburgh
Culbin Sands, Highland	Axe	*ibid.*	Edinburgh
Culbin Sands, Highland	Axe	*ibid.*	Edinburgh
Culbin Sands, Highland	Knife	(C.M. 21)	Edinburgh
Strathconan, Highland	Axe	*Proc. Soc. Antiq. Scotl.*, **38**, 492	Edinburgh
Fouland, Grampian	Axe	*ibid.*	Edinburgh
New Deer, Grampian	Axe	*ibid.*	Edinburgh
Ferintosh, Highland	Axe	*Proc. Soc. Antiq. Scotl.*, **63**, 12	Edinburgh
Glenn Rinnes House, Dufftown, Grampian	Axe	*Proc. Soc. Antiq. Scotl.*, **64**, 14	Edinburgh
Burredales, Fyvie, Grampian	Axe	*Proc. Soc. Antiq. Scotl.*, **89**, 458	Edinburgh
Kilmailie, Highland	Ring	*Proc. Soc. Antiq. Scotl.*, **9**, 248	Edinburgh
Marnoch, Grampian	Axe	*Proc. Soc. Antiq. Scotl.*, **38**, 492	Banff Museum
Cutties Hillock, Elgin, Grampian	Axe	None	Elgin Museum
Pitdowlzie, Grampian	Axe	*Proc. Soc. Antiq. Scotl.*, **40**, 35	Untraced
Wallington, Cambo, Northumberland	Axe and ring	*Archaeol. Aeliana*, 1855, **4**, 107	Untraced
Longden Common, Shropshire	4 sided, all flat axes	Chitty Fest.[11]	Shrewsbury
Hurbuck, Lanchester, Durham	Axe	Later Prehistoric Antiqs. 1953, p. 37. (WG 2267)	Brit. Mus.
Hurbuck, Lanchester, Durham	Axe	—	—
Fens, Suffolk	2 flat axes 1 bar and ring	(0. 6923)	Manchester
'Lough Gall, Antrim'	4 flat axes and 2 ingots		Pitt Rivers Museum, Blandford
Lough Scur, Leitrim	2 flat axes 1 flanged axe	Evans[12] p. 430 (W. 83)	Dublin
Ballynahinch, Down	4 flat axes	(1911, 1121)	Belfast
Ballyglisheen, Carlow	Open mould for 8 axes	*J. R. Soc. Antiq. Irel.*, 1958, **88**, 139. (1957, 342)	Dublin
Carrickfergus, Antrim	4 flat axes	Evans[12] p. 430	Belfast
Kilchronat, Cork	Flat and flanged axe	Coghlan and Raftery[13]	Cork
Doonour, Cork	4 flat axes	O'Kelly[14] (1966.33)	Dublin
Lyre, Cork	2 flat axes	O'Kelly[15] (G. 680)	Cork
Ballymena, Antrim	Flat axe	Evans[12] p. 428	Belfast
Ballymoney, Antrim	Knife	Evans[12]p. 433	Belfast
Broughshane, Antrim	Halberd	Evans[12] p. 433	Belfast
Caernarvonshire [Gwynedd]	Flat axe	Hodges[16]	Bangor
Cornwall	Flat axe	Hencken[17]	Truro

of Bronze Age castings are of this type and that true feeding was hardly ever present. When feeding does occur it is far more efficient in an alloy than in a pure metal, and one would not expect the density of a copper artifact, as cast, ever to exceed 95% of the theoretical density. The density could, of course, be improved by hammering after casting, and there is much evidence to show that this was done in the case of flat copper axes, but mainly to harden the edge.

Where the mould and core are made from clay, the clay would have to be dried thoroughly, or on filling the mould with molten metal, steam would be formed which might enter the surface layers of the metal as hydrogen. During drying, a clay mould would crack and craze; these cracks would be useful in allowing air and also any steam from water remaining in the clay to escape during filling. Metal does not normally penetrate small craze-like cracks, owing to its surface tension. On the other hand, the fact that the imprint of wooden patterns has sometimes been transferred to the metal from the mould shows how faithfully the features of the pattern may be taken up by the casting. The actual composition of the metal has a marked influence in this respect, some elements such as lead improving the fluidity of the metal.

The need to remove the wooden patterns makes a two-piece mould a necessity. However, when two-piece moulds are used, arrangements must be made to enable the two pieces to be fitted together accurately after removal of the wooden pattern. This can be most

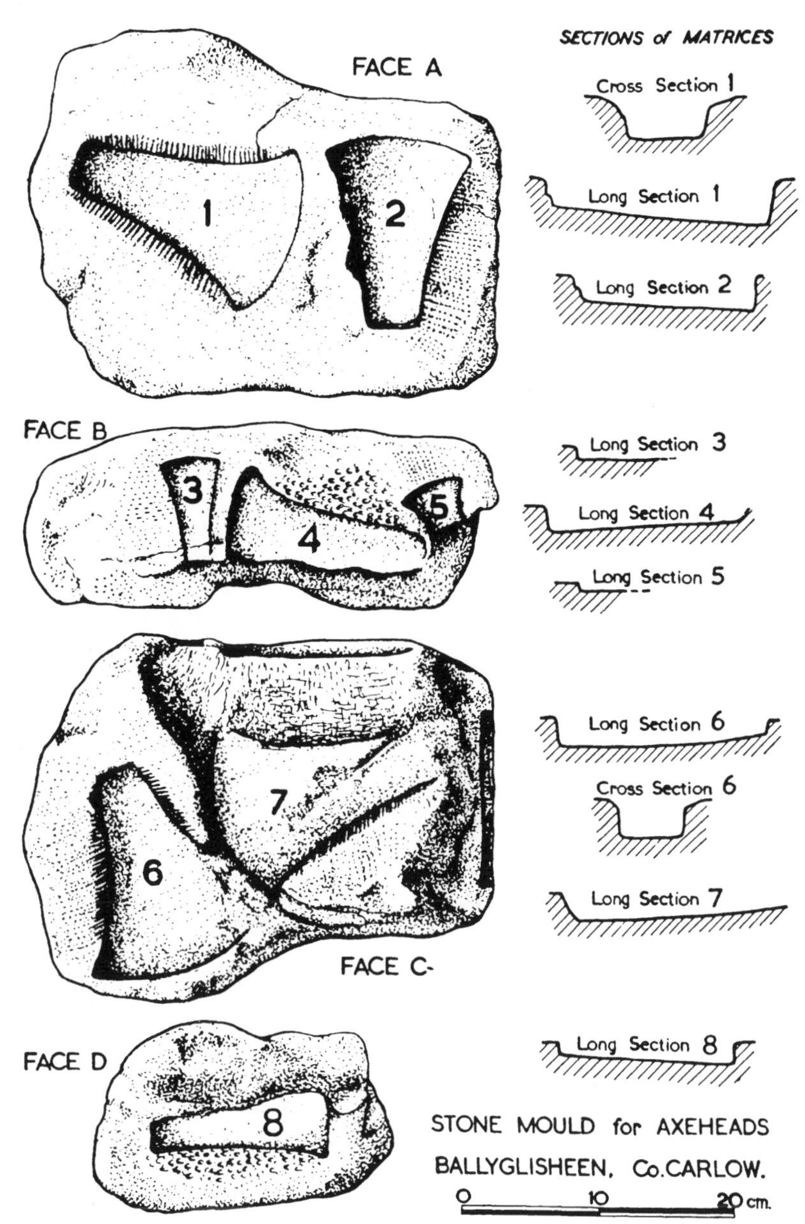

41 Stone axe mould from Ballyglisheen, Co. Carlow (by courtesy of the Royal Society of Antiquaries of Ireland and the National Museum of Ireland)

easily done in clay moulds by first making a half-mould and imprinting a pebble into the soft clay at opposite corners, so leaving two or more depressions. After drying, the second half mould is made by replacing the pattern in the first half and allowing the clay to penetrate the depressions made by the pebble. This will ensure accurate registration of the two half moulds when they are assembled ready for casting.

Core-prints, i.e. holes for the location of cores, are easily made in the moist clay before firing, and additional support for the core obtained by means of chaplets, as in modern practice. These chaplets are usually small dumb-bell shaped pieces of metal and should be made of an alloy rather purer than the metal to be cast, so that they do not completely melt; their surfaces should just be raised to the melting point by the liquid metal during casting. If they were completely melted by the heat of the liquid metal, they would fail in their purpose of maintaining the position of the core and hence the casting thickness. Whether the early smiths appreciated this fact has still to be determined by sectioning suitable objects and looking for the chaplets, which should be visible under the microscope. If these are not visible it may be concluded that the early smiths either did not use them, or made them from the same metal as the casting.

There are other ways in which this may be achieved. At Dainton[3] the two halves fitted onto each other and were sealed by an outer 'wrap' which held the two halves together during casting and stiffened the whole unit so that wooden reinforcements were not required.

The use of the two-part mould at Dainton suggests that investment or 'lost-wax' casting was not used in the Bronze Age, but by the Iron Age its use was well understood, as may be seen from the discarded mould material from Gussage All Saints.[4] By investment casting is meant the surrounding of a wax or low melting point metal pattern with the moulding material, so that by heating the mould, the pattern can be run out in liquid form. In this way the necessity for a two-part mould and loose cores is avoided. The *cire perdue* or

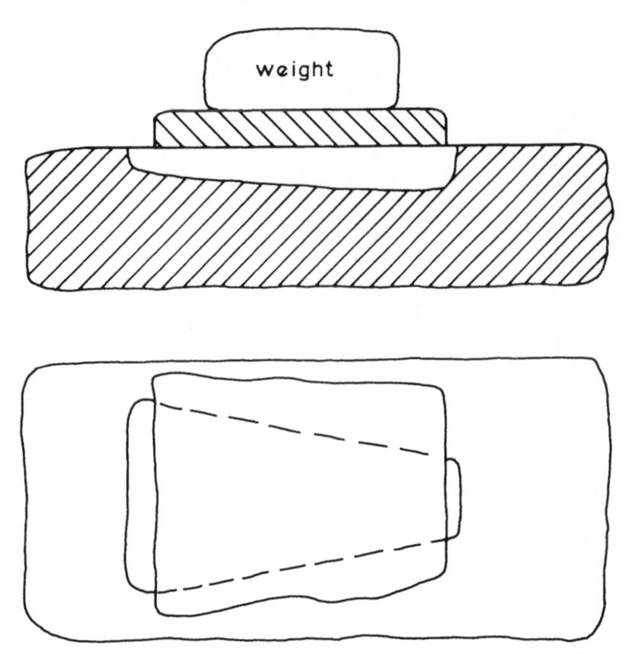

42 Use of weighted stone cover on an open axe mould

lost-wax process was used from Bronze Age times abroad, but the use of a low melting point metal pattern is thought to date from very recent times.

There is conflicting evidence for the use of lead as a pattern material. Experiments have shown that lead is not a very efficient pattern material for investment casting (i.e. lost-lead casting) as it tends to oxidize when melted out of the mould, and the oxide is retained resulting in cavities in the final metal casting.[5] It is mentioned by neither Theophilus[6] nor Cellini,[7] two art founders who are likely to have considered it, but the evidence for its use in some way is strong. We find pieces of lead socketed axe in some of the bronze moulds and it is possible that lead patterns were widely used. Lead patterns would be cast in bronze or stone two-part moulds and used in the normal two-part clay moulding process in which the pattern was removed before casting. They would be easier to make than bronze or wood patterns. In a way this approximates to the modern process of using aluminium patterns.

Strong evidence of lead casting has been found in the Irish LBA site of Rathgall but this was associated with clay moulds and not stone or bronze.[8]

The commonest moulding material today is a permeable mixture of moist sand and about 4% clay. This, however, was not used until the eighteenth century. Previous to this all moulding was done with clay, sometimes strengthened and made more permeable with the aid of chopped straw or dung. The methods of moulding in use in the 11th century AD, are described by Theophilus[6] and it is clear that in principle they cannot have changed much over the previous 3 000 years. It is only by attention to detail that such impermeable material could have been used with such satisfactory results as may be seen by the examples of bell founding given by Theophilus. Much the same methods are used for bells and propellers today.

Stone moulds

Moulds of the Copper and Early Bronze Age, mostly made for casting flat axes, were of stone. Fine-grained stones, such as the fell sandstones of the Northern Pennines were mainly used (Table 48).

Hodges[9] has examined many stone moulds and remarks that they show little variation in manufacture, all appear to have been cut from natural tabular blocks, using the technique of hammer dressing, sometimes with later grinding. In many cases the moulds for flat axes are of uniform depth throughout, but in two cases at least (Borough Moor and Burredales, Grampian, Table 48) the impressions become shallow towards the cutting edges; some others, however, have precisely the opposite tendency (Fig. 41). Many moulds have cavities for two or more axes and some have cavities for casting rectangular bars and other shapes. Since only single halves of many of these stone moulds have been found, they have been described as 'open', implying that the casting was made by pouring metal into the single cavity and allowing it to freeze with one side in contact with the air. This would seem to be unnecessarily wasteful since a good deal of the metal would be lost by oxidation and the cooling rate would be very much faster than has been observed on the axes examined, such as the Irish flat axe reported by Coghlan.[18] Hodges[9] presumes that these 'open' moulds would be closed by means of a flat movable cover. It is possible that the cover was so placed over the cavity before casting as to leave a space at one end into which metal could be introduced (Fig. 42).

Some of the flat axes are enormous. That found at Lawhead in the Pentland Hills, Scotland,[19] is 32 cm long and weighs about 2·7 kg. The wastage that would result from the use of an open mould would have led to the use of a cover of some sort, or a closed mould.

Moulds of the Middle Bronze Age were mainly used for casting socketed spearheads with side loops, and, in a few cases, palstaves. The moulds are mostly two-part moulds made of steatite or chlorite schist. Unlike those of the Early Bronze Age, these moulds were carefully faced to produce a tapering block of stone, and the cavity is far more carefully finished. In the spearhead moulds shown by Hodges[9] from Scotland and northern England, in every case the cavity for the central rib of the spearhead protrudes to the end of the mould and in many cases the mould makes no provision for a point to the spearhead. As a result Hodges concludes that the spearheads cast from these moulds were poured from the point as shown in Fig. 43*a*. The main reason for this method is the easy location of the core; the method was also used for the knives and rapiers of the period.[12]

On the other hand, the majority of the Irish spearhead moulds found near Omagh in 1883[20] have the point complete and must have been filled through the socket end (Fig. 43*b*). The cores had been supported by means of core prints made in the two halves of the stone moulds. Some of the moulds had been made for tanged blades (Arreton Down type) and these had also been filled through the tang.

Usually all four surfaces of a stone mould were cut for different objects. Presumably, this did not indicate a shortage of stone, but a need for portability. Since many bronze smiths were itinerant, economy in weight would be important.

Some of the stone moulds for palstaves, such as that from Lough Gur contain rough and apparently unnecessary cavities. It is suggested that these cavities

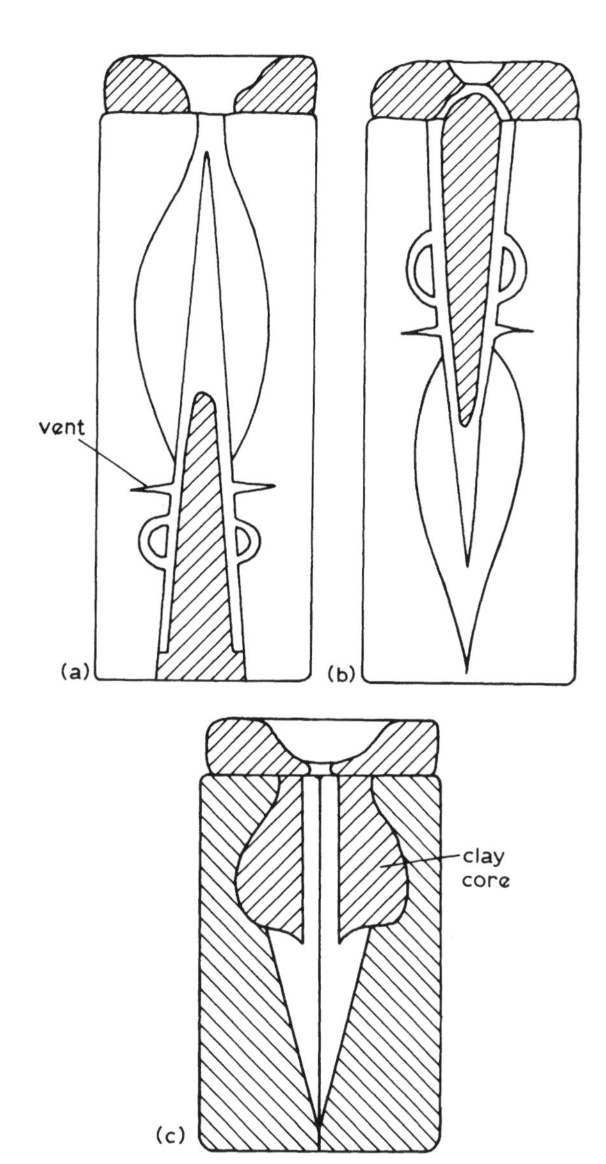

43 Stone moulds and cores

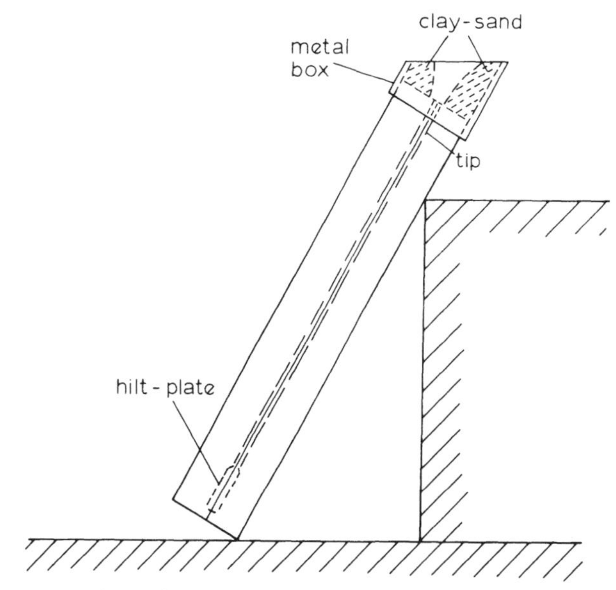

44 Method of casting a rapier in a stone mould

were made in the moulds for the location of clay cores which would enable the casting of heavy undercuts, such as those to be seen in a small class of palstaves (Fig. 43*c*).

The cavities for the loops of some of the spearhead moulds are semicircular, while the spearheads themselves always have loops which are rectangular when seen on the face and lozenge-shaped when seen on edge,[21] showing that the loops have been finished by hammering. This, together with the absence of the end of the point in those examples shown by Hodges, argues strongly against the use of these moulds for the manufacture of patterns for the lost-wax process. Coffey draws attention to the blackening in the cavities of the Irish moulds and attributes this to their use for hot-metal casting.[20]

Many spearhead moulds and at least one rapier mould have rather crude incisions in the surface of the mould (Fig. 43*a*). As Evans[12] correctly commented with regard to the rapier mould from Knighton, these are vents to let out air and gases within the cavity of the mould. The parting line itself will do this to a limited extent, since the fit of the two surfaces is never hermetic. These incisions do not have to reach the outside of the mould – it is sufficient to make them big enough to take the gases a short distance from the mould cavity itself for efficient dispersal, as in the spearhead mould from Lough Gur illustrated by Evans.[12] Further diffusion will occur by way of the parting surface. The incisions are so shallow that molten metal does not usually flow into them, nor even plaster of Paris as is shown by a model spearhead cast in such a mould in the National Museum of Scotland.

Experiments carried out on the making of rapiers in a pair of stone moulds show that the impermeability of stone does cause problems.[22] Even with the type of grooves shown in the Knighton mould the air does not leave fast enough to avoid defects. Besides casting in a sloping mould heated to 300°C with a clay sand runner bush (Fig. 44) it was necessary to wedge the mould open about 0·65 mm to enable the air to escape as the metal rose in the mould. This caused quite an extensive 'flash' which could be trimmed off and which would supply 'plate' metal for remelting. The 'opening' of the mould by 0·65 mm provided enough excess metal for working, which causes thinning, especially of the edges.

While later two-piece moulds were fitted together by means of dowels, the moulds of this period seem to have been tied together. Registration was obtained by deep scratches on the outside surfaces across the two halves of the mould. This can be clearly seen on the rapier mould from Killymaddy in the Dublin Museum (Table 49).

By contrast, many of the moulds of the Late Bronze Age are of clay, but some made of stone have been found, such as the two socketed axe moulds of mica-schist found at Rosskeen, Highland (Table 50). Here we find that in one case, the two halves were located by means of dowel pegs which fit into opposed holes, and in the other with dowels fitted into holes in one piece which register with slots in the other. The runner is provided with gates (channels) for filling the mould round the core (this being presumably of clay).

The permanent nature of the stone mould is such an advantage over the need to make a fresh clay mould for each casting, that one wonders why it went out of fashion in the Late Bronze Age. Although a good quality sandstone would last a considerable number of castings, it would have to be carefully heated before

Table 49 Stone moulds of the MBA for casting palstaves, socket looped and base looped spearheads and rapiers (mainly after Hodges[1, 10] and Collins[23])

Provenance	Materials	Implements	Publication (Museum Register No.)	Present whereabouts
England				
1 Knighton, Devon	Green Mica-schist	Rapiers	Evans[12] p. 434	—
2 Yorkshire Moors	Steatite	Spearhead Palstave	—	Br. Mus.
3 Croglin, Cumbria	Steatite	Spearhead	*Trans. CWAAS*, 1884, **7**, 279	Carlisle (cast only)
4 Gwithian, Cornwall	Chlorite-talc schist	Socketed axe	Megaw[24]	
Wales				
1 Bodwrdin, Anglesey	'Hone stone'	Spearheads	*Archaeol. J.*, 1846, **3**, 257	Tolson Museum, Huddersfield (? lost). Casts in Ashmolean Museum Oxford, and elsewhere
Scotland				
1 Campbeltown, Strathclyde	Steatite	Spearhead, razor	*Proc. Soc. Antiq. Scotl.*, 1947, **81**, 171	Edinburgh
1a Campbeltown	Steatite	Spearhead	*ibid.*	Edinburgh
2 Cromar, Grampian	Steatite	Spearhead	(C.M. 22)	Edinburgh
3 Culter, Grampian	Sandstone	Spearhead	—	Ashmolean
4 Eildon Hills, Borders	Sandstone	Palstave	*Proc. Soc. Antiq. Scotl.*, 1923, **57**, 142	Ashmolean
5 Glengyre, Dumfries and Galloway	Steatite	Palstave	*ibid.*, 105	Edinburgh
6 Orkney	Sandstone	Palstave	*Proc. Soc. Antiq. Scotl.*, 1909, **43**, 10	Edinburgh
7 Aberdeenshire [Grampian]	Mica-schist	Spearhead	(C.M. 31)	Edinburgh
Ireland				
1 Armoy, Antrim	Steatite	Spearhead	Evans[12] p. 435	Br. Mus.
2 Ballycastle, Antrim	Sandstone	Palstave	(B. 17.24) (R. 140)	Dublin
3 Ballymena, Antrim	Steatite	Spearhead	(WG/1639)	Br. Mus.
4 Ballymena	Granite	Spearhead (2 pt)	Collins[23] (543.1924)	Belfast
5 Near Ballymena	Schist	Spearhead	Collins[23] (536.1924)	Belfast
6 Near Ballymena	Schist	Spearhead	(533.1924)	Belfast
7 Near Ballymena	Schist	Spearhead	(534.1924)	Belfast
8 Near Ballymena	Schist	Spearhead	(537.1924)	Belfast
9 Near Ballymena	Schist	Spearhead	(539.1924)	Belfast
10 Near Ballymena	Schist	Ingot and spearhead	(535.1924)	Belfast
11 Near Ballymena	Schist	Razor	Collins[23] (540.1924)	Belfast
12 Killymaddy, Antrim	Schist	Rapier, spearhead, razor	*Proc. R. Ir. Acad.*, 1913, **30C**, 83 (1911.73, 76–88)	Dublin
12a Killymaddy, Antrim	Sandstone	Sickle (Continental type)	(18, 1161) (1911, 80)	Dublin
13 Rasharkin, Antrim	Steatite	Spearhead	(1924/542) Collins[23]	Belfast
14 Whitepark Bay, Antrim	Sandstone	Palstave (1 pt) and flat blade	(1930/644)	Belfast
15 Co. Carlow	Sandstone	Palstave (1 pt and flat blade	(W. 84 and 84A)	Dublin
16 Lough Ramor, Cavan	Sandstone	Spearhead	Wilde[25] p. 91 (W. 90)	Dublin
17 Inchnagree, Cork	Sandstone	Rapiers	—	Dublin
18 Maghera, Derry	Steatite	Spearhead	Evans[12] p. 435	Dublin
19 Maghera, Derry	Steatite	Spearhead	(W.G. 1640)	Br. Mus.
20 Ardvarnock Glebe, Donegal	Steatite	Palstave	Rep. Nat. Mus. Ireland, 1932–33, p. 14	Dublin
21 Ballyliffin, Donegal	Steatite	Spearhead	*J. R. Soc. Antiq. Irel.*, 1927, **57**, 64	Dublin
21a Ballyliffin, Donegal	Steatite	Palstave	*Proc. R. Ir. Acad.*, 1935, **42C**, 156 (1925, 27)	Dublin
22 Inch Isle, Donegal	Sandstone	Spearhead	(1926, 4)	Dublin
23 Dromore, Down	Steatite	Palstave	(1876, 1257)	Dublin
24 Clarinbridge, Galway	Steatite	Spearhead	*Archaeologia*, 1807, **15**, 394	Dublin

Table 49 –*continued*

	Provenance	Materials	Implements	Publication (Museum Register No.)	Present whereabouts
25	Culfin, Galway	Steatite	Palstave, spearhead	*J. Galway Hist. Archaeol. Soc.*, 1945, **21**, 107	Dublin
26	Lough Corrib, Galway	Sandstone	Palstave	*ibid.*	Edinburgh
27	Moonbawn, Laois	Sandstone	Palstave	—	Dublin
28	Lough Gur, Limerick	Sandstone	Part palstave	*Proc. R. Ir. Acad.*, 1954, **56**, 401	U.C. Dublin
29	Dundalk, Louth	Steatite	Transverse palstave*	Evans[12] p. 431 (W. 96)	Dublin
30	Toorglass, Mayo	Steatite	Spearheads	Raftery, 'Prehist. Ireland', 1951, p. 127	Dublin
31	Omagh, Tyrone	Steatite	Spearheads	*J. R. Soc. Antiq. Irel.*, 1907, **37**, 181	Dublin
32	Loughash, Tyrone	Schist	Palstave	*Ulster J. Archaeol.*, 1939, **2**, 257	Belfast
33	Moylisha, Wicklow	Sandstone	Spearhead	*J. R. Soc. Antiq. Scotl.*, 1946, **76**, 125	Dublin
34	River Bann	Steatite	Palstave	Evans[12] p. 431 Br. Mus.	Br. Mus.
35	N. of Ireland	Steatite	Spearhead	(1903, 236)	Dublin
36	N. of Ireland	Steatite	Palstave	*ibid.*	Ashmolean
37	Ireland	Stone	Rapier	(1882, 86)	Dublin
38	Ireland	Stone	Palstave	(P.2)	Dublin
39	Ireland	Stone	Palstave	(P.438)	Dublin
40	Ireland	Stone	Palstave	(1882, 89)	Dublin
41	Ballyliffin, Donegal	Chlorite schist	Palstave	Collins[23]	Buncrana
42	N. Ireland	Chlorite schist	Part palstave	Collins[23] (3788/11.55)	Belfast
43	Ballyliffin, Donegal	Chlorite schist	Blank for spearhead	*ibid.*	Buncrana
44	Ballyliffin	Chlorite schist	2 blanks for ? palstave	Collins[23] (3788/11.55)	Buncrana
45	N. Ireland	Chlorite schist	Spearhead and razor	Collins[23] (3785/6.54)	Belfast
46	N. Ireland	Chlorite schist	Spearhead and razor	Collins[23] (on loan)	Belfast
47	Maghera, Derry	Chlorite schist	Spearhead (1 pt)	Collins[23] (on loan)	Belfast
48	N. Ireland?	Chlorite schist	Spearhead	Collins[23] (on loan)	Belfast
49	Mistyburn, Antrim	Chlorite schist	Socketed chisel? (1 pt, frag.)	Collins[23] (538. 1924)	Belfast

*Blade at right angles to parting line and septum.

Table 50 LBA moulds for casting socketed axes, leaf-shaped swords, leaf-shaped spearheads, and chisels (mainly after Hodges[10])

	Provenance	Material	Implements	Publication (Museum Register No.)	Present whereabouts
England					
1	Bulford Water, Wilts.	'Greenstone'	Socketed axes	*Antiq. J.*, 1937	Salisbury
2	Burderop, Wilts.	Keratophyre	Socketed axes	Needham[26]	Salisbury
3	Gwithian, Cornwall	Chlorite-talc schist	Socketed axes	Megaw;[24] *Br. Archaeol. Rep.*, 33	Truro
4	Ham Hill, Somerset		Socketed axes		Taunton
5	Egham, Surrey	Keratophyre	Socketed axes	Needham[26]	Brit. Mus.
6	Helsbury, Cornwall	Chlorite-schist	Socketed axes	*J. R. Inst. Corn.*, 1936, **24**, 303	Truro
7	Milton, Dorset	Chlorite-talc	Socketed axes	*Archaeologia*, 1820, **28**, 451	Dorchester
8	Linkinshorn, Cornwall	Sandstone	Socketed chisel	Hodges[16]	Private
Wales					
1	Afon Mâd, Dyfed	Fine gritstone	Chisel	*Archaeol. Cambrensis*, 1944, **98**, 146	Cardiff
Scotland					
1	Aberdeenshire [Grampian]	Steatite	Spearheads	*Proc. Soc. Antiq. Scotl.*, 1926, **60**, 19	Edinburgh
2	Ardrossan, Strathclyde	Mica-schist	Socketed axe	—	Ashmolean, Oxford

Table 50 –*continued*

Provenance	Material	Implements	Publication (Museum Register No.)	Present whereabouts
3 Strathnaver, Highland	Steatite	Spearhead	*Proc. Soc. Antiq. Scotl.*, 1906, **40**, 129	Edinburgh
4 Rosskeen, Highland	Mica-schist	Socketed axe	*Proc. Soc. Antiq. Scotl.*, 1938, **72**, 157	Ashmolean
4a Rosskeen, Highland	Cast	Socketed axe	—	Edinburgh
Ireland				
1 Mistyburn, Antrim	Steatite	Socketed axe	Hodges[10]	Belfast
2 Ballydag, Kilkenny	Sandstone	Socketed axe	*J. R. Soc. Antiq. Irel.*, 1863, **7**, 30	Dublin
3 Fethard, Tipperary	Mica-schist	Socketed axe	*J. R. Soc. Antiq. Irel.*, 1889, **19**, 290	Br. Mus.
4 Lough Gur Limerick	—	Palstave	—	Dublin
5 N. of Ireland	Steatite	Spearhead	Cat. Nat. Mus. Antiq. Scotl., 1892, p. 125	Edinburgh
6 Ireland	Steatite	Socketed chisel	Hodges[10]	Belfast
7 Ireland	Mica-schist	Socketed axe	Wilde[25] p. 91	Dublin
8 Ireland	—	Leaf-shaped swords	*Mém. Antiq. Nord*, 1872–77, 145	—
9 Ireland	Stone	Socketed axe	(W. 85)	Dublin
10 C. Kilkenny (?)	Steatite	Socketed axe	(1901, 56)	Dublin

Table 51 Clay two-piece moulds of Bronze Age character

Provenance	Implement	Site	Publication (Museum Register No.)	Present whereabouts
England				
1 Grimes Graves, Norfolk	Basal looped spearheads etc.	Flint mine	Burgess[27]	Br. Mus.
2 Fimber, Yorks.	Sword, spearhead, chape	Settlement	*Naturalist*, 1930, 347	Hull
3 Dainton, Devon	Swords, spearheads, ferrules, rings	Settlement	Needham[3]	Br. Mus.
Scotland				
1 Jarlshof, Shetland	Socketed axe, knife, sword, pin	Dwelling	*Proc. Soc. Antiq. Scotl.*, 1934, **68**, 279	Edinburgh
2 Loanhead of Daviot, Grampian	Sword	Cemetery	*Proc. Soc. Antiq. Scotl.*, 1936, **70**, 300	Edinburgh
3 Traprain Law, Lothian	Spear-butt, sword	Dwelling	*Proc. Soc. Antiq. Scotl.*, 1920, **54**, 80; 1922, **56**, 213	Edinburgh
Wales				
1 Breiddin	Unidentified	Hill fort	Musson[28]	
Ireland				
1 Whitepark Bay, Antrim	Sword, spearhead	Inhabited dwelling	*Ulster J. Archaeol.*, 1953, **16**, 39 (654, 30)	Belfast
2 Bray, Dublin	Sword, knife	Cemetery	*J. R. Soc. Antiq. Irel.*, 1894, **24**, 54; 1895, **25**, 106	Dublin
3 Dalkey Isle	Swords etc.	Settlement	Liversage[29]	Dublin
4 Boho, Fermanagh	Sword		Hodges[10] Fig. 26; (1902. 24, 25)	Dublin
5 Lough Gur	Spearhead	Dwelling	*Proc. R. Ir. Acad. (C)*, 1954, **56**, 403	Dublin
6 Rathgall, Wicklow	Swords, spearheads, axes, etc.	Settlement	Raftery[30]	Dublin

1–5 from Lough Eskragh, Co. Tyrone; *6* from Boho, Co. Fermanagh (by courtesy of *Ulster J. Archaeol.*, A. E. P. Collins and H. W. M. Hodges)

45 Construction of clay sword moulds with wooden reinforcement

use each time to avoid cracking due to thermal shock. Since accidents will happen, the enormous amount of work in making the more complex mould of the Middle and Late Bronze Ages gradually led the bronzesmith to discard stone and use expendable clay moulds. It is possible that this process was assisted by the use of lead patterns made in permanent bronze moulds. Clay moulds were almost certainly being made by the Middle Bronze Age, or even earlier, but none has survived.

Clay moulds of the Late Bronze Age

Clay moulds (Table 51), being fragile, tend to come from excavated sites, while the majority of stone moulds are chance finds. Remarkable consistency has been noted in the fragments of clay moulds so far found, and it would appear that a uniform technique of clay moulding was in use during the Late Bronze Age. All were two-piece moulds often provided with dowels and where necessary with a core.

In many cases the clay used was of two grades. That in contact with the metal had been carefully worked and then deliberately grogged with sand to minimize shrinkage on drying. The outer layer was a coarser clay, often containing some vegetable matter and supported in the case of some sword moulds with long thin pieces of wood to prevent warping (Fig. 45).

The material from Dainton[3] shows the use of three fabrics for the inner part of the mould. These were all well prepared while the outer wraps were made of clay of similar provenance but of minimal preparation. The moulds were fired to a temperature of about 650°C and probably poured at this temperature or one a little lower.

The surviving fragments of moulds for casting swords allow a working reconstruction of the method used for a Late Bronze Age sword. Casting was from the hilt end by means of a small gate cut in the mould itself; and the tangs of many bronze swords show the runner to have been broken off at this point – no effort being made to clean it up since it would be covered by the hilt plates. No attempt was made to core the holes in the tang to receive the rivets of the hilt plates, but slight projections were often made in the mould so as to leave hollows in the tang. This seems surprising as one would have thought that it would have been much easier to core the hole than drill it. The probable explanation is that a series of fully cored-out rivet holes would interfere too much with the filling of the mould.

It would be expected that the casting of such a long thin object as a sword would result in a considerable amount of segregation of certain alloy constituents from one end to the other if the cooling rate was slow. This point has been investigated by Coghlan[31] for the hilt and part of the blade of an Irish sword and the results are shown in Table 52*b*. It may be seen that

Table 52 Degree of segregation in copper-base artifacts

(*a*) EBA flat axe (after Richards and Blin-Stoyle[32])

Element, %	Top to bottom Centre	Top to bottom Edge	Through thickness One surface	Through thickness Centre	Through thickness Oppos. surface
Cu	96·4	96·2	—	—	—
As	2·89	3·15	2·97	2·97	2·84
Ni	0·021	0·021	0·021	0·021	0·021
Ag	0·194	0·207	0·206	0·193	0·189
Sb	0·247	0·266	0·259	0·262	0·254
Pb	0·31	0·31	0·30	0·30	0·30

(*b*) Irish LBA sword (after Coghlan[31])

	Position							
	Half-way down blade						Hilt	
Element, %	Test no. 1	3	5	7	8	9	10	12
Pb	13·7	13·5	13·2	13·2	13·0	13·0	12·7	11·8
Sn	9·0	9·5	9·4	9·3	9·4	9·4	9·5	9·3
Cu	76·2	76·1	76·4	76·3	76·6	76·5	77·0	78·0

while some downward segregation of the heaviest element, lead, has occurred, the tin and copper are of fairly uniform composition throughout. This absence of major vertical segregation must have been caused by fairly rapid cooling due to the small volume-to-surface area ratio of the clay mould. On the other hand the flat axe of arsenical copper solid solution, which was probably cast flat, shows very little segregation indeed [32] (Table 52*a*).

Cope[33] reports the degree of silver–lead segregation in a strip of vertically cast Cu–Sn–Pb coinage alloy and found lead to vary from 10·35% at the bottom to 14·62% at the top. This was clearly due to gravity segregation in the crucible. The silver content of the lead was more or less constant; the overall composition gave 1·15% Ag at the bottom and 1·57% at the top.

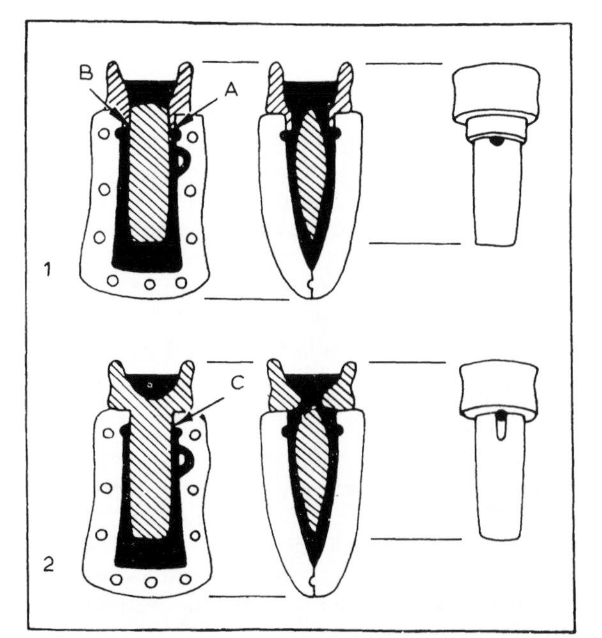

46 Socketed axe moulds showing methods of coring and gating (by courtesy of H. W. M. Hodges)

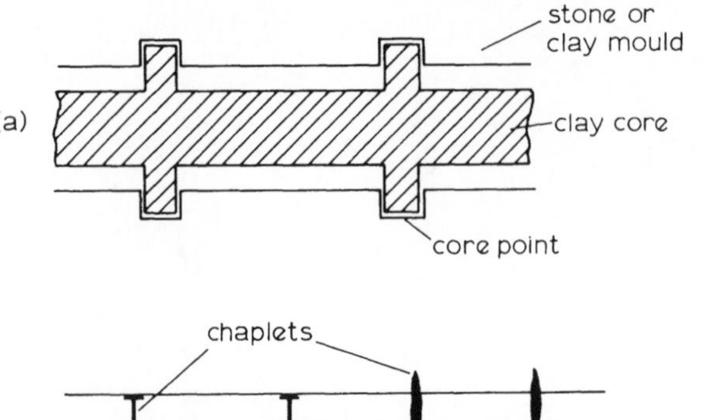

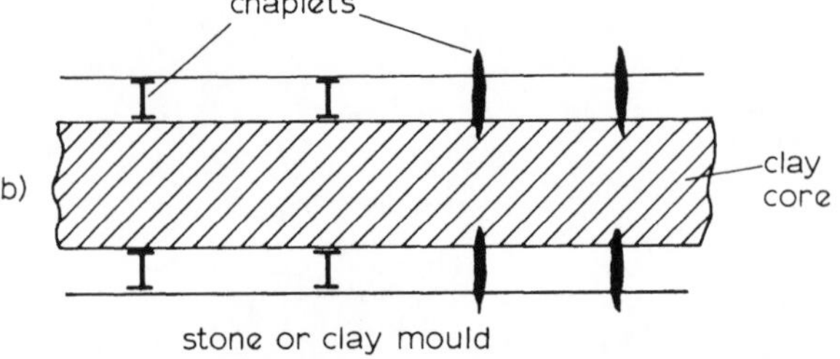

47 Location of cores by means of trunnions and chaplets

There is no reason to doubt that clay moulds were used for direct casting of metal since analytical work on the inner face of one of the Irish clay moulds shows the presence of large quantities of copper, tin, and lead.[10]

The surviving fragments of clay socketed axe moulds show how the moulds were made but not the coring and gating practice. It would seem that the core and gates were made in one piece as shown in Fig. 46. The flash that can be observed on the axes themselves points to the more common use of method 2 (Fig. 46), which leaves a single flash all round the periphery of the cored socket, rather than method 1 which would mean flashes at A and B.

Cores can also be supported by means of chaplets or trunnions (Fig. 47), or by side pieces attached to the upper parts of the mould (Fig. 49, below) and it is possible that a combined core and runner was made by paring down a piece of clay first roughly shaped in the two-piece mould, or more conveniently, by pressing clay or a mixture of clay and sand into a mould in which a finished axe had been placed. The sockets of most axes are so tapered that this method could have been used. Gates would still have to be cut into the side of the core after moulding. After casting, of course, the cores would be destroyed by their removal, and only in cases of defective castings are cores found inside.

An unfinished spearhead of Late Bronze Age type was found in a hoard from Guilsfield, Powys.[34] This had the flash still left on and had been scrapped because the core for the socket had been accidentally omitted. Instead of a socket there was a solid shaft of bronze, short bronze bars now occupying the position of the rivet holes, showing how the core was held in place by 'prints'. This method serves two purposes, that of coring out the rivet holes and locating the core for the socket.[35] The fact that such a method of making rivet holes was practised on spearheads suggests that there must have been some very good reason for not using it on the swords of the period.

The fact that casting was not always without its difficulties is shown by the scrapped Late Bronze Age swords from Ballycroghan,[36] which were found with their flash on and 'dimples' for rivet holes. These had been scrapped owing to large blowholes near the top (hilt) of the casting. They had been made from a leaded bronze and cast in clay moulds for which the same pattern may have been used. This shows how ineffective is the feeding of such objects, and how important it is to cast gas-free metal.

The normal pattern material for clay moulds was wood, and the impression of the grain of the wood can be clearly seen in the pieces of the sword mould from Jarlshof.[9] Wooden patterns for socketed axes, spearheads, and a hammer are known from Ireland.[10] The facets of many bronze axes may be due to some extent to the use of wooden patterns, such faceting being typical of wood carving. A wooden sword capable of being used for a pattern was discovered in Orkney.[37] There would be no objection to using the bronze artifacts themselves as patterns, or lead models, but such a heavy metal is always difficult to remove neatly from the mould.

No trace of core-boxes which might have been made from wood have been found, but the insides of the socketed axes sometimes show the use of a moulded core with a 'parting line' which is transferred to the bronze axe itself and shows on the inside. This is almost certainly proof of moulding with a two-part core box.

According to Clarke,[38] the core of the lead socketed axe found at Anwick, near Sleaford, Lincs., was made in a jointed core-box, as there is a groove down the centre of the two sides of the socket.

In addition, there are often additional ridges or ribs which have been intentionally moulded in by cutting grooves into the core. The distribution and pattern has been extensively studied[39] and it is possible that these denote a smith's personal mark or a workshop group.

Towards the end of the Late Bronze Age, axes appeared with 'waisted' sockets, presumably as a better method of hafting. The cores for these, of course, could not be made in a single-piece core-box but could have been made in a two-piece box or pared down from a lump of clay.

One of the most important requirements of a core is its permeability to gases. Gases evolved on the solidification of molten metal will find their way out relatively easily through the surface of a mould but this is not the case with a core, which must be far more permeable. A good core could be made by applying a thin coating of clay to a bundle of straw, and the resulting fragility could be a reason for the complete absence of unused cores.

Modern cores are made in mixtures of sand and clay, or sand bonded with oil or a cereal, which gives them very high permeability. A solid lump of clay or sandy clay with little sand would not be expected to give a good casting, and it would seem more likely that a

Table 53 Bronze two-piece moulds (mainly after Hodges[10] and Smith[41])

Provenance	Implement	Publication or present whereabouts
Middle Bronze Age		
1 Llyn Mawr, Gwynedd	Palstave	Br. Mus.
2 Danesfield, Gwynedd	Palstave	Br. Mus. Grimes: 'Pre-history of Wales', 1951, p. 192. *Archaeol. Cambrensis*, 1856, **2**, 128
3 Danesfield	Palstave	*ibid.*
4 Wiltshire Downs	Palstave	Evans[12] p. 440. (B.M. 55, 5–3, 1)
5 South London	Palstave	Smith.[41] Hull Museum
6 Stoke, Rochester (alias Hoo)	Palstave (fragment only)	Smith[41] Jessup: 'Archaeol. of Kent', 1930, p. 108
Late Bronze Age		
1 Arkesden, Essex	Socketed axe	Fox: 'Archaeol. of Cambridge Region', 1923, p. 324
2 Beddington, Surrey	Socketed axe	Evans[12] p. 447
3 Cambridge, New St	Socketed axe	Fox: 'Archaeol. of Cambridge Region', 1923, p. 58 (Cambridge)
4 Cleveland	Socketed axe	Evans[12] p. 447
5 Donhead St. Mary, Wiltshire	Socketed axe	*Wilts. Archaeol. Nat. Hist. Mag.*, 1927, **43**, 373
6 Norwich, Unthank Rd	Socketed axe	Castle Mus., Norwich (161.946)[42]
7 Heathery Burn, Durham	Socketed axe	*Archaeologia*, 1904, **54**, 99
8 Hotham Carr, Yorks.	Palstave	Evans[12] p. 439
9 Isle of Harty, Kent	(*a*) Socketed gouge	*Inventaria Archaeol. GB*, 18
	(*b*) Socketed axe	*ibid.*
	(*c*) Socketed axe	*ibid.*
	(*d*) Socketed axe (1 pt)	*ibid.*
10 Quantock Hills, Somerset	Socketed axe (faceted)	Evans[12] p. 447 (BM.OA 116, 117)
11 Roseberry Topping, Yorks.	Socketed axe	Elgee: '*Archaeol.* of Yorks.', 1933, p. 94. (Sheffield Mus.)
12 Southall, Middlesex	Socketed axe	Vulliamy: 'Archaeol. of Middlesex and London', 1930, p. 110 (*Inventaria Archaeol. GB*, 51)
13 Charnwood Forest, Leics.	Socketed axe	Hodges[16] (Leicester Mus.)
14 Beeston Regis, Norfolk	Part socketed axe	Craddock[43] (Norwich Mus.)
15 Washingborough, Lincs.	Socketed axe	Evans[12] p. 447
16 Wickman Park, Surrey	Socketed axe	Whimster: 'Archaeol. of Surrey', 1931, p. 79
17 Wilmington, Sussex	Socketed axe	Curwen: 'Archaeol. of Sussex', 1937, p. 206
18 Beacon Hill, Leics.	Socketed axe	Clarke[38] (Leicester)
19 Unlocalized, Yorks. (?)	Socketed axe	Evans[12] p. 447 (Br. Mus.)
20 Isleham, Cambs.	Palstave	*Antiquity*, 1960, **34**, 281
21 Ireland	Palstave (half only)	Dublin (p. 745), Burgess[27] Plate XVI

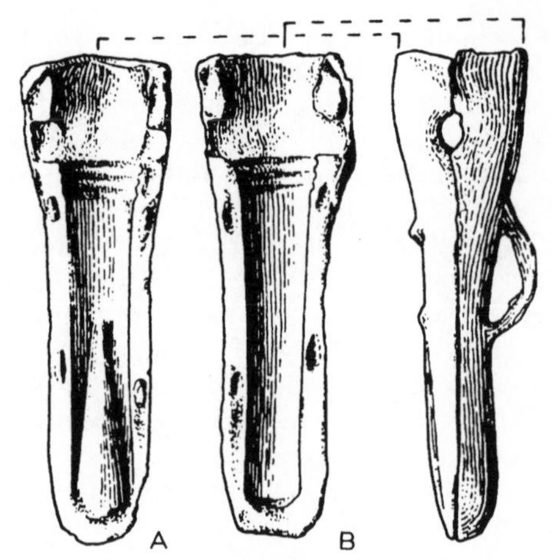

48 Bronze mould with trunnion support from Isle of Harty, Kent (by courtesy of *Inventaria Archaeol*.)

mixture of sand and clay, with a higher proportion of sand to clay than normal, was used. The socket of an unfinished Breton-type axe in the museum at Belfast was found to contain such a mixture.

After the clay mould had been assembled with its core in position, it would be kept in a suitable position for casting in a sand-filled pit, such as that found by Curle[40] at Jarlshof. Since the pit was only 28 cm deep, a sword mould would probably have been cast in a sloping position as shown in Fig. 44.

Bronze moulds

At some time during the Middle Bronze Age, permanent bronze moulds were introduced (*see* Table 53).

Hodges[9] notes that these moulds are mainly found in Lowland England and therefore one assumes that the idea was imported from the Continent. Certainly there is only one example recorded from Ireland, which seems to indicate that Irish founders were losing their hold on the English market. These objects show the advanced state of the founder's craft and would only be considered today when the quantity of castings required was large enough to justify the expense. It is therefore natural that they should only appear in the latter half of the Bronze Age when the demand for palstaves and socketed axes became very great. The bronze mould is a direct reproduction of the clay mould, which fact suggests that clay moulds were used in the Middle Bronze Age. The bronze mould would still need a clay core and it is strange that a core box for such cores has never been found. This lends support to the possibility that they were often used for wax or lead patterns and not always for the direct casting of bronze axes.

That it is possible to cast bronze direct into such moulds cannot be doubted, for it is done today, but usually with cast iron or steel moulds. Coghlan[44] has made two 7% tin bronze castings in a bronze palstave mould, pouring the metal at 1150°C into the mould which was heated to 150°C. The mould was quite undamaged, and the 'chill' effect would have given a stronger casting than a thick clay mould. But the important question is how many castings could be made in such moulds? Would it not be cheaper to use the mould for making wax or lead patterns when it would have an indefinite life rather than for bronze castings when it would need replacing every 50 or so

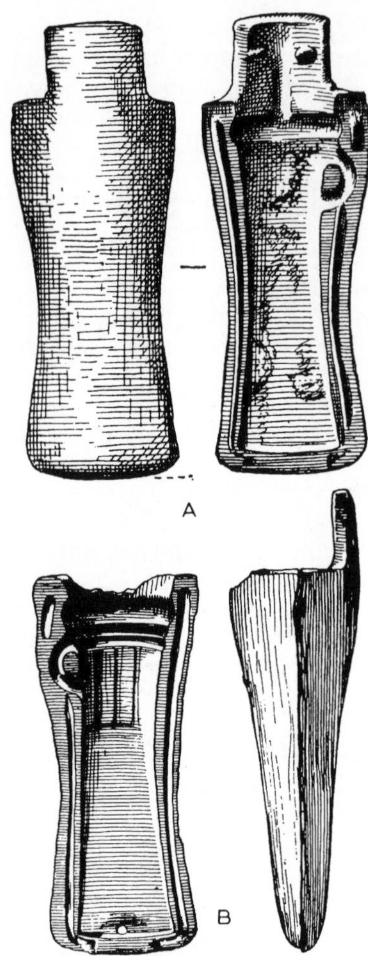

49 Bronze mould from Southall, Middlesex, probably an exact copy of a clay mould (by courtesy of the Trustees of the British Museum)

castings? The mould from the Isle of Harty[45] shows that the core has been supported by trunnions fitted into holes made in two vertical projections above the main part of the mould (Fig. 48). Such moulds must have been used for direct casting. On the other hand the mould from Southall[46] does not have holes for trunnions and the core must have been supported by clamping it between the vertical projections. (Fig. 49.)

The absence of identical objects implies that bronze moulds were comparatively short lived. The Stibbard hoard[47] contains a number of looped palstaves of which five appear identical. These could have been made either by casting directly in a bronze mould or by using the lost-wax process.

From time to time a number of bronze moulds have been found bearing traces of lead, and in the one found in New St, Cambridge, there is half of a badly fitting lead socketed axe. Complete lead axes have been found at Anwick, near Sleaford, Lincs., and on Seamer Moor, Yorks.[38] Lead was found in bronze moulds from Beacon Hill, Southall, and Roseberry Topping,[38] and furthermore a bronze mould from the Isle of Harty had a 'lead lining', according to Evans, and still contains an accretion of lead[45] (Table 54).

Further afield, in the north of France, many hoards of lead socketed axes or chisels have been found.[48] About forty were found near Donges (Loire Atlantique), one axe weighing 2·1 kg and containing 99·49% Pb. In Moellan (Finistère), lead and bronze axes

Table 54 Lead axes and lead in bronze moulds

Provenance	Object	Remarks	Ref.
Anwick, near Sleaford, Lincs.	Lead socketed axe	Now at Leeds	38
Seamer Moor, Yorks.	Lead socketed axe	Br. Mus.	38
New St, Cambridge	Half a lead socketed axe in bronze mould (ill fitting)	Now in Cambridge Museum of Archaeol. and Ethnol.	38
Beacon Hill, Leics.	Lead in bronze mould		
Isle of Harty, Kent	Lead in bronze mould	Accretion of lead according to *Inventaria Archaeol. GB*, 18	12, p. 447
Roseberry Topping, Yorks.	Lead in bronze mould	Lead now no longer evident (Sheffield Museum)	38
Southall, Middlesex	Lead in bronze mould	Gowland found this to be lead carbonate derived from decay of metallic lead (*Inventaria Archaeologia GB*, 51)	46

were found together, but do not appear to have been the same size. All these are of 'Breton' type. It is not thought that these can have had any votive significance since they were not found near tombs or funerary sites. The hoards of socketed bronze axes found show that they were made by mass production methods and it is possible that the lead axes found at Donges[48] had been cast in a bronze mould and the products were ready for use as patterns. Since bronze casting was a mass production process, pattern making and moulding would have to be geared to the final process.

It is not possible to tell from the reports how many axes of the several hoards of French lead axes are identical, but many of these, although outwardly socketed axes, had their sockets filled with lead. This would seem to confirm that it was the founder's intention to use these as patterns for clay moulds with separate cores. After moulding, a runner would be made with the necessary gates and the prepared clay–sand core inserted.

Some of the Breton bronze axes have been exported in the unfinished condition with their flash left on. Since they appear to contain as much as 43% Pb,[49] and little tin, they would be virtually useless as cutting implements. It has been suggested that they were frauds, but it is far more likely that they had some ritual significance, since if they were intended to deceive, the deceiver would obviously have finished them. The same argument probably applies to at least some of the Breton hoards of pure lead axes, but the small number of English type found in England must have served some other purpose, perhaps as patterns. Lead brooches and rings are also known, and it is possible that some of these also served as patterns.

Later moulding materials

Stone moulds did not go out of use entirely with the Late Bronze Age, but were still used in the late Roman period and up to the 18th century. Parts of two-piece moulds of Bath stone for casting pewter dishes were found at Camerton,[50] Nettleton,[51] Lansdown,[52] and Silchester[53] and these can be dated to the 3rd–4th century AD when large scale pewter manufacture started in south-west and south-east England with the revival of the working of Cornish tin ores. Pewter took the place of silver tableware in an increasing number of Roman households (Table 55).

Stone moulds were also used for casting small items or 'trinkets'. But probably most of this type were used for casting patterns for straight two-part moulds or for investment moulding.[62]

Sandstone moulds for casting ingots, which date from the early years of the 1st millennium AD, were found at Traprain Law.[63] At Garranes, Co. Cork[64] (6th century AD) a bronze bar 15 cm long fitted exactly into a stone mould. Other stone moulds and a large number of pieces of clay moulds were found on this site.

The largest deposit of clay moulds in a post-Bronze Age context was found at Gussage All Saints, and clearly most of the material was used for lost-wax investment moulding. The parts made were mostly for horse harness or chariot fittings. All the objects could be paralleled in chariot burials.[4] The moulds were made from selected clays with the deliberate and measured additions of fillers in the form of quartz or sand.

The use of the lost-wax process continued into Roman times as the 2nd–3rd century AD material from Caistor-by-Norwich shows.[65] Clay investment moulds have been used for pins, awls, fibulae, and bracelets. They had been fired before use at a fairly high temperature (probably above 600°C). The material cast was leaded tin bronze with zinc contents not exceeding 5%. Investment moulding has also been used for small statuettes about 35 cm high as may be seen from the Gestingthorpe example.[66] It seems that in this case the inner layer was heated after moulding, presumably to melt out the wax, and the outer layer was applied later (details are given in Table 56).

Evidence for the migration period from Sweden shows the use of the two-part mould for square-headed brooches.[74] Medieval material from Southampton probably relating to bronze cauldrons and cooking pots suggests the use of two-part moulds with wooden patterns.[75]

A clay mould for a bronze pin was found on the site at Lough Faughan Crannog,[76] probably to be placed in the Early Christian period, and pieces of clay

Table 55 Iron Age and later stone moulds

Provenance	Material	Object	Publication, Reference, or (Museum Register No.)	Present whereabouts
Canonbie, Dumfries & Galloway	Cast (1 part)	Circle and bar	—	Edinburgh
Trochrigg, Strathclyde	Cast (1 part)	Odd shapes	*Proc. Soc. Antiq. Scotl.*, 1851–4, 45	Edinburgh
Ardifuar, Strathclyde	Stone (1 part)	Sickle? Bar	(G.R. 19)	Edinburgh
Traprain Law, Lothian	Sandstone (2 single moulds)	Ingot	—	Edinburgh
Dun Beag, Skye	Stone	—	(G.A. 1068–70)	Edinburgh
East Lomond, Fife	Sandstone	—	(H.H. 385)	Edinburgh
Cumlins, Shetland	Steatite	Ring	(H.D. 740)	Edinburgh
Garranes, Cork (6th cent. AD)	Stone	Bars lamp (?)	O'Riordain[64]	Edinburgh
Camerton, Somerset (Roman)	Stone	Pewter dishes	Cunnington[54]	Edinburgh
Lansdown, Somerset (2nd–5th cent. AD)	Oolite and white lias.	Various objects of unknown use	*Proc. Soc. Antiq. Scotl.*, **22**, 34–8	Bath
St. Just in Penwith, Cornwall (3rd–4th cent. AD)	Decomposed granite	Pewter dishes	Brown[55]	
Worms Head, W. Glamorgan (unstrat.)	Red sandstone	For disc or mirror (1 of 2 parts)	Cunnington[54]	Swansea
Dinorben, Clwyd–Powys, Wales (EIA?)	Igneous	For disc (2 parts)	Guilbert[56]	Cardiff
Kiondroghad, Isle of Man (7–8th cent. AD)	Red sandstone	Ingots	Gelling[57]	Isle of Man
Langton, E. Yorks. (Roman Villa)	Limestone	Pewter dishes	Goodall[58]	Malton, Yorks.
Nettleton, Wilts. (Roman)	Limestone	Pewter dishes	Wedlake[51]	Bristol and Devizes
Thetford, Norfolk (Saxon)	Chalk	Ingot bars	Bayley[59]	
Hunsbury, Northants.	Sandstone	Small disc	Fell[60]	Northampton
Silchester, Hants.	Oolite Limestone	Dishes and plates	Blagg[53]	Reading
Whitcombe, Glos.	'Stone'	Jug	Blagg[53]	—
Perth (13th cent.)	'Stone'	Brooches	Bogdan[61]	Perth

moulds have been found on Iron Age sites in Scotland. Clay continued to be the chief moulding material of the Anglo-Saxon and medieval period, but until the work of Biringuccio[77] little was recorded showing the methods used. By the early medieval period, large castings were made for bells and memorial brasses.

Of the other objects, perhaps the commonest was the three-legged bronze cauldron. The size of these varies from about 15 cm to 34 cm in diameter at the belly. There are many examples of this type in the museums at Dublin and Edinburgh and some in Belfast, England, and Scandinavia. Illustrations of them can be seen in illuminated manuscripts of the end of the 12th century and they remained in common use until the 17th century when they were replaced by

Table 56 Later sites yielding clay moulds

Site	Date	Objects	Process	Reference
Gussage All Saints	Iron Age	Chariot fittings	Investment	Foster[4]
South Cadbury	Iron Age	Bridle bit		Spratling[67]
Breiddin	Iron Age			Musson[28]
Stibbes, Leicester	Roman	Plaque	2-part	Blank[68]
Gestingthorpe, Essex	Roman	Statue	Investment	Frere[66]
Caistor by Norwich	Roman	Brooches	Investment	Tylecote[65]
Mucking, Essex	Migration			Jones[69]
Elgin (Nicholson's Garage)	14th cent.	Pins etc.		Cracknell[70]
Birsay, Orkney	Pre-Norse	Brooches	2-part	Curle[71]
Much Park St, Coventry	14th cent.	Buckles	Stack	Bayley and Wilson[72]
Copthall Ave., City of London	Late med.	Buckles	Stack	Armitage *et al.*[73]

Table 57 Analyses of crucibles, crucible slags, and a red opaque glass

Element or compound, %	York, Feasegate[86] (Roman or (12th–13th cent.)		York, Petergate[87] (Medieval)	Craigy-Warren[85] (EC)	Wilderspool, Lancs.[88] (Roman)	Meare, Somerset[84] (Iron Age)		Gussage All Saints[89] (Iron Age)	Dalkey Isle[29] (EC)		Beckford[90] (Iron Age)	Antrim[83] (EC)	Cavan[83] (EC)	Dunshaughlin[83] (EC)	Irish[83] No. 158	Kirkstall Abbey[91]	Tara[92]
	Crucible	Slag	Slag	External slag	Slag	Crucible	Slag glaze	Red slag	Slag	Slag	Slag	Slag	Slag	Slag	Slag	Slag	Red glass
SiO_2	62·6	57·1	64·70	46·2	41·3	76·7	66·8	56–61	21–32	4–8	58	46·2	45·0	43·9	43·2	73·58	43·28
Fe_2O_3	2·6	8·6	2·86	—	(9·4 incl. alkali)	—	—	—	0·7	pres.	—	—	—	—	—	—	—
FeO	—	—	—	11·2	—	2·7	7·3	3·2	—	—	6·0	11·2	7·7	8·7	15·2	5·78	2·75
Al_2O_3	30·6	19·6	8·84	10·6	—	2·7	7·3	5–10	2–4	—	16·0	10·6	2·2	18·2	18·2	14·28	—
MnO	nil	tr.	—	tr.	—	—	—	—	—	—	—	—	—	—	—	—	—
CaO	1·2	3·0	0·40	12·8	—	1·5	6·2	4–10	3–4	0·7	7·5	12·8	1·0	3·7	4·9	1·66	1·44
MgO	0·1	1·73	0·50	6·5	—	1·5	2·0	—	2	—	—	6·5	0·5	0·8	0·5	1·08	—
Cu_2O	nil	1·8	13·20	1·0	39·1	tr.	2·0	4–9	6–11	—	6·0	1·0	26·2	31·6	—	2·83	9·86
PbO	nil	6·70	0·60	0·2	—	nil	tr.	—	0·5	0·04	—	0·2	—	0·3	—	0·22	32·85
Sn	—	—	0·63	—	4·3	—	—	2·4	3–5	0·01	—	—	—	—	—	—	—
Ni	nil	tr.	—	—	—	—	—	—	—	—	—	—	—	—	—	—	—
ZnO	nil	nil	—	—	0·75	—	—	5·8	—	—	5·0	4·7	—	—	1·2	—	(with CaO)
K_2O	—	—	1·30	8·8	—	3·2	n.d.	—	1·0	—	—	4·1	—	—	1·0	—	9·82
Na_2O	—	—	—	—	—	—	—	n.d.	4–10	2·0	—	1·2	tr.	—	0·5	—	—
P_2O_5	—	—	—	1·2	—	—	—	—	—	—	0·5	—	—	0·9	—	—	—
As_2O_5	—	—	—	0·5	—	—	—	—	—	—	—	0·3	—	2·6	—	0·018	—
S	—	—	0·20	0·3	—	tr.	0·3	—	—	—	—	—	—	—	—	0·018	—
TiO_2	—	—	—	tr.	—	1·0	tr.	—									
Cone refractoriness, °C*	1 500–1 600																

*Cone refractoriness is temperature at which clay softens. Maximum working temperature would be 100–200°C below this.

— = not sought; n.d. = not detected.

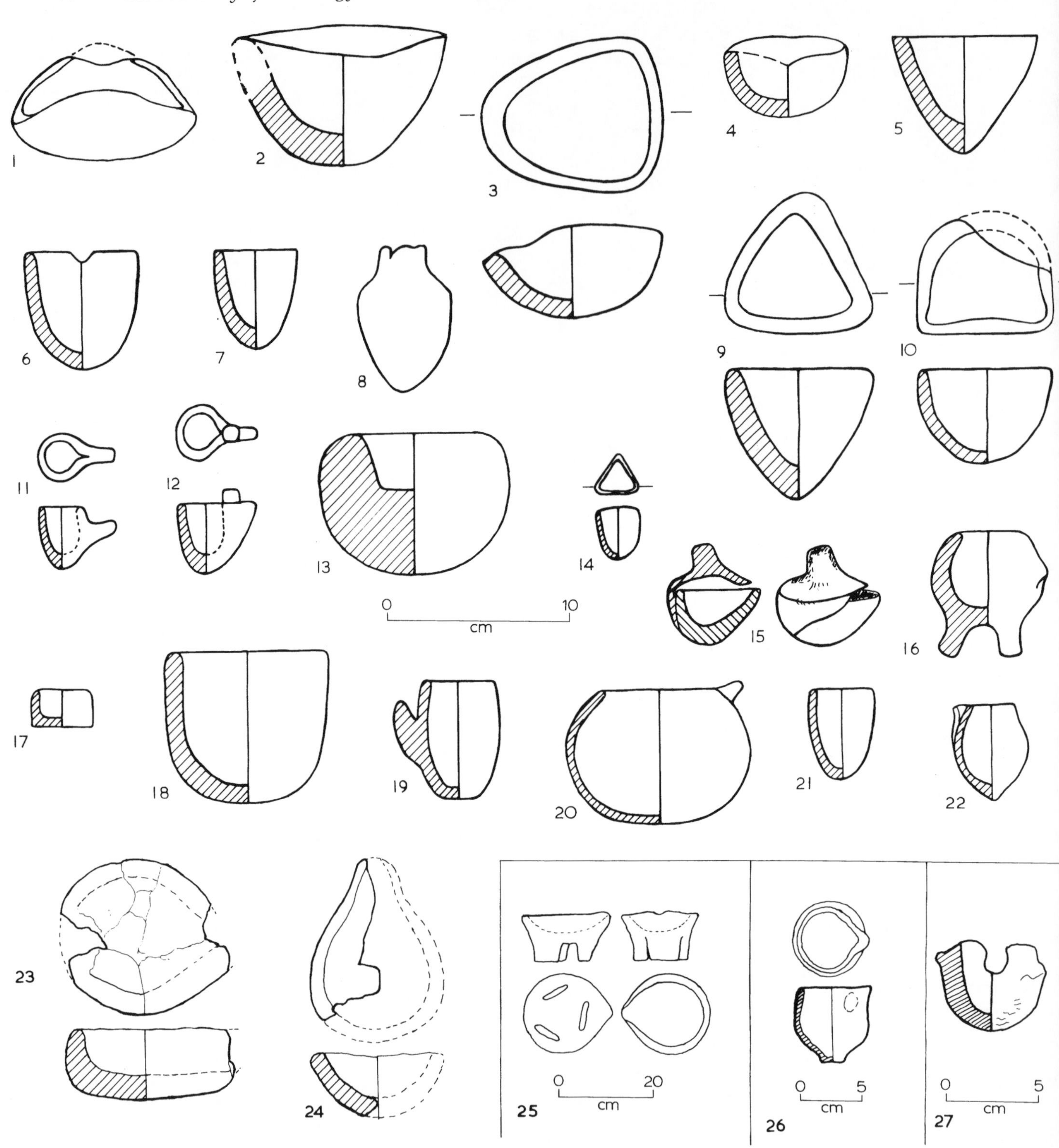

iron. The Belfast examples examined by Marshall[78] are all short-legged but it would appear that this has no significance and short-legged ones also occur in England and Wales.[79] The cast cauldrons were probably made by bell-founders and have been cast in a two-part mould with a central core and cored handles. Scottish cauldrons had a wall thickness of 0·8–2·5 mm and a hardness of 72–122 HV. They contained less than 11% Sn and were basically in the 'as cast' state with a variable degree of final cold work due to planishing. Marshall[78] connects their appearance with the increased exploitation of the Cornish ores by the Earls of Cornwall during the latter part of the Anglo-Saxon period. There is no doubt that they could only be afforded by the richer members of medieval and later families.

Stone continued to be used for small ingot moulds on the Viking site at Coppergate, York[80] and we have a Tudor coiners' mould of graphite.[81] One suspects that all types of moulding technique were being used from the Roman period onwards, apart from the more modern 'green sand' process.

Stack moulding, as described by Biringuccio[77], was in use in medieval times for making buckles, as we see from examples found at Much Park Street, Coventry[72] and Copthall Avenue, City of London.[73]

50 Crucibles
1 Glastonbury; shallow triangular; Type A1 (100 BC–AD 50)
2 Maiden Castle; 'sub-angular'; Type A1 (25 BC–AD 1)
3 Sutton Walls; D-shape; Type A3 (1st cent. BC–3rd cent. AD); contained a trace of bronze
4 Exeter; 'sub-angular'; Type A1 (Roman)
5 Great Casterton; conical; Type B3 (Roman)
6 Wilderspool; hemispherical; Type B1 (Roman)
7 Heronbridge; hemispherical; Type B1 (1st–2nd cent. AD)
8 Wakefield (?); necked; Type C1 (Roman)
9 Lough Faughan Crannog; deep triangular; Type A2 (Dark Age)
10 Nendrum; D-shape; Type A3 (6th–13th cent. AD)
11 Nendrum; pinched; Type D1
12 Corraneary, Co. Cavan; pinched, with knob; Type D2 (Early Christian)
13 Garranes; globular, stone or clay; Type B4 (5th–6th cent. AD)
14 Aghaloghan; hemispherical bottom, triangular top; Type A/B (Early Christian)
15 Dinas Powys; lidded; Type E1 (7th cent. AD)
16 Lagore Crannog; three-legged; Type B5 (7th–11th cent. AD); third leg is behind one shown on right
17 Lagore Crannog; flat bottomed; Type B2 (Dublin Reg. No. E.14.415)
18 Dunadd; hemispherical; Type B1 (6th–11th cent. AD)
19 Dunadd; hemispherical with handle; Type B1
20 Oxford; bag-shaped; Type F1 (Late Saxon)
21 Jarlshof; hemispherical; Type B1 (9th–10th cent. AD)
22 Wadsley; pointed bag-shaped; Type G1 (Medieval); Sheffield Mus. Reg. No. J.1928.14
23 Rathgall; Type J1 (Late Bronze Age)
24 Breiddin; Type J1 (Late Bronze Age; 1005–110 BC)
25 Dainton, Devon; Type J2 (Late Bronze Age)
26 Northampton; Type B6 (Roman); No. 292–3, Ref. 94
27 Wilderspool; side hole; Type B1A (courtesy John Hinchcliffe)

CRUCIBLES

Crucibles could either be used for collecting the products of smelting, or for melting metals for castings. Of the first method of use there is no evidence on British sites and it would seem that a clay-lined hollow below the fire was a more satisfactory way of collecting the products of smelting. Gowland[82] came to the conclusion that many of the early crucibles from abroad had been used for melting and had been placed below the fire because the external surfaces bore little or no trace of the action of heat, and they were thought too thick to be efficient for melting metal within a fire. However, an examination of the majority of British crucibles provides sufficient evidence for the external application of heat.[64, 83] Furthermore, Early Iron Age and Migration Period crucibles were relatively thin and could be efficiently heated within a fire. For example the thickness of the wall of the crucible found in Meare Lake Village, Somerset,[84] was 5·0 mm; one from Craigywarren[85] was as little as 2 mm at the top and 4 mm half-way down. A modern clay foundry crucible is never thinner than this, and the reason for the greater thickness of some of the more primitive crucibles is the poor quality of the clay.

In using a crucible for melting, some of the metal forms a slag with the material of the crucible, a typical composition being that found in the crucible from York[86, 87] (Table 57). The outside, which is only in contact with charcoal or charcoal-ash, does not slag easily although, for example, the crucible found at Dinas Powys[93] and other sites had vitrified to a limited extent owing to the fluxing action of wood ash. On pouring, some of the contents will drip down the outside, leading to local slagging on reheating the crucible for a further melt.

In some cases it would seem that the same hearth was used for smelting and melting. In this case it would be possible for the crucibles used for melting to pick-up on the outside traces of ore or slag from smelting. In this way traces of material on the outside of crucibles may give us a valuable indication of the ores used for smelting.

We now have crucibles from at least three Bronze Age sites (Table D, Appendix) and these are mostly of the large shallow, oval, type (Fig. 50). But the crucible from Dainton[3] has three legs, which reminds us of Schliemann's crucible from Troy III (No. 469).[95] Although this type is rare it reappears in an Early Christian context at Lagore. The problem of pouring the Bronze Age crucible was a difficult one and it is possible that the legs assisted tipping. They would also help to heat the underside, as pieces of charcoal could be pushed beneath, with difficulty.

The small, pinched crucible from Iron Age Cullykhan[96] is the first of a type that was to become very common in the Roman and post-Roman periods. The piece found in a Late Bronze Age level at Jarlshof, Shetland, has been described as 'boat-shaped' and may be a piece of a thick oval crucible.[97,98]

The large Early Iron Age crucibles are in the main triangular, although for the smaller sizes, cylindrical or globular types are found. The main type-sites for the triangular crucibles are in the lake villages of Meare[84] and Glastonbury,[99] and at Gussage All Saints, Dorset.[89] The Glastonbury triangular crucibles are all fragmentary, with no complete examples. All of them, however, seem to be shallow (Fig. 50, 1) compared with the rather deeper ones from the neighbouring site at Meare. This would lead to high heat losses and would appear to necessitate a lid, although fragments of lids have not been reported from these sites. Maiden Castle has yielded a curious assortment. One described by Wheeler[100] as sub-angular is of the Meare type (Fig. 50, 2); but the large saucer-shaped dish would hardly have been considered a crucible unless bronze had been found in it. It certainly has none of the usual attributes of a crucible, since it has a large surface area and a thickness in places as great as 3·8 cm. It is very probable that the bronze deposit is accidental. Another of circular shape (9·5 cm diameter) came from Belgo-Roman levels and possibly shows Roman influence.

The D-shaped crucible from Sutton Walls[101] (Fig. 50, 3), which is either pre-Roman or Roman, is a natural variant of the triangular type. It would be poured by means of one of the sharper corners.

With few exceptions the Roman period crucibles are circular, the exception being that from Exeter,[102] which, lying on the edge of the Roman province, was probably more influenced by native types. For some reason the Roman world seems not to have taken to

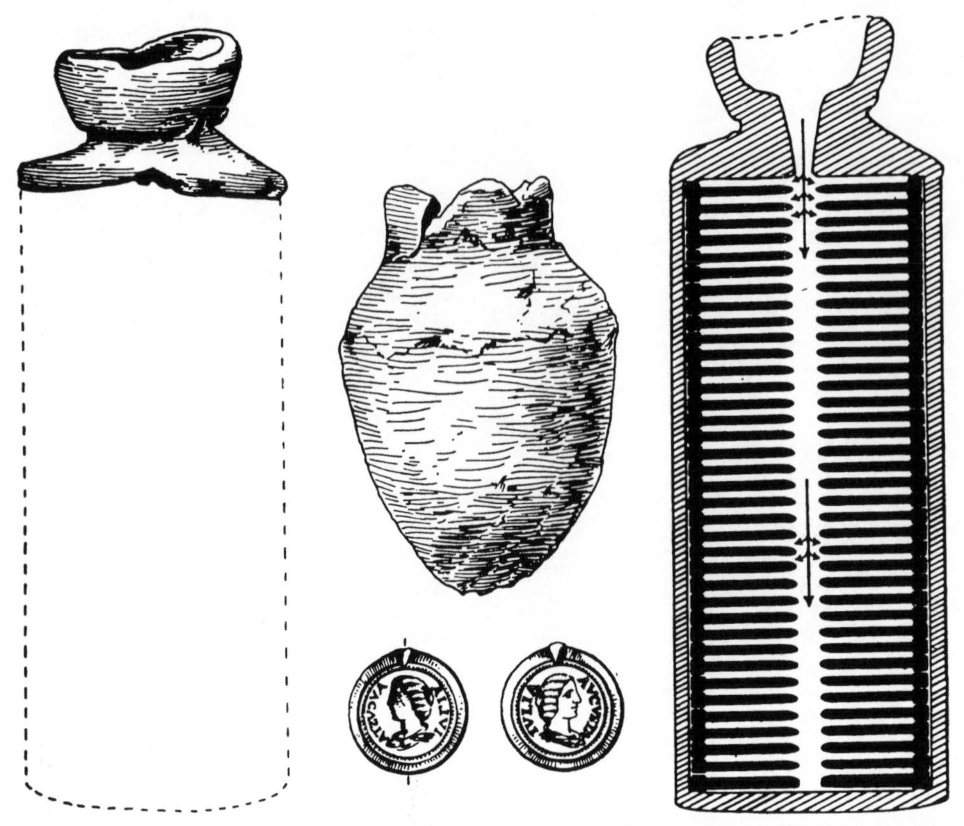

51 Crucible and coin mould from near Wakefield, Yorks (by courtesy of the Trustees of the British Museum[103])

the more easily held and poured triangular crucible. While most of these found are hemispherical or conical (Fig. 50, 5 and 6) we have two examples of the narrow-necked type (Fig. 51). This type has the advantage of not needing a lid, but would be harder to make and the neck would tend to slag up after several heats. The specimen shown in Fig. 51 is thought to have been found near Wakefield and is now in the British Museum.[103] Artis[104] shows metal being poured from a similar crucible into coin moulds of the same type as those shown in Fig. 51, and it is possible that he got the idea for the crucible and mould shown in his drawing from these.

Among the round crucibles found at Wilderspool is one with a side hole (Fig. 50, 27), not unlike some from the German site of Nida-Heddernheim.[105] The advantages of this type were well-known in the Eastern Mediterranean in the Bronze Age. During pouring, the slag is held back and clean metal can be passed through the side hole which can be plugged during melting.[106]

It is clear that the native sites continued to use the triangular type during the Roman period, except for the very small sizes, which were circular and in the main used for jewellery. The triangular type persisted in Scotland and Ireland during early Christian times, and only in Saxon England does it seem to have been superseded by the bag-shaped type[107] (Fig. 50, 20).

Nendrum, Co. Down, has produced another D-shaped specimen (Fig. 50, 10), and two very interesting examples of the 'pinched' type (Fig. 50, 11), which are made by pinching a larger crucible before firing in order to produce a handle. One variant of this is that from Corraneary Crannog, Co. Cavan, which has a square knob on top (Fig. 50, 12). These knobs or projections allow the crucible to be easily held with iron tongs and poured. The position of the knob on the Corraneary crucible makes pouring easier than the horizontal handles on the Nendrum crucibles; similar crucibles were also found at Dunadd, Strathclyde.[108] Nendrum also produced a knob which appears to be that of a crucible lid. A large number of knobs have been found on other sites and it would appear that some of these, at least, belong to lids.

The crucibles' heat loss problem was in part solved by the integral-lidded crucibles found at Dinas Powys, Glamorgan[93] (Fig. 50, 15), and at Dunadd. While the Dinas Powys lids were moulded around an already formed hemispherical crucible, that at Dunadd was moulded onto the rim of the bowl to give a similar result.

Towards the end of the 1st millennium AD, the large circular crucible seems to have made its appearance in Scottish and Irish native sites. This is clearly shown by the examples from Dunadd,[109] one of which has a handle (Fig. 50, 19). The three-legged crucible from Lagore Crannog (Fig. 50, 16) is unique in this context (but see above).[110] It would appear that its maker was about to form a handle (now to be seen in embryo), when he realized that one of the legs would do as well.

Viking levels at Jarlshof[98] (AD 800–1000) have yielded a lengthened hemispherical crucible without handles, not unlike the Roman type (Fig. 50, 21). Saxon levels at Oxford[107] have produced the characteristic bag-shape with lugs, the largest of which is shown in Fig. 50, 20. Those from the Danelaw site at Flaxengate, Lincoln are very similar.[111] Less baggy types are known from Coppergate, York.[111] Both those from the last two sites are of one fabric, Stamford-type ware, which is also used in a wide range of domestic

Table 58 A typology for clay crucibles

Type	Shape	Period	Examples
A1	Triangular; shallow	EIA	Glastonbury
2	Triangular; deep	EIA EC	Meare Garranes Lough Faughan Crannog
3	D-shaped	EC	Nendrum
B1	Hemispherical	Roman Roman SIA	Heronbridge Templeborough Dundadd
1A	Hemispherical + side hole	Roman	Wilderspool
2	Cylindrical (flat bottom)	EC	Lagore
3	Conical	Roman	Lullingstone Great Casterton
4	Globular	EIA EC	Glastonbury Garranes
5	Globular + 3 legs	EC	Lagore
6	Pedestal	Roman	Northampton
C1	Necked	Roman	Wakefield Castor (Durobrivae)
D1	Side-lugged	EC	Nendrum
2	Top-lugged	EC	Corraneary
E1	Integral-lidded	EC	Dinas Powys
F1	Bag-shaped	Saxon	Oxford
G1	Pointed bag-shaped	Roman	Gestingthorpe
		Medieval	Wadsley
H1	Truncated cone	Medieval Modern	Northampton
J1	Oval	Bronze Age	Rathgall and Breiddin
2	Oval + 3 legs	Bronze Age	Dainton
3	Oval; flat bottomed	Bronze Age	

pottery. They were shaped on a wheel and hand finished from a white-firing fabric with fine quartz temper. The wall thickness was only 2–4 mm but the crucibles were often coated on the outside with some less refractory clay which served to increase the mass and the resistance to thermal shock.[112]

The crucible with the pointed base (Fig. 50, 22) is a medieval example from Wadsley, Yorks. of a Roman type. A tentative typology is given in Table 58, and dimensions are given in Table D of the Appendix.

It is extremely difficult to reconcile the sizes of crucibles (20–100 cm³) with the size of some of the artifacts found. The small flat axes of the Early Bronze Age weigh about 100 g and could be cast from a crucible with a maximum capacity of 20 cm³. But the palstaves weigh from 0·34 kg to 0·7 kg and the larger size would need a 100 cm³ crucible, of which very few have been found. The flat axe from Lawhead, in the National Museum of Scotland, weighing about 2·7 kg, would have needed a crucible with a capacity of 350 cm³ to cast it. Equally difficult is the question of the large pieces of sheet incorporated into the Late Bronze Age buckets and cauldrons, some of which must weigh 2·3 kg. These would need a crucible with a capacity of 300 cm³. It would seem that either much bigger crucibles than have yet been found have been used, or that some of the larger pieces of metal are the result of the simultaneous pouring of several crucibles.

The method of holding crucibles in early times is supposed to have been by means of green withies, which would not carbonize in the short time at high temperatures owing to the boiling off of the sap within them. This method can be seen in use in the many reproductions of the drawings in the tomb of Rekh-mi-re dated to the 15th century BC.[114] The method is undoubtedly possible provided that the hands are protected by wet rags or held at a distance greater than about 15 cm. The heat radiated from a crucible containing bronze at 1 000°C is such that one's hands are soon burnt if held any closer.

Copper or bronze tongs were found in the Late Bronze Age cave at Heathery Burn,[115] and iron tongs on the Early Iron Age site at Llyn Cerrig Bach,[116] and on other sites, and so there would be no need for such a primitive method as the use of green withies by the Late Bronze Age. The Glastonbury type of triangular crucible could have been held by gripping the edge with iron tongs (Fig. 52). It would only be possible for the bronze tongs from Heathery Burn to have been used for holding crucibles if the ends were tipped with clay envelopes. These would prevent the heat from melting the bronze for long enough to enable the crucible to be poured.

So it may be assumed that by the time of the Late Bronze Age tongs would normally be used for holding crucibles. Even so, holding them by the rim would be a clumsy method for some of the small spherical crucibles of the post-Roman era. Here we notice an attempt to build onto the crucible a handle or projection which could be gripped by a pair of tongs (*see* Fig. 50, 19 from Dunadd). Later, no doubt, the method in use today of putting a slightly tapered crucible into a ring fastened to the end of a rod would be used (Fig. 52). The bag-shaped medieval crucible could be used in such a manner, but it would not be ideal. Crucibles for use with this method would have to be made to a fairly close dimensional specification. Yet the large crucible from Dunadd looks as though it was designed for just such a method. Perhaps some of the inexplicable pieces of iron found on Roman and Migration sites have been used for this purpose.

The contents of a crucible melt at about 1 000°C, when removed from the furnace would very soon solidify owing to radiation of heat. While this effect is not too marked in the case of a modern crucible holding 14 kg of bronze, pouring has to be done without too much delay. In the case of crucibles holding as little as 50 cm³ the effect would be very marked and therefore necessitate pouring the contents in a matter of seconds. Indeed one wonders how it was done at all. It would appear that the founder only melted enough for

top ring-shanks gripping modern crucible, height 22 cm *bottom* simple tongs gripping triangular crucible
52 Methods of holding crucibles

one artifact, and super-heated it sufficiently to enable him to pour it before it solidified.

One way in which the cooling process can be delayed is to use a lid. Unfortunately lids will not stay on during pouring unless stuck on with slag, in which case they are unlikely to be removed without breaking. Alcock's discoveries at Dinas Powys[93] show one way in which the 7th century founder got over this problem (*see* Fig. 50, 15). His lid was part of the crucible, fixed on with clay after the first firing but leaving a narrow slit through which he could charge small pieces of solid metal and pour liquid metal. Furthermore, he would be able to hold the crucible with tongs by means of the lid, which has a handle almost exactly the same shape as that on a modern crucible lid. Now that we have certain evidence for the use of lids in the 7th century one wonders how often these were used in earlier periods, but so damaged as to be unrecognizable by the excavators.

Copper slags often look like red enamel and this has led many to assume that some of the crucibles have been used for enamelling. It is certain that their red colour gave early people the idea of enamelling and some crucibles have clearly been used for melting red enamel. That such material was made in large quantities is shown by the finding of a 15 cm diameter ball of red enamel at Tara, Co. Meath.[117] Pieces were broken off this lump and either cut to shape or placed in a prepared cavity and melted into shape. Such red enamels contain copper and lead silicates and can easily be confused with copper melting slags.

A considerable number of crucibles and crucible fragments have been reported from Irish sites. All these crucibles are small, showing that small amounts of metal were remelted and cast into individual items. Two triangular crucibles were found at Craigywarren Crannog:[83, 85] one 3·8 cm high and 4·76 cm across the mouth, and the other 5·4 cm high. This latter crucible had a capacity of 40–50 cm^3. On the 6th century site at Garranes, Cork,[64] a large number of crucibles and fragments were found, of the usual triangular and spherical shapes. The triangular ones all seemed to have been used for copper and bronze, and had capacities between 20 and 100 cm^3. Some spherical stone crucibles were found, one measuring 10 cm across (Fig. 50, 13), but these were thought to be too thick for melting copper or bronze and it is possible that they have been used for glass or enamel. A stone crucible was also found at Nendrum.[113]

Crucible slags

Early crucible slags were formed by the reaction between the clay of the crucible and the ash of the fuel and are therefore more likely to be sodium or potassium aluminium silicates rather than cuprous silicate which has a high formation temperature and which does not form easily. The latter, if formed at all seems to be amorphous and thus escapes phase analysis. But most crucible slags contain iron which can be introduced in three ways:

(i) as smelting slag entrapped in the metal, i.e. as the ferrous silicate, $2FeO.SiO_2$
(ii) as metallic iron, present as a second phase in the original metal
(iii) iron introduced by the use of unprotected iron tools.

The second is more common than was hitherto realized, as raw copper can often contain very high iron contents.

The sulphur present in some slags (Table 59) arises, no doubt, from the oxidation of the residual sulphide or matte phase from the smelting process. Phases detected by X-ray diffraction are iron oxides and double oxides such as the spinel type phase $CuFe_2O_4$, or the oxide delafossite ($CuFeO_2$). It seems that these

Table 59 Crucibles with analysed deposits

Ref.	Site	Date AD	Museum Register No.*	Capacity, cm^3	Shape	Deposits
83	Craigywarren Crannog		1902–76	40–50	Δ	Analysis given in Table 57
83	Craigywarren Crannog		1902–77	19	Δ	External slag + Cu
83	Moylarg, Co. Antrim	9–16th cent.	1905–224	70	O–Δ	Cu–Cu_2S
83	Moylarg		1905–234	30–40	Δ	Cu–Cu_2S
83	Toneymore, Co. Monag		F.235	33	Δ	Cu
83	Co. Cavan		191	9	Δ–O	Cu–Cu_2S
83	Co. Cavan		192	10	Δ	Slag on bottom: 23·4% Cu, 45% SiO_2, 9·7% S, 7·7% Fe_2O_3
83	Ballyjamesduff, Co. Cavan		1682	33	Δ	Cu. No Slag
83	Dunshaughlin, Co. Meath		1881 E	33	Δ–O	Cu, Sn, Pb. Slag: 28% Cu, 5% S, 4% CaO, rest SiO_2
110	Dunshaughlin		468.124	—	Δ	85% Cu, 11·6% Sn
83	Lough Rea, Galway		L.R.I 18 15	12	O	Slag: 20% Cu, 20% SiO_2, 23% FeO, 10% S, 4·8% Zn. Bead: 15·52% Zn, 3·1% Sn, 1·0% Ni, rest Cu
83	Ex R. Soc. Antiq. Irel. (find spot not known)		158	93	O	Contents: Cu. Slag: 10% Cu_2O
83	Lagore Crannog, Co. Meath	7–11th cent.	378	30–40	Δ	Contents: 4·2% Sn, 2·2% Ni, 1·3% Fe, rest Cu
110	Lagore Crannog	7–11th cent.	205	—	Δ	Inside Cu_2S, outside Cu_2S + Fe + SiO_2
110	Lagore Crannog		1634	—	Δ	Inside FeS_2 + tr. Cu_2S
110	Lagore Crannog		582	—	Δ	Cu + FeS_2; $CuCO_3$, Cu + Fe(tr.) + S(tr.)
110	Lagore Crannog		263	—	shallow	FeS_2, SiO_2 + Cu(tr.)
110	Lagore Crannog		232	—	shallow	Vitreous SiO_2 + Fe + Cu + S (? reduced pyrites)
110	Lagore Crannog		32	—	shallow	SiO_2 + Fe; no Cu
89	Lagore Crannog		42	—	bag-shape	Cu_2S + $CuCO_3$ + Fe(tr.)
88	Wilderspool, Lancs.	Roman	—		O	Contents: 88·7% Cu, 9·73% Sn, 1·53% Zn; for slag see Table 57

*For specimens in the National Museum of Ireland, Dublin.

slags are essentially alkali silicates or alkali aluminium silicates with some iron and copper. But the main components that give the reddish colour to relatively unoxidized examples of these slags are metallic copper and Cu_2O which are not in solution. These red glassy slags resemble ruby glasses in their composition.

Modern crucible slags are not very different, although they may often contain elements introduced from cover fluxes and other compounds added to protect the melt during melting down. Slags so formed by melting a leaded gunmetal (85% Cu, 5% Sn, 5% Zn, and 5% Pb) in a graphited clay crucible were found to contain major amounts of Zn, Cu, Mn, Fe, Pb, and Sn. Trace amounts of barium were also found. The only crystalline phases were alpha-quartz, cassiterite, and willemite (Zn_2SiO_4). The matrix was an amorphous glass.

Moss[83] examined a large number of crucibles from Irish sites with their deposits, and I have attempted to summarize his results in Table 59. The majority of these crucibles are triangular, which facilitates pouring. Some are triangular at the top and circular towards the bottom. Only three out of the thirteen examined were wholly circular. The capacity varied from 9 to 93 cm^3.

In some cases the slags show contamination with wood ash and sulphides, and one is forced to conclude that the crucibles had been used for melting after smelting in the same hearth and that the outside had come into contact with slags and ore remaining after smelting. The external deposits and internal contents show that sulphide ores were smelted, and a wide range of copper-base alloys were produced, including simple tin bronzes, leaded bronzes, and brasses.

While copper is found in 15 Irish counties, principally Cork, Waterford, Wicklow, and Tipperary, two-thirds of the crucibles came from Antrim, Monaghan, and Cavan where there are no recorded economic deposits of copper ores. Copper ores are however, found in the neighbouring counties of Down and Tyrone. The sites in Meath are about 20 km from the nearest known deposits at Beauparc and Brownstown.

At Kirkstall Abbey, Yorks., (*c.* AD 1200), 'crucible' residues have been found. These residues have been analysed by Haynes,[91] and are much the same as those found on the Irish crucibles and seem to be products of either smelting or melting. One was a bright green powder, looking rather like malachite but containing charcoal and entrapped metal particles. Analysis of the powder showed it to contain copper and tin oxides and when the powder residue was reduced it produced a metal containing 92·5% Cu, 5·96% Sn, and 0·49% Pb. A further residue was almost certainly a crucible slag; its composition is given in Table 57.

The fabric of the crucibles shows a general improvement with the period. The piece in the Late Bronze Age levels at Jarlshof was coarse and thick.[98] The clay used for the small triangular crucibles from Aghaloghan, Antrim, was very fine and light grey. The cylindrical crucibles from Lagore were two-layered, the thin layer being of very fine clay while the outer was thick and coarse. A similar structure was observed in the triangular crucibles from Buston, Strathclyde.[118]

The composition reflects this improvement (Table 57). The clay used for the Meare[84] crucibles can hardly be described as a fire-clay since it only contains 13·1% Al_2O_3 and has a good deal of fluxing material. On the other hand, the Roman or medieval crucible from York[86] is made of a high-class fire-clay (30·6% Al_2O_3) with very little alkali, and consequently has a very high softening temperature (1 500–1 600°C) which would enable it to be used up to temperatures of 1 400°C.

METAL FORMING AND PLATING PROCESSES

The technical methods of smiths are a very important part of our subject since there is a limit to what can be achieved by casting, and a good many articles have to be assembled out of various pieces made in different ways.

The Late Bronze Age and Early Iron Age smiths were capable of producing surprisingly large pieces of sheet metal. The maximum size of casting must have been about 320 cm^3 but only in the most ductile metals was it possible to beat cold cast metal into thin sheet. This, of course, could easily be done with gold and the thickness of the lunula from Tullynasole (Belfast Museum) is only 0·15 mm. This would be beaten by means of a bronze hammer, which would consist of a square-shaped bronze casting about 8 cm long provided with a handle like a socketed axe. A bronze anvil from Sligo, also in the Belfast Museum, measures about 8 × 8 × 5 cm and has a small piece of gold hammered into its working surface.

Coffey[119] shows examples of socketed bronze hammers, the largest of which was found at Abbey Shrule, Co. Longford. He also shows a small bronze anvil with a beak and a pointed base which was intended to be inserted into a piece of wood and which is one of the few specimens of Bronze Age anvils found in the British Isles. Another with its built-in swage-block comes from Sutherland [Highland]. Such a pattern is not uncommon amongst jewellers today (Fig. 53).

In order to make wire,[121] narrow strips were cut from sheet, and these were made round by twisting and then rotating them between two flat surfaces of bronze or stone, rather like making butter pats today. The wire had first to be annealed, i.e. softened, by heating it above its recrystallization temperature, which varies with the metal or alloy used. For pure copper and silver, 200°C would suffice, but for more complex alloys such as low-tin bronzes, 650°C would be required.

By the Middle Bronze Age it would appear that gold wire of semi-circular section was being made by swaging using the swage blocks evident on the Bronze Age anvils (Fig. 53). Swaging is the process of hammering progressively into a prepared groove of the right size and shape, and with a pair of swages or with the addition of a grooved hammer it is possible to make circular wire.[122, 123] The anvil from Sutherland was found[120] to contain 26·4% Sn and should have a hardness of 360 HV (Fig. 53).

The next stage in the wire-making sequence is drawing, by which a larger and often rougher wire can be thinned down by pulling through a smaller hole. There is, as yet, no evidence for this until the Migration period in Jutland,[124] if we ignore the small triangular bronze plates found in the Isleham hoard.[125] The draw plates from Jutland were made of steel with hardness reaching 900 HV, and could be used for making iron wire.

The fact that rivet holes are often not 'cored-in' by suitable mould design means that some method had to be found for drilling them. Examination of a rivet hole from a sword hilt from the Gilmonby hoard showed signs of longitudinal drifting rather than rotation, which suggested that a hard pin or punch had been hammered through to open up the hole which may have been started by drilling.[126] Allen believed that a hollow tube drill fed with sand was used to drill holes in arsenical copper halberds.[127] This process has been discussed at length by Coghlan[1] who also claims that with a well work-hardened bronze punch it would be possible to make holes through thin sheet metal such as cauldrons but not the thicker sword hilts. Holmes[128] suggests that pointed 'bits' of flint were used for the Irish trumpets as the holes are so uniform in size.

Rivets were comparatively easy to make. However, working down to small sizes could result in the loss of easily volatilized elements such as arsenic, as may be seen when the compositional relationship between blades and rivets is examined.[129] On the whole this would have been an advantage as it is better to attach an organic hilt plate with a softer rivet than a hard one. The use of a bronze rivet in the arsenic blade from Dorchester is an anachronism which could either indicate the use of a small piece of metal that the smith happened to have handy or the re-riveting of the weapon at a later date.[130]

Dividers or compasses were used, but none have survived from the Bronze Age. Saws can be made of notched flint but Coghlan found these useless for wood.[1] Bronze saws, such as those found at Grimes Graves and Felixstowe[131] were used in the Bronze Age, almost certainly for organic materials, as their hardness is unlikely to be greater than 200 HV. In fact

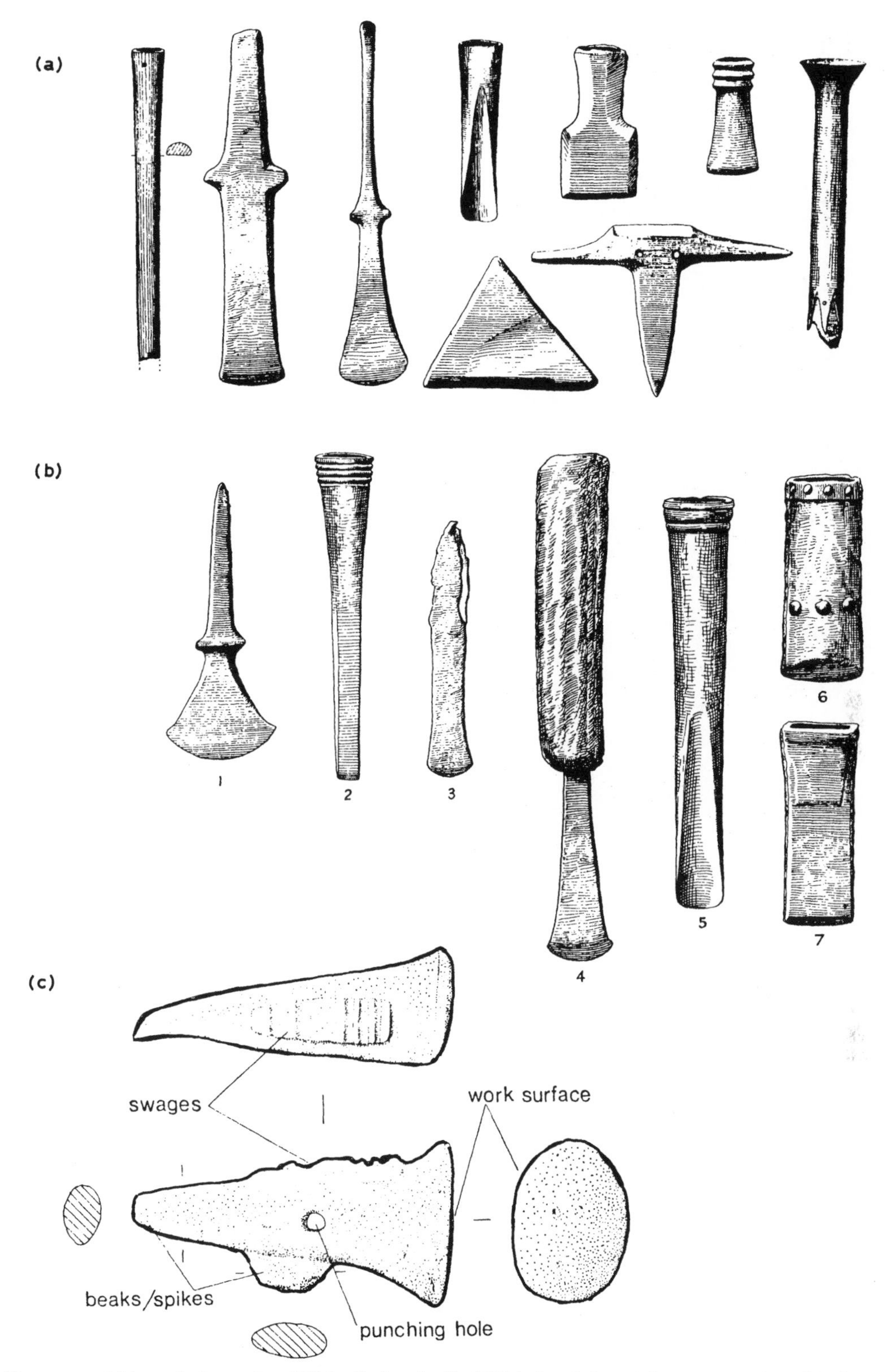

a set of bronzesmith's tools from Co. Offaly, Ireland, all of LBA date *left to right* two trunnion chisels, socketed gouge, two hammers, anvil *lower centre* polishing stone
b *1* tanged splayed chisel from the Thames (8·6 cm long); *2* socketed mortise chisel from the Thames (10 cm long); *3* chisel from Irthington, Cumbria (8·1 cm long); *4* chisel in antler handle from the site of the Tate Gallery, London (16·3 cm long); *5* socketed gouge from the Thames at Richmond (13 cm long); *6, 7* socketed hammers (both 6·35 cm long)
c anvil with broken spike and swages, from Kyle of Oykel, Highland (half actual size)

53 Examples of bronze smithing tools (*a* and *b* courtesy Trustees of the British Museum; *c* anvil,[120] courtesy of Society of Antiquaries, London)

the saw from Felixstowe which was made from a Cu–Pb alloy had a hardness of only 95 HV. Another saw, from Burgess Meadows, was made of a 10% tin bronze and had a hardness of 170 HV.

The bronze chisels and gouges were no harder than expected. Gouges from the Yattendon hoard[132] had edge hardnesses of 145 and 172 HV. The tanged chisels had lower hardnesses but were badly corroded.

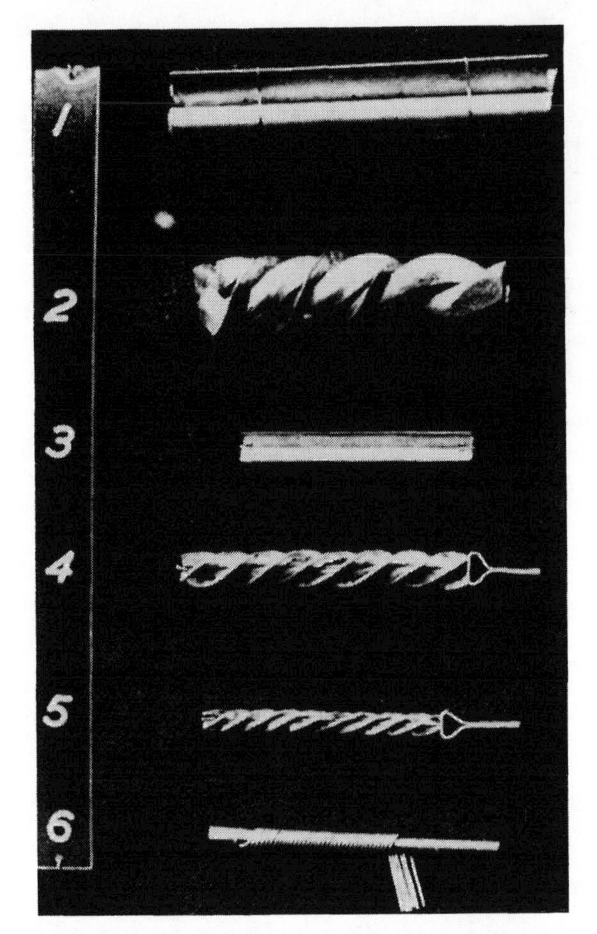

1 to twist a thin strip of gold without breaking, it must be placed between two plain bars and bound in position *2* the whole is then twisted and the ribbon of gold, now torc-shaped is unscrewed from between the bars *3* when a cruciform shape is forged from the solid *4* this may be twisted without danger of tearing; *5* plain rod with lapping of six wires to illustrate manufacture of necktorc from near Aughrim, Co. Galway

54 Methods of constructing torcs (photo: National Museum of Ireland)

Iron or steel saws appear in the Roman period but they are barely hard enough for cutting iron. Some found at Wanborough, Wilts., had a maximum hardness of 470 HV (1 kg).[133] The same site produced a file not unlike that found at the 1st century BC/1st century AD site of Waltham Abbey.[134] This had a surface hardness of 535 HV (1 kg).

The technique of making tubes was well developed. A gold torc from Broighter was made of two tubes hinged at the back. The tubes were made of sheet bent round a rod and soldered along the edges. They were then filled with sand, pitch, or resin and bent. If the radius of the bend was small, i.e. about a centimetre, sometimes small corrugations were formed on the inside, as in the gold tubular penannular rings from Late Bronze Age sites which have been called 'ring money'.[119] These are probably dress fasteners since many have complicated terminals like torcs, and are far too elaborate to be intended primarily for currency. But the annular 'crescentic ornament' used for the hilt of the Whittingham sword is clearly a bronze casting and almost certainly a 'lost wax' one[135] (Fig. 67 below).

Torcs were often made from twisted gold rod or ribbons. While being twisted, the rods or ribbons were

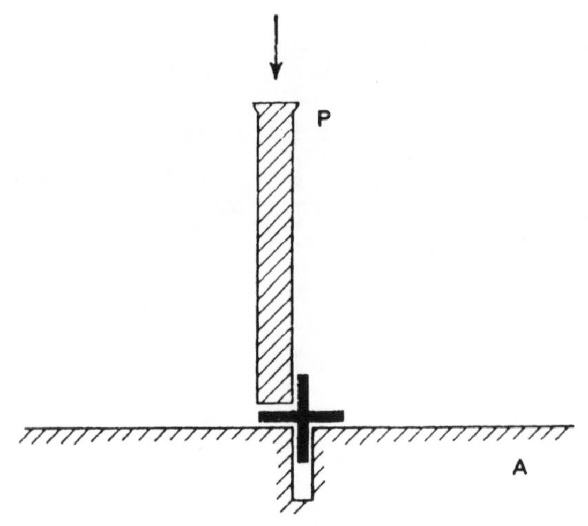

55 Method of making a gold torc of cruciform type (after Taylor and Shell[136])

supported on each side with rods or ropes to maintain their relative positions (Fig. 54). A gold torc from Bracks Farm[136] was made in a most interesting manner by forging a solid bar along its length, by grooving with a chisel and then finishing with a square punch and slotted anvil (Fig. 55). After forging-in the cruciform section, the torc was twisted as shown in Fig. 54. The terminals could be forged from residual metal left at the ends, or cast or soldered on. The Yeovil torc was 43 cm long and shows the skill attained by European goldsmiths in the Late Bronze Age[137] (torc sections and other references are given in Fig. 56).

A number of tubular gold-alloy 'torcs' from Snettisham[139] had two iron cores rectangular in section 10 × 6 mm) which terminated in a neat fastener. The space between the iron cores and the gold tube was filled with wax. The Broighter 'torc' also has a neat 'bayonet-type' fastener but no iron core.[140]

The Snettisham hoard was followed by the finding

No.	Section	Examples
(A) Solid-hammered or swaged (MBA or LBA)		
1		Llanwrthwl
2		Towednack
3		Llanwrthwl
4		Heyope, Powys
5		Brack, Cambs. Castlemount, Kent Boynton, Suffolk Tara, Meath
6		Haxey, Lincs. Kilmutt, Mayo
(B) Tubular (EIA)		
1	iron core; filling of wax or sand	Snettisham Broighter, Londonderry

56 Methods used in making gold torcs (after Taylor,[136] Maryon,[121] and Hawkes[138])

one-tenth actual size

57 Irish cast bronze trumpets (horns) of the Late Bronze Age (*c*. 800 BC): six from Clogherclemin and one from Derrykane, both in Co. Kerry; one from Dromabest, Co. Antrim; and two unprovenanced (photo: National Museum of Ireland)

in 1968–70 of six torcs near Ipswich, and these finds inspired a considerable amount of technological work on the methods of making the rod and wire, and casting-on the terminals.[141] Further torcs were found at Snettisham between 1964 and 1968.[142] Striations on one of these have been put down to drawing through a stone die but it is more likely that these are due to swaging or even drawing through a steel die as stone tends to split owing to the high pressures applied in drawing such hardenable, highly alloyed gold.

Thin gold foil was often formed into hollow objects, such as the heart-shaped 'bullae', which were then filled with lead or tin.[143]

One of the most accomplished articles metallurgically is the Irish trumpet (Fig. 57). This has been described by Coffey[119] and exists in three types:

1 In the shape of a horn open at both ends, having mouthpiece and trumpet cast in one piece.

2 Similar in shape to the above, but closed at the narrow end; cast, with an aperture for the mouth in the side near the closed end. Both of these are predominantly Late Bronze Age types.

3 Also horn-shaped, with a long straight tube attached to the narrow end of the curved portion; both parts made from sheet metal. In some cases the end of the tube has four rivet holes, which may have been used for attaching a cast mouthpiece. This type resembles the Roman *litunus*, and all types were probably derived from the South since they are similar to those of Gaul. These are predominantly Iron Age.

The Portglenone trumpet, now in the museum at Belfast, is an example of type 2. The closed end still contains its clay–sand core. The thickness of the cast metal is about 1·3 mm and the ends of the chaplets that served to position the core in the mould are still visible. They are spaced at 4–5 cm intervals over the whole internal surface of the casting.

Coles [144] has proposed a typology for the BA horns which recognizes their slightly different provenances within Ireland and their decoration. Holmes[128] has investigated and reproduced their manufacturing technology. This involves the production of a mould (cope and drag) and a core, and the provision of 'prints' in the moulds, and chaplets to locate the core accurately. It was not a case of a lost-wax process but approximates to modern techniques of sand moulding. A pattern would be made in wood or clay resembling the outside of an existing trumpet and a core of sandy clay would have the dimensions of the inside. All this follows directly from normal Bronze Age casting techniques, as used on socketed axes for example. But such a large core as was used in these horns would have to be made oversize to compensate for shrinkage in drying. The technique used for the design and placing of the chaplets has been carefully studied by Holmes and shows a very high level of foundry technology.

Various rings and mounts have been 'cast-in' by first making them in sheet metal, making a hole in the horn wall and in the core and running molten bronze to 'braze' the horn and the mount together. Techniques similar to that shown in Fig. 65 (below) were probably used.

Examination of all the single-piece cast horns shown in Fig. 57 shows that the apertures for blowing were all on the same side, in such a position as to have the wide opening to the left.

The Iron Age horn from Ardbrin Crannog, near Annaclone, Co. Down, is unique and was formed from two pieces of wrought bronze sheet, each piece about 1·2 m long.[121] The tubes were first formed from bent pieces of annealed bronze sheet about 1·3 mm thick riveted to 1 cm wide strips (Fig. 58); 658 rivets were needed to make this horn! After being riveted into the shape shown in Fig. 58 using an iron mandrel, it was filled with sand and finally bent into the shape required. Local annealing would probably be necessary on the outer surface of the radius to facilitate the stretching and thinning which would take place.

English and Scottish examples are rare. The Caprington horn from Ayrshire [Strathclyde][145] consists of a lead-free tin bronze casting. A trumpet found in the River Witham, Lincs., was nearly straight, curving upwards to an irregularly shaped expanding mouth.[145] The tube was formed from hammered 12% tin bronze sheet and had been soldered with tin. This specimen probably belonged to a late period of the Early Iron Age.

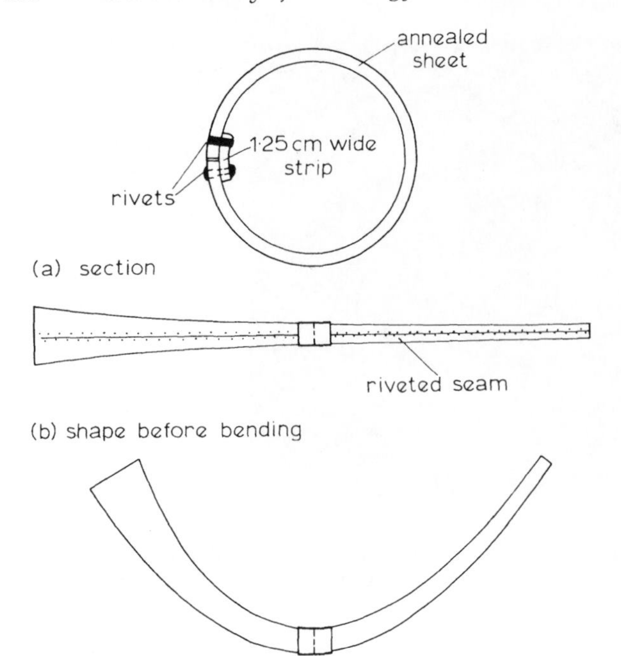

58 Construction of the Ardbrin horn (not to scale)

Childe[146] has said that 'all the main species of manual tools employed in handicrafts and husbandry today [1944] were already in vogue by the beginning of our era'; this is true of tools with the possible exception of scissors and screws. In the case of techniques there seems to be little left to discover but wire-drawing and rolling.

It is now proposed to deal with some of the major metal-working techniques in rather more detail.

BEATING AND TURNING

The need for buckets, cauldrons, and bowls to replace the skin and pottery vessels previously used for cooking, water storage, and carrying, spurred on the Late Bronze Age bronze workers to supply the demand as their technique improved.

One of the prototypes of the British Bronze Age bucket is probably that found at Nannau, in Wales.[147,148] This is believed to be an import adapted from a Kurd design which had a distribution embracing central and north-west Europe and Italy, and which belongs to the first half of the 1st millennium BC. This bucket is made of three sheets, the two sheets forming the upper portion being riveted together along two vertical seams, which were then riveted into the bottom sheet, which has been formed as shown in Fig. 59. This latter must have been a very complex forming operation. The base was reinforced with four sheet-metal base-plates and a foot ring, and the staples for supporting the handles were also made of sheet-metal. The rivets were 4·6 mm in diameter, with 9·2 mm diameter flat heads on the inside of the horizontal seam and on the outside of the vertical side seams.

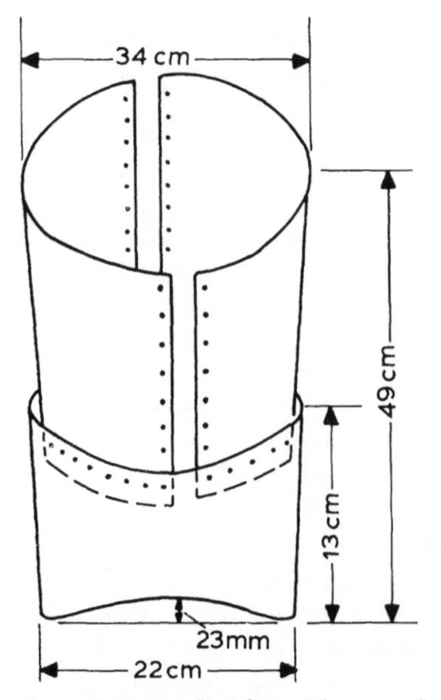

59 Construction of the bucket from Nannau, Wales

The lower part was raised by means of horn or wooden tools since there are considerable traces of 'pleating' or slight gathering of the sheet metal. The rim is curved inwards and reinforced with a bronze wire as in many examples of modern sheet-metal work.

The Whigsborough or 'Dowris' (Ireland) bucket is modelled on this design, but the foot ring is reinforced by a set of six sheet-metal angle plates, positioned radially. The original rivets are hammered flat on the outside and have conical heads on the inside. The handles are missing but would probably have been cast ring handles like those on the Nannau bucket. Hawkes[148] thinks that this represents the first large-scale sheet-metal work in Ireland, dated 750–700 BC, and was made by Irish smiths who had learned the process on the Continent.

The next stage is represented by the base-plates found in a founder's hoard from Bagmoor, Lincs.[149] These are cast bronze and represent the only surviving parts of a Late Bronze Age bucket, probably of the Whigsborough–Nannau type, but with six cast bronze segmental base-plates instead of six sheet-metal plates as used in the Whigsborough example. The Bagmoor hoard is dated to the 7th century BC.

The final stage is shown by the Heathery Burn bucket,[115,150] which differs in two respects from the previous examples. First, the base-plate is made from a single wheel-shaped bronze casting and the handle staples are also cast. The staples are cast in two parts; the outer part consists of an external vertical strap which lies on the neck and shoulder. The inner strap and the rest of the staple were cast and joined to the outer strap probably by 'burning-on' through a hole cut in the rim of the bucket, so joining the two halves of the staples and fastening the staple to the bucket in one operation.

These buckets give us some idea of the maximum size of casting that could be made into a single sheet in the Late Bronze Age. The dished bottom sheet of the Nannau bucket was made from a piece of bronze with a volume of about 130 cm^3 (weight: 1·17 kg), and the side plates each had a volume of about 160 cm^3 and must have weighed about 1·4 kg. The bottom sheet from a similar bucket from Flanders Moss at Edinburgh was calculated to weigh 1 kg.[151]

The Whigsborough bucket and the Heathery Burn bucket are similar in their component parts, but the wheel-shaped base casting of the latter must weigh about 1·14 kg.

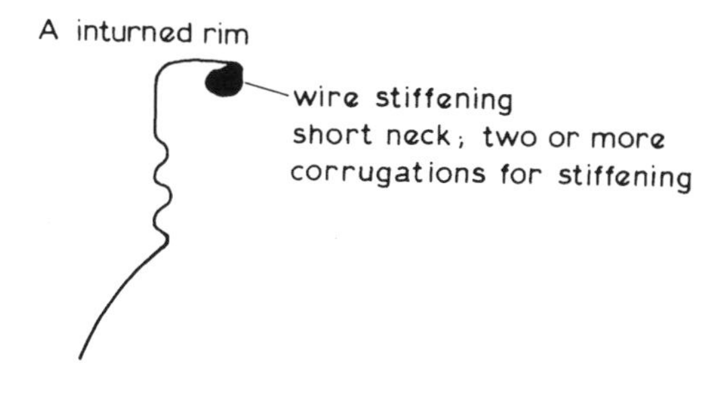

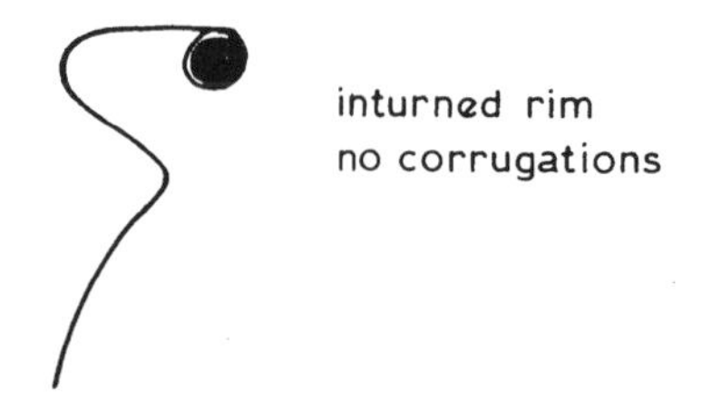

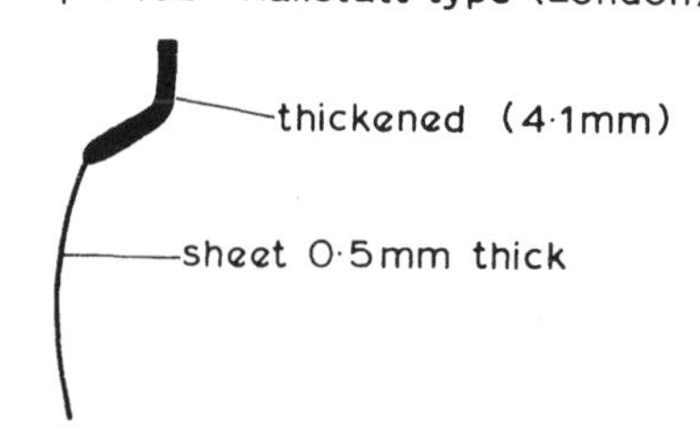

60 Types of cauldron rim (after Hawkes and Smith[148])

The cauldrons differ from the buckets in having a spherical bottom and for this reason must have been somewhat easier to make. Otherwise they are made in the same way, with a single hemispherical bottom sheet consisting of about the same weight of metal as the bottom sheets of the buckets, with one or more side sheets.[149] These also date from the Late Bronze Age. There are two main classes[148] (Fig. 60): *A* has a short neck with two or more horizontal corrugations and a rim bent into a lip turned inwards. The second type, *B*, has a rim everted at a slant, and the neck is a mere constriction without corrugations.

The Colchester cauldron is an example of a transitional type without corrugations and with a turned-in rim. This is a particularly good piece of workmanship: 0·673 m maximum diameter and 0·49 m high with no creasing at the rim. The upper part is made from a single sheet 0·28 m wide and 2·13 m long, which makes it one of the largest pieces of British Late Bronze Age bronze sheet known (weight about 2·27 kg). The bronze handles are triangular in section and 0·14 m in diameter externally. The supporting staples have been cast with the rings in place.

The cauldron from the River Cherwell, Oxford[147] is made of three sheets; even so, the base-sheet is 0·02 m high and 0·055 m in diameter. The 0·43 m diameter Late Bronze Age shield from the Trent near Newark[152] is a comparatively simple piece of work compared with this. The Llyn Fawr[153] cauldron, on the other hand, has sides consisting of four zones of riveted plates and is dated by the Hallstatt-type sword found with it to 600 BC.

Another type of cauldron was found in London, probably in the Thames.[148] This has been made from a single sheet of bronze which must weigh about 4 kg. It has a simple turned-in rim which is reinforced so as to give it a total thickness of 4 mm, while the body is 0·5 mm thick. This type has a distribution confined solely to Central Europe (Hallstatt area) with this one exception, and therefore is most probably an import. It seems to show a rather higher standard of workmanship than that prevailing in this country, since all the others found in the British Isles have been made of two or more sheets.

Such cauldrons and buckets must have been extremely difficult to make, both as a casting and as a panel-beating operation. They can only have been owned by the few, and it is therefore fortunate that so many have come down to us.

A chemical analysis of the Heathery Burn bucket base-plate shows that it contains 5·2%Sn and only 0·66%Pb.[154] A side-sheet was found to contain 12·2%Sn and 1·6%Pb.[154] This shows a careful choice of metal, since Late Bronze Age casting alloys usually contain about 5·0%Pb. The 'Dowris' base-plate was found to contain 6·2%Sn and 1·0%Pb.[154]

The EIA Welwyn bowl contained ~5%Sn, 1%Pb, 1%Fe, and 0·2%Sb. It had been worked and annealed and finally reworked.[155] A small Belgic style bowl (No. 689), 20 cm in diameter and weighing 231 g, from the Whitfield, Northumberland, hoard was found to consist of a cold-worked bronze with low lead and some residual eutectoid. The hardness was 168 HV. Since the thickness was only 0·25 mm, the wrought structure is not surprising.

The average thickness of the sheet-metal of the cauldrons is usually of the order of 0·50 mm which shows that the smiths had attained a considerable degree of skill in order to beat out a bronze containing as much as 12%Sn into such thin sheets.

Some of the Irish cauldrons such as that from Castlederg, Co. Tyrone (Fig. 61) have conical rivets projecting at least 6 mm from the surface. It would seem that a necessary feature has been intentionally turned to good account, since the projection of these rivets beyond the surface of the cauldron would assist the transfer of heat from the fire to the cauldron and its contents.[156] This would be most effective in the case of the multiple plate cauldrons and its high efficiency might have been noticed in use. The original intention would have been merely a decorative feature, but the design would be maintained once its superiority

61 LBA cauldron from Castlederg, Co. Tyrone (c. 700 BC); max. dia. 0·52 m; rivets project 6 mm from sheet metal surface (photo: National Museum of Ireland)

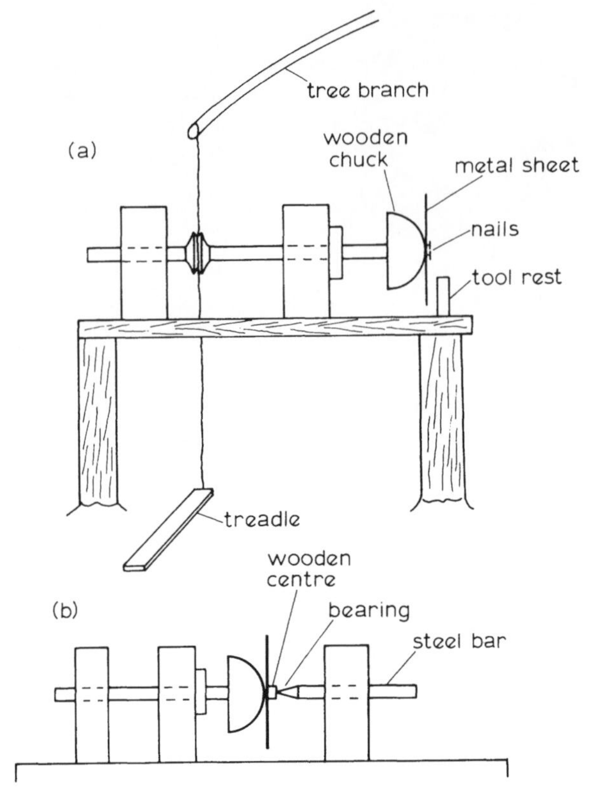

62 Primitive pole lathe adapted for metal planishing

had been established. The same principle is used in some modern power station boilers and heat exchangers.

Compared with the cauldrons, the manufacture of the bowls must have been comparatively simple since they are much smaller. The designs of many of these were based on pottery forms. Two Belgic bowls (1st century BC–1st century AD) have been found at Felmersham, Bedfordshire[157] and are thought to have been made in the same workshop as the bowl found at Welwyn.[155]

Many of these bowls have holes in the bottom, and various explanations have been put forward to account for them, such as their use as water clocks or 'killing' i.e. piercing for some ritualistic reason, before burial with their owner.[158] The bottom of one of the bowls from Felmersham contains a circular patch 14 cm in diameter over a central hole. This patch has been riveted and soldered. The bowl has been worked after casting but shows no signs of hammering. However, the second bowl shows signs of spinning marks where the metal tended to break. Voce[155] noticed the same defect on the bowl from Welwyn and came to the conclusion that the cracks are caused by stress corrosion due to the cold-work left in after spinning.

It is important to distinguish at this stage between spinning and polishing. True spinning involves cold-working a flat sheet of metal over a bowl-shaped former on a lathe and requires considerable leverage on the tool and considerable power from the machine. It results in a fully worked metallurgical structure. Polishing, on the other hand, is merely a finishing operation and will produce circumferential striations but does not work the material completely throughout its thickness. Unfortunately a combination of hammering and polishing on a lathe can be difficult to distinguish from true spinning.

At Thealby, Humberside[159] bronze bowls were found of 3rd century AD date with an iron rim. These also had a hole in the centre and one had a patch made rather like a brass paper clip. Dudley was also of the opinion that they had been finished by spinning, and that the central hole was a necessary part of the process. Smith[158] points out that in the case of a bowl from Wotton, Surrey, a neat round hole 0·50 cm in diameter had been covered with a patch that had been finally embossed.

Whereas most of the 'spun' bowls have been assigned to the Early Iron Age, Hawkes (in Watson[157]), thinks that the one-piece bowls may be Late Roman or Anglo-Saxon. The bowls made of two or more pieces, like the cauldrons, are earlier and probably date from about AD 50.

The pole lathe was certainly well-known by Belgic times and there can be no objection on technical grounds to its use for a metal-finishing operation. However, it is probable that true 'spinning' was beyond its capabilities. The bowl was probably beaten to shape by hammering and finished on the lathe by polishing.

The main difficulty is to decide how the bowl was supported or 'chucked' in the lathe, since the Belgic lathe is thought to have consisted of a shaft in two bearings with the chuck outside these bearings (Fig. 62). It would appear that holes were made in the centre, and the bowl nailed or otherwise fastened to a face plate as shown. This would explain the holes found. After working, the holes would be filled with rivets and washers or patched. In some cases the patch was soldered on but has corroded off after loss, leaving the hole open.

The circumferential marks on one of the Felmersham bowls on the inside at the bottom and on the outside near the shoulder, suggest that lathe turning was only used as a finishing operation to remove the hammer marks. Two shield bosses found in the Polden Hills, Somerset[155] also show signs of lathe turning and it would be worthwhile to examine a number of Early Iron Age artifacts that appear to have been spun, in order to elucidate the details of the technique used.

A number of thin metal bowls are known from the

63 Gold boxes of the Late Bronze Age, about 5 cm dia., constructed by hammer welding; details of joints are given in Fig. 64 (photo: National Museum of Ireland)

Anglo-Saxon period and are similar to those shown in illuminated manuscripts. Six specimens found in England are mentioned by Allen[160] and since their ornament has similarities to sculptured monuments in Ireland and Scotland, they have probably been manufactured in these countries. A lead pattern suitable for a 5·3 mm patch or plaque was found at The Brough of Birsay, Orkney.[71]

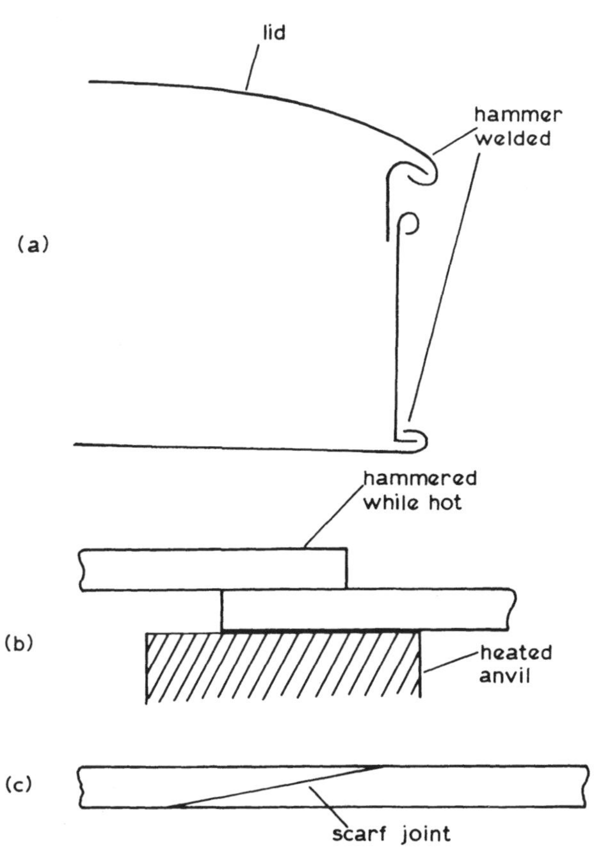

64 Hammer welding: *a* gold box hammer welded (after Maryon[121]); *b*, *c* silver sheet from Traprain Law hot hammer welded against a heated anvil

The making of sheet-metal cauldrons, as well as bowls, continued into the Anglo-Saxon period, and it is possible that the former were only finally displaced by the cast bronze cauldron of the early medieval period. Coghlan[161] cites Oldeberg in concluding that some of the Roman and Migration period (600 AD) bowls were cast. The thinness of castings made under primitive conditions is shown by two Korean bowls dated to the 12th–14th centuries AD, which were certainly not spun as their composition was about 20% Sn, which cannot be cold-worked.[161] Yet their maximum thickness was only 0·5 mm, and they show circumferential marks. It would appear that such bowls are thin castings finally finished by grinding on a lathe. While the Felmersham bowl examined by Voce[155] was definitely not a simple casting so finished, but made of beaten sheet, the possibility exists that some early bowls could have been finished by this process and show marks similar to those obtained by spinning.

A good many vessels of the Early Iron Age and later periods consist of metal-mounted wood. Many buckets and tankards have been found in which this technique was used.[162, 163] This does not necessarily indicate a retrogression, but more probably a cheaper article more widely used than the expensive buckets and cauldrons of the Late Bronze Age.

The medieval cast cauldrons varied in thickness from 0·8 to 2·5 mm. Three Scottish cauldrons from the Dumfries Museum varied from 'as-cast' bronzes with 9–10% Sn and some lead, to homogenized bronzes with severe cold-work. The hardness varied from 72 to 122 HV and all showed some cold-work, no doubt as part of the finishing process.[164]

Welding and soldering

The welding of iron dates from the beginning of the Iron Age because it is implicit in the smithing and working of iron by the direct process. Today this process would be known as smith-welding or pressure welding to distinguish it from the large group of welding processes based upon fusion.

Gold, like iron, can also be pressure welded.[165] While the blacksmith-welding of iron requires a high temperature (more than 1 000°C) and the presence of slag as a flux, gold, on account of its ductility and freedom from oxide films can be welded at room temperature. This used to be done in the gold filling of a tooth.[166] Thin gold foil was first heated in a spirit flame to remove adsorbed films of gas (water vapour and oxygen), and then tooled and packed into the cavity in the tooth with a probe. The layers of chemically clean foil weld to each other under the pressure of the probe. A group of small Late Bronze Age gold boxes, now to be seen in the National Museum, Dublin, has been made in this way, probably by using a light hammer[166] (Figs. 63 and 64).

At least one piece of the Traprain Law silverware hoard (4th–5th century AD) has also been made in this way.[167] It appeared to be the longitudinal seam of a vessel or tankard. The edges had been lapped over and hammered. As silver requires a temperature of 500°C for this operation and a heated anvil (Fig. 64), the use of this technique implies a long standing tradition which perhaps had its origins in the cold welding of the gold boxes above. The hardness of the silver sheet adjacent to the weld was 92 HV and that along the

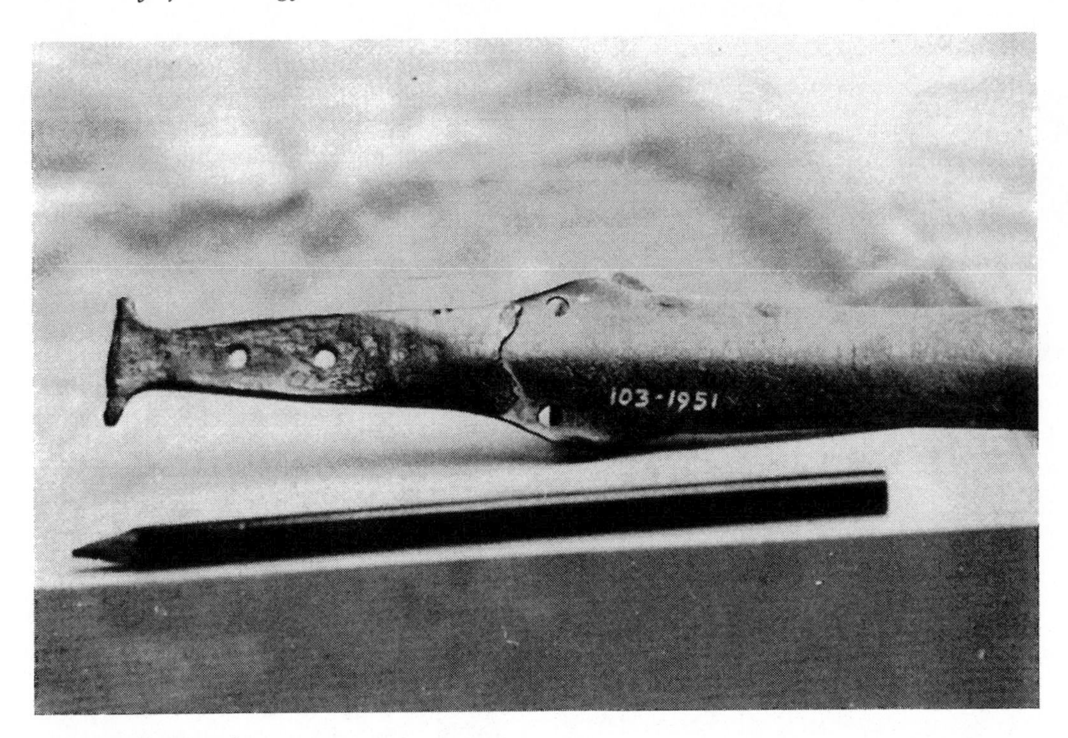

pencil is 15 cm long

65 Welded repair to a sword hilt, Late Bronze Age (photo by permission of the Ulster Museum)

weld line was 118 HV. This shows slightly increased hardness due to welding. A square section gold torc from Co. Roscommon, now in the Belfast Museum (No. 173, 1937), has been welded to give a similar lap-joint, but of course, being gold, would not need a heated anvil.

The nearest process to modern fusion welding is the technique of pouring molten metal between two pieces of heated solid metal. Known as 'burning', this process has been used for the repair of an Irish Late Bronze Age sword hilt (Fig. 65), and in the construction of a trumpet.

This may be contrasted with 'casting on', 'running on', or 'burning on' in which a new piece is both made and attached in one operation. In Dublin there is a sword from Canbo, Co. Roscommon, on which this has been attempted unsuccessfully. In principle these processes are the same as 'Thermit' welding today. The additional metal is first melted in a crucible. A mould is made round the piece to be repaired (Fig. 66) and the edge to be joined to the new casting, and the mould, are heated by building a small charcoal fire round the mould. The additional metal should be as hot as possible: well above its melting point, as it has to supply the heat to fuse the broken edges for a short distance to obtain a good join. If the edges have been preheated by the charcoal fire round the mould, this is not so important, but the main trouble in this case will be avoiding oxidation of the edges, which would prevent them being properly welded. Failure to weld the Canbo sword was either due to insufficient preheating or to oxidized edges.

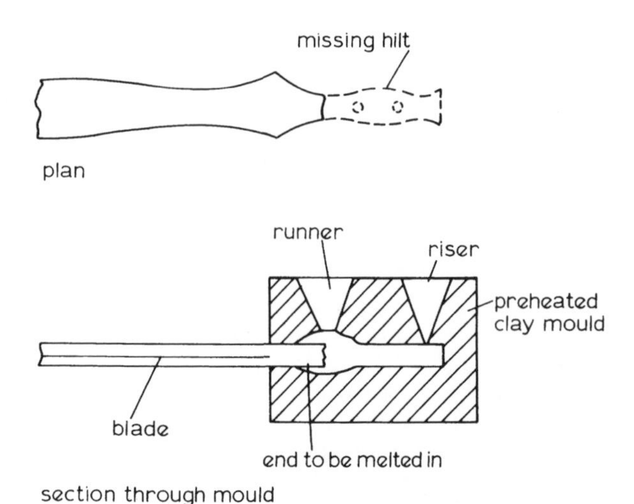

66 Casting on a new sword hilt

The above processes can be used for joining fairly massive pieces, but if thin sheets and foils have to be joined, soldering and brazing are normally used.

Soldering of gold and silver is mainly based on the fact that alloys of copper and gold, and silver and copper have melting points below those of the pure metals. In other words, a gold–20% copper alloy melts at 884°C compared with 1 063°C for pure gold. An alloy of 70% silver and 30% copper melts at 780°C compared with 960°C for pure silver. There is little evidence of lead–tin soldering in Western Europe before the beginning of the Iron Age, and the use of gold soldering has not been confirmed in the later Bronze Age torcs.[136]

In the case of gold the required alloy may be formed by diffusion of the copper into the gold. Gold does not form an oxide film so that diffusion of gold to form alloys is relatively easy at elevated temperatures. In the case of the baser metals this is largely prevented by the formation of an oxide film which has to be dissolved or dispersed with the aid of a 'flux'.

If gold is heated to 1 000°C (60°C below its melting point) in contact with a small amount of copper, a gold–3% copper alloy is quickly formed which is molten at this temperature and which fuses the two

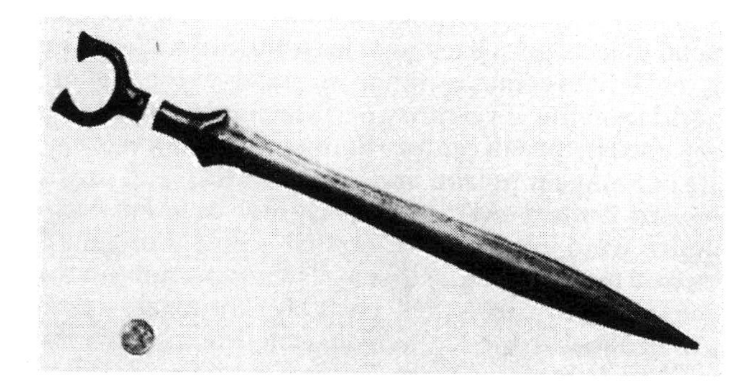

coin is 2·5 cm in diameter

67 Sword from Whittingham, Northumberland; penannular ring fitted as pommel to Late Bronze Age socketed hilt sword (by courtesy of the Museum of Antiquities, University of Newcastle upon Tyne)

surfaces together. It is not then detectable to the naked eye as the colour is almost the same.

The fact that a base metal (copper) enters into this process makes a flux necessary, and goldsmiths use chrysocolla ($CuSiO_3.2H_2O$). In the 11th century Theophilus[168] describes the manufacture of a copper solder for gold by making a soap from lye and pig fat, heating copper and salt and adding them to the soap. The soap acts as a flux and the copper chloride decomposes at high temperatures to give copper which alloys with the gold, and chlorine, which probably reacts with the alkali of the soap to form potassium chloride.

Maryon[121] points out that this technique is very similar to that outlined in Littledale's patent (1934.415181), where the sequence of reactions appears to be:

at 100°C $Cu(OH)_2$ + glue → CuO + glue
600°C glue → C
850°C C + CuO = Cu + CO(gas)
900°C Cu + Au = Cu–Au alloy.

The Whittingham sword[135] is an interesting example of the use of a complex series of casting and brazing processes (Fig. 67). It appears to be an attempt by a Late Bronze Age bronzesmith to copy the 'antennae' swords which either he or his patron had seen on the Continent. A leaf-shaped sword has been provided with a socket instead of the more normal tang. The socket, cast in one piece with the blade, has been pierced with two holes parallel to the axis of the cutting edges. The pommel, also socketed, consisted of three pieces of metal, a penannular ring, a socket, and a patch which had been used in joining the ring to the socket. The ring was an intricate cored casting which had had a hole cut into it to which the socket had been hard-soldered. Technically it is superior to the socket and was certainly not made by the same hand. The socket is a 9% tin bronze and had been hard-soldered to the ring with the aid of a bronze of higher tin content, and therefore, of lower melting point. This shows that by the Late Bronze Age the principle of using a more alloyed metal to join or solder a less alloyed metal was understood in this country.

From the Roman period (before AD 380), Friend and Thorneycroft[169] report a very interesting case of the copper brazing of an iron ring. Today it is common practice to join steel components together with pure copper at about 1 100°C in an atmosphere of hydrogen to prevent oxidation. The ring was 4·3 cm in diameter and 1·5 cm wide with a thickness of 4 mm. It was made of two pieces welded together at one place, and copper-brazed at the other. Why two different methods should be used in a simple ring is uncertain. It is possible that the copper braze is a repair, and that the smith who repaired it did not want to weld it as he would have had difficulty in restoring the object to its original shape if he did. More probably, the weld was merely incidental, and the braze was the intended method of joining the two ends.

So far, we have little evidence of hard silver soldering using copper–silver alloys as used today for the joining of silver articles.[170] In the Mildenhall[171] and Sutton Hoo[172] silverware, the handles have been soft-soldered with lead–tin solders; a very crude technique, and one which has led to preferential corrosion and separation of the handles from the rest of the dishes.

Useful conclusions can be drawn regarding the state of metal working in the early 5th century AD from the hoard of silverware found at Traprain[167] which represents native loot from a Roman site. In this hoard, as at Mildenhall, silver-soldering is notably absent, handles being attached to silver dishes by means of soft solder containing 15% Pb and 85% Sn. The stems of wine-cups are riveted to the bottoms of the bowls, and in some cases the rims of the bases of bowls are formed by beating rather than by soldering or riveting. As already noted, pressure welding has been used to join silver sheet into cylinders. There is plenty of evidence for lathe finishing. Gilding and niello have been used for decoration, and Curl[71] assumes that the former has been carried out by brushing on an amalgam and heating to evaporate the mercury and fix the gold. The niello technique must have been primitive, and much of it has perished, which tends to support Moss's conclusion regarding the technique used at this time.[173]

Tinning, gilding, and plating processes

A number of authorities have reported the apparent tinning of bronze axes.[174,175] Smith and Macadam[174] report a number of flat bronze axes with high tin contents from Scotland. Three of these have tin contents of 35·8, 37·3, and 32·9%.

As there are several ways in which tin-rich layers may be formed on bronzes a considerable amount of work has been done on this subject.[176–179] It is now generally accepted that in most cases the tin has been applied intentionally either by rubbing or dipping and that continued heating after application leads to the inter-diffusion of tin and copper so that the tin layer is now a tin-rich layer as found by Smith and Macadam. This layer contains enough tin to give it a silvery colour (i.e. more than 30%) and should have a high hardness in the range 350–500 HV.

From the Roman period onwards copper-base cooking vessels seem to have been generally tinned as indeed they are today. A bronze patera or skillet from Kyngaddle, Dyfed, has been tinned internally.[180]

By the 10th century AD the tinning of iron articles was an accepted practice, and Jope[181] has reported on the tinning of iron and bronze spurs. Many of these have been found in or near Oxford and range over seven centuries. Tinning has rarely been detected on medieval ironwork other than spurs, and not even on stirrups which are of similar workmanship. It appears only to have been introduced after the time when the making of stirrups and spurs became separate crafts, i.e. in medieval times.

The tin had been fusion plated, probably not by dipping but by rubbing sticks of tin onto heated bronze or iron which had previously been coated with a thin layer of resin to prevent oxidation. We have, as yet, no proof of the use of resin, and fat or tallow may have been used. In 13 out of 15 specimens examined by Jope the coating was pure tin, but in the other two cases it was a tin–lead alloy.

Tinning can also take place by electrodeposition, not necessarily by an externally applied current but by an internally produced current by galvanic dissolution in a suitable electrolyte. Diderot[182] shows how brass pins were tin-plated in the 18th century by boiling them in a copper vessel containing an aqueous solution of argol (lees of wine – crude bitartrate) and tin plates 0·5 mm thick. The tin dissolves slowly in the solution and by some complex electrochemical process is plated out onto the pins in contact with the tin plates. This process was certainly applied to pins found in Britain from the 18th century and probably much earlier. Thouvenin[183] believes that it was used in Gallo-Roman and Merovingian times.

Depletion gilding and silvering can be used where the matrix contains enough noble metal to provide a coating when the base metal is removed by oxidation and leaching. But recent work on Peruvian copper-base alloys has clarified the mechanism of electrochemical replacement gilding.[184] Very thin gold layers (1 μm) were obtained by immersion in a warm aqueous solution containing an 85% Au–15% Ag alloy. The solution contains NaCl, KNO_3, and alum and it is clear that local cells are set up, with the small areas of Cu-alloy acting as the anode and the majority as the cathode where there is deposited a thin layer of Au–Ag alloy. The active constituent is probably $AuCl_3$. It would seem that this is the mechanism by which tin is plated onto brass pins. Naturally the tin must be cathodic to the brass.

It has been observed that small iron bells were coated with bronze in Roman and Early Christian times. 'Bronzing' would be a far more difficult process than tinning and a resin flux would be quite inadequate. It would require either a dipping process or heating the article in charcoal to maintain a non-oxidizing atmosphere and melting on the bronze under the charcoal. An iron spur covered with thin bronze sheet was found in the River Ray at Islip;[181] covering by hammering thin bronze sheet on the iron would be an alternative but still difficult process. However, the steel links of the bridle bit from Gussage All Saints have been bronze-plated by dipping, which has also been used as a process of 'brazing' several pieces of iron together. Owing to the high melting point of bronze (950–1 000°C) a considerable amount of iron will dissolve in the bronze during the dipping process. Apparently this was prevented at Gussage by previously coating the iron with tin.[89]

The same principle of coating and joining has been used in the assembly of medieval padlocks from Goltho where brass has been used as the brazing medium and appears to have run over the outside of the casing.[185]

The application of gold in some form to wood or metal objects goes back to at least the 3rd millennium BC at Ur. It becomes common on metals in the classical world from the 5th century BC. Mercury or fire gilding was used in the 6th century BC on the Late Bronze Age site of Rathgall, Ireland and was in widespread use by the 3rd century AD.[186] Two slightly different techniques were involved. In the first a gold amalgam is applied to the cleaned surface. The proportions in the amalgam are 1 part Au to 7 parts Hg. The object is then heated causing the Hg to evaporate leaving only the gold, ready for burnishing. In the second the Hg is first applied to the surface and then gold leaf is pressed on top in layers until no more will adhere. The object is then heated as before.[187,188]

Gilding can be applied to non-leaded bronzes and brasses and also to iron. Charles II's suit of armour was gilded, but iron has to be copper-plated first.[189] The technique is described in 'Mappae Clavicula',[190] the anonymous 12th century manuscript; vinegar was boiled in a copper pot for 1 h and the iron plate stood in the acetate solution for some time before mercury gilding.

There are other methods of obtaining the appearance of gilding; one such, called 'pseudo-gilding' by Oddy and Meeks[191] consists in applying a layer of gold-coloured chalcopyrite ($CuFeS_2$) to a tinned layer with the aid of a tin–lead solder. This was used on a Roman bust consisting of 13% Zn, 3·9% Sn, and 4·7% Pb, rest Cu. The lead content would have rendered it unsuitable for normal mercury gilding.

Techniques of decoration

Since the beginning of the Bronze Age non-functional decoration has been an important aspect of the metallurgist's craft. At the beginning, much decoration was forged in with a punch and a chisel. Later, as in the case of socketed axes it was cast in to give shield and other patterns. Lowery *et al.*[192] have discussed the shaping and decorating techniques available (Fig. 68). Bowls and vases were made by sinking i.e. with blows on the inside of the vessel, or by raising, i.e. by blows on the outside of the vessel, both done with special hammers and anvils.

Decoration can be applied by *repoussé* methods in which the design is in relief and the background is depressed by means of a hammer or punch. Alternatively the pattern is worked from the back into a soft base of bitumen or lead. Detail may be applied with the graver or the tracer. The former tool is pushed by the hand and cuts a groove, producing ahead of the tool a small curl or swarf which is eventually removed. The outline may be completed and emphasized by means of short narrow marks with the chisel-like tracer which is hammered onto the metal.

In the Roman and later periods when the repetition of decorative motifs became more common, dies were used. At Wroxeter[193] there was found a block of wrought iron 28·7 × 8·4 × 1·5 cm which had been engraved with recessed (intaglio) designs on both sides. It was presumably used for embossing designs on copper-base alloy plates. The designs indicate that it could hardly be earlier than the close of the 1st century BC and is most likely to be Roman. No opinion has been expressed as to the method used for engraving it, but it could have been cut with hardened steel

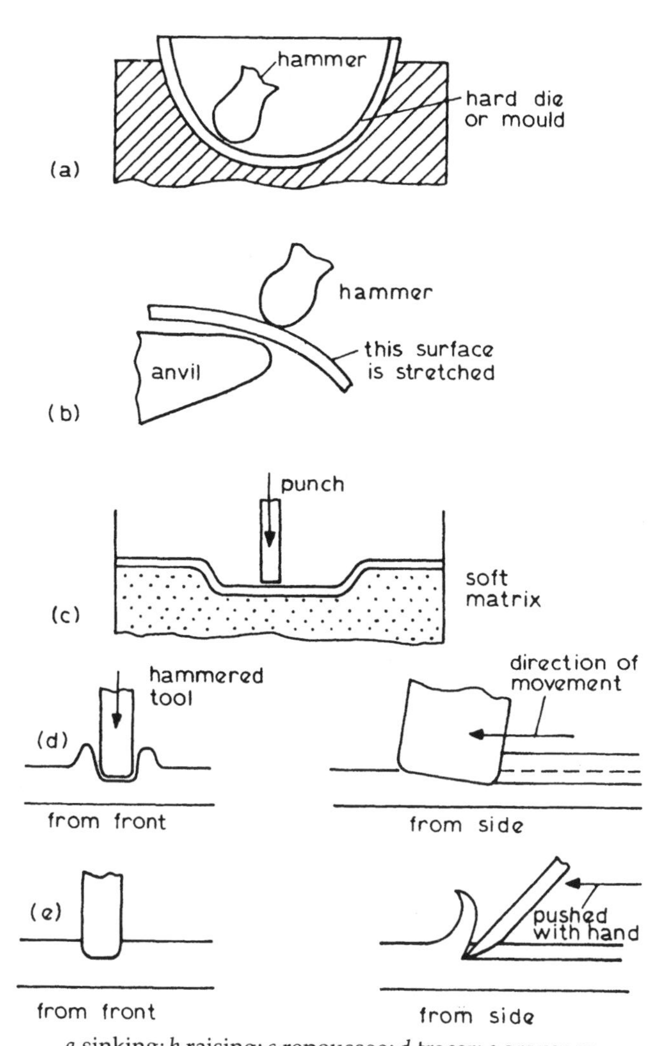

a sinking; *b* raising; *c* repoussee; *d* tracer; *e* graver or scraper

68 Metal shaping and engraving techniques (after Lowery, Savage, *et al*.[192])

tools. On the Sheepen, Colchester, site a bronze die was found which seems to have been intended for a belt-plate decoration.[194]

How such die plates may have been used has been shown by Northover.[195] A small 7th century AD gun-metal die plate was found at Rochester, Kent. This could have been used by pushing annealed soft metal foil into the hollow areas of the die by hammering a sheet of a softer metal such as lead. It is possible that this die was used for making gold mounts for drinking cups which are found on a number of Anglo-Saxon and medieval sites.

The Sheepen die contains a large peg on the underside and since this is in good condition it is not likely to have been hammered but more likely to have supported the die in a hole in a vice or anvil.

PRODUCTION AND COMPOSITION OF COINS

The majority of pre-Roman coins found in Britain arise from normal trade contact with the continent which preceded the Roman invasion. Many are Gallo-Belgic from north-west France and were not minted in Britain although they were in current use here for a long period.[196]

These coins were modelled on the gold staters of Philip II of Macedon which were struck about the middle of the 4th century BC. The Celts of Gaul probably learnt the art of coining from these examples by way of the Danube.[197]

One group, Gallo-Belgic A (Table 60), came into this country by way of east Kent and Clacton, and is widely distributed in south Britain. Another group, those of the Gaulish Atrebates (Gallo-Belgic C), towards the beginning of the 1st century BC, gave rise to the first series of locally made gold staters, British A. These spread north and south-west from an area centred on Sussex. Several coins of this type were found in the Westerham hoard in conjunction with Gallo-Belgic types and their composition is given in Table 60.

Other hoards of Gaulish coins show evidence of trade between Britain and France. The hoard from Le Câtillon in Jersey contained over 765 coins of copper–silver alloy and 41 of gold. There are Gallo-Belgic, British B, and Durotrigian together with thin silver coins of the Hengistbury type. But the majority belong to the Armorican Corisolites tribe, although other Armorican tribes are represented.

Hengistbury Head has also provided us with a large number of Armorican coins together with Durotrigian and other British coins.[199] Mount Batten, Plymouth, which may have a claim to be an export port for tin, has also provided many Armorican coins.[200]

The Corisolites had a port at Alet near St. Malo not unlike Hengistbury Head.[201] It is an interesting fact that this site yielded the remains of cupellation debris which contains Cu, Pb, and Ag in metallic form, and Cu and Pb as oxides just as at Hengistbury Head.[202] There is a strong concentration of silver in the copper. It would appear that, just as at Hengistbury, attempts were being made to remove copper from silver by means of cupellation with lead.[201] The composition of the Corisolitan coins was in the range 55–70% Cu, 25% Ag, 0·3–3% Pb.

Table 60 Gold coins of Gallo-Belgic and British origin (after Allen[197] and Brooke[198])

	Westerham Hoard			Chute Hoard	Separate finds				
	Gallo-Belgic A	Gallo-Belgic C	British A	British B	Gallo-Belgic Xc	British Qc (40–20 BC)	Tincommius 97	Tincommius 104	Verica 122
Au, %	69·02	69·70	59·51	37·92	51·75	57·3	47·37	48·55	72·2
Ag, %	22·83	18·62	28·22	40·06	34·6	16·4	12·91	13·56	7·6
Cu, %	8·15	11·68	12·27	22·02	13·65	23·9	39·72	36·74	17·2
Sn, %	—	—	—	—	—	2·4	tr.	1·15	—
Weight, g	7·5–7·8	6·5–6·7	6·41	6·10	—	—	5·2–5·4	5·2–5·4	5·3
S.G.	15·0–16·5	14·5–15·5	13·85	12·19	13·01	13·22	10·88	10·64	12·6

Table 61 Composition of bronze coins of the Iron Age: 'tin money' and Durotrigian coins from Hengistbury Head

	'Tin money'		Hengistbury Head[203]			
	Sunbury[204]	Brentford[205]	*A*	*B*	*C*	*D*
Cu, %	72·4	73	73·1	71·7	85·0	76·0
Sn, %	22·1	27	11·0	17·4	2·3	11·7
Pb, %	4·6	—	7·78	0·37	—	9·6
Oxygen etc., %	—	—	8·14	10·6	11·4	2·7
Ag, %	—	—	—	—	1·31	—
Fe, %	0·4	—	—	—	—	—
Estimated hardness, HV	260	410				

Another type of gold stater, British B, appears to have originated in the Hampshire–Dorset area about 70–80 BC, and specimens found in a hoard at Chute, Wilts., have been analysed (Table 60). According to Allen[197] the reduction in gold content indicates that B is later than A. These types gave rise to a large number of silver and later, bronze staters which are generally ascribed to the tribe of the Durotriges. Some of those found at Hengistbury[203] (Table 61) are typical of this group.

The British Q coins are the product of the latest of the pre-Roman invasions from the Continent, that under Commius, which probably took place about 40–20 BC. In the succeeding reign of Tincommius (20 BC–AD 10) a reduction in the gold content has taken place (Table 60). Revaluation occurred during the time of Verica (AD 10–40) when the gold content of the pre-Roman coinage reached its peak of 72%.

Another early British produced coinage is the so-called 'tin money', 'Potin', or speculum coinage of the south-east of England, which was minted around 50 BC. Finds of these speculum or high-tin bronze coins have been made at Sunbury-on-Thames,[204] Brentford, Reculver, and Snettisham,[139, 206, 207] and at other places in south-east England. Mack divides the coins into two groups: Class I, in which there is an attempt at portraiture, facing sometimes left and sometimes right, with a degenerate bull on the reverse side. This is typical of those found at Brentford, Reculver, and Snettisham; Class II are generally smaller but about the same weight as Class I, but the face is no longer recognizable and the bull takes the form of a meaningless rectangle; examples of the latter were found at Great Chesterford, Essex.

Opinion is divided as to the method of manufacture of these coins. It is certain that the coins were cast, mainly because of the high hardness of their composition (Table 61) and therefore the difficulty of striking. The hardness of one such coin was 370 HV; this would give it considerable wear resistance. Modern bronze coins have a hardness of 110 HV. The structure is typical of a cast high-tin bronze. It is possible that two-piece carved wooden pattern moulds were made, and that wax was cast in these; the wax patterns were then invested with clay, after the parting line was carefully removed from the wax. The wax was run out by warming the mould, and metal poured in. The wax patterns would be so positioned in the clay moulds that many could be cast together. After casting, if the runners were thin enough they could be easily broken by hand, or alternatively, cut with a chisel. This is of course a variation of the lost-wax process.[208]

It is much more likely that the wooden patterns were used to impress designs into clay moulds. The 'strings' of coins which were thus obtained were then cut up. This, however, still leaves the absence of a parting line unexplained.

Verulamium (St. Albans) was one of the main mints of the Catuvellauni in the time of Tasciovanus (20 BC–AD 10). The 1956[209] and 1957[210] excavations on this site have yielded clay moulds which show that the coin blanks or flans were cast into small depressions, about 10–17 mm in diameter at the mouth. The clay plates measured about 20 × 20 cm and contained as many as fifty depressions. One of these holes contained a bronze 'pellet'. It is suggested that Epaticcus minted here in about AD 25–35. The blanks cast in these moulds were obviously struck.

A section of this type of mould is shown in Fig. 69. Under certain conditions it is possible to form small globules of metal in them, rather than cylinders, which would not be as easy to coin. Allen records that a hoard of fifty gold globules was found, marked only with a cross, in Peebleshire [Borders]. It is probable that they were the direct result of melting in this now rather

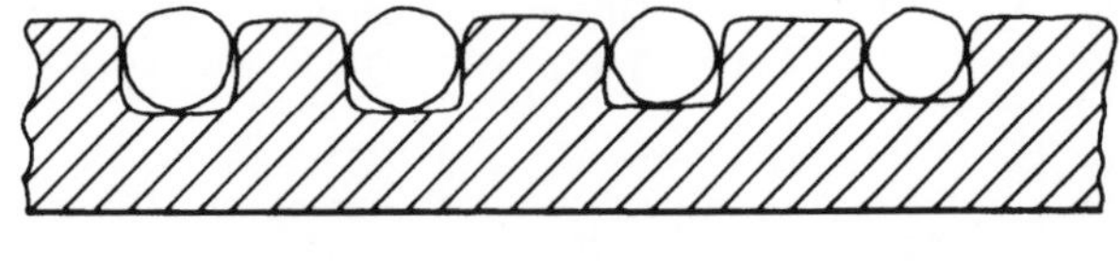

69 Section of Iron Age coin mould

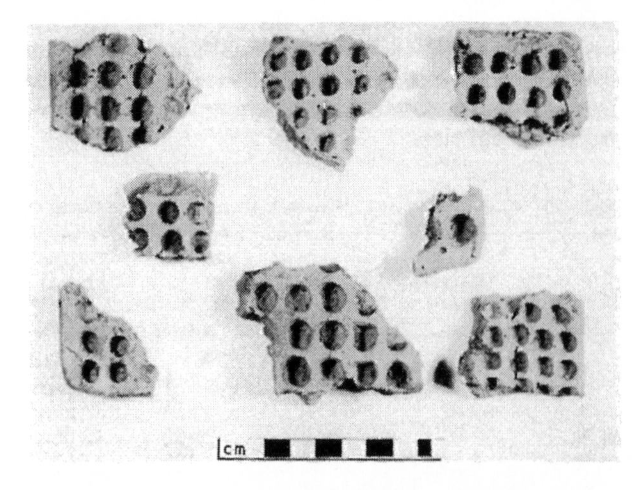

70 Pieces of clay coin blank moulds of the Early Iron Age from Old Sleaford, Lincs. (photo by W. T. Jones, A.R.P.S. Crown Copyright Reserved)

Table 62 Roman copper and copper-base alloy coins (found in England[216]

Emperor	Date	Composition, % Sn	Pb	Fe	Ni	As	Sb	Zn
Caligula	AD 37–41	0·10	0·46	0·2	tr.	—	—	nil
Vespasian	AD 60–79	—	—	0·27	0·2	0·2	—	nil
Vespasian	AD 60–79	0·22	tr.	tr.	0·33	—	0·32	nil
Augustus	30 BC–AD 14	0·72	tr.	0·43	tr.	tr.	tr.	11·8
Vespasian	AD 71	—	0·14	0·12	—	—	—	18·7
Trajan	AD 98–107	0·39	—	0·27	—	—	—	20·7
Sabina	AD 100–137	0·43	tr.	0·38	—	—	—	16·8

common type of mould, but that for some reason no attempt had been made to strike them.

Similar moulds have also been found at Braughing, Camulodunum, Needham, Silchester, Soham Toney, Bagendon, and Old Sleaford (Fig. 70). Spectrographic analysis[211] of traces of metal in some of these moulds showed the following results:

Site	*Probable metal or alloy*	*Ref.*
Old Sleaford	20Ag–30Cu	212
Bagendon	Brass (or impure Cu) Copper–silver	211
Verulamium (Batch I)	Bronze Bronze Copper–silver	211
Verulamium (Batch II)	Copper Copper–silver Copper–silver–gold Silver Silver–copper Silver–gold	211

It would appear, therefore, that coins from a wide range of metals and alloys had been made at these places, and that this technique was current from immediately pre-Roman times to well into the Roman period.

Experiments have shown that it is easy to make sessile drops (globules) of accurate weight from compositions containing large amounts of noble metals in these moulds.[212] Small pieces may be weighed out and placed in each of the 50 cavities and the moulds heated to the melting point under charcoal. Base metals such as copper tend to stick to the walls of the cavities unless about 10–20% of charcoal is added to the clay to provide a reducing atmosphere within. Two per cent of residual charcoal was found to be present in a used mould from Sleaford.

An Icenian Ag–Cu coin (ANTED) had a hardness of 101 HV. An experimentally cast globule of the same composition cast in a coin mould had a hardness of 105 HV (1 kg) and cold striking raised this to 153–185 HV on the surface. It would appear therefore that the original coin had been struck hot.

There is little doubt that these moulds were used for producing globules which were hot struck to produce flans or blanks which were further struck to produce the coins themselves. They were also common on the Continent.[213]

Excavations at Hengistbury Head[203] in 1911–12 yielded the remains of cupellation hearths (*see* Chapter 4). This process may have been used in connection with minting. A block of 50/50 silver–copper alloy was probably the result of unfinished cupellation of material destined for silver coinage. No coin moulds were found however, but several ingots and over 3 000 coins, many being cast coins of bronze and some of silver. Of these 1 660 were cast bronze issue of the Durotriges and 1 308 were struck bronze.[214] More than 100 were Roman, ranging from Republican issues to those of Antoninus Pius, and a handful of others included Gaulish and British issues. The silver coins were almost completely converted into silver chloride by the saline soil and it was not therefore possible to prove their method of manufacture.

The composition of some of the bronze coins is given in Table 61. *A* and *D* resemble Roman coins of the time of Augustus and Agrippa (*c.* 30 BC) which had 78·5% Cu, 13% Sn, and 8·6% Pb. All other Roman coins later than the Republic (28 BC), except an issue of AD 268–270 were richer in copper than any of those found, and usually contained zinc.

In addition to coins of the above compositions, some were found made of pewter (tin–lead alloy) containing a trace of silver and copper, which were thought to be counterfeits, but could have been patterns for casting.

The proportion of cast to struck coins on this site probably reflects the change in methods of coining during the period of its occupation. The moulds found at Verulamium show that striking, as a means of production, was being introduced there during the early years of the 1st millennium AD.

Hengistbury yielded no definite evidence of Roman occupation, but its native occupation went on into the mid-4th century AD. It is probable that it was the site of a pre-Roman mint.

For a long period after the invasion the Romans in Britain imported their currency from mints at Rome and elsewhere in the Empire.[215] After AD 287 a small proportion was made in temporary mints at London and Colchester, and the London mint continued until the mid-4th century. Analyses of Roman coins found in England are given in Tables 62–64.

Finding an explanation for the small amount of silver so often found in an otherwise bronze coinage (Table 63) has presented some difficulties. Clearly, 2·0% is inadequate to form a silver coating by depletion, and Cope[218] feels that it is added for economic reasons. On the other hand the presence of tin as well as silver will enhance the whitening effect of a given amount of silver either in a Ag–Cu alloy, or when the surface is enriched by depletion, as the tin displaces the Ag from solution in the alpha phase and causes a

Table 63 Analysis of copper-base alloy coins of the Roman period from the London mint (after Cope[217–222])

Type	Emperor	Date of issue AD	Composition, % Cu	Sn	Ag	Pb	Fe	Ni	Co	Zn
Reduced follis	Maximian	307	36·07	5·56	1·85	6·44	0·01	0·02	0·01	0·01
	Constantine I	322–3	95·2	0·21	1·83	1·71	0·41	0·02	0·06	0·01
	Galerius	300	92·86	2·58	2·23	2·19	0·01	0·02	0·01	nil
Large follis	Maximian	300	86·30	5·02	1·70	6·43	0·02	0·03	0·02	0·01
	Severus	306–7	90·49	3·56	2·07	3·50	0·04	0·05	0·02	0·02
Reduced follis	Constantine I	310	86·78	5·54	1·76	6·01	0·01	0·02	0·01	nil
	Constantine I	310–2	85·12	3·39	1·51	9·56	0·03	0·02	0·04	0·01
	Lucinius I	313–4	90·12	3·35	1·42	5·00	0·03	0·03	0·02	0·01
	Constantine I	316–7	86·57	4·24	1·32	7·60	0·04	0·03	0·03	0·02
	Constantine I	318	89·50	4·27	1·27	4·47	0·02	0·04	0·04	0·01
	Constantine I	318	87·89	4·33	2·11	5·57	0·02	0·01	0·01	0·01

larger proportion of the silver–copper eutectic to form.[32]

Methods of coin production included casting and striking. Wootton[223] has examined a Hadrianic *dupondius* (AD 117–138) made by striking brass containing about 20–25% Zn. He found that it was made from a cast blank, presumably cast in an insulating mould of baked clay. There was a considerable number of zinc oxide inclusions, showing that it had been oxidized during melting. After casting, the blank was struck too cold, producing a severely distorted structure and numerous cracks.

Many more coins have now been examined by Sellwood[224, 225] and Cope.[33] It is clear that many of the blanks were cast vertically. But it is clear that before striking they were heated for a short time, probably in a brazier and then struck hot with bronze or steel dies. Homogenization of the cast structure is almost non-existent, and the degree of cold-work is in most cases minimal. The temperature would be falling as striking was taking place and it is clear that in some cases the blank was below its recrystallization temperature when struck.

Similar methods were used for making 3rd century copper–silver coins such as those found at Dorchester (Oxon.). Smith[226] found that these coins had either been struck while still hot after casting, or in the cold state. In any event they had received no long-term heating and the cast structure was very clear. The composition of some silver coins is given in Table 64, and since new coins of compositions below 40% Ag would be yellow, they would be dipped in acid to dissolve away the copper from the surface to make them look like pure silver. The chief impurities were lead and gold. The lead would be left over from the cupellation of the silver.

Silver coating of copper-base coins was practised during the Roman period either by counterfeiters or by official moneyers – probably by both. Two techniques were used: (*a*) enhancement of the silver content of a copper-base alloy containing sufficient silver (15–40%), by leaching out surface-oxidized copper; (*b*) the wrapping of a non-silver-containing blank with silver sheet.[227–229] Naturally this latter process would form its own Cu–Ag solder as in Sheffield plate. There is also some evidence for the use of silver 'washes' or thin coating which could have been applied either by dipping or electrochemical plating using silver or silver chloride. Naturally in many cases these have worn off and it is now difficult to decide how they were applied.

Some Continental and Greek staters were gold plated copper-base alloys. These had been mercury gilded – a process known since the 7th century BC if not before.[230]

We now have several finds relating to a method of coining thought to be used by counterfeiters. Two of these come from Somerset. Coin moulds found at Edington in the Polden Hills were used for making silver denarii in the reign of Commodus (AD 180–192). Separate clay two-piece moulds were used for casting the finished coin, each piece of the mould being 3–6 mm thick. The two halves were put together, probably luted with a thin wash and stacked in three tiers of 12 moulds with runners and ingates arranged so that 36 of these could be made in one cast. From time to time since 1891 similar moulds have been found around Whitchurch.[231] An excavation in 1961 revealed leaded

Table 64 Composition of Roman imperial silver coins issued between AD 70 and 360 (after Cope[33])

Emperor	Date AD	Mint	Composition, % Ag	Au	Cu	Sn	Pb	Zn
Vespasian	70–2	Rome	95·2	0·7	3·81	tr.	0·25	0·005
Vespasian	70–2	Rome	86·9	0·35	10·8	n.d.	0·37	1·5
Trajan	112–7	Rome	81·0	0·35	17·9	tr.	0·64	0·002
Hadrian	132–4	Rome	82·8	0·28	16·5	tr.	0·3	0·002
Lucius Verus	161	Rome	68·36	0·16	25·3	n.d.	6·08	0·004
Maximus I	235–6	Rome	34·08	0·14	65·1	n.d.	0·58	0·015
Gordian III	241–3	Rome	38·06	0·07	61·4	n.d.	0·48	0·04
Volusian	253	Rome	21·8	0·14	76·9	n.d.	1·03	0·04
Julian II	360	Lyons	93·26	0·32	5·6	n.d.	0·61	0·01

Table 65 Composition of Anglo-Saxon silver coins (after McKerrell and Stevenson[234])

	Date, AD	Composition, % Ag	Au	Cu	Sn	Pb	Zn
Mercia (Coenwulf)	769–821	98	0·11	2·0	0·0	0·1	0·02
Mercia (Burgred)	852–874	89	10·9	5·1	2·9	1·5	1·2
Northumbrian (C'nut)	Viking	91	0·5	5·3	0·2	2·5	0·2
Wessex (Aethelwulf)	839–858	64	0·39	30·0	0·0	0·9	5·4
Wessex (Eadweard Elder)	899–925	92	1·4	5·5	0·1	0·9	0·1
England (Aethelstan)	925–939	95	0·7	3·1	—	1·0	0·2
England (Aethelred II)	979–1016	91	0·3	4·5	—	1·9	2·9

bronze *antoniniani* and *sestertii* together with mould fragments and crucibles. These moulds would fit together in the same way as those from Edington and Castor, Northants. (Fig. 51). This figure comes from Artis's book on Durobrivae[104] and is probably based on the find from Wakefield which is now in the British Museum together with some of the coins cast in it.[103] Tudor counterfeiters are known to have used moulds of graphite, probably because of its ease of working.[81]

The separate two-piece moulds are presumably made with the aid of a wood or wax pattern. There is no reason why a lead pattern should not be used, and this makes one wonder if the pewter coins from Hengistbury[203] were, in fact, patterns for such a purpose. In the University of Amsterdam there is a lead pattern for an Athenian tetradrachm,[232] which suggests that such a method had been used.

Official minting was done on horizontal or vertically cast flans cast into two-part moulds of steatite or baked clay. In some cases the blanks were faced-up with a turning tool like a carpenter's bit, which left a hole near the centre of one or both sides even after striking.[225]

Another process was introduced to demonstrate the genuineness of a silver coin and prove that it was not plated. This involved the making of chisel cuts on the face and the making of radial cuts round the whole edge before striking (serrati). This tended to be ineffective and was eventually dropped.[225]

A totally different process was in use at Lydney, Gloucestershire, in the later Roman period.[233] Blanks had evidently been made by cutting off short lengths of round copper alloy bar with the aid of a cold chisel. This method resulted in very well rounded coins, the edges of which in many cases form a distinct angle with the face. It is clear that the cut-off pieces had been flattened to the required thickness, before striking with a die. A proportion of these coins possessed holes, roughly through the centre, which had been caused by 'piping' in the original casting. Usually piping is confined to the top part of the casting and may be cut off before use. In this case, insufficient metal was removed so that some of the coins were made from defective material. The composition was found to be: copper, rather less than 97%; lead, about 2%; tin, under 1%; and iron, a slight trace. They contained no zinc. These coins were made in the 4th century and some as late as 390 AD, and seem to represent a retrogression in technique. Traces of a bronze smelting hearth containing pieces of smelted copper and slag were found on the site, but no analyses were made of these remains. The copper–lead alloy is a comparatively common alloy in the Roman period, and would appear in this case to be a natural product of smelting and not an intentional alloy.

The departure of the Romans was followed by a decrease in the amount of currency, and since the Saxons brought no coinage with them, Roman coins circulated for almost two centuries. In the 7th century, Anglo-Saxon mints were established at London, Canterbury, and Winchester. A national coinage was not established until the 10th century, when the London mint supplied the dies used by the numerous provincial moneyers.[215] Some examples of compositions used in the Anglo-Saxon period are given in Table 65.

In 1280 the coinage was made from cast square-section bar from which pieces were cut which were then forged into round blanks.[215] These were pickled (i.e. steeped in dilute acid to clean the dirty surface) and coined between upper and lower dies as in the Roman period.

Balog[232] suggests that in the 13th century silver coins were struck from pellets made by pouring molten silver into water. This method produces very small pellets, and could only have been used for the smallest coins such as the minims, weighing only 0·25 g. Machinery was not introduced into the London mint until 1628 when rolling mills were used.[215] Screw presses followed in 1662.

Silver plates, which were used for the verification of the purity of the coinage, are in existence dating from 1279. In addition to this use, most of them have been used at Goldsmith's Hall in connection with the assay and hall-marking of wrought silverware. A plate was cast and hammered, and pieces were distributed to various authorities. Some were made by the Royal Mint, others by the Goldsmith's Company, which had long held the responsibility for the periodic testing of the coinage known as the Trial of the Pyx.

A spectrographic analysis of these plates shows how the standard of purity has changed through the centuries.[235] The percentage of impurity elements in the plates made before 1600 is roughly of the same order as that found in coinage and wrought silverware of similar date. The nominal silver content of this sterling silver is 92·5%, the bulk of the remainder being copper. The reason for the reduced silver content is that pure silver is normally considered too soft, and 7·5% copper has the effect of causing an appreciable increase in the hardness and still retaining the colour of silver. The main impurities up to the year 1600, besides the intended addition of copper, were: 0·11–0·38% Au; 0·6–1·25% Pb; 0·02–0·087% Sb; and 0·011–0·16% Bi. The purity should be symptomatic of the gradual improvement in metallurgical technique, but the plate produced in 1477 is very much purer than the others,

Table 66 Composition of Hadrianic die (AD 134–8)[239]

Element, % Cu	Sn	Pb	Zn	Ni	Mg	Total
73·8	14·5	6·55	2·81	tr.	tr.	97·7

which suggests that material made abroad, in a country using superior refining techniques, may have been used.

In 1601, the impurities showed a big fall, owing, no doubt, to improved refining techniques. The gold, which previously had not been removed – cupellation does not remove gold from silver – has now dropped to 0·003%. In 1600, a trial plate was made of a 25% silver alloy, the rest being copper, in which the main impurity is 0·29% arsenic. This is one of three plates made for a debased coinage. The others are: 1526(?), 88·5% silver; and 1542, 76·4% silver.[215]

It appears evident that special refining techniques were employed in the manufacture of the silver trial plates between 1600 and 1829 since these were considerably purer than the alloys actually used for coinage and wrought silver wares.

Silver coins can be embrittled in the course of time by grain-boundary precipitation processes. Thompson and Chatterjee[236] have investigated this, and state that it is particularly bad in the presence of copper, i.e. sterling silver with 7·5% Cu is more likely to be affected than pure silver. Brittle silver coins, therefore, show that the metal itself is not of high purity.

The principal cause of embrittlement would seem to be the precipitation of lead and copper in the silver–lead–copper alloys used. The solubility of copper in silver decreases with temperature from about 8·8% at 750°C, the striking temperature, to less than 1% at room temperature. After striking, the coins are cooled quickly in water and the precipitation of copper takes place slowly over 2 000 years. This raises the hardness from about 70 HV to 163 HV with consequent embrittlement.[33] If reheated, the hardness[237] falls to 80 HV and ductility is regained. Embrittlement has been observed mainly in Saxon and English coins which contain added copper and residual lead (*c.* 1%) as a result of poor cupellation. The silver trial plate of AD 1279 contained 6·2% Cu and 0·81% Pb.[215] A coin of similar purity would probably suffer embrittlement.

The first gold trial plate dates from 1477 and contains 99·35% Au and 0·515% Ag.[238] A gold trial plate used in 1649 contained 5·1% Ag and 3·59% Cu.[215]

Coining dies

It is not yet certain when coining dies were first used in Britain, but by the Roman period official issues were being struck rather than cast. A top die or trussel found at Verulamium[239] is made of 15% tin bronze (*see* Table 66 for analysis) and was found in a deposit dated to the second half of the 2nd century AD. The die is that of a reverse of a *denarius* of Hadrian, whose coinage was struck in fairly pure silver and shows Hadrian being received by Roma within the legend ADVENTS AVG. This type is dated by Mattingly and Sydenham[240] to *c.* AD 134–8. Since Verulamium had no mint in Roman times, it is possible that this die was used for counterfeiting or was brought there and lost after earlier use elsewhere. A similar bottom die (or anvil) for striking Helvetian or Gaulish coins has been described by Keller.[241]

The face of the top die has a hardness of 150 HV which is not sufficient to coin metals harder than gold or pure silver in the cold state. But, since it was clearly intended that the blanks should be struck hot, a die with the hardness of the Hadrianic specimen would be adequate. Steps would have to be taken to see that the die itself remained fairly cool, but this could be done by dipping the trussel in water occasionally, and sprinkling water on the anvil.

According to Balog[232] the large bronze coin of Caligula was struck with cast dies as it shows the marks of casting. Since the coin is not itself cast but struck, these marks must have been transferred from a cast die. We are not, however, told what these 'marks' were, but the evidence is certainly very strong for cast bronze dies. One cannot help feeling, however, that a die with the impression cast into it would need some finishing before being capable of giving a sharp impression. There is no reason why this should not have been done with the aid of an iron punch or engraving tool.

We must now consider how such dies were produced. Hill[242] suggests that they were produced by a 'hub' or master punch being pressed into them to give them their negative or recessed image. This would be made of iron or steel, and if so would not be cast. It would be made from wrought material and engraved.

Since the bronze dies themselves would wear relatively quickly (as indeed steel dies do today), some method is necessary to reface them or produce new ones fairly often. Thus the idea of a steel hub or 'master' is an attractive one if the technique of the time is of a sufficiently high standard to make such a 'master'. It is known that in pre-Roman times gem cutting was carried on by using a drill with a soft end and cutting a groove by using a mixture of oil and abrasive, usually particles of the gem itself. The abrasive particles stick into the soft end – copper is very suitable – and when the drill is rotated, wear away the hard gem-stone as and where required. There is no objection to the use of this process on iron or steel. On the other hand, if the processes of hardening steel by carburization and quenching were understood, there is no reason why a soft steel master should not be engraved by a hard steel engraving tool and the master finally hardened.

In fact, there is evidence to show that it was possible in Roman times to engrave a piece of wrought iron to a considerable depth.

Of course, wrought iron need be no harder than cast bronze containing 15% Sn, but an iron die could have been impressed into heated bronze or, alternatively, the surface of the iron could have been carburized, and therefore hardened, after the engraving of the design.

One such die was found in the River Moselle at Trier.[243] It was much worn and corroded and was probably thrown away because of its useless condition. However, enough remained to show that it consisted of a phosphorus-containing wrought iron shaft with a carburized steel head containing pearlite and cementite, which reached a maximum carbon content of 0·94%. The head consisted of a different iron from the shaft, with a lower phosphorus content. It is possible that the face of the die had been quench-

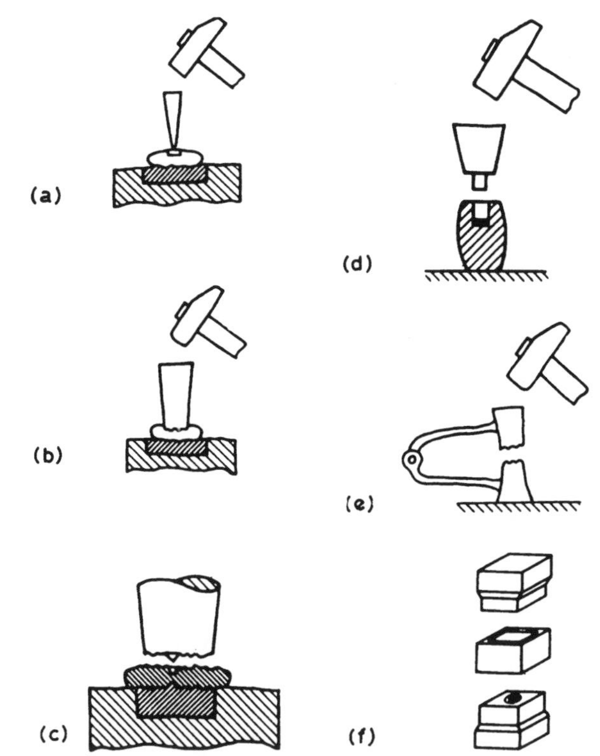

71 Coin stamping by hand (by courtesy of *Met. Treat. Drop Forg*.)

hardened but all evidence for this had been destroyed by corrosion.

This implies a very high level of technique and at the moment it is not certain whether such a level had been reached before the Roman era. It is clear, however, that iron or steel dies were in a minority. Vermeule[244] lists 38 Greek and Roman bronze dies and only eight of iron.

Bronze dies could not have been used economically for cold-striking the brass coin examined by Wootton.[223] It appears, therefore, that iron or steel dies must have been used for at least some of the brass coinage of the Hadrianic period.

The simplicity of the early designs on some of the more degenerate coins suggests that die designs could have been built up with a small number of iron or steel punches. These could have been driven directly into the bronze dies, and used to reface them when worn. If made of iron, the punches would deform and harden with wear and could be sharpened in the usual way on a stone. Iron dies could have been made in a similar way using hardened steel punches. Shirley-Fox (in Macdonald[245]) was of the opinion that the die used for the production of the short-cross coinage of Henry III was made with the aid of about a dozen differently shaped iron or steel punches.

In a discussion of the subject of Islamic die manufacture, Balog[232] believes that the Arabs did not cut the design into the hard surface of a die, but made a pattern in lead, which was impressed into fine clay. This was then used for making dies by casting either bronze or iron. Since no cast iron dies have been found, this process is more likely to have been applied to bronze if used at all. No Byzantine or Arab steel dies exist and the only iron die (at Algiers) is one probably made in more recent times to produce antique (12th century) coins for jewellery. Little evidence exists for the use of steel dies before the 18th century (although Cellini[246] claims to have made them) and it is possible that dies made for 18th century coins are, in fact, of later date.[247]

Some coins bear parts of two impressions on one side, and it would seem that the anvil or pile contained at least two obverse dies, and that the coin was inadvertently placed between them. Hill[242] suggests that in some cases this is due to two obverses being accidentally hubbed into the same anvil, and in others to the obverses of two different coins being put into the same anvil, so that coins of two different denominations could be struck merely by changing the top dies and selecting the appropriate bottom die.

The Coppergate site at York has recently produced clear signs of Saxon minting activities including an iron die and a lead trial piece.[111]

Die alignment

If one examines the two sides of modern coins it will be found that the designs have a fixed relationship to one another. This has not always been so. Early coins were struck free-hand with no system of die alignment (Fig. 71*a*,*b*). This makes the repeated striking of a coin difficult. During the period of the Roman Empire some method of ensuring die alignment was clearly used, and von Wedel[248] suggests a toggle guide or hinge as shown in Fig. 71. Another arrangement is shown by a pair of dies in the Museum at Lyons.[242] The dies are for an *aureus* of Faustina II (AD 130–175), and are said to be cut in steel. They have been inserted into square section iron bars, the top being hollowed out to a depth of 6 cm forming a square socket, into which the bottom die or anvil fits. In this way, a consistent alignment could be obtained. A similar arrangement, shown in Fig. 71*f*, was used in Gaul in the 2nd century BC and was used in England in Anglo-Saxon times until AD 875, according to von Wedel.[248] At this date, coining in England appears to have reverted to the free-hand system, which was in use to the end of the Middle Ages.

Dolley[249] claims that there is some degree of regularity in the later Anglo-Saxon coinage, consistent with the use of a square die. With a square die working in a square guide as shown in Fig. 71*f*, it is possible to get four die alignments. Dolley shows that in the case of 640 coins ranging from AD 975 to 1017, the die axes were evenly spread over the four possible alignments, only 25 being irregular. It is possible that a closer examination of die axes would reveal a similar regularity in much of the medieval coinage and hence prove the use of square dies.

Although free-hand or square die striking became general in Europe during this period, certain exceptions exist in the case of the coinage of Sicily and in Eastern Europe where a fixed axis system was in use.[247] Balog[232] concludes that a fixed alignment by means of pegs fitting into holes or grooves in the opposite die was not used until the 18th century.

NOTES AND REFERENCES

1 H. H. COGHLAN: 'Notes on the prehistoric metallurgy of copper and bronze in the Old World', 53–6, 136; 1975, Oxford.

2 L. H. COPE: *Numismatic Chron.*, 1971, (Nov.), 402–4.
3 S. NEEDHAM: *Proc. Prehist. Soc.*, 1980, **46**, 177–215.
4 J. FOSTER: 'The Iron Age moulds from Gussage All Saints (Dorset)', Brit. Mus. Occas. Paper No. 12, 1980.
5 E. FOLTZ: *Archaeol. Korrespondenzbl.*, 1980, **10**, 345–9.
6 THEOPHILUS: 'Diversarum artium schedula'. (Trans. C. R. DODWELL: 'The various arts. De divers. artibus'; 1961, London, Nelson.
7 B. CELLINI: 'Due trattati'; 1568.
8 P. T. CRADDOCK: 'Scientific examination of metallurgical samples from Rathgall Co. Wicklow', Brit. Mus. Res. Lab. File No. 4552, 6 Oct. 1981.
9 H. W. M. HODGES: *Sibrium*, 1958–9, **4**, 129–37.
10 H. W. M. HODGES: *Ulster J. Archaeol.*, 1954, **17**, 62–80.
11 N. THOMAS: in 'Prehistoric man in Wales and the West; essays in honour of Lily F. Chitty', (ed. F. Lynch and C. Burgess), 161–66; 1972, Bath.
12 J. EVANS: 'Ancient bronze implements'; 1881, London.
13 H. H. COGHLAN and J. RAFTERY: *Sibrium*, 1961, **6**, 223–44.
14 M. J. O'KELLY: *J. R. Soc. Irel.*, 1969, **99**, (2), 117–24.
15 M. J. O'KELLY: *J. Cork Hist. Archaeol. Soc.*, 1970, **75**, 25–8.
16 H. W. M. HODGES: *Sibrium*, 1960, **5**, 153–62.
17 H. O'NEILL HENCKEN: 'Archaeology of Cornwall and Scilly'; 1932, London.
18 H. H. COGHLAN: 'Notes on prehistoric copper', 133.
19 National Museum of Antiquities, Edinburgh.
20 G. COFFEY: *J. R. Soc. Antiq. Irel*, 1907, **37**, 181–6.
21 J. EVANS: 'Ancient bronze implements', 321.
22 R. F. TYLECOTE: *Bull. Hist. Metall. Group*, 1973, **7**, (1), 1–5.
23 A. E. P. COLLINS: *Ulster J. Archaeol.*, 1970, 33, 23–36.
24 J. V. S. MEGAW: in 'Settlement and economy', (ed. C. Burgess and R. Miket), 70, Br. Archaeol. Rep. 33, 1976.
25 W. WILDE: 'Catalogue of the Museum of RIA'; 1857.
26 S. NEEDHAM: 'The Bulsford-Helsbury manufacturing tradition', Brit. Mus. Occas. Paper No. 13, 1981.
27 C. BURGESS: 'The age of Stonehenge'; 1980, London, Dent.
28 C. MUSSON: Breiddin, 1970–1, private communication.
29 G. D. LIVERSAGE: *Proc. R. Ir. Acad.*, 1968, **66**, (C2), 53–233.
30 B. RAFTERY: *Antiquity*, 1970, **44**, 51–4; 1971, **45**, 296–8. *J. R. Soc. Antiq. Irel.*, 1970, (C2), 200–11.
31 H. H. COGHLAN: *Man*, 1953, **53**, 97–101.
32 E. E. RICHARDS and A. E. BLIN-STOYLE: *Archaeometry*, 1961, **4**, 53–5.
33 L. H. COPE: in 'Methods of chem. and met. investig. of ancient coinage', Special Publication No. 8, 1–47, R. Numismatic Soc., London, 1972.
34 W. F. GRIMES: *Archaeol. Cambrensis*, 1931, **86**, 358.
35 J. EVANS: 'Ancient bronze implements', 437.
36 E. M. JOPE: *Ulster J. Archaeol.*, 1953, **16**, 37–40.
37 R. B. K. STEVENSON: *Proc. Soc. Antiq. Scotl.*, 1957–8, **91**, 191–2.
38 E. K. CLARKE: *Proc. Soc. Antiq. Lond.*, 1905, **20**, 258–61.
39 M. EHRENBERG: *Antiquity*, 1981, **45**, 214–18.
40 A. O. CURLE: *Proc. Soc. Antiq. Scotl.*, 1932, **67**, 82–136.
41 M. A. SMITH: *Proc. Prehist. Soc.*, 1959, **25**, 144–87.
42 'Bronze Age metalwork in Norwich Castle Museum', 2 ed.; 1977, Norfolk Museums Service.
43 P. T. CRADDOCK: 'Beeston Regis', Rep. No. 4453, Br. Mus., 1981–2.
44 H. H. COGHLAN: 'Notes on prehistoric copper', 55, 137.
45 *Inventaria Archaeol. GB*, 18.
46 *Inventaria Archaeol. GB*, 51.
47 *Inventaria Archaeol. GB*, 50.
48 P. DE LISLE: *Rev. Archéol.*, 1881, 42, 335–43.
49 S. S. CRICHTON MITCHELL and M. E. CRICHTON MITCHELL: *Proc. Soc. Antiq. Scotl.*, 1934–6, **69**, 424–30.
50 W. J. WEDLAKE: 'Excavations at Camerton, Somerset', 83, Pl. 17; 1958, Camerton Excavation Club.
51 W. J. WEDLAKE: 'The excavation of the shrine of Apollo at Nettleton, Wilts. 1956–1971', Res. Rep. No. 40, Soc. Antiq., London, 1982.
52 T. S. BUSH: *Proc. Soc. Antiq. Lond.*, 1908, **22**, 34–8.
53 T. BLAGG and S. READ: *Antiq. J.*, 1977, **57**, 270–6.
54 M. E. CUNNINGTON: *Archaeol. Cambrensis*, 1920, **20**, (6th set), 251–6.
55 P. D. C. BROWN: *Corn. Archaeol.*, 1970, **9**, 107–10.
56 G. GUILBERT: Dinoben Hill Fort; personal communication. (This stone may be part of a cheese press.)
57 P. S. GELLING: *Medieval Archaeol.*, 1969, **13**, 67–83.
58 I. H. GOODALL: *York Archaeol. J.*, 1972, **44**, 32–8.
59 J. BAYLEY: in: A. Rogerson and C. Dallas 'Excavations in Thetford, 1948–59 and 1974–7'; East Anglian Archaeol. (forthcoming).
60 C. I. FELL: *Archaeol. J.*, 1936, **93**, 57–100.
61 N. Q. BOGDAN and J. W. WORDSWORTH: 'The medieval excavations at the High St, Perth', Perth High St Excavation Committee, 1978.
62 L. B. HUNT: *Gold Bull.*, 1980, **13**, (2), 63–79.
63 A. O. CURLE and J. E. CREE: *Proc. Soc. Antiq. Scotl.*, 1915–16, **50**, 64–144.
64 S. P. O'RIORDAIN: *Proc. R. Ir. Acad. (C)*, 1941, **47**, 77–150.
65 R. F. TYLECOTE: *Bull. Hist. Metall. Group*, 1969, **3**, (2), 46–7.
66 S. FRERE: *Britannia*, 1970, **1**, 266–7.
67 M. G. SPRATLING: 'An Iron Age bridle bit mould fragment from South Cadbury Somerset', Rep. 11. Sept. 1974.
68 E. BLANK: 'Ratae Coritanorum'; 1971, London.
69 M. U. JONES: *Antiq. J.*, 1975, **55**, (2), 407–8.
70 S. CRACKNELL: personal communication, 4 Feb. 1977.
71 C. L. CURLE: *Proc. Soc. Antiq. Scotl.*, 1972–4, **105**, 301–7.
72 D. L. C. WILSON and J. BAYLEY: personal communication. AM Report, 2953. Publication forthcoming in *Trans. BWAS*, **92**.

73 K. H. ARMITAGE, J. E. PEARCE, and A. G. VINCE: *Antiq. J.*, 1981, **61**, (2), 362–4.

74 K. LAMM: *Bull. Hist. Metall. Group*, 1973, **7**, (2), 1–7.

75 R. F. TYLECOTE: Material examined from Southampton and reported 2 Apr. 1974.

76 A. E. P. COLLINS: *Ulster J. Archaeol. (3rd ser.)*, 1955, **18**, 45–82.

77 C. S. SMITH and M. T. GNUDI (eds.): 'The Pirotechnia of Vannoccio Biringuccio'; 1959, New York.

78 K. MARSHALL: *Ulster J. Archaeol.*, 1950, **13**, 66–75.

79 C. S. BRIGGS and G. A. WARD: *J. Kerry Archaeol. Hist. Soc.*, 1979, **12**, 5–8.

80 J. BAYLEY: 'Non-ferrous metal and glassworking in Anglo-Scandinavian England; an interim statement', **7**, 487–96; 1982, PACT.

81 G. L. BOON: *Trans. CWAAS*, 1976, **76**, 97–132.

82 W. GOWLAND: *Archaeologia*, 1899, **56**, 267–322.

83 R. J. MOSS: *Proc. R. Ir. Acad. (C)*, 1924–7, **37**, 175–93.

84 H. ST. G. GRAY and A. BULLEID: 'Meare Lake Village', Vol. I; 1948, Taunton.

85 G. COFFEY: *Proc. R. Ir. Acad.*, 1906–7, **26**, 109–18.

86 J. DYER and P. WENHAM: *Yorks. Archaeol. J.*, 1958, **39**, 419–26.

87 P. WENHAM: *Yorks. Archaeol. J.*, 1972, **44**, 65–113.

88 T. MAY: *Iron Coal Trades Rev.*, 1905, **71**, 427.

89 M. G. SPRATLING, R. F. TYLECOTE, *et al.*: in 'Proc. 16th Internat. Symposium on Archaeometry, Edinburgh 1976', (ed. E. A. Slater and J. O. Tate).

90 W. BRITNELL: *Current Archaeol.*, 1974, **45**, 293–7.

91 R. HAYNES: *J. Iron Steel Inst.*, 1956, **183**, 359–61.

92 C. H. READ: 'A guide to the antiquities of the Early Iron Age', 89; 1905, Br. Mus. Since this was published many additional analyses have been made on red and other glasses and enamels. One of the most complete is M. J. HUGHES: Proc. Prehist. Soc., 1972, **38**, 98–107.

93 L. ALCOCK: *Archaeology*, 1964, **17**, (2), 104–11.

94 J. H. WILLIAMS: 'St. Peter's St, Northampton; Excavations 1973–76', Archaeol. Monograph No. 2, Northampton Development Corp., 1979.

95 H. SCHLIEMANN: 'The city and country of the Trojans', Ilios, The burnt city, Ch. VII, Moulds of mica schist, Crucibles; 1880, London.

96 C. GREIG: *Current Archaeol.*, 1972, **32**, 227–31.

97 A. O. CURLE: *Proc. Soc. Antiq. Scotl.*, 1933–4, **68**, 224–319.

98 J. R. C. HAMILTON: 'Excavations at Jarlshof, Shetland'; 1956, Edinburgh, H.M.S.O.

99 A. BULLEID and H. ST. G. GRAY: 'Glastonbury Lake Village', (2 Vols.); 1911 and 1917, Glastonbury Antiquarian Soc.

100 R. E. M. WHEELER: 'Maiden Castle', Res. Rep. No. 12., Soc. Antiq., London, 1943.

101 K. M. KENYON: *Archaeol. J.*, 1963, **110**, 1–87.

102 A. FOX: 'Roman Exeter. Excavations in the war-damaged areas', 1952, Manchester University Press.

103 'Antiquities of Roman Britain', 78, Fig. 39; 1951, London, Trustees of the Brit. Mus.

104 E. T. ARTIS: 'The Durobrivae of Antoninus'; 1828, London.

105 H. G. BACHMANN: *Germania*, 1977, **55**, (1–2), 85–107.

106 R. E. TYLECOTE: in 'Archaeological ceramics', (ed. A. D. Franklin and J. Olin), 231–56; 1982, Washington DC, Smithsonian Institution Press.

107 E. M. JOPE: *Oxon.*, 1952–3, **17–18**, 77–111.

108 D. CHRISTISON, J. ANDERSON, and T. ROSS: *Proc. Soc. Antiq. Scotl.*, 1905, **39**, 259–322.

109 J. H. CRAW: *Proc. Soc. Antiq. Scotl.*, 1929–30, **64**, 111–46.

110 H. HENCKEN: *Proc. R. Ir. Acad. (C)*, 1950, **53**, 1–247.

111 J. BAYLEY: *PACT*, 1982, **7**, 487–96.

112 K. KILMURRY: 'The pottery industry of Stamford, Lincs., AD 850–1250, Br. Archaeol. Rep. 84, 1980.

113 Now in the Museum at Belfast.

114 H. H. COGHLAN: 'Notes on prehistoric copper', 69.

115 W. GREENWELL: *Archaeologia*, 1894, **54**, 88–114.

116 C. FOX: 'A find of the Early Iron Age from Llyn Cerrig Bach, Anglesey', 1947, Cardiff, National Museum of Wales.

117 Now in the National Museum, Dublin. (For analysis see Table 57.)

118 R. MUNRO: *Archaeol. Hist. Collect. Ayr Wigtown*, 1882, **3**, 19.

119 G. COFFEY: 'The Bronze Age in Ireland'; 1913, Dublin.

120 M. R. EHRENBERG: *Antiq. J.*, 1981, **61**, (1), 14–28.

121 H. MARYON: *Proc. R. Inst. Ir. (C)*, 1938, **44**, 181–228.

122 W. A. ODDY: *Masca J.*, 1980, **1**, (4), 110–11.

123 W. A. ODDY: *Gold Bull.*, 1977, **10**, (3), 79–87; *Masca J.*, 1970, **1**, (2), 44–5.

124 F. K. NEUMANN: 'Metallkunde; Untersuchungen an drei Wikinger-zeitlichen Zieheisen aus Haithabu', in: Berichte 5, Untersuchungen zur Technologie des Eisens, 84–98 Neumünster, 1971.

125 D. BRITTON: *Antiquity*, 1960, **34**, 279–82, Pl. 36 (4).

126 R. F. TYLECOTE: Metallurgical examination of the Gilmonby, Co. Durham Bronze Age hoard. Forthcoming.

127 I. M. ALLEN, D. BRITTON, and H. H. COGHLAN: 'Metallurgical reports on British and Irish Bronze Age implements and weapons in the Pitt Rivers Museum', Occas. Paper No. 10, Pitt Rivers Museum, Oxford, 1970.

128 P. HOLMES: in 'The origins of metallurgy in Atlantic Europe', (ed. M. Ryan), 165–88; 1978.

129 H. McKERRELL and R. F. TYLECOTE: *Proc. Prehist. Soc.*, 1972, **38**, 209–18.

130 H. CASE: *Antiquity*, 1965, **39**, 219–22.

131 C. BURGESS and D. COOMBS (eds.): 'Bronze Age hoards; some finds old and new', Br. Archaeol. Reps. (Brit. Ser.) 67, Oxford, 1979.

132 H. H. COGHLAN and G. PARKER: 'A report upon the hoard of Bronze Age tools and weapons from Yattendon nr. Newbury, Berks'; 1970, Newbury.

133 R. F. TYLECOTE: Metallurgical examination of iron work from Wanborough, Wilts. Forthcoming.

134 W. H. MANNING: in 'Aspects of early metallurgy', (ed. W. A. Oddy), 87–96; 1977, London.

135 J. D. COWEN and H. MARYON: *Archaeol. Aeliana*, 135, **12**, 280–309.

136 J. J. TAYLOR: 'Bronze Age goldwork of the British Isles'; 1980, Cambridge.

137 H. ST. G. GRAY: *Proc. SANHS*, 1909, **55**, 66–84.

138 C. F. C. HAWKES: *Man*, 1932, **32**, (222), 177.

139 R. R. CLARKE: *Proc. Prehist. Soc.*, 1954, **20**, 27–86.

(Recent examination by P. T. Craddock of the Brit. Mus. Res. Lab. has shown that the fragments of 'tin' torcs referred to in this report are in fact made of silver.)
140 A. J. EVANS: *Archaeologia*, 1897, **55**, 391–408.
141 J. BRAILSFORD and J. E. STAPLEY: *Proc. Prehist. Soc.*, 1972, **38**, 219–34.
142 J. E. BURNS: *Proc. Prehist. Soc.*, 1971, **37**, 228–9.
143 A. G. G. LEONARD and P. F. WHELAN: *Sci. Proc. R. Dublin Soc.*, 1929, **19**, 55.
144 J. COLES: *Proc. Prehist. Soc.*, 1963, **29**, 326–56.
145 J. EVANS: 'Ancient bronze implements', 362 *et seq.*
146 V. G. CHILDE: *J. R. Anthropol. Inst.*, 1944, **74**, 20.
147 E. T. LEEDS: *Archaeologia*, 1930, **80**, 1–36.
148 C. F. C. HAWKES and M. A. SMITH: *Antiq. J.*, 1957, **37**, 131–98.
149 H. E. DUDLEY: *Archaeol. J.*, 1946, **103**, 8–11; *Inventaria Archaeol. GB*, 23.
150 *Inventaria Archaeol. GB*, 55.
151 The bottom plate of the Hatton Knowe cauldron (D.U.8) in the National Museum of Antiquities, Edinburgh, weighs about 1·6 kg.
152 J. W. BAGGALEY: *Antiq. J.*, 1950, **30**, 195.
153 C. FOX and H. A. HYDE: *Antiq. J.*, 1939, **19**, 367–404.
154 M. A. BROWN and A. E. BLIN-STOYLE: *Proc. Prehist. Soc.*, 1959, **25**, 188–208. (Analyses in *Archaeometry*, 2, Supp.)
155 H. H. COGHLAN, W. WATSON, and E. VOCE: *Man*, 1951, **51** (234), 37.
156 This was suggested by Dr J. Raftery of the National Museum, Dublin.
157 W. WATSON: *Antiq. J.*, 1949, **29**, 37–61.
158 R. SMITH: *Proc. Soc. Antiq. Lond.*, 1914–15, **27**, 76–80.
159 H. DUDLEY: 'Early days in north-west Lincolnshire,; 1949, Scunthorpe.
160 J. R. ALLEN: *Archaeologia*, 1898, **56**, 39–56.
161 H. H. COGHLAN: 'Notes on prehistoric copper', 62.
162 G. BALDWIN BROWN: 'The arts in Early England', Vol. 2, 'Saxon art and industry in the Pagan period'; 1915, London.
163 A. J. EVANS: *Archaeologia*, 1890, **52**, (2), 357–88.
164 R. F. TYLECOTE: Result of the examination of three vessels from Locherbie, Lochmaben, and Appin. By courtesy of the Dumfries Museum and J. Williams; 8 Nov. 1967.
165 R. F. TYLECOTE: *Br. Weld. J.*, 1962, **9**, 512–22.
166 R. F. TYLECOTE: 'The solid phase welding of metals'; 1968, London, Edward Arnold.
167 A. O. CURLE: 'The treasure of Traprain'; 1923, Glasgow.
168 J. G. HAWTHORNE and C. S. SMITH (trans. and eds.): 'On divers arts: the Treatise of Theophilus', 1963; The University of Chicago Press.
169 J. N. FRIEND and W. E. THORNEYCROFT: *J. Inst. Met.*, 1928, **39**, 61–2.
170 J. LANG and M. J. HUGHES: in 'Aspects of early metallurgy', (ed. W. A. Oddy), 169–77; 1977, London.
171 J. W. BRAILSFORD: 'The Mildenhall Treasure', (2 ed.); 1955, London, Br. Mus.
172 R. L. S. BRUCE MITFORD: 'The Sutton Hoo ship burial'; 1947, London, Br. Mus.
173 A. A. MOSS: *Conservation*, 1953, **1**, 49–61.
174 J. A. SMITH: *Proc. Soc. Antiq. Scotl.*, 1873, **9**, 435–43.
175 J. ANDERSON: *Proc. Soc. Antiq. Scotl.*, 1900–1, **35**, 266–75; *Inventaria Archaeol. GB*, 26.
176 W. A. ODDY: in 'Aspects of early metallurgy', (ed. W. A. Oddy), 129–34; 1977, London.
177 S. NEEDHAM and I. KINNES: *Antiquity*, 1981, **55**, (214), 133–4.
178 I. A. KINNES, P. R. T. CRADDOCK, S. P. NEEDHAM, and J. LANG: *Antiquity*, 1979, **53**, 141–3.
179 J. CLOSE-BROOKS and J. M. COLES: *Antiquity*, 1980, **54**, 228–9.
180 J. R. ALLEN: *Archaeol. Cambrensis*, 1901, **1**, 20–44.
181 E. M. JOPE: *Oxon.*, 1956, **21**, 35–42.
182 DIDEROT and D'ALEMBERT: 'Encyclopédie', Recueil de Planches, Vol. 24, 662–5; 1762, Paris.
183 A. THOUVENIN: *Rev. Hist. Mines Métall.*, 1970, **2**, (1), 101–9.
184 H. LECHTMAN *et al.*: *Am. Antiq.*, 1982, **47**, (1), 3–30.
185 G. BERESFORD: 'The Medieval clay-land village', Monograph No. 6, 85, *Soc. Med. Archaeol.*, 1975.
186 P. T. CRADDOCK: 'Report on material from Rathgall', (exc. B. Raftery), Brit. Mus. Res. Lab., 1981.
187 P. A. LINS and W. A. ODDY: *J. Archaeol. Sci.*, 1975, **2**, 365–73.
188 W. A. ODDY: *Gold Bull.*, 1981, **14**, (2), 75–9.
189 A. R. WILLIAMS: *Gold Bull.*, 1977, **10**, (4), 115–17.
190 C. S. SMITH and J. G. HAWTHORNE (eds.): *Trans. Am. Philos. Soc.* (New Ser.), 1974, **64**, (4), 64.
191 W. A. ODDY and N. D. MEEKS: *Masca J.*, 1981, **1**, (7), 211–13.
192 P. R. LOWERY, R. SAVAGE, and R. L. WILKINS: *Proc. Prehist. Soc.*, 1971, **37**, (1), 167–82.
193 D. ATKINSON: 'Report on excavations at Wroxeter (The Roman city of Viroconium) in the County of Salop, 1923–7', 217; 1942, Oxford, Birmingham Archaeol. Soc. (The die is now in Rowley House Museum, Shrewsbury.)
194 G. WEBSTER: Bronze die from Sheepen, Colchester; private communication.
195 S. C. HAWKES, G. SPEAKE, and P. NORTHOVER: *Frühmittelalterliche Stud.*, 1979, **13**, 382–92.
196 D. NASH: in 'Cross-Channel trade between Gaul and Britain in the pre-Roman Iron Age', (ed. S. Macready and F. H. Thompson), 92–107; 1984, London, Society of Antiquaries.
197 S. S. FRERE (ed.): 'Problems of the Iron Age in Southern Britain', Occas. Paper No. 11, University of London Institute of Archaeology, 1960. D. F. ALLEN: 'The origins of coinage in Britain: a re-appraisal', 97–308.
198 G. C. BROOKE: *Numismatic Chron.* (5th ser.), 1927, **1**, 371–7.
199 B. CUNLIFFE: in 'Cross-Channel trade between Gaul and Britain in the pre-Roman Iron Age', (ed. S. Macready and F. H. Thompson), 3–23; 1984, London, Society of Antiquaries.
200 L. SELLWOOD: *Oxford. J. Archaeol.*, 1983, **2**, (2), 199–211.
201 L. LONGOUËT: in 'Cross-Channel trade between Gaul and Britain in the pre-Roman Iron Age', (ed. S. Macready and F. H. Thompson), 67–76; 1984, London, Society of Antiquaries. (The work on the cupellation material was done by the BNF Metals Technology Centre, Wantage, on behalf

of J. R. Maréchal and H. H. Coghlan, Report DB/DI/TE77/263.)

202 J. P. BUSHE-FOX: 'Excavations at Hengistbury Head, Hampshire, in 1911–12', Rep. No. 3, 1915, London, Society of Antiquaries. App. by W. Gowland.

203 *ibid.*

204 R. A. G. CARSON: *Numismatic Chron.* (6th ser.), 1950, **10**, 148–9.

205 G. F. HILL: *Numismatic Chron.* (4th ser.), 1917, **17**, 316–18.

206 D. F. ALLEN: in 'Trans. Internat. Numismatic Congress of 1936', 351–7; 1938, London.

207 R. P. MACK: 'The coinage of ancient Britain'; 1953, London.

208 J. EVANS: 'The coins of the Ancient Britons', 124; 1865, London.

209 S. S. FRERE: *Antiq. J.*, 1957, **37**, (1, 2), 1–15.

210 S. S. FRERE: *Antiq. J.*, 1958, **38**, (1, 2), 1–24.

211 E. E. RICHARDS and M. J. AITKEN: *Archaeometry*, 1959, **2**, 53.

212 R. F. TYLECOTE: *Numismatic Chron.* (7th ser.), 1962, **2**, 101–9.

213 J. TOURNAIRE, O. BUCHSENSCHUTZ, J. HENDERSON, and J. COLLIS: *Proc. Prehist. Soc.*, 1982, **48**, 417–35.

214 B. CUNLIFFE: 'Hengistbury Head, Dorset'; 1978, London.

215 W. A. C. NEWMAN: 'British coinage'; 1953, London, R. Inst. Chem.

216 J. GOWLAND: *J. Inst. Met.*, 1912, **7**, 23–49.

217 L. H. COPE: *Bull. Hist. Metall. Group*, 1967, **1**, (9), 1–6.

218 L. H. COPE and H. N. BILLINGHAM: *Bull. Hist. Metall. Group*, 1968, **2**, (1), 51–3.

219 L. H. COPE and H. N. BILLINGHAM: *Bull. Hist. Metall. Group*, 1968, **2**, (2), 71–2.

220 L. H. COPE and H. N. BILLINGHAM: *Bull. Hist. Metall. Group*, 1969, **3**, (1), 30–2.

221 L. H. COPE: *Bull. Hist. Metall. Group*, 1969, **3**, (2), 62–4.

222 L. H. COPE and H. N. BILLINGHAM: *Bull. Hist. Metall. Group*, 1970, **3**, (1), 33–6.

223 A. WOOTTON: *Seaby's Coin Medal Bull.*, 1960, (Jun.), 223.

224 D. SELLWOOD: MSc thesis, University of Durham, 1976.

225 D. SELLWOOD: in 'Roman crafts', (ed. D. Strong and D. Brown), 63–73; 1976.

226 E. A. SMITH: *J. Inst. Met.*, 1939, **65**, 271–5.

227 L. H. COPE: 'Surface-silvered ancient coins. R.M.S. Symposium', Special Report No. 8, 261–78, R. Numismatic Soc.

228 L. H. COPE: *Metallurgia*, 1967, (Jan.), 15–20.

229 E. KALSCH and U. ZWICKER: *Mikrochim. Acta*, 1968, **Supp. III**, 210–20.

230 U. ZWICKER: in 'Jahrbuch für Numismatik und Geldgeschichte', No. 23, 115–16; 1973.

231 G. C. BOON and P. A. RAHTZ: *Archaeol. J.*, 1966, **122**, 13–51.

232 P. BALOG: *Numismatic Chron.* (6th ser.), 1955, **15**, 195–202.

233 R. E. M. WHEELER and T. V. WHEELER: 'Report on the excavation of the prehistoric, Roman and post-Roman site at Lydney Park, Gloucestershire', Rep. No. 9, Soc. Antiq., London, 1939.

234 H. McKERRELL and R. B. K. STEVENSON: in 'Methods of chemical and metallurgical investigation of ancient coinage', (ed. E. T. Hall and D. M. Metcalf), Special Publication No. 8, 195–210, R. Numismatic Soc., London, 1972.

235 J. S. FORBES and D. B. DALLADAY: *J. Inst. Met.*, 1958–9, **87**, 55–8.

236 F. C. THOMPSON and A. K. CHATTERJEE: *Conservation*, 1954, **1**, (3), 115.

237 F. C. THOMPSON: in 'Methods of chemical and metallurgical investigation of ancient coinage', (ed. E. T. Hall and D. M. Metcalf), Special Publication No. 8, 67–8, R. Numismatic Soc., London, 1972.

238 J. H. WATSON: *Bull. Inst. Mining Metall.*, 1959, **68**, 475–85.

239 R. E. M. WHEELER and T. V. WHEELER: 'Verulamium, a Belgic and two Roman cities', Res. Rep. No. 11, Soc. Antiq., London, 1936.

240 H. MATTINGLY and E. A. SYDENHAM: 'Roman Imperial coinage'; 1923–33, London.

241 F. KELLER: *Archaeol. J.*, 1862, **19**, 252–8.

242 G. F. HILL: *Numismatic Chron.* (5th ser.), 1922, **2**, 1–41.

243 G. BECKER and W. DICK: *Arch. Eisenhüttenwes.*, 1967, **38**, 351–4.

244 C. C. VERMEULE: *Numismatic Circ.*, 1953, **61**, (10–12); 1954, **62**, (1–3).

245 G. MACDONALD: 'The evolution of coinage', 69–70; 1916, Cambridge University Press.

246 B. CELLINI: 'The life of Benvenuto Cellini', (trans. A. Macdonald), 73; 1907, (reprinted 1960), London.

247 P. GRIERSON: *Numismatic Chron.* (6th ser.), 1952, **12**, 99.

248 E. VON WEDEL: *Met. Treat. Drop Forg.*, 1960, **27**, 401–8.

249 R. H. M. DOLLEY: *Br. Numismatic J.*, 1952, **27**, 167–74.

6 *The coming of iron*

The use of iron was introduced into Britain around the 7th century BC during the late Hallstatt period, but it was not until the La Tène period that its use increased significantly and there is evidence for local production. The European background is complex; dates in northern Europe are similar to those of Britain while those from the south suggest an earlier date around the turn of the millennium.

The reasons behind the gradual replacement of copper by iron in Europe are obviously a complex admixture of sociological, economic, and technological factors. Its gradual acceptance in Britain would have been heavily influenced by trading links with late Hallstatt and subsequently La Tène Europe where iron was becoming the predominant metal. It seems probable that iron was imported from the continent in some quantity, especially into the Thames Valley where it was worked into local forms and distributed. With increasing local production smiths would be able to acquire a cheaper product and eventually achieve self-sufficiency by smelting their own ore – so much so that Strabo[1] lists iron as one of Britain's mineral resources and exports before Caesar's campaign.

The furnace technology does not differ significantly from that of copper, and iron could easily be produced in copper smelting furnaces. It has been suggested that iron did not become a viable alternative until carburization and consequent hardening of the metal had been mastered, thus producing a metal harder than bronze. However, iron from the bloomery process is often heterogeneous with varying amounts of carbon which, together with small amounts of phosphorus, would result in a metal as hard, if not harder, than much of the leaded bronze available at that time.

One other decisive factor in the transition would have been the abundant nature of iron ores; unlike non-ferrous ores, iron ore has been available in nearly all British counties. Much of our information on ore sources can be dated to the 19th century when there was a notable development of the industry. Relying on these relatively modern sources for information has its drawbacks, since it is only a record of sizeable economic deposits. Recent archaeological evidence points to early iron smelting in areas not usually considered to be ore-bearing, e.g. Chinnor in Buckinghamshire[2] and Mucking in Essex,[3] and it is evident that small, workable deposits of iron ore were readily available throughout Britain. (Examples of British ores are given in Tables 67 and 68.)

IRON ORES

Three main types of ore are found in Britain: carbonate, hematite, and limonite ores.

Carbonate ores are by far the most common and occur in two forms: (*a*) as nodules, found in the Wealden Series and in the Coal Measures; (*b*) as sedimentary deposits in Northamptonshire, Lincolnshire, Oxford (Lias), and the Cleveland Hills, which are sometimes called siderite or spathic iron ore. In both forms, the iron is present essentially as ferrous carbonate ($FeCO_3$). When the ores are roasted the carbon dioxide is driven off, as in lime burning, and ferrous oxide (FeO) is left. If roasting is taken to a further stage, the ferrous oxide will be converted to magnetite (Fe_3O_4) or to ferric oxide (Fe_2O_3) which is chemically the same as hematite.

The nodular form is sometimes called clay-band, argillaceous, or clay-ironstone, and its manner of formation is not clear. The nodules can be any size from a few centimetres in diameter up to a metre. It is very common in the Coal Measures, where it appears to have formed by the crystallization onto suitable nuclei of the iron in the surrounding shales. Normally it will underlie the coal, and where seams outcrop, the beds of iron ore nodules may also appear. In the medieval period this material was mined by digging 'bell-pits' (not always bell-shaped) through the shale and coal. In many cases no attempt was made to use the coal, and the coal that had to be moved to get at the ore below was back-filled with the shale.[4]

The other carbonate ores are found along the Jurassic scarp from the Cleveland Hills into Oxfordshire. They were deposited in the Lias, and although always carbonate ores at depth, the surface deposits have weathered in some places to limonite, magnetite, or hematite.

The second group, the hematite ores (Fe_2O_3), were much sought after during the Industrial Revolution, as they contain very little of the phosphorus which gave difficulties in the early days of steelmaking. Such ores occur mainly in West Cumberland and in the Furness district of Lancashire (both areas are now in Cumbria). Now that these are exhausted, most of our indigenous ore comes from the Jurassic scarp and contains an appreciable amount of phosphorus.

Table 67 Composition of some British iron ores

Chapter Ref.	105	6	105	105	105	105	105	105	106	107	105	29	108
%	Forest of Dean limonite	Lydney, Forest of Dean limonite	Llanharry, S. Wales limonite	Furness, Cumbria hematite	Brendon Hills limonite	Seend, Wilts. limonite	Fawler, Oxon. limonite	Westbury, Wilts. siderite	Ashburnham, Sussex nodules	Northumberland, Coal Measures, nodules	Towcester, Northants.	Marlborough, Wilts. limonitised pyrite	Crawley, Sussex 'Horsham stone'
FeO	—	—	—	—	—	—	1·04	47·25	42·00	44·05	—	—	—
Fe_2O_3	90·05	91·20	74·71	84·47	73·71	64·61	46·93	1·32	6·85	6·71	64·62	97–99	73·80
SiO_2	1·07	2·2	3·86	6·95	1·56	18·02	10·55	14·72	6·46	8·62	13·52	2–3	8·52
CaO	0·06	0·5	1·90	0·25	0·15	0·64	11·79	0·74	3·87	0·94	0·90	0·3–0·5	1·80
MnO	0·08	—	0·19	0·22	10·80	—	0·51	—	2·32	—	} 3·91	—	4·44
Al_2O_3	—	—	1·90	—	—	3·85	7·86	5·46	2·64	2·89	}	—	5·54
MgO	0·20	0·3	0·94	0·41	2·66	0·20	1·12	0·36	1·76	0·77	0·25	0·2–0·4	1·86
P_2O_5	0·09	0·4	0·01	0·03	—	0·64	0·66	tr.	0·65	—	2·15	—	0·75
CO_2	—	—	4·90	—	0·14	—	8·55	30·01	32·70	28·35	} 14·60	—	0·12
H_2O	9·22	—	7·88	8·48	11·06	11·85	10·15	—	0·15	4·85	}	—	1·86
S	—	—	0·02	—	—	—	tr.	0·05	0·10	—	0·02	0·2	—
TiO_2	—	—	—	—	—	—	—	0·05	0·21	0·20	—	—	0·35
Alkali	—	0·1	—	—	—	—	—	—	—	2·62	—	—	0·48
FeS_2	—	—	—	—	—	—	—	—	0·11	—	—	—	—

tr. = trace; — = not sought

The third and largest limonite group of deposits are found in the Forest of Dean and along the edge of the South Wales coalfield between Rudry and Llanharry. Limonites are hydrated iron oxides and consist mainly of the crystalline oxide goethite ($Fe_2O_3.H_2O$ or FeO.OH) with varying amounts of absorbed water.

Most deposits of non-ferrous metal yield iron ore of one type or another. For example, iron ore occurs in the copper deposits of Alderley Edge, and was smelted nearby;[5] there are siderite and limonite deposits in close proximity to the lead ore (galena) of Weardale (Durham); hematite has been found at Priddy in the Mendips; and an enormous range of minerals, including siderite and hematite, occurs in the Leadhills area of Lanarkshire.

In more recent times the mining of most of these ores, except for those of the Jurassic scarp, has had to be carried out by deep mining methods, and the deeper ores would not have been available to early people. However, all deposits have outcropped somewhere or other at some time. Those in the Coal Measures outcrop near the coal. Valleys intersecting deposits are usually the sites of outcrops elsewhere, and there is no doubt that most of the deposits mentioned above could have yielded sufficient outcrop ore for primitive peoples. At Lydney[6] (Forest of Dean) a small underground working has been located, sealed by the floor of a 3rd century AD hut. This is perhaps an exception, since both ferrous and non-ferrous ores were mainly mined by open-cast methods up to early medieval times.

Bog iron ore is widespread in the northern and western parts of the British Isles. As its name implies, it is not a rock but a deposit formed under wet conditions. In many ways it resembles peat and, in areas where it contains manganese, has a definite indigo colour. In Scandinavia it is often deposited on the bottom of lakes and is recovered by cutting holes through the ice and dredging with sieve-like spades. In Britain it can be found just under the turf on moorland, and in Weardale it occurs in thicknesses of 8–10 cm at depths of about 20 cm. Where iron-bearing surface waters meet organic material, conditions suitable for the precipitation of iron oxides are created. Deposition is often continuous and once the ore has been removed, more will form at the rate of 5–10 cm in a period of 30 years.

The Tertiary strata of the Hampshire area have yielded deposits worked in the pre-Roman Iron Age and in the 19th century, and there are deposits in similar strata in the Home Counties. In fact there is no part of the British Isles where iron ore of some sort or another is not fairly accessible. This is the prime reason why the use of iron advanced so rapidly in comparison with the use of copper-base metals, once the more difficult technique of production had been mastered.

Composition of iron ores

Table 67 gives the chemical composition, as far as it is known, of ores typical of those described in the previous section. Table 68 gives the analysis of ores found on excavated sites, some of which have been roasted, and therefore are not completely representative of the natural ore.

The analysis of ore minerals is based on elemental analysis supplemented by physical tests, which give

Table 68 Composition of iron ores from excavated sites

Chapter Ref. / %	4 Harthorpe Mill, Durham, 15th cent. roasted nod.	19 Ashwicken, Norfolk, Roman roasted nodules	109 Weardale, 12th cent. bog ore	110 Templebro', Yorks., Roman	48 Dean Moor, Devon, LBA hematite	111 Ballyvourney, Co. Cork, E-Christ, bog ore	108 St. George's Hill, Weybridge, siderite	112 Brookfield Cottage, Northants., limonite	113 Panningridge, Sussex, Period 1 siderite	114 Tintern, Gwent, hematite
FeO	9·18	nil	—	—	—	—	—	0·25	28·9	0·39
F_2O_3	62·48	72·5	39·2	61·00	90·91	20·11	65·0	77·7	19·4	97·40
SiO_2	7·60	20·02	21·5	9·64	5·02	20·1	10·5	6·6	15·3	0·78
CaO	0·84	1·28	1·2	1·0	0·26	—	0·6	1·4	1·87	0·35
MnO	0·88	0·86	9·9	1·46	0·81	22·7	—	—	—	0·03
Al_2O_3	4·72	4·21	2·1	9·4	1·06	7·29	2·0	3·4	3·47	0·26
MgO	0·80	1·10	1·3	1·0	0·14	—	0·31	0·3	1·56	0·16
P_2O_5	0·22	0·31	—	2·5	0·03	—	0·07	0·27	0·45	—
CO_2	6·70	nil	—	} 18·2	} 1·68	—	26·98	} 9·3	22·50	0·04
H_2O	5·60	3·0	21·8			—	2·22		1·8	—
S	0·023	—	—	0·36	0·02	—	0·1	0·1	0·07	0·005
TiO_2	0·05	—	—	—	0·02	0·36	0·28	—	0·38	—
Alkali	—	—	—	—	—	—	0·42	—	—	—

Table 68 – *continued*

Chapter Ref. / %	115 Coed Ithel, Gwent, limonite	116 Stamford, Lincs., hematite	117 West Runton, Norfolk, nodules	118 Hartfield, Sussex, nodules	119 Muncaster Head, Cumbria, hematite	120 Berkhampstead, Herts. nodules	121 Allensford, North'land. nodules	52 Rannoch Moor, Upper Dall, bog ore	118 Hartfield, Sussex, roasted ironstone	122 Shifnal, Salop., CM nods.	14 Odell, Beds.
FeO	—	—	57·8	45·5	—	28·3	35·75	0·5	—	0·8	66·8
Fe_2O_3	87·95	80·51	—	—	91·7	—	14·44	41·5	80·6	66·4	—
SiO_2	<0·5	10·70	11·0	8·9	2·9	49·0	8·51	17·9	5·57	14·8	7·7
CaO	<0·5	0·6	<0·5	6·2	0·6	} c. 16·0	4·74	0·5	1·73	0·3	0·4
MnO	<0·1	0·1	0·7	1·28	0·02		0·85	7·9	1·99	1·5	—
Al_2O_3	<0·1	5·4	3·6	3·25	1·0		8·73	4·1	3·39	3·2	2·3
MgO	—	<0·1	<0·1	1·84	0·1	—	0·99	0·2	2·44	<0·2	0·1
P_2O_5	0·08	0·34	1·32	0·37	n.d.	0·8	—	—	1·05	<0·1	0·3
CO_2	<0·1	1·25	14·16	25·40	—	c. 6·0	(32·7)	15·3+	} 2·93	(12·3)	} (23·4)
H_2O	10·07	—	—	4·1	3·7	—	—	9·8			
S	0·01	0·03	0·08	0·09	n.d.	nil	—	0·1	0·05	—	—
TiO_2	<0·1	0·20	0·20	—	—	—	—	—	0·29	0·2	—
Alkali	—	—	—	—	—	—	3·29	—	0·1	—	—

Table 68—*continued*

Chapter Ref.	146	10	10	—	29	147	148	29
%	Bourne Pool, Staffs.	Rushall, Staffs.	Rockley Smithies, W. Yorks.	Shalbourne, Wilts. bog ore	Burrow Hill, Suffolk SA 178	Alstead, Surrey	Minepit Wood, Sussex	Hengistbury Head, Dorset
FeO	6·90	1·80	27·14	—	—	—	0·5	—
Fe_2O_3	55·71	63·48	2·57	34·9	69·10	59·2	64·6	78·60
SiO_2	14·40	13·48	37·10	31·4	10·00	3·0	10·6	1·58
CaO	0·50	1·40	3·6	0·6	0·8	—	0·4	3·2
MnO	0·18	0·93	0·58	9·3	n.d.	8·3	2·7	tr.
Al_2O_3	8·46	5·92	10·06	4·3	3·2	—	5·3	—
MgO	4·68	1·22	2·23	0·1	0·5	—	0·7	1·0
P_2O_5	0·12	0·6	0·41	1·29	2·3	—	—	0·07
CO_2	} 8·00	} 11·50	} 16·25	} 14·36	—	10·5	} 11·8	} 15·5
H_2O						11·5		
S	0·01	0·01	—	0·04	—	—	0·03	—
TiO_2	—	—	—	0·20	—	—	—	—
Alkali	—	—	—	—	0·3	—	n.d.	—
FeS_2								

+ = Mostly organic matter; ( ) = by difference; n.d. = not detected; — = not sought

some indication of the compounds in which the analysed elements are present. Experience based on other methods of analysis also plays a part in identifying compounds. In some cases the results as tabulated do not give a complete picture. For example, although only two iron oxides are given in Table 67, there are in fact three, the missing one being magnetite (Fe_3O_4). From the chemical point of view this consists of the two other oxides, FeO and Fe_2O_3, and therefore is not normally given in analytical results. Whenever these two oxides are present together, it is possible that some magnetite is also present. Its presence can be checked magnetically, the other iron oxides being non-magnetic. Phosphorus is usually given as the compound P_2O_5, although it may be present in other forms. Sulphur may occur as iron pyrites (FeS_2) or as a sulphate such as gypsum ($CaSO_4.2H_2O$). The insertion of a dash means that the element is either not present or has not been sought. In some cases the total adds up to more than 100%, which means that not all the iron is present as the oxide Fe_2O_3 but a small amount is present as FeO.

In the tables the carbonate $FeCO_3$ is split up into FeO and carbon dioxide (CO_2), while limonite is expressed as its components, Fe_2O_3 and H_2O. Where the tables return iron in the form of Fe_2O_3, it may exist in the ore as hematite or the hydrated form, limonite. Where the tables give FeO, the iron is normally present as the carbonate, $FeCO_3$. The ratio of FeO to CO_2 for ferrous carbonate is about 48:30, and it will be seen that the Westbury siderite gives this ratio very accurately. Some of the CO_2 is present as limestone ($CaCO_3$ = CaO + CO_2), which accounts for the departure from this ratio in the case of the ore from Ashburnham, Sussex (Table 67).

The best ore for smelting purposes is not necessarily that with the highest iron content but one with a high iron to gangue ratio. The gangue minerals: silica, calcia, and alumina, are slagged off during the smelting process and take a proportion of iron (as a flux) with them. Hence the lower the proportion of these elements, the less iron required to flux them during the formation of the slag.

Certain elements such as manganese are of value in the process, because they replace iron in the slag, thereby increasing the yield of iron. In almost all cases smelting has been carried out on pre-roasted ore, and during the roasting process certain non-metallic fractions are removed, in particular moisture and carbon dioxide. For this reason the carbonate and some limonite ores appear to contain less iron than the hematite ores.

Some of the elements in the ore such as manganese go only into slag and hence are a good indication of the ore used, while other elements such as calcium, potassium, silica and even phosphorus can originate from the ore and the fuel ash. Phosphorus enters both the slag and metal, and while it is more common in clay ironstones/siderites its presence in the metal depends to a certain extent on the smelting conditions and may be influenced by contributions from the fuel ash.

Some ores have an appreciable sulphur content (up to 0·9%). Marcasite (iron sulphide) is commonly found in a nodular form widely distributed over the southern chalklands. The presence of sulphur in iron is detri-

Table 69 Smelting and melting conditions for various metals

Metal	Melting point °C	Minimum smelting temperature °C
Gold	1 063	Only requires melting at 1 063°C
Silver	960	800 (as lead)
Copper	1 083	400
Lead	327	800
Tin	232	600
Iron	1 540	1 100

mental to the malleability of the metal, and consequently it is generally considered that sulphidic ores were not used in the production of iron. However, many of these nodules are so extensively weathered (limonitized) that they retain only a small percentage of sulphur which is easily removed by roasting, so rendering these ores a rich source of iron. Experimental work has been successful in producing iron with an acceptable sulphur content from marcasite nodules, some of which have been found on iron-working sites such as Swallowcliffe Down. Small quantities of nickel, arsenic, and cobalt have also been found in early iron objects. (A currency bar examined by Gowland[8] had a nickel content of 0·23%.) Such elements are rare and with our present information it is impossible to say where this material could have originated. Our knowledge of characteristic trace elements in iron and iron ores is minimal, although research into this problem is presently being carried out at Oxford by Salter.[9]

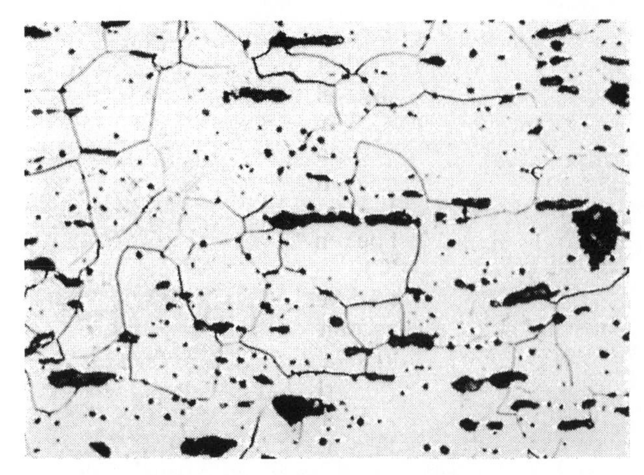

72 Wrought iron with slag made in an experimental furnace (after Wynne and Tylecote[25])

PRINCIPLES OF PREHISTORIC AND EARLY IRON SMELTING

Iron production came late in the history of metallurgy and was preceded by over two millennia of copper and copper alloy-based technology. The reasons lie not in the fact that iron has a much higher melting point than the non-ferrous metals, but that it requires much stronger reducing conditions in the furnace.

It can be seen from Table 69 that some metals can be

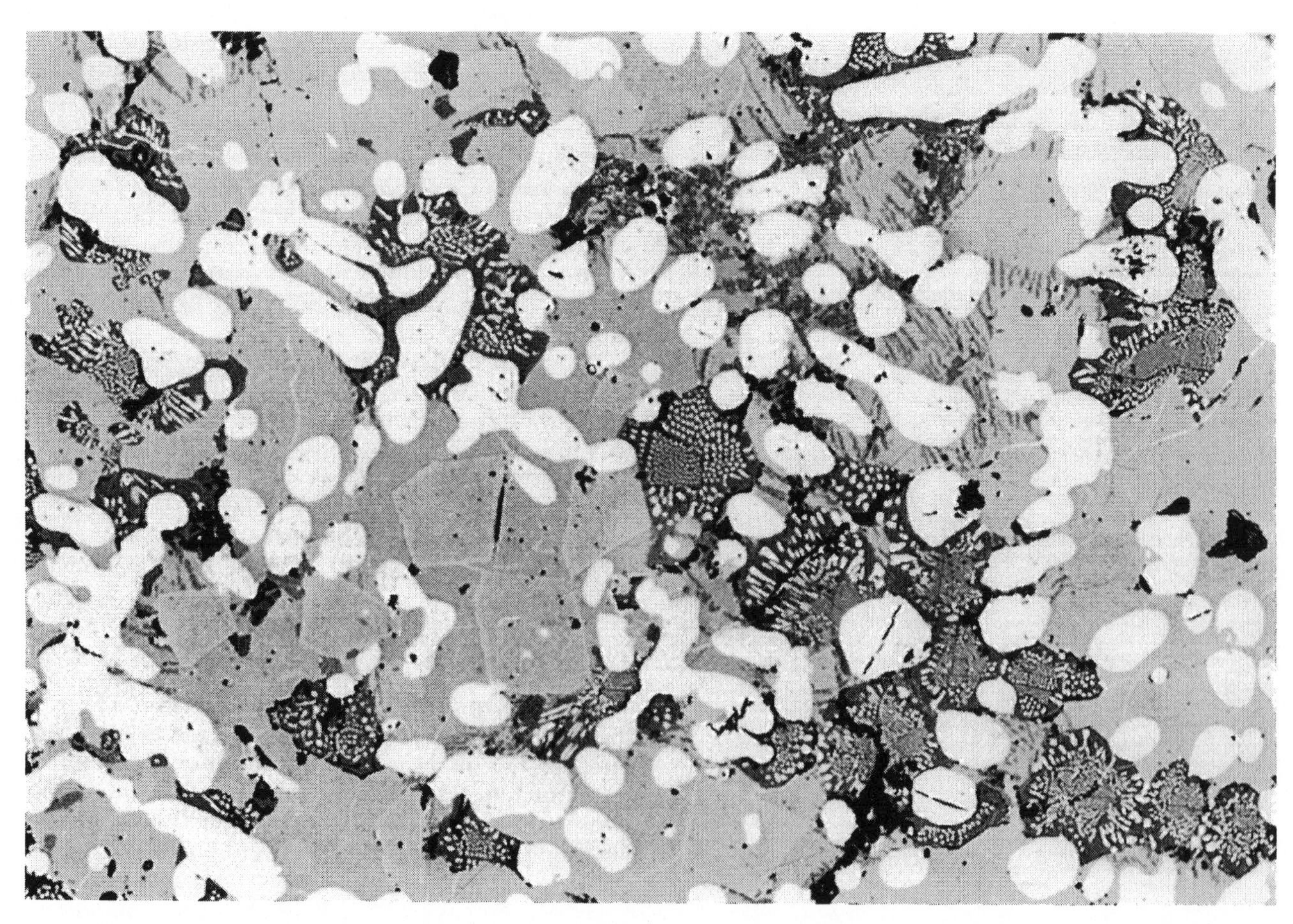

Lightest phase; wüstite (FeO) dendrites
Dark phase; glass with eutectic
Grey matrix, euhedral crystals of silicate (fayalite)

73 Photomicrograph of slag from Garden Hill, Sussex (courtesy of D. S. Butler)

reduced (i.e. oxygen removed from the oxide ore) at temperatures well below their melting points. Iron can be reduced at approximately 800°C, well below its melting point of 1540°C. Unfortunately, iron ores are rarely pure and are composed of the various iron oxides along with varying proportions of gangue minerals such as quartz. Where these cannot be removed by washing, they must be removed by slagging at high temperatures (>1 100°C).

As the iron is produced below its melting point it is in the solid state, as a sponge or raw bloom composed of a mixture of iron and slag (Fig. 72). Hence the term commonly applied to the early production technique is the 'bloomery process'. Early iron slags are predominantly fayalite (iron silicate) and other minerals forming a glassy phase. Because the iron is in the solid phase it is necessary to remove the entrapped slag by liquation, and therefore the smelting process must take place at or above the temperature that allows the slags to drain away from the solid iron. This occurs (depending on composition) at about 1 150°C, well above the minimum temperature required to reduce the oxide, but substantially below the melting point of the metal. The slag remaining in the bloom is gradually removed during the forging process, when it is squeezed out by a cycle of reheating and hammering.

The separation of the growing iron crystals from the slag is critical for the formation of the bloom and serves to highlight the importance of the slag in the bloomery process. To achieve the correct viscosity enabling it to flow from the iron at the temperatures reached in the process, the slag must have a limited range of compositions. Ancient slags are rich in iron for this very reason. The iron has been lost to the slag in order to flux the gangue minerals, silica and alumina. These minerals both have high melting points – in excess of 1 700°C, well above the temperature attained in the early furnaces. The addition of iron to these minerals in the correct proportions can reduce their melting points to below 1 200°C.

Thus, by modern standards the process could be seen as inefficient; so much so, that from the 17th century onwards there was extensive trade in ancient slags as a source of ore for the blast furnace. Blast furnaces could achieve much greater temperatures and hence were able to use large quantities of lime, replacing iron as the flux, and resulting in a slag with a much higher melting point (*c.* 1 400°C).

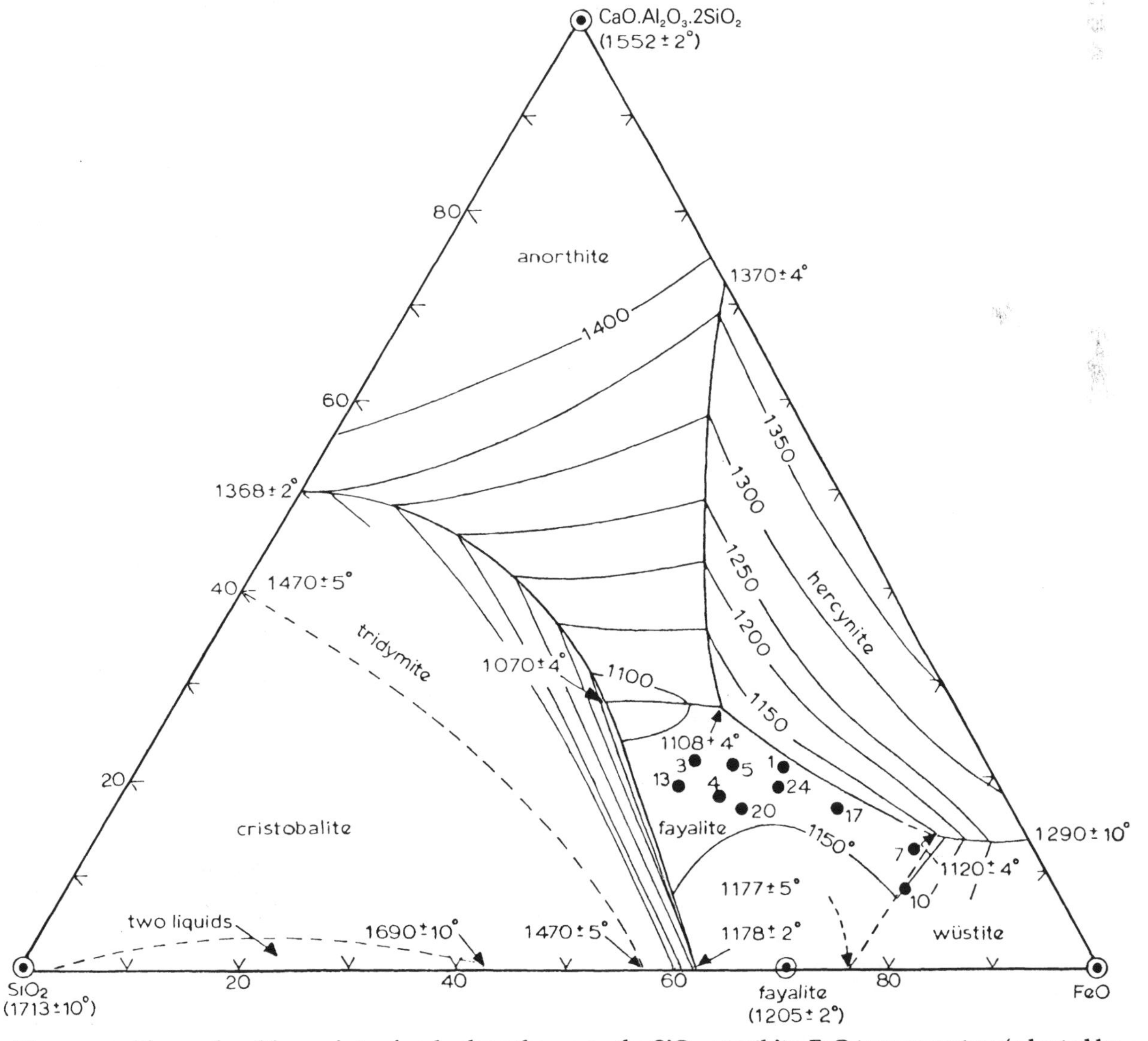

74 The composition and melting points of early slags shown on the SiO_2–anorthite–FeO ternary system (adapted by Morton and Wingrove[10] from Levin *et al*.)

Table 70 Free running temperatures of slag

Composition %			Free running temperatures °C
FeO	SiO_2	CaO	
78	22	0	1 180
70	30	0	1 200
62	38	0	1 180
60	40	0	1 400
59	29	12	1 115
40	40	20	1 150
35	35	30	1 200
30	30	40	1 500
0	50	50	1 540

From 'Phase Diagram for Ceramicists', ed. E. M. Levin *et al.*, 204; 1964, American Chemical Society.

Slags in the bloomery process consist mainly of the mineral fayalite ($2FeO.SiO_2$) with variable quantities of wüstite 'FeO' and a glassy matrix (*see* Fig. 73). The fayalite–wüstite–glass eutectic has a melting point of around 1 170°C and can absorb small quantities of MnO, CaO, MnO, Al_2O_3, FeO, etc., which do not raise or lower its melting point by more than 50°C (Table 70). Increasing the quantities of these elements, particularly lime, alumina, and magnesia, results in a sharp increase in the flowing temperature of the slag. Modern blast furnace slags that have as much as 35% CaO and less than 1% FeO have melting points in the region of 1 400°C. Lime was not used as a flux in the early bloomery smelts but up to 6% has been recorded in ancient slags (Table 76 *below*). These quantities could easily originate from the ore or the fuel ash.

Morton and Wingrove[10] extensively studied the chemical and mineralogical composition of ancient Roman and medieval slags. The results enabled them to assess the efficiency with which smelters controlled the composition of ores and furnace conditions to produce a slag with a low free-running temperature and optimum iron extraction. They plotted composition on ternary phase diagrams which show the relationship between the composition of the slag and the process (*see* Fig. 74). The majority of ancient slags can be seen to fall within narrowly defined fields on the phase diagram, again suggesting a close control over the smelting conditions by the ancient smelter.[11]

The main principle in iron smelting, as in primitive copper smelting, is the reduction of the metal oxide with carbon monoxide:

$$Fe_2O_3 + 3CO \rightarrow 2Fe + 3CO_2$$

The carbon monoxide is formed by partial combustion of charcoal with the air from the tuyere, and it requires a certain distance (about 12 charcoal diameters) before the gas coming from the tuyere contains sufficient carbon monoxide to reduce the iron ore to iron. Any iron ore passing too near the tuyere is likely to remain unreduced and to become slag. However, in the higher levels of a furnace, provided that sufficient charcoal is present at all times, the ore will be fully reduced but not hot enough to separate the gangue as slag. Some of this reduced ore may fall near the tuyere, but if it consists of cementite, i.e. iron carbide (Fe_3C), short oxidation in this zone will not slag it but merely burn out more or less of the carbon and convert it to iron.

The significance of this is that if the iron ore is in contact with carbon monoxide-containing gases for a long time, there is less likelihood of it being slagged, and a more efficient operation will result. However, this requires a container or shaft above the furnace. The superiority of such furnaces is attested by the fact that most primitive furnaces used in the world today are of this type. They have, however, one disadvantage. When the time comes to remove the solid bloom of iron, this has to be done through the side of the furnace, which may require the destruction and replacement of a considerable amount of the furnace. The open bowl or hearth has the advantage that the bloom can be removed easily, and smelting of the next charge begun at once.

Another advantage of the shaft furnace is that it can operate as a natural draught furnace if the chimney or shaft is tall enough. This saves the very considerable labour of blowing bellows. The principle of the induced draught is the same as that of the domestic chimney, i.e. hot air rises because of its low density compared with cold air. The greater the difference in temperature and the taller the column of hot air, the more rapid will be the flow of air up the shaft. Haynes[12] found that a temperature of 1 150°C could be produced at the bottom of a 22 × 22 cm square section furnace only 25 cm high. However, the filling of a shaft with ore and charcoal sets up a resistance to the flow of air, and the empty height has therefore to be increased to maintain the same flow of air. The majority of induced draught furnaces used today by primitive peoples appear to have a height of 2–3 m, which agrees with those found on the Continent and in the UK dating from early times. It is never easy to prove that shaft furnaces have been operated by induced draught, but in some cases their design strongly suggests it.

Some writers are very concerned with the orientation of the furnaces, in the belief that the wind velocity has a marked effect on their operation. The orientation of those found seems to be very variable. Since some shaft furnaces found have been at the bottom of a pit about 2 m deep, wind cannot have had much effect on their operation. It is doubtful whether wind can contribute much to the draught in a shaft furnace over 2 m high.

To date, Britain has failed to provide evidence for iron production in the pre-Roman period on a large scale, but there are three areas where a reasonable scale of production is likely: the Weald, the Forest of Dean, and Northamptonshire. These have all been subject to intensive iron-working well into the blast furnace period. For example, the evidence of Lydney and Ariconium shows that the Romans exploited the Forest of Dean, and they also used the ores in the Weald extensively, but it is only in the Weald that extant remains suggest the magnitude of such operations.

Slag-tapping furnaces have previously been seen as the hallmark of iron production during the Roman period. The only case of tapped slag from a pre-Roman context was, until recently, that found at the site of All Cannings Cross,[13] although the dating of this specimen must be questioned. However, late Iron Age/ Romano-British evidence from the sites of Odell, Beds.,[14] Gussage All Saints,[15] Wakerley,[16] and Pip-

75 Tap slag from High Bishopley, Co. Durham

pingford[17] suggests that slag-tapping was also a pre-Roman feature. These slags do not have the ropey flow structure of the typical Roman furnace (Fig. 75), but their distorted structure suggests that the slag has been raked out of the taphole.

PREPARATION OF IRON ORE FOR SMELTING

Iron ores have a considerable range of compositions varying in both iron and gangue content. To assess whether an ore could have been used, the iron content must be considered relative to the other minerals. Iron should be considered together with manganese and small amounts of magnesia and calcia relative to the quantity of silica and alumina present. However, to simplify the discussion, the slag will be treated as pure fayalite and the ore as a mixture of iron oxides and silica.

The weight ratio of FeO to SiO_2 in the slag will be approximately 3:1. As iron in the roasted ores is usually in the higher oxidation state as Fe_2O_3, the ratio required in the ore to produce a pure fayalitic slag is approximately 3·5:1. Therefore, a considerably greater ratio of iron oxide to silica is required to produce a significant quantity of iron. Iron content alone is not a good indicator of the value of the ore. For example: the Seend limonite (Table 67) has an Fe_2O_3 content of 65% in a ratio ~3·6:1, while the Northumberland nodules have an FeO content (includes 6% Fe_2O_3) of ~50% but in a ratio of almost 6:1. These nodules could be smelted without further preparation, while the Seend ore would require considerable dressing by winnowing or washing to reduce the silica content before charging in the furnace. It should be noted that slags are rarely pure fayalite and the FeO/SiO_2 weight ratio can be as low as 2:1, suggesting slags more viscous than desired, but also a furnace temperature well above 1 200°C.

The size of the ore charged in the furnace is quite important since the reducing gas, carbon monoxide, will take longer to penetrate thicker pieces, in contrast to fine ore which will be reduced rapidly but will also impede the flow of gases in the furnace.

Some ores such as the carbonates, and the sulphides – if indeed they were used – benefit by first roasting to drive off the carbon dioxide, or water and sulphur, respectively. This increases the porosity and so allows better penetration of the reducing gases during smelting. Iron carbonates decompose in the temperature range 500–700°C,[18] but some other compounds such as calcium sulphate require temperatures as high as 1 200°C. While we have no evidence for roasting in the pre-Roman period, it is very likely that it was used, possibly in the simplest way by making a 'raft' of timber and branches, and setting it alight with the ore on top. The high oxygen content of the fire of raw wood is no detriment. Not only does roasting increase the porosity but even if it is not taken to the chemical endpoint shown by the equations:

$$FeCO_3 \rightarrow CO_2 + FeO$$
$$2FeO + \tfrac{1}{2}O_2 \rightarrow Fe_2O_3,$$

it results in a much more friable charge which is more readily broken up to give a consistent size.

With some ores such roasting can be effected in the upper levels of a shaft furnace, but with sulphidic ores it is preferable for the roasting procedure to be separated from the smelting, leaving only residual sulphur to be driven off in the furnace.

The size of the ore used in the bloomery process seems variable and evidence from sites is limited. Samples from the 2nd century AD site of Ashwicken[19] seem to have been 6 × 2·5 × 2·5 cm, while those from the Anglo-Saxon levels of Odell (Beds.) were rough cubes of ~4–6 cm. Ethnographic evidence from the African and Asian continents[20] gives examples of ore ranging from sand to fist size.

FUELS

The fuel of the early iron smelter was charcoal, although there is some evidence for the use of peat. Raw wood does not provide a satisfactory fuel for iron smelting, owing to its oxygen and moisture content. Traces of it have been found in some furnace slags but these probably originated from the preheating of the furnace when wood was used.

The early smelter preferred the use of hardwood charcoals which are denser and longer-burning and so somewhat more economical. Oak was the main source of fuel in Britain until the 13th century; after this, shortages forced the smelter to turn to whatever timber was available.

Charcoal contains 3–20% of ash, depending on whether it originated from hard or softwoods and also on the part of the tree it came from. Hardwoods generally have a lower percentage of ash than softwoods and bark has a higher content than heartwood. The ash content can play a significant role in the process, because it is incorporated into the slag. Ash contains a high proportion of alkalis and alkaline earths, such as potash and lime, both of which act as fluxes, lowering the free-running temperature of the slag.

The composition of some charcoal ash is given in Table 110 (Chapter 11) which shows that the alkali content can be as high as 22%. This could contribute

76 Slag accretion, furnace lining and tuyere from Chelms Combe, Somerset (30 cm wide × 30 cm high) (courtesy of Prof. M. J. O'Kelly)

more than 3% of alkali to the slag, which, together with the lime, could lower the free-running temperature of the slag by as much as 100°C. Yet very few analyses show much alkali even when these elements are sought, and it is probable that they have been leached out by groundwaters after formation. Analysis of slags from Garden Hill (Sussex)[21] demonstrates a considerable loss of these elements from the structure of the slag. This corrosion gives a misleading value to the free-running temperature of the slag and suggests a higher temperature than the original operating temperature of the furnace.

The charcoal size is also important, since the smaller the charcoal the faster the conversion of the oxygen in the blast to carbon monoxide. However, fine charcoal also impedes the flow of gases in the furnace and a greater blast will be required to expel it from the furnace. At Ashwicken[19] (2nd century AD), the size appears to be 3–8 cm, which is similar to that of charcoal entrapped in slags from other sites. By the time the charcoal reached the tuyere zone of the furnace it would be much smaller than that originally charged.

IRON SMELTING FURNACES

There have been numerous attempts to construct a typology of ancient furnaces from as early as Percy,[20] and Gowland,[22] through to Forbes,[22a] Schubert,[23] and Cleere.[24] Percy, being a metallurgist, viewed furnace structures from the ethnographic record with a

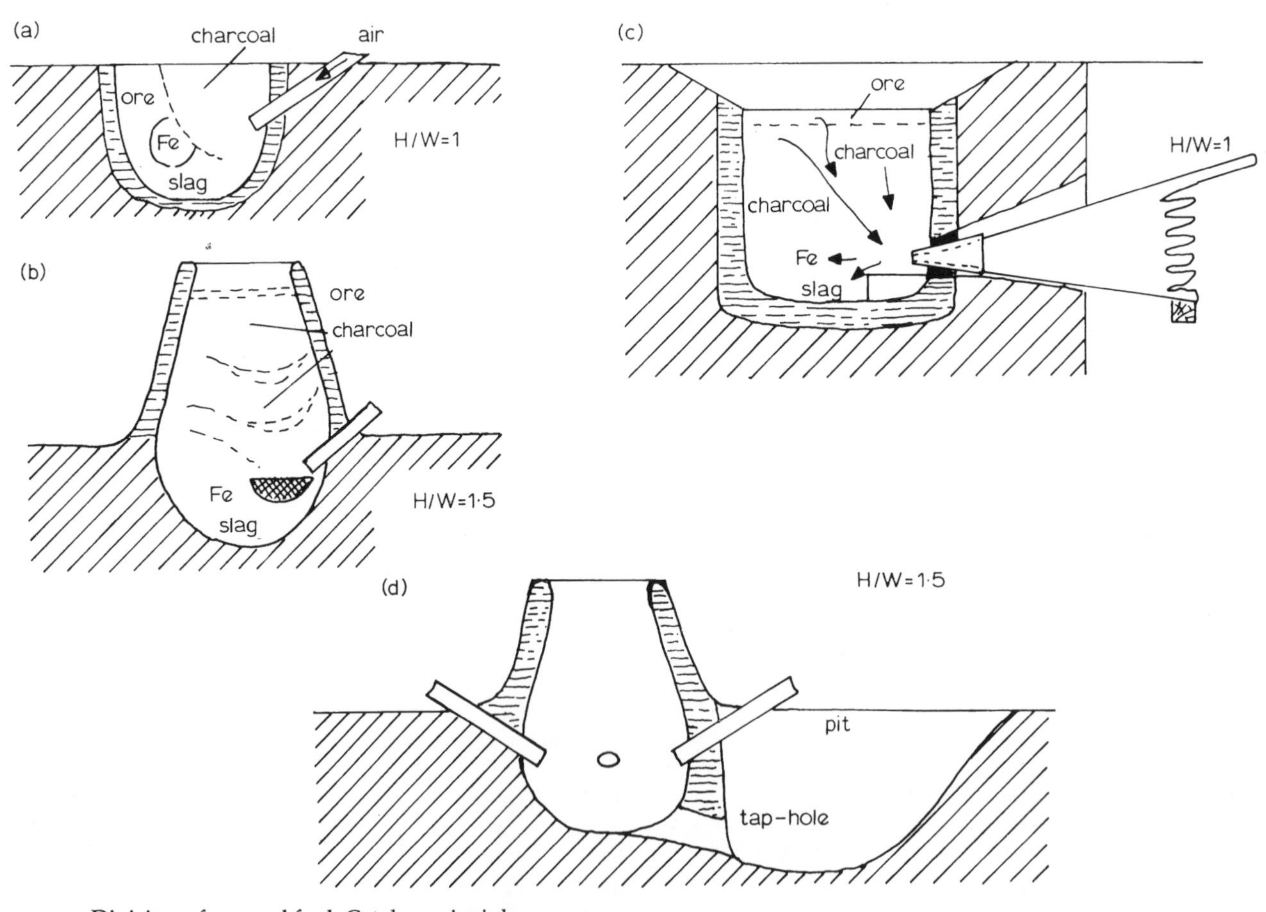

a Division of ore and fuel; Catalan principle
b Layered charge; needs reservoir = low shaft
c Manipulation technique (after Busch[27] and Evenstad[28])
d Developed, or slag-tapping bowl furnace

77 Types of bowl furnaces

technological bias. Others, being archaeologists, have tried to use typologies to trace movements of ancient metal-workers and as a means of dating industrial sites. The major drawback encountered by all is the paucity of the archaeological record. There is certainly no evidence to support complex furnace typologies. The superstructure of all but a few furnaces has been destroyed, with only the furnace base remaining (Fig. 76). Without a slag-tapping pit, this is invariably interpreted as a 'bowl' furnace, whether its use was for smithing, smelting, or melting, or even perhaps non-metallurgical purposes.

In recent years numerous smelting experiments have been carried out in Europe and Britain in an attempt to understand the ancient processes more fully. The results of these experiments, in conjunction with ethnographic and recent archaeological evidence, enable us to propose a different approach to early furnace structures.

The 'bowl' furnace

The bowl furnace is something of an enigma, since experimental work has shown how difficult it is to produce iron from a true 'bowl', i.e. a furnace with a height about equal to its diameter (H/W = 1). When one compares the results first with the historically documented Catalan hearth which fits into this category, and then with African furnaces recently seen at work by anthropologists, 3 types are recognizable:

1 A furnace in which the ore and fuel are to a large extent separated, the fuel going down the tuyere side and the ore down the opposite side so that the ore is blasted with a current of carbon monoxide. This is essentially the principle of the Catalan hearth and the one found to work best, as shown in Wynne's experiments[25] (Fig. 77*a*).

2 The same type of furnace, but the charge is mixed or layered, and reduction takes place in the upper parts of the furnace. During its passage past the tuyere it tends to get re-oxidized, giving a lot of slag and a poor yield. The metal is found mixed up with charcoal in the slag at the bottom. This type of furnace can be improved by adding a 'chimney' or reservoir, whereupon it becomes a shaft furnace with H/W > 1·0. Both Wynne[25] and Gilles[26] found this by experiment (Fig. 77*b*).

3 The re-oxidation of the reduced iron in the tuyere zone can be avoided by removing the metal from this area at the right time. This is what is meant by 'manipulation'; it is a technique seen quite recently by many anthropologists, mainly in East Africa, and it is well described by Busch[27] and Evenstad[28] (Fig. 77*c*). The technique used here explains why some bowl furnaces seem unnecessarily large, since it is quite clear that the blast from a tuyere supplied with air from a primitive bellows system cannot penetrate as much as the width of a 1 m diameter furnace. In some cases more than one tuyere has been used in such a large furnace. In these cases the area around each tuyere must be considered as a separate furnace, although it may be possible for the slag to drain into a common receptacle.

The manipulated bowl furnace must be wide enough at the top to enable it to be worked manually with primitive tools such as pieces of sapling. For this reason it cannot be permanently covered over and is more likely to be flared than domed. Blowing will be directed downwards, and burning from the top with consequent waste of fuel may be prevented by dampening the charcoal.

There are no early furnaces in Britain known to be of the Catalan type. From experiments in a low-shaft furnace[29] it has been found that a layered charge produces characteristic slag and furnace remains. As the burden descends through the combustion zone of the furnace, liquated slag collects under the tuyere. Such a structure can be seen in the remains of the Chelms Combe[30] and Kestor[31] furnaces (Fig. 76). Also typical of a layered charge are runlets of slag entrapping charcoal in the base of the furnace, which would tend to leave a mixture of pieces of slag and charcoal. Associated with these smaller pieces of slag would be a larger cake of slag or furnace bottom (FB) which might be removed to extract the bloom. The West Brandon furnace[32] had pieces of charcoal and slag remaining in the base of the furnace with the superstructure collapsed on top. The remains of this site suggest a low-shaft furnace superstructure rather than the bowl furnace originally proposed.

One of the problems in the bowl furnace is its inefficiency in retaining both the heat and a reducing atmosphere. As Wynne discovered,[25] the yield was poor until the furnace was covered, and even then a large quantity of unreduced ore remained in the furnace. The amount of ore that can be pre-packed into such a furnace is small – at most 5–7 kg. This, together with the inherent inefficiency of the bloomery process as far as its consumption of ore is concerned, precludes it being a common mode of production.

The product of the bowl furnace is the furnace bottom (FB) consisting mainly of slag with the bloom on top of it (Fig. 78). In some very primitive furnaces the two may be intermixed and the iron must be recovered by breaking up the slag. Considering that the slag in a furnace bottom amounts to 80% of the total product, the remains are very much in evidence and, where compact, will be large. For this reason they can be distinguished from the smaller smithing hearth bottoms (SHBs) which, because of their small size, are not likely to represent smelting furnace bottoms since the 20% metal product would be too small to be worth making.

There will of course be exceptions to this argument: SHBs often contain layers of non-ferrous metals showing that the smithing hearths were often used for a wide variety of purposes. Smelting furnace bottoms would represent the slag from one smelt and not the accretions of many operations.

The shaft furnace

This type (Fig. 79), in which H/W > 2, almost certainly developed from the second type of bowl furnace. The problem with shaft furnaces is to distinguish them from bowl furnaces when only the bottom part remains. To help in this respect it must be remembered that the yield of the bloomery process is poor – 10–20% of the iron in the ore – therefore the slag volume is high. If the furnace does not provide for the tapping of slag, it will stay in the bottom and take up a very large proportion of the total furnace volume, leaving little room for the metal and the burning charcoal above it. Also the slag 'furnace bottom' must always be below

78 Furnace bottom from Type B bowl furnace (after R. E. Clough[29])

the tuyere, since in primitive furnaces it is not possible to obtain enough pressure to blow into the slag for very long.

On the other hand, the bowl of the bowl furnace is *thermally* very efficient since if care is taken in its building, heat conduction through the ground can be minimized, while the loss through the walls of an above-ground shaft furnace is high. For this reason some shaft furnaces such as those from Igołomia, Poland, have a wide bowl and a narrow shaft or chimney, all below ground.[33]

A small 'bowl' – say only 20 cm in diameter – is unlikely to be the remains of a bowl furnace but more likely to be the bottom of a shaft furnace, since the metal represented by a small bowl furnace of this size would hardly be worth smelting. The shaft furnace works in a different manner to the bowl furnace. Even if the charge is layered and not mixed to begin with, by the time it arrives in the reduction zone it will be well mixed and reduced to iron and slag.

Reduction must go beyond the pure iron stage to give an iron with a considerable amount of carbon, so that it may survive the high oxygen content of the tuyere region without being converted back to iron oxide. Some iron will be oxidized, however, and will join the slag and lower its viscosity. The slag will drain away to the bottom of the furnace and may be tapped, while the solid iron will stick to the sides and will gradually slide down under its own increasing weight.[34]

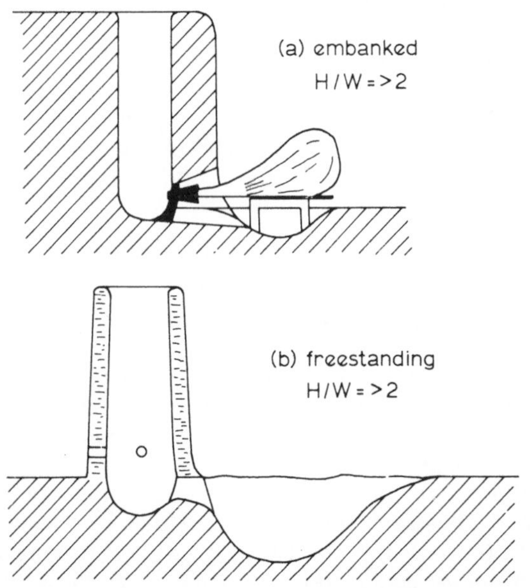

79 Shaft furnaces: (*a*) embanked; (*b*) freestanding

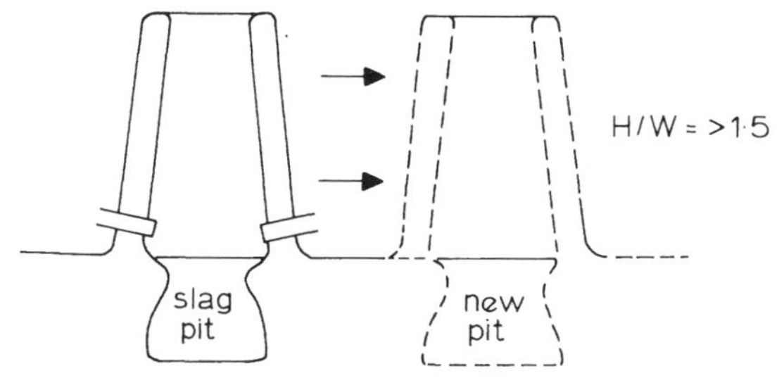

80 Slag-pit furnace

Once we know the manner of extracting the 'bloom' we can determine whether the furnace is bowl or shaft. In a bowl the bloom is extracted through the top; in a shaft this cannot be done, and it must be removed through a hole in the side. Both types may be buried in the ground, but if a shaft furnace is buried it will need an adjacent pit for bloom and possibly slag removal.

The best-authenticated shaft furnaces in Britain belong to the Roman period and there are many of them, stretching in a belt from the Weald of Sussex and Kent through Northamptonshire to Norfolk. Most of these were built above ground; some were built into the sides of a bank. Some of the remains of pre-Roman furnaces (50 BC–AD 50) could well be the bottoms of shaft furnaces, since it is most probable that a widely distributed type in the 1st and 2nd century AD should have had antecedents.

Increasing the height of the shaft increases the efficiency of the furnace, so much so that there is a danger of the conditions being too reducing, resulting in the production of cast iron. As the burden descends down the shaft, several stages of the reducing reaction occur.

If the ore has not been pre-roasted, then this occurs in the upper levels of the furnace just before the initial stages of reduction. It is important at these levels not to expose the ore to temperatures which will cause slagging, as this will decrease the inward diffusion of the carbon monoxide and inhibit the formation of crystals of iron and subsequently the bloom. As the ore descends, reduction proceeds and slagging of the gangue mineral begins. As the slag drops through the combustion zone, around the tuyere, the slag and metal fall towards the base of the furnace, and the small particles of iron gradually accumulate and weld to form the porous bloom, allowing the slag to draw away from it. The effect of the shaft is to keep the ore in a reducing atmosphere, and compared with the bowl furnace less fuel is required. The longer the ore is kept in a reducing atmosphere, the more iron will be reduced. It follows that a higher fuel/ore ratio will have the same effect by increasing the residence time. While an increase in yield may seem desirable, any attempt to obtain it has two negative effects:

(i) by removing too much iron from the ore there will be insufficient to slag the gangue minerals, resulting in a viscous slag difficult to separate from the bloom during smelting or smithing;
(ii) too much carbon will be absorbed into the iron, which in the extreme case could lead to the production of cast iron.

81 Slag-pit furnace bottom from Aylsham (36 cm wide) (courtesy of Norfolk Museums Service)

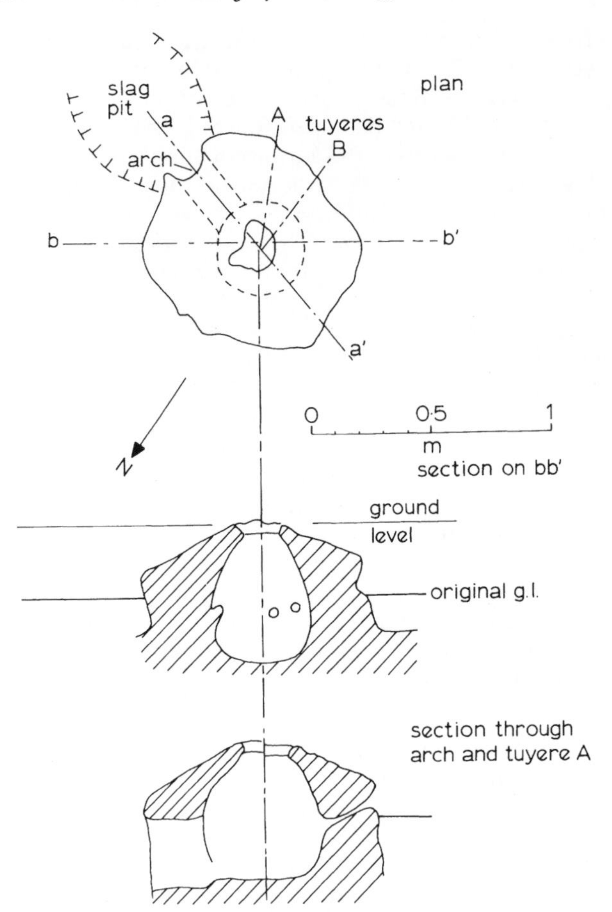

82 Domed furnace from Levisham, Cleveland (based on drawing by M. Davies-Shiel)

There have been several examples of cast iron from Romano-British and Roman sites. The piece from Hengistbury Head (Table 78) is dated to the 1st century AD[35] and could be pre-conquest. The accidental production of cast iron in China led to its intentional production, and cast iron became the main ferrous metal from the Han Dynasty onwards.

Slag pit furnaces

One development for separating metal from slag was the slag-pit furnace, which is really a shaft furnace with H > 2 W but with a deep pit directly under the shaft into which the slag was encouraged to go at or towards the end of the smelt (Fig. 80). After the metal was removed from the top of the slag through a hole in the side, the shaft was lifted *in toto* and moved sideways to rest on top of a previously excavated empty pit. By this means there was no trouble in removing a heavy slag lump from within the furnace – it was merely left in the hole in which it had formed.[3]

This type of furnace must have been very common in the pre-Roman Iron Age judging by evidence from Northern Europe and sub-Saharan Africa.[3, 3a] The earliest dated specimens are in Africa (400 BC) but the ^{14}C dates given for some of the examples in Northern Europe span the pre-Roman and Roman periods so that it is possible that they were in use in Britain also. The examples of slag bottoms from such furnaces found in Britain are unstratified, but the 24 kg furnace bottom from Mucking, Essex, is probably Migration period,[3] although like the others from Hevingham and Aylsham, Norfolk (~50 kg) (Fig. 81, *above*) it is not well stratified. The fact that the only examples in Britain come from East Anglia and that it was a type common in Jutland and North Germany during the Roman period suggests post- rather than pre-Roman influence.

The evidence for the moving of the slag-pit furnace shaft after use lies in the lightness, slender lines, and thin wall of the restored specimens in the museum at Harburg. Also, Nicole Echard has shown in her film on the smelting of iron in the Sahara that such furnaces can be moved when dry and, furthermore, that they are extremely strong after use.[149] But it is not claimed that all slag-pit furnaces were moved in one piece over a new pit. Some of them, like the Polish ones, probably have walls too thick to make this possible.

The variety of furnaces from Britain is quite considerable and because of the doubts discussed above, it is not possible usefully to classify all those found. Generally, however, there are above-ground and below-ground types; bowl, dome, or shaft furnaces; slag-tapping and non-slag-tapping (Figs. 77, 79, 80, 82). Most, but not all, of these are represented by finds in Great Britain. Specific examples of these types will now be discussed.

EXAMPLES OF SMELTING FURNACES AND SMITHING HEARTHS

Any discussion of the archaeological evidence of the early industry must take into account problems of dating and identification. Often we are confronted with the remains of smelting without the evidence for relating it to period or culture. Alternatively, the samples of slag and vitrified products can be dated but are difficult to relate to a specific process such as smelting or smithing. In many cases where the furnace remains cannot be dated by stratigraphical associations they are dated by furnace type, but this can be misleading as furnace types often cannot be identified with a particular period.

Some pieces and cakes of slag are commonly found in datable levels but without furnaces or large quantities of tap slag (Fig. 75), and it is difficult to assign these to either smithing or smelting. Numerous attempts have been made to distinguish these by analytical techniques, but this has proved difficult, as smithing slags, particularly those from reheating and the consolidation of the bloom, have a similar chemical and mineralogical composition to smelting slags. The characteristics of the principal finds are given in Table 71.

Thus, consideration of slag characteristics must be supplemented by further field evidence. The presence of large quantities of slag, or tapped slag run from the furnace, and prepared or roasted ore, suggest smelting on site; likewise, the absence of these suggests smithing only. The hillfort of Danebury, for example, produced a few kilograms of small pieces of slag and vitrified products.[36] The lack of tap slag and slagged furnace structure leads to the conclusion that these pieces are the remains of forging and other pyrotechnological processes. A summary of the principal sites producing iron-working evidence is given in Table 72.

While the first evidence for iron in Britain appears in conjunction with Late Bronze Age bronzes at Llyn Fawr[37] and possibly with the LBA hoard at Gilmonby,[38] at the moment we have no furnace or slag remains which can be dated before the 5th century BC.

83 Furnace slag from Great Weldon, Northants.

At this time the site of Brooklands, near Weybridge,[39] produced the first of 21 furnaces in use up to the Roman period. This site produced about 44 kg of slag, quite a moderate amount for the five centuries of its existence; clearly a much greater quantity must have been removed at some time, perhaps for medieval and later re-working. (Analysis is given in Table 76.)

At Brooklands the most complete furnace had a bowl-shaped base and a superstructure of dense red clay. On one side of the base was a red stain 1·5 m long and 15 cm wide, which probably indicates the remains of a slag-tapping channel similar to some from the Northants site of Wakerley. Very little tap slag was found on the site, the bulk being small pieces and furnace bottoms. This furnace and an FB from a pit have been dated to the 5th century BC. Twelve furnaces had internal diameters of 30–35 cm and were probably bowl furnaces with low shafts (Fig. 75*b*). Another six 'furnaces' were over 60 cm in diameter and may have been reheating hearths for forging. The only ore found on the site was a pyritic nodule, but St. George's Hill, which contains sideritic ironstone of

Table 71 Characteristics of principal finds on iron-working sites

Charcoal:	Before the 13th century mainly oak.
Slag:	Before the Roman period rarely tapped. Mostly furnace slag and furnace bottoms; rarely magnetic.
Furnace slag: (Fig. 83)	Porous and not as dense as tap-slag (SG = 3–4). Often contains charcoal and/or wood. May be magnetic as it may contain partially reduced ore as well as slag and fuel. If in large pieces (over 20 kg) likely to be Roman or later.
Tap slag or run slag: (Fig. 75)	Dense black. Has been tapped from the furnace in semi-molten condition. Smooth top surface with solidified trickles which are often hollow due to gas (SG=4). Bottom surface is rough owing to tapping onto sand or soil. All slags have very much the same composition.
Soft cinder:	Blue-black, friable; usually *in situ* in bowl furnaces. A mixture of semi-reduced ore and charcoal. Often magnetic.
Hard floor:	The smithy or smelter working floor. Consists of a hard mixture of fuel, iron ore or hammer-scale and any other debris. Usually so hard that it can only be broken with a pick.
Furnace bottoms: (Fig. 78)	Plano-convex lumps of dense slag. The smaller ones (*c.* 500 g) are smithing hearth bottoms (SHBs) which are an accretion of hammer-scale, fuel and smelting slag. The larger ones which may weigh up to 25 kg consist mostly of slag formed during smelting but may contain some wood or charcoal. They are the main product of the bowl furnace when tapping proves difficult.
Hammer-scale: (Fig. 118)	Oxides of iron with some silicate. Product of heated iron (700–1 200°C) shed when reheating or on quenching. Usually found around the anvil but can be put back and re-smelted.

Table 72 Summary of evidence for iron-working in the pre-Roman Iron Age

Site	Date	Furnace	Slag or cinder	Charcoal	Iron*	Ore	Chapter ref.
Chelms Combe (Som.)	250 BC	+	+				129
Merthyr Mawr Warren (Glam.)	4th–1st cent. BC	+	+				49
Glastonbury	150 BC–AD 43	?	+				54
Rowberrow Warren Cave (Som.)	2nd cent. BC	+	+	+		+	53
Rudh' an Dunain (Skye)	1st cent. BC	+	+	+			51
Kestor (Devon)	After 400 BC	+	+	+			31
Mynydd Bychan (Glam.)	150 BC–AD 100		+	+		+	61
Ewell (Surrey)	200 BC–AD 150		FB		+		62
Hunsbury (Northants.)	3rd cent. BC		+			?	43
All Cannings Cross (Wilts.)	400–250 BC	?	T	+		+	13
Hengistbury (Dorset)	500 BC–AD 100		+		+	+	35
Chysauster (Cornwall)	50 BC–AD 350		+			+	47
Bac Mhic Connain (N. Uist)	After AD 200 (?)		+			+	64
Sedlescombe (Sussex)	1st cent. BC		+	+			132
Crowhurst Park (Sussex)	1st cent. AD		+	+			69
Saxonbury Camp (Sussex)	1st–2nd cent.	?	+	+			133
Wookey Hole (Som.)	Pre-Rom. or Roman	+			+		79
Swallowcliffe Down (Wilts.)	EIA		+	+	+	+	7
Little Woodbury (Wilts.)	400–250 BC		+				42
Sudbrook (Gwent)	1st cent. BC–1st cent. AD		+	+			134
Hascombe Camp (Surrey)	Before 50 BC	?	+			+	135
Camerton (Som.)	1st cent. AD		+				131
Camerton (Som.)	200 BC		+				131
Chinnor (Bucks.)	4th cent. BC		+	+			2
Loanhead of Daviot (Aberdeen)	Scottish IA		+	+			136
Bagendon (Glos.)	*c.* AD 10–50		+	+	+	+	41
Wiltrow (Shetland)	*c.* 100 BC–AD 100	?	+	+		+	55
Ffridd Faldwyn (Gwent)	EIA	+	+	+			137
Trevisker, St. Eval (Cornwall)	160±90 BC		+	+		+	46
Gussage All Saints (Dorset)	1st cent. BC–AD	+	+	+	+		104
Levisham, Cleveland	AD 50	+	+	+			58
Roxby, Cleveland	EIA	+					57
Berkhamsted (Herts.)	AD 15	+	+			+	120
Pewsey Vale (Wilts.)	4th cent. BC		+				138
West Brandon (Durham)	3rd–2nd cent. BC	+	+	+			32
Brooklands (Surrey)	5th cent. BC–1st cent. AD		FB	+		+	39
Wakerley (Northants.)	2nd cent. BC	+	+		+	+	16
Broadfield (Sussex)	190±80 BC	+	+	+			60
Hardwick (Northants.)	AD 1–50		+			+	139

Table 72 *–continued*

Site	Date	Furnace	Slag or cinder	Charcoal	Iron*	Ore	Chapter ref.
Huckhoe (Northumb.)	7th–6th cent. BC	+	+	coal	+		66
Odell (Beds.)	1st cent. BC– 1st cent. AD	+	+			+	14
Longbridge Deverill, Cow Down (Wilts.)	4th cent. BC						40
Hod Hill (Dorset)	?		FB		+		
Meltham (Yorks.)	EIA		+	+		+	140
Dry Hill (Surrey)	1st cent. BC– AD		+	+			141
Pipers Copse	1st cent. BC– AD		+			+	141
Ructstalls Hill (Hants.)	2nd cent. BC– 3rd cent. AD		+				142

FB = Furnace bottoms; T = Tap slag
* = Iron showing evidence of smelting or smithing on site

high quality (*see* Table 68) has been suggested as the source of the ironstone actually used. However, this deposit would give a slag of much lower phosphorus content than that actually found on the site (i.e. 1·6% P_2O_5 shown in Table 76 *below*).

The site of Cow Down at Longbridge Deverill in Wilts.[40] has also produced material dated to the same period. Several furnace bottoms were found in pits of this 6th–5th century settlement, but no furnace structures and only one piece of ore, again a pyrite nodule. The smaller plano-convex lumps of slag weighing only about 500 g are almost certainly all SHBs. Analysis of one such specimen from the later Iron Age site of Bagendon[41] confirmed this interpretation. A green corrosion product detected in the cross-section proved to be from a tin bronze, showing that the forge had been used for working copper-base metals as well as iron. This is not an unusual occurrence; similar bottoms have been found in Greece and Carthage.

The simultaneous existence of slag and ore, provided there is enough ore to prove that it has not been used merely for colouring, implies the existence of a bloomery. The presence of large quantities of slag itself implies a bloomery, since the primitive smithy process does not produce much slag. Tap slag (i.e. slag that has leaked or been allowed to run from a furnace in a semi-liquid state) indicates smelting. The position of the find also indicates its purpose to some extent. Until the late medieval period smelting was invariably carried out near the source of ore and fuel, i.e. in woods or on iron deposits near woods.

The forge fire, unlike the smelting furnace, need not be placed in a hole in the ground. Highly reducing conditions are not needed, only a pile of charcoal and a tuyere and bellows. Naturally, if the forge is more than temporary, some structure or a hole will be used to prevent the charcoal blowing away. In many cases it is very difficult to distinguish forge slag from smelting-furnace slag of the more primitive type. Furthermore, there is no reason why smelting furnaces should not be used for reheating blooms and forging purposes.

At Hengistbury Head[35] (400 BC–AD 50) there is little doubt that smelting was carried out since ore, slag and a piece of iron were found as well as ore outcrops along the cliffs. This site was also the scene of non-ferrous metallurgical activities of an advanced type (*see* Chapter 4), but it is a surprise to find a piece of cast iron (composition in Table 78 *below*) which was previously thought to be a product only of the medieval Iron Age, at least in Western Europe. Since only a few pieces of well-authenticated cast iron dating from prehistoric times have been found in the British Isles, it is very probable that this was an accidental product. It would be formed as a result of moderately high temperatures (~1 250°C) and highly reducing conditions, and since these two requirements rarely occur together in primitive furnaces, it would be infrequent. As cast iron is not forgeable it would be quickly thrown away as an unwanted product.

Of the classic Iron Age 'A' sites (500–300 BC), All Cannings Cross[13] (400–250 BC) yielded tap slag, ore from the Lower Greensand (probably from Seend), and charcoal consisting of a variety of trees including oak, holly, ash, and elder. No evidence of furnaces other than small pits was found, but the remains clearly indicate smelting.

Little Woodbury[42] yielded little more than slag, but Hunsbury[43] proved to be the richest site as far as Iron Age 'A' objects were concerned. The site was riddled with 300 or more pits of varying sizes, 1·6–3·3 m in diameter by 1·8–2·1 m deep, which were probably grain storage pits. Some of these, however, penetrated as far as the Northamptonshire ironstone, and a considerable amount of iron slag was found, suggesting smelting operations on the site.

Iron slags have been found on the 3rd–2nd century BC tin-smelting sites at Chûn[44] and Trevelgue[45] in Cornwall, and in the Iron Age settlement at St. Eval.[46] Slag from the latter has been analysed (*see* Table 76 *below*). Another Cornish site, Chysauster,[47] has yielded iron slag, a piece of tin, but no tin slag. The presence of limonite strongly suggests that this was an iron-smelting site. On the other hand, a Late Bronze Age hut on Dean Moor, Devon,[48] was found to contain about 23 kg of unused hematite built into the walls. It appeared that this had been separated from tin ore collected for smelting.

Merthyr Mawr Warren,[49,50] an Early Iron Age site in Glamorgan, produced a mound containing iron slag of the primitive type which had not been tapped. A trench cut across the centre of this mound revealed two small hearths, the largest of which was 76 cm in

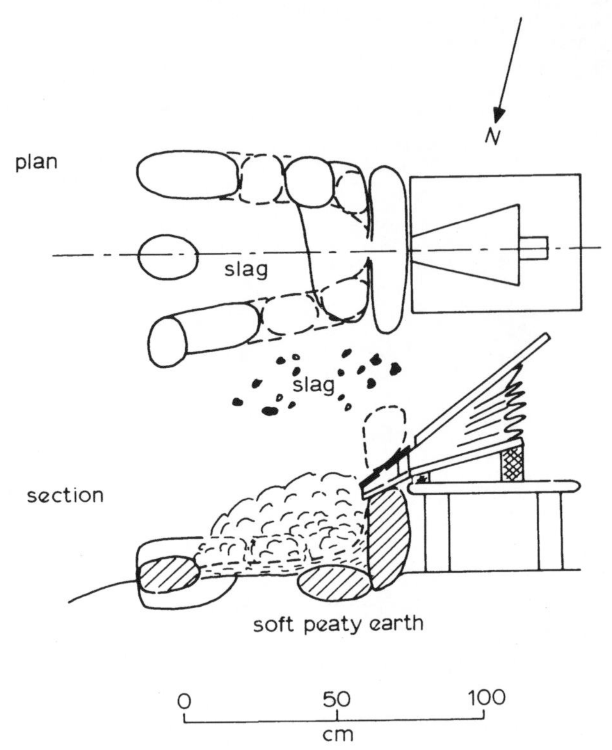

84 Reconstruction of furnace from Rudh' an Dunain, Skye (after Scott[51])

diameter. The smaller hearth was 30 cm in diameter, lined with clay, and contained a small piece of iron slag of high iron content. The site is dated from about the 4th century down to the 1st century BC and, besides iron-working remains, it contained copper-base alloy working hearths and crucibles of Glastonbury type (*c*. 1st century BC).

Perhaps one of the best Early Iron Age smelting sites yet found is that at Kestor, near Chagford in Devon.[31] Unfortunately, this site is not closely dated and may have been occupied at any time after 400 BC. Here, in the dark occupation soil inside a hut, were found the remains of a furnace about 30–45 cm in diameter containing furnace slag and a mixture of charcoal and soil. The furnace was 22 cm deep, and seems to be a typical bowl hearth. There was a stone to one side which might have served as a rest for the bellows. The slag was typical high iron-containing material (Table 76 *below*). Unfortunately, it is not possible to identify the source of the ore, as the analysis is not complete. It could have come either from bog ore or from one of the many other sources in the area.

A larger pit was found some 0·60 m away from the bowl furnace containing ash and charcoal but no slag. The bottom was reddened by burning. This could have been used for reheating the bloom for forging. No slag was found in it, which shows that it was not used for smelting.

The remains of a small bowl furnace were found in a rock shelter at Chelms Combe, Cheddar. This furnace was dated to about 150 BC–AD 50 by the presence of Iron Age 'B' sherds within it.[30] It appears to have the remains of a semi-permanent tuyere built into the side, which is an unusual feature of early bowl furnaces. The remains of the furnace are shown in Fig. 76; in many respects it is very similar to the smelting furnace from Kestor. The slag found in the furnace has been analysed and the result is given in Table 76 (*below*). There is no doubt that this furnace, like the one from Kestor, had been used for smelting.

The only complete 'bowl' furnace from the Early Iron Age is that from Rudh' an Dunain cave in Skye.[51] This furnace is composed mainly of stone slabs; a reconstruction is shown in Fig. 84. The space between the slabs measured about 38 × 23 cm and was unpaved. These structures have close affinities with those found on Rannock Moor,[52] which belong to the medieval period (*see* Chapter 8). Clay would have filled the spaces between the stones. Some slag was found within the furnace but most was piled up on the north side. Most of the slag was in an Iron Age 'A' level, which suggests that the use of the furnace dates from about the 4th century BC. The slag, which totalled about 8 kg in weight, was the typical fayalite–wüstite variety. Charcoal and wood were found in all levels. In the iron-smelting level, birch and willow or poplar charcoal were present and some hawthorn wood. A pinewood shovel was also found which could have been used for putting on charcoal.

Remains of iron smelting were also found in a Somerset cave at Rowberrow Warren.[53] Here, a pit 0·50 m deep had a floor of flat stones to which iron cinder was adhering. The pit also contained charcoal, and slag and hematite were found in the vicinity. This site probably belongs to the 2nd century BC.

Slag was also found at Glastonbury Lake Village[54] in the same county, which was dated to the 1st century BC. Here a 45 cm diameter hollow was found filled with charcoal ash and some fragments of iron slag. The absence of ore, however, may point to the use of the furnace for smithing rather than smelting.

At Wiltrow, in Shetland,[55] the remains of another very primitive unit were found. The description and plans given by Curle are not metallurgically convincing and one doubts whether he has interpreted the evidence correctly. The facts are that iron-smelting slag, furnace slag, and bog iron ore were found on this site, which may be dated to about 100 BC–AD 100. The fuel used seems to have been largely peat, since a considerable amount of peat ash was found. In spite of the high sulphur content of raw peat, we know that this can be reduced by charring to as little as 0·16%. So there is no reason to ignore it as a fuel for iron smelting.[56]

In the north of England the remains of Iron Age furnaces have been found at Roxby (Cleveland)[57] and at West Brandon in Durham.[32] The former was a bowl furnace, and since no slag was found adhering to the sides it is just possible that this was a smithing hearth.

The rock-hewn furnace in the palisaded site dating from the 7th–6th century BC site at West Brandon has been referred to as a bowl furnace, but consideration along the lines suggested earlier indicates that it is too small for this purpose (only 20 cm in diameter) and that it is the bottom of a small shaft furnace as shown in Fig. 77*b*.

Cleveland has produced the only domed furnace of the pre-Roman Iron Age. This was found at Levisham[58, 58a] by J. G. Rutter and R. Hayes (Fig. 82). There are problems with regard to this type of furnace: bloom removal is difficult; it was either through the top or, more likely, along the horizontal passage. Gilles[59] has reconstructed this type, which is common

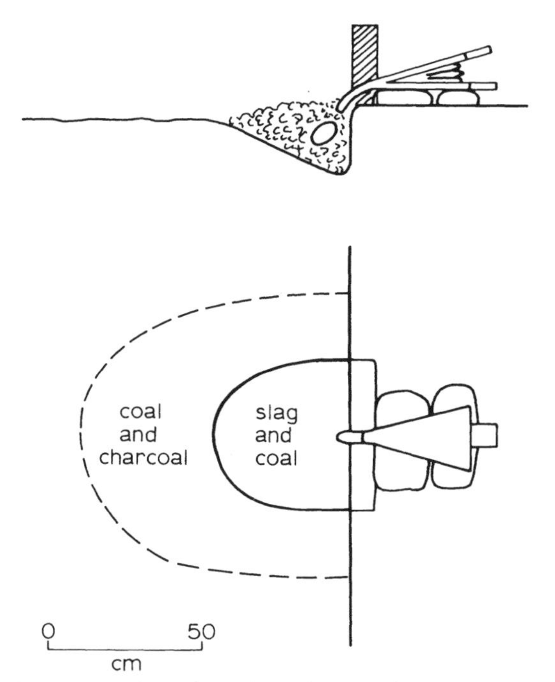

85 Reconstruction of smithing furnace from Huckhoe, Northumberland (after George Jobey[66])

in the Siegerland region of Germany, and has produced FBs containing weldable pieces of low-carbon iron in a fayalite slag. But his furnace was larger (1·7 m tall inside) and he used natural draught through a 60 mm diameter tuyere.

Shaft furnace finds tend to cover the period of the 1st century BC into the Roman period, suggesting that this type was not introduced by the Romans but had developed or been introduced by contact before the Roman invasion. Two sites show evidence of earlier, pre-Roman remains; these are Broadfield[60] in Sussex (190–80 BC) and Wakerley in Northants.[16] The finds at Broadfield are 30–45 cm in diameter shaft furnaces about 1·25 cm high with tapping pits in front. As there are no signs of tuyere holes, it is possible that the tuyere was in front and blowing was stopped immediately before tapping. In this respect they are like the Roman furnaces at Ashwicken and Pickworth.

At Wakerley there seem to have been several types, but since these are only extant to ground level it is difficult to be sure of the type. Some are narrow and therefore the bases of shaft furnaces, while others are wider and could be the bottoms of bowl furnaces. Together with these are the 'channel' hearths, which are long, narrow, trench-like features containing only charcoal. They might have been used for smithing or charcoal manufacture.

Besides the above remains of furnaces dating from the Early Iron Age, numerous pieces of evidence indicating iron smelting have been found on many other sites. At Mynydd Bychan, near Llysworney in Glamorgan,[61] ore, slag, and charcoal were found, some of it in an Iron Age 'B' context. The ore was analysed, and found to contain 8·24–28·7% silica and 0·08–0·244% phosphorus, the ferric oxide content being 70–89%. The phosphorus content indicates a hematite of the Mwyndy type rather than a limonite from Llanharry which is very much lower in phosphorus. The charcoal consisted of the usual Early Iron Age mixture including oak, willow, elder, and hawthorn.

Excavations at Purberry Shot near Ewell in Surrey have yielded FBs and a small bloom.[62] These remains date from 200 BC to AD 150, but the FBs measure only 7–10 cm in diameter and are about 2·5 cm thick; they must be the result of smithing. In many cases these bottoms are heavily contaminated with charcoal, since the conditions in the bottom of the furnace are often not sufficiently oxidizing to burn it. The small 'bloom', which weighed 340 g and measured 7·5 × 4 cm, had been forged.

An early site on Swallowcliffe Down, Wilts.[7] produced slag, ore from the Lower Greensand, charcoal, and a piece of iron. This piece seems to have been broken off the corner of a squared block, possibly a smithed bloom or billet; its weight was 1·1 kg. This site also produced marcasite nodules, but there is no evidence to show that they were used for ironmaking.

Another site, on Boscombe Down East,[63] produced some ferruginous material which may have been a very well weathered slag. This was found in a Late Bronze Age–Early Iron Age context and is probably the earliest piece of man-made ferruginous material found in the British Isles.

Two more Scottish sites have yielded remains of early iron smelting. Callander[64] found some iron slag and hematite in an earth house at Bac Mhic Connain (N. Uist, Hebrides). This site, however, is not closely dated and is probably after 200 AD. An Early Iron Age bowl barrow at Pityoulish, Strathspey, produced 'two small lumps of fused iron' and much charcoal.[65]

One of the most interesting sites to have produced a hearth is another palisaded site at Huckhoe in Northumberland for which new ^{14}C dating has given the very early date of 7th–6th century BC.[66,67] The hearth was rock-hewn in the shape of a semicircle, and was probably used as shown in Fig. 85. Some of the slag was similar to smelting products, but in view of the presence of a large quantity of coal, the author believes that this hearth was finally used for smithing purposes. A piece of piled iron consisting of twelve laminations, 2 mm thick and welded together, was found in the centre of a mass of cinder. This piece had been cut off a longer piece and clearly it was intended to carburize the laminations by prolonged heating in red-hot charcoal or coal in the relative absence of air. In this way wrought iron, low in carbon, can be converted into a piece of steel containing a considerable amount of carbon. A microexamination, however, showed that the carburization had gone only a very little way, and in spite of the elaborate treatment given, the desired result had not been achieved.

An early type of tanged knife and some smithing scales were also found on this site and are discussed below.

Tuyeres

A number of tuyeres have been found on various sites, many of which have not been closely dated. It is not known whether all of them relate to ironworking, but there is no reason to suppose that a tuyere used for bronze working is significantly different from that used for iron. For this reason it is now intended to discuss the structure of the tuyeres found.

The most perfect of those known to be primitive in

86 Tuyere from Arnbathie (courtesy of National Museum of Scotland)

type is that from Arnbathie, 6 km ENE of Perth, which is illustrated in Fig. 86. It consists of fired yellowish clay and is unused. It was unstratified and is a typical single tuyere, only large enough to take a single tube from a pair of bellows. Another, very similar, has been found in a 14th century context in Much Park St, Coventry. The bore tapered from 24 mm to 9 mm at the furnace end.[68]

A more complicated, but probably more primitive, double tuyere was found on the site of the bloomery at Crowhurst Park, Sussex.[69] The remains are in the museum at Lewes, although there are also some remains at Bexhill that may well be a part of the same tuyere. Both parts are heavily slagged, and it is difficult to be certain of the dimensions of the original tuyere. There is no doubt that this was designed for the reception of two tubes from the bellows, which suggests that the bellows themselves were of the most primitive type without valves. The two bellows would be operated alternately, as suggested by Straker,[69] the air being sucked into the bellows through an opening around the poorly fitting bellows-tube.

More double tuyeres have been found at Bardown in Sussex;[70] these are undoubtedly Roman and are illustrated in Fig. 87. They seem to be designed to receive separate streams of air from two bellows, but instead of uniting them, they enter the furnace separately and in a divergent manner.

Another example of the simpler type of tuyere belonging to the early 1st century AD was found at the Sussex bloomery of Chitcombe.[69] A tuyere of a slightly different pattern, oval in section and therefore probably designed to take the two bellows-tubes, was found on the Lake Village site at Glastonbury[54] (shown in Fig. 87*a*); it would seem that a part of the narrow end is missing.

The ends of tuyeres of the 6th–10th century AD were found at Garranes[71] and Lagore[72] in Ireland; these are shown in Fig. 87*b*, *c*. Also found in Ireland, at Ballyvourney[73] were a number of circular discs of clay, slagged on one side, with holes about 25 mm in diameter. These represent the inside ends of tuyeres set into the furnace wall (Fig. 87*d*). Their survival is due to slagging and the high temperature to which they were exposed. The minimum diameter of the holes in the tuyeres varies from 9 to 25 mm. These diameters should be compared with the size of 12 mm found by Wynne to be necessary in his experimental work on a small bowl furnace 23 cm in diameter.[25]

'Block tuyeres' consisting of square blocks of fired clay in which suitable holes have been moulded were often used in the Roman world. One has been found on the legionary fortress at Carpow in Perthshire and is shown in Fig. 87. The smithing furnace in which it was inserted had sloping sides, as shown by the line of vitrification.[74]

Summary

The results of excavations allow us to reach firm conclusions only about the two centuries before the Roman invasion. The period from 700 to 100 BC is still very obscure, but it would seem that continuous contact with the Continent led to the existence of two types of furnace: the bowl and the shaft. In the south-east and east of England the evidence is stronger for the shaft furnace, while the bowl furnace is the main type found in the rest of the country, with one example of the domed furnace.

IRON AND STEEL OBJECTS FROM THE EARLY IRON AGE

The introduction of iron into the British Isles was a gradual process. Now that the invasion hypothesis is no longer in fashion we can suggest, with more likelihood of acceptance, a gradual diffusion of technology across the English channel using the shortest sea route across the Straits of Dover. There can be no doubt that contact was established in the LBA, as may be seen from the evidence of a cargo of scrap bronze deposited by wreck at the bottom of the Dover cliffs.[75] Ideas would no doubt have travelled the same way, and bronze and iron were used together for a period before the widespread establishment of iron smelting. But there are signs of an increase in trade from the Rhine to the Thames estuary, and this route might have been more influential in spreading new ironmaking techniques.

Perhaps the value attached to iron in this period is best shown by the finding of an iron torc weighing 0·4 kg at Spettisbury and similar specimens on other sites.[76] Iron was used in Denmark as a decorative inlay on bronze razors at the end of the BA, and for finger rings in Mycenaean Greece.

a Glastonbury Lake Village (100 BC–AD 50)
b Garranes, Co. Cork (AD 600)
c Lagore Crannog (7th–10th century AD)
d Ballyvourney, Co. Cork (6th–13th century AD)
e Arnbathie, Perth (unstratified)
f Carpow, Perth (Roman or post-Roman)
g Bardown, Sussex (Roman) (after H. F. Cleere)

87 Clay tuyeres from various British and Irish sites.

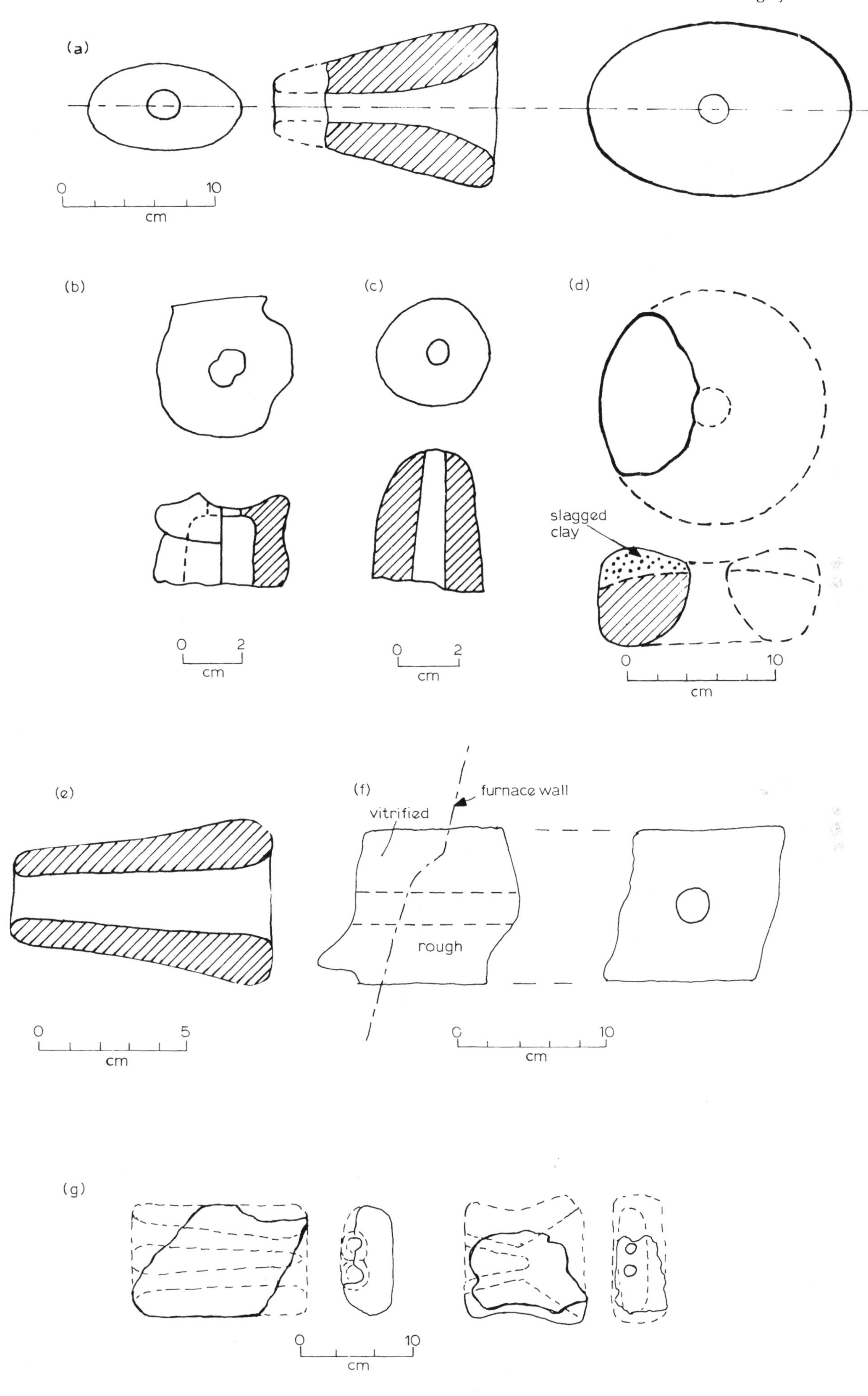
(a)
0
10
cm
(b)
0
2
cm
(c)
0
2
cm
(d)
slagged clay
0
10
cm
(e)
0
5
cm
(f)
vitrified
furnace wall
rough
0
10
cm
(g)
0
10
cm

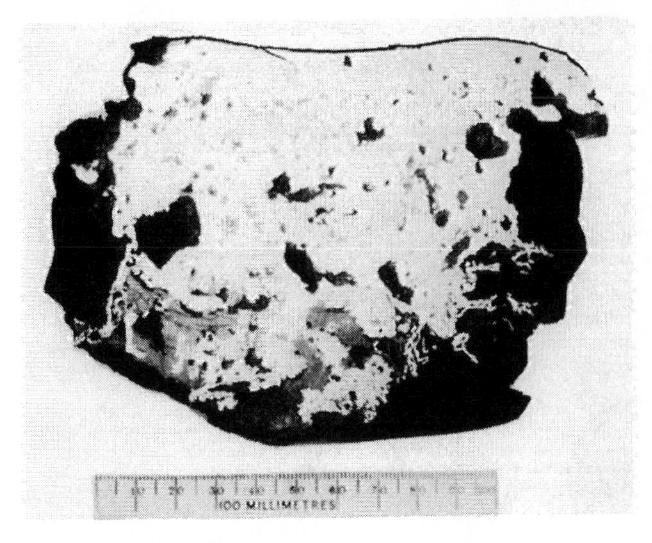

88 Section through iron bloom from No. 27 experimental smelt (weight of complete bloom = 6·5 kg) (Tylecote *et al*.[34])

Blooms and the effect of carbon and phosphorus

The iron produced in the bloomery process is remarkably pure except for the entrapped slag (Fig. 88). This is due to the low temperature (1 200°C, compared with the 1 300–1 400°C used today) and the fact that the iron was solid rather than liquid. As a consequence very few impurities are reduced into the iron. Unless the conditions are exceptionally reducing, all the impurities, except for phosphorus and sulphur, go into the slag. Under strongly reducing conditions, small amounts of silicon and manganese will be reduced from the silica and manganese oxides in the ore, but then carbon will be present as well, making the product virtually a steel. It is, however, very rare for the silicon and manganese contents to approach that of a modern steel (0·2%). Some authors. confronted with ores high in manganese, often refer to them as good ores from which steel can be made. Thermodynamically this is unlikely to occur, and over 99% of the available manganese will enter the slag. But the presence of manganese in the slag will favour a higher carbon content in the iron and so produce a more 'steely' iron.

On the other hand, phosphorus can be reduced and taken up by the iron although a proportion of it finds its way into the slag; the Corbridge beam, for example, contains 0·063% and the Chesterholm chisel 0·076%P (*see* Table 78). These values are a little higher than those of modern steels. However, wrought iron made from high phosphorus irons in the pre-Roman and late medieval period contained appreciable quantities of phosphorus, which made it stronger but more brittle in the cold state (*see* Table 75). For example, several pieces of iron from Odell[14] had as much as 1%P, but the metal was extensively hot-forged.

Sulphur is a very detrimental element in iron. Iron ores – particularly when roasted – are sometimes, but not always, low in sulphur which is usually introduced by fuels such as coal. Charcoal irons rarely show more than 0·05% sulphur. Kendall[105] reports that Forest of Dean ores contain about 0·1% sulphur. However, these amounts are relatively innocuous and may be removed during roasting.

Much larger amounts of sulphur have been found in primitive irons believed to have been smelted with coal. For example, one of the few pieces of cast iron made before the late medieval period was found in the Roman site at Wilderspool[78] together with coal. Its composition (0·485%S) showed conclusively that it had been worked with sulphur-containing fuel, probably the coal found on the site. The smelting furnace either produced a number of small pieces of iron which were separated from the slaggy furnace bottom and welded in a smithing furnace, or a more homogeneous piece of metal relatively free of slag. The latter would be the product of the shaft furnace.

The bloom found at Wookey Hole[79] is probably the largest pre-Roman bloom found, and has a high carbon content. Its weight when found was 3·12 kg and it measured about 20 × 9 × 6·4 cm. It was, however, completely oxidized. Analysis gave the following results (%): Fe_2O_3 57·2; FeO 30·4; SiO_2 0·61; CaO 0·52; C 0·48; MgO 0·34; P_2O_5 0·24; S 0·12; insoluble 0·55.

Besides the oxides of iron that form 87% of the specimen, the high carbon content is of some interest as it could not have been picked up from the ground. It seems that this piece of rusty iron had been a piece of fairly pure steel with a carbon content of about 0·7%. In addition, it would have originally contained about 0·15% phosphorus, but the rest of the material would be derived from soil entrained by the growing rust. The original weight of this bar would have been about 2·0 kg. It is not known whether it was made from more than one piece, but it probably represents the upper limit of size of the smelted blooms of the period. It is not possible to tell whether the carbon content was intentional.

Two small blooms have also been found in Sussex and examined metallographically by Smythe.[80] The larger bloom (1·25 kg) from Foreward bloomery, Crowhurst (EIA or Roman in date), varied in carbon content from 0 to 0·3%, and was undoubtedly the direct product of smelting. The smaller bloom from Nanny's Croft, Arundel Park, which is probably late Roman, weighed 0·33 kg and had a carbon content of about 1·6%. Smythe believed that this was also the direct product of reduction, but was probably accidental.

Neither of the Sussex blooms had been carburized in the smith's fire, which shows that it is quite possible to obtain very variable carbon contents during reduction in the bowl hearth. The product would be tested, and if found harder than required might be thrown away or, more likely, kept to one side and used for 'steeling' wrought iron by welding. There is no doubt that the majority of iron required and produced was of the low-carbon type. The fact that most of the artifacts examined contain hardly any carbon, whereas these blooms contain appreciable amounts of carbon, suggests that the latter may have been accidental. Usually when the smith wanted carburized iron he took low-carbon wrought iron and carburized it himself under his own controlled conditions.

The diffusion of carbon into iron, like most solid-state diffusion processes, is slow. Figure 89 gives some idea of the rates at 900°C when iron is embedded in various carbonaceous media. The easiest way to do

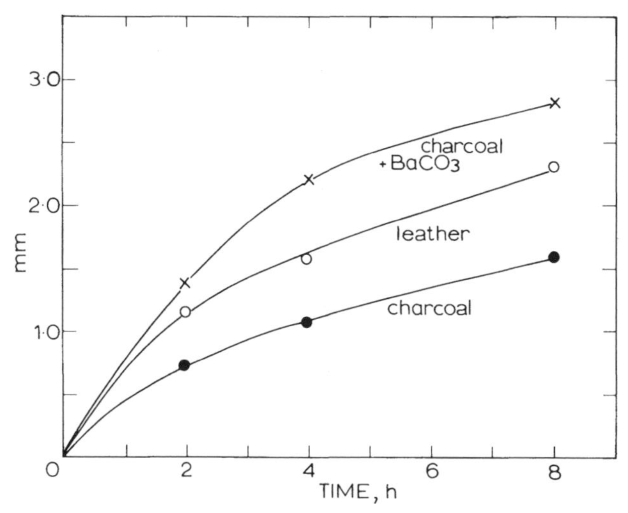

89 Rates of carburization of iron in different media at 900°C (after Shaw Scott[143]; courtesy of the Council for British Archaeology)

this is to put the pieces into a pot with charcoal and heat the pot externally with whatever fuel is available. The carbon from the charcoal will combine with the small amount of available oxygen in the system to form carbon monoxide which will dissociate on the surface of the iron, leaving the O_2 free to return to the system. The elemental carbon will then diffuse interstitially through the iron, building up a high concentration on the surface, decreasing to zero in the centre of the bar. The actual concentration gradient will depend on temperature, time, iron composition, and the carbonaceous medium. Naturally, to maintain controlled conditions over a large surface area is extremely difficult and much variation is bound to occur. Welding of ready-made steel is one way of minimizing this variation.

Very few pre-Romano-British iron tools have been examined for evidence of quench-hardening but even those of the Roman period rarely show evidence of this process. There are two possible reasons for this. The first is the widespread use of high phosphorus iron which can be work-hardened to give hardnesses as high as 300 HV. The second is the high technological level required to obtain a satisfactory end-product. Not only must the iron be carburized to turn it into steel (i.e. not less than 0·5%C) but the minimum quenching temperature depends on the carbon level reached. If the carbon level and the quenching temperature are too high the steel will be unduly brittle.

A thick piece of piled iron can rapidly be built up by carburizing strips of low phosphorus iron and welding them together. In fact, a single strip of reasonable surface area can, with many reheatings, be built up into a small but thick piece of steel. This has been demonstrated by Whitaker and Williams[81] and is probably the technique used to form the steel chariot tyres found at Llyn Cerrig Bach.[82] A pile of ten strips, about 2·8 cm wide and 15 cm long and each 1 mm thick, was heated to between 950°C and 1 000°C in an atmosphere of carbon monoxide, then quickly withdrawn and hammer-welded. After welding there was a very slightly carburized layer at the junction of each strip with the next.

In a second test small quantities of finely-powdered cast iron were placed between the layers, and in the third test larger quantities were used. This produced a layer of high-carbon iron between each strip which could be uniformly dispersed by heating for 8 h at 900°C.

It is not known whether cast iron was used for this purpose as early as the pre-Roman Iron Age, but carburization for 2–3 h before piling would have the same effect.

The reaction to carburization in the forge fire would vary with the phosphorus content, as shown in Table 73. Therefore, on the whole, we could not expect the high phosphorus irons to carburize so readily. The figures in Tables 77 and 78 (*below*) show that all the high phosphorus irons have low carbon, while the one example of a smithed high-carbon iron has a low phosphorus content. There is no reason why the smith working a high phosphorus iron should be dissatisfied

Table 73 Effect of phosphorus on the diffusion of carbon in iron (after Stead[124–5])

Starting material	%C absorbed in a given time
Pure iron	0·94
Iron + 0·6%P	0·60
Iron + 1·25%P	0·53

Table 74 Effect of carbon and phosphorus on the hardness of annealed iron

%	HARDNESS (HV) Carbon[B]	Phosphorus*
0	75	75
0·05	—	78
0·15	100	91
0·31	—	113
0·23	110	145
0·60	180	216
0·90	200	250
1·20	260	287
1·50	—	270

[B] After E. C. Rollason, 'Metallurgy for Engineers', 150; 1961, London.
* First four values after Hopkins and Tipler[126] on pure iron and phosphorus. Remainder based on experimental work by J. G. McDonnell of the University of Aston, using less pure materials.

Table 75 Effect of cold-work on the hardness of iron and phosphorus-containing iron

Cold-work % reduction in thickness	HARDNESS (HV) Iron + 0·1%C (Williams[127])	Iron + 0·5%P (McDonnell*)	Iron + 1%P
0	120	216	250
10	150	300	320
20	170	320	340
30	180	—	—
40	188	—	—
50	192	—	—

* Based on experimental work by J. G. McDonnell of the University of Aston.

Table 76 Composition of Early Iron Age iron-smelting slags (%)

Ref.	128 Maiden Castle, Dorset	31 Kestor, Devon	129 Chelms Combe, Som.	130 Blewburton Hill, Berks.	131 Camerton, Som.	41 Bagendon, Glos.	14 Odell, Beds.	16 Wakerley, Northants.	46 St. Eval, Cornwall	120 Berkhamstead, Herts.	108 Brooklands, Surrey	Shalbourne, Berks.	Crown End, N. Yorks.
	AD 25–45	After 4th cent.	After 3rd cent.	Pre-Rom. or RIA	Pre-Rom. IA	AD 10–50	AD 1–50			50 BC–AD 40	Pre-RIA	?	EIA
FeO	53·0	48·3	50–57	61·54	14·34	59	64·2	63·8	50·0	68·8	68·0	65·9	72·3
Fe_2O_3	22·87	17·4	8–11	19·69	6·44	18	—	13·0	14·3	—	—	—	—
SiO_2	15·95	21·4	23–27	10·15	66·01*	23	18·1	15·7	19·3	21·6	19·0	23·1	18·7
CaO	2·75	0·39	4·4–5·6	5·70	2·10	0·5	1·4	1·1	1·3		0·68	0·6	0·2
MnO	tr.	—	0·1	tr.	—	—	—	—	—	c. 8·8	—	1·3	0·5
Al_2O_3	1·47	—	—	1·1	9·16*	2·6	—	—	—	—	2·3	7·1	1·9
MgO	0·45	0·11	1·1	0·40	0·33	—	—	—	—	—	0·19	0·5	—
P_2O_5	0·40	—	—	0·96	0·16	—	0·4	—	—	0·46	1·63	0·65	—
C	}	}	—	—	}		}	}	—	0·04	—	—	—
H_2O	} 2·97	} 1·21	2·45	—	} 0·11		} 12·8	} 0·41	—	—	2·21	—	—
CO_2	}	}	1–2·6	—	}		}	}	—	—	—	—	—
TiO_2	—	0·25	0·30	—	0·84	—	—	0·34	0·5	—	0·15	0·3	0·1
SO_3	—	—	—	—	0·09	—	—	—	—	0·024	tr,	0·1	—
Alkali	—	—	—	—	—	—	—	—	1·3	—	0·58	—	—

— = not sought; * = slagged with clay

with his product, for phosphorus will harden iron almost as much as carbon, as the figures in Table 74 show.

Like most other metals, wrought iron can be work-hardened by cold-hammering. Table 75 gives some indication of the effect of cold-work on iron containing a small amount of carbon (i.e. low-carbon steel). In the same table are shown some estimated values for iron containing 0·5%P. As have already been shown in Table 74, phosphorus, being a solid solution element in iron, can exert a considerable effect on the hardness in the annealed state. This fact was well recognized in later times when iron with substantial phosphorus content merited the term 'cold short', i.e. somewhat hard and brittle in the cold state. Today amounts not exceeding 0·1% are deliberately added to mild steel to increase its strength.

Currency bars

At various times, mainly in the south of England, objects commonly known as 'currency bars' have been found (Fig. 90). These appear to consist of two types: (*a*) tapered sword-shaped bars; (*b*) parallel-sided bars, almost all having a socketed end. Many of these have been found in hoards, comprising in one case 394 bars and in another about 140, but many have been found in ones and twos.

The term 'currency bar' derives from a much debated interpretation of a passage in Caesar's *De Bello Gallico*, Bk V, Ch. 12. In this book Caesar is giving a description of the customs of British tribes. The disputed passage runs as follows: *utuntur aut aere, aut nummo aureo, aut taleis ferreis ad certum pondus examinatis pro nummo*. Handford[83] renders this: 'for money they use either bronze, or gold coins, or iron ingots of fixed weights.' But archaeologists have always accepted the meaning of *taleis ferreis* as 'bars of iron'. The meaning has been discussed at length by many scholars;[8,84,85] most assume that the sword-shaped iron bars found on native sites are what are referred to in this passage. It is pointed out how difficult it is for a smith to make a bar to a fixed weight and it is suggested that the bars are 'moods' or semi-finished swords.[84] As such they would still have some sort of value dependent on their weight, but they need not be made to an accurate weight or size. Most of the bars can be safely dated to between the beginning of the 1st century BC and the end of the 1st century AD, and therefore agree with the period referred to in *De Bello Gallico*.

Attempts have been made by various exponents of the currency theory to classify the bars into 'units'. One of the difficulties here is to decide whether to take a unit of weight or length, since the bars vary so widely in both these measurements. The heaviest bar yet found weighs 1·93 kg and was found at Littleton (Bredon Hill), Worcestershire in 1882. However, this bar is only 32·3 cm long and is therefore one of the shortest, the longest being one from Hod Hill which is 87·6 cm long.

More than half the bars are sword-shaped, i.e. tapered, and although all but a few of the handles are socketed, this may be an unfinished stage, and they could easily be forged into the standard handle or tang of the La Tène sword. The sword-shaped bars are, in the main, about 76 cm long and therefore are roughly the length of the swords of the time. None of the short

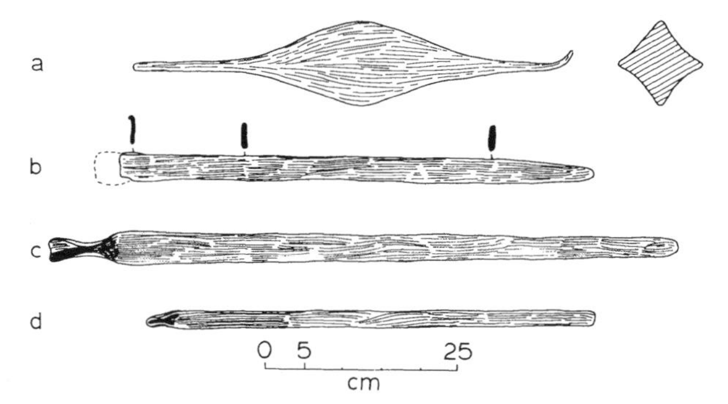

90 Types of currency bars

bars is tapered. Even so, the heavier parallel-sided bars could have been forged into such implements.

Most of the bars are heavily rusted, and one of those from Llyn Cerrig Bach has rusted into three pieces across the smithed welds. The loss (or gain) by rusting makes the accurate recording of the weights somewhat meaningless. This should be one difficulty facing those trying to classify the bars into 'units', but it seems to have been ignored, and many have gone out of their way to record the weight to the nearest gram! The other difficulty is one of statistics. Are the four recorded from the hoard of 140 from Bourton-on-the-Water representative of the whole or just odd ones picked out at random? The same question may be asked about those from Hod Hill. It would appear that little attention was given to the measurements of the early hoards before they were scattered, and many bars from them were destroyed.

On the basis of weight, which seems to be the best way of recognizing material intended to be a sword-mood, only two groups can be distinguished, if single specimens are discounted. These groups are 390–650 g and 170–310 g. Since many bars are incomplete, the original weights may be regarded as being about 650 g and 310 g respectively.

As far as size is concerned it is found that the thickness varies by a factor of three. Therefore it is difficult to see how the bars could be valued by length. However, when the length is considered one very large group in the range 70–90 cm is noticeable. This group will include a number that are incomplete – handles or tips missing – and it would appear that it contains bars well on their way to being finished swords. Furthermore, the fact that this group was approximately the correct weight for such a purpose was shown by plotting weight against length when it was seen that well over half the currency bars known fell into a closely defined group. The remainder are too few to allow any useful conclusion and may have been intended for making into the other implements and weapons of the period. Perhaps one way of resolving the problem would be to examine metallurgically the tangs of some of the contemporary swords. If they were made by forging the almost tubular handle of the 'currency' bar, this might be clear from the structure.

Chemical Composition

So far, about 14 bars have been analysed, or partly analysed, and the chemical compositions of these, given in Table 77, show that five out of six were made from phosphorus-containing ores, and one from an ore containing an appreciable amount of nickel. The latter is unusual in British ores; the Northamptonshire orefield, for one, is remarkably free from it. There are, as yet, insufficient data on the nickel content of British iron ores to enable its origin to be determined. However, it can be definitely said that this is not an example of the use of meteoric iron, as Gowland thought.[8] No meteoric irons contain less than 4·5% Ni (as shown in Fig. 1, Chapter 1). In fact a considerable number of continental iron objects have been found to contain layers of high-nickel iron and it is known that deposits of such iron ores exist in Greece, and other Balkan countries.

The geographical distribution, given by Fox,[86] shows that most of the bars were found in an area centering on the Cotswolds, and it has often been stated that these bars must have been made from Forest of Dean ores. This is usually based on the erroneous idea that all early iron came from a few well known deposits of high-grade ores. However, as has been explained previously, iron is very widespread and there is no doubt at all that local deposits were worked where there were any; early iron-workers tended to go to the iron, provided there was sufficient fuel to smelt it.

The Forest of Dean ores have very low phosphorus contents (*see* Table 67) and therefore cannot have produced the five bars analysed with phosphorus contents from 0·35 to 0·954% (*see* Table 77). Phosphorus-containing ores are very common and it is quite possible that if the currency bars excavated in the Cotswold region had been made there, they would have been made from ores closer at hand than the Forest of Dean – from Oxfordshire ores similar in type to those from Fawler and Towcester (*see* Table 67). Furnace bottoms found on the site of Hod Hill, Dorset, which has yielded currency bars, contained 2·5%P and could have been connected with the production of high-phosphorus currency bars.

Metallurgical structure

One of the bars examined was one of 147 found at Bourton-on-the-Water, Gloucestershire, in 1860,[87,88] and now in the museum of the University of Newcastle upon Tyne. When examined it was fractured 16·5 cm from the bottom end. The total length was 78·7 cm, the width of the blade at the tang 4·0 cm, at the bottom 2·2 cm, and at the fracture 3 cm. The bar was approximately rectangular in section, with a thickness of about 2·5 mm.

A section, 3 mm wide, was removed just below the fracture. In thickness, it varied from 3·1 to 1·65 mm. The surface of the section showed a typical steel structure. Hardness measurements on this surface varied from 145 to 182 HV; this was in keeping with the composition, which varied from about 0·2% to 0·8% carbon. The ferrite had a Widmanstätten distribution, which suggests that the bar had been finally cooled from a high temperature (~1 000°C) at a fairly fast rate, but not enough to harden it. The fact that the ferrite areas showed none of the Neumann bands so common in early wrought irons indicates that no attempt was made to forge the bar at a low temperature while it was cooling.

The section indicated that the bar was made of a number of pieces of thin carburized iron containing high- and low-carbon areas. Slag lines were common

91 Bundle of currency bars from Beckford, Worcs.: left, section; right, photomicrograph (light ferrite; dark pearlite) (courtesy of W. Britnell)

in the centre and the thick end, but the thin end was fairly homogeneous with a mean hardness of 158 HV. The almost pure ferrite area at the thick end was unduly hard for ferrite (i.e. 142 HV) and since the bar showed no signs of cold-work, this hardness is probably owing to the fairly high phosphorus content (*see* Table 74).

One of the bars in the bundle from Beckford was examined and found to contain a high-carbon steel band in the centre, sandwiched between two layers of low-carbon steel (Fig. 91). It is unlikely that this was done intentionally, but is more likely to be the result of a decarburization like that on the bar from Worthy Down (*see below*). Bars from Danebury were merely heterogeneous bloomery iron and it would appear that no effort was made to maintain or control the carbon content of the currency bars and that they represent the normal products of the bloomery, reheated in a smithing hearth.

Modern methods of examination now permit the analysis of slag inclusions in the iron, and it was hoped that the use of EPMA might be able to provenance the currency bars, since, during smelting, the unwanted elements such as Ca, Al, Si, Mn, etc., typical of the orebody, are concentrated in this slag. While this technique does lead to certain conclusions regarding the *type* of ore used, the possibility of changes in slag composition during smithing would seem to render this technique of less value than the analysis of trace elements in the metallic iron itself. Such elements are Cu, Ni, Zn, Cr, As, and P. This work was begun by Haldane,[90] and his results are given in the analyses shown in Table 77.

Some trace elements tend to concentrate on the surface of the iron while the iron is undergoing heating under oxidizing conditions during smithing.[103] When two such surfaces are welded together a 'white line' is shown in the metallographic section which is ferritic and contains a high concentration of one or other of the trace elements. In this way iron containing as little as 0·04% As may show enrichment bands containing over 1% As. Such bands allow us to locate the weld lines and thus estimate the degree of smith-welding used in the production of the object.

Myers[91] examined one of the bars from Worthy Down and found it had a similar structure to the one from Beckford. He felt that the bar had first been carburized and then decarburized in the smith's fire. Chemical analysis gives only the average, but the low value of the average carbon suggests that local carburization was not intentional. The Bourton bar had quite enough strength to be used as a weapon, the average hardness being 158 HV. In modern steels the strength is largely due to the carbon which is the normal hardening element for steel.

One of the long bars from the Dorset Museum at Dorchester was very heterogeneous, with carbon contents varying from 0 to 0·8% C. It had been fairly rapidly cooled but not quenched. The hardness varied from 124 to 194 HV (20g).[92]

A complete examination was made on one of the bars from the Gretton hoard;[93] the composition is given in Table 77. The carbon content was variable, alternating between zones of carbon-free iron and regions containing grain-boundary cementite with a carbon content equivalent to modern mild steel. Both regions appeared to be high in phosphorus. The low-carbon areas had a coarse grain size and contained nitride needles. The hardness varied from 210 HV(0·5 kg) in the carbon-free areas to 171 in the carbon-containing areas, the difference probably owing to the phosphorus content.

Electron-probe microanalysis was carried out on the slag inclusions and the result 66% SiO_2, 20% FeO, and 13% Al_2O_3 was obtained. Those near the edge had a rather higher iron content (67%) and lower SiO_2 and Al_2O_3. Comparison with local iron ores showed that this bar could have been smelted locally.

The production of semi-finished sword moods seems to be confined to England and Wales[86] as, so far, few currency bars have been found outside this region. The equivalent Continental type of semi-finished product was the 'Spitzbarre' – a pointed bloom, similar to those found in Sargon's Palace,

Table 77 Composition of currency bars (%)

Chapter Ref.	Worthy Down, Winchester 91	Bourton on the Water, Glos. —	Beckford, Worcs. 89	Ham Hill, Dorset 1 90	Ham Hill, Dorset 2 90	Gretton, Northants. 93	Danebury, Hants. 9	Meare, Som. 90	Little Solsbury, Glos. 90	Read's Cavern, Som. 90	Worlebury, Som. 90	Wookey Hole, Som. 90	Glastonbury, Som. 90	Provenance unknown 1 8	Provenance unknown 2 8
C	0·06	0·02–0·8	0–0·8	—	—	0·25	—	—	—	—	—	—	—	0·08	tr.
Mn	tr.	<0·05	<0·01	0·003	0·003	0·04	—	0·01	0·003	0·005	0·007	0·004	0·024	nil	nil
P	0·954	0·35	0·029	—	—	0·39	—	—	—	—	—	—	—	0·35	0·69
S	0·014	0·014	0·028	—	—	0·029	—	—	—	—	—	—	—	—	—
Cu	—	—	—	0·015	0·017	<0·01	—	0·032	0·019	0·007	0·041	0·004	0·016	—	—
Ni	tr.	<0·05	<0·003–0·006	0·024	0·014	0·014–0·1	<0·004–0·04	0·017	0·019	0·042	0·048	0·026	n.d.	—	0·23
Co	—	—	—	0·015	0·004	—	—	0·017	n.d.	0·006	0·062	n.d.	0·004	—	—
Si	0·11	0·2	—	—	—	0·31	—	—	—	—	—	—	—	0·02	0·09
As	—	tr.	<0·014–0·056	—	—	0·01	<0·017–0·086	—	—	—	—	—	—	—	—
V	—	—	<0·011–0·028	—	—	<0·01	—	<0·1	<0·1	<0·1	<0·1	<0·1	<0·1	—	—

tr. = trace; n.d. = not detected

where they were dated to 721–705 BC. Two of this type have been found at Portland, Dorset[93b] and one is illustrated by Grinsell[93a, b] (Fig. 90*a*). It is possible that these are confirmation of the increasing evidence for cross-channel trade which has been noticed from the LBA onwards into the Iron Age as shown by the material found at Hengistbury Head and in other sites in the south of England.[94]

In conclusion, the distribution of the flat bars seems to indicate that they were made and used by people occupying the south-west central region of England centred on the Cotswolds. These could be the Belgae or the Dobunni. There is no evidence for stating that they were made mostly from Forest of Dean ores. The best conclusion that one can reach is that in the south-west of England there was a fully developed industry for the manufacture of semi-finished bars of iron which could be distributed to local smiths for working up into tools or weapons, and have a value as currency.

Early Iron Age swords

A number of Early Iron Age swords exist in fairly good condition, although many are rusted into their scabbards. The total lengths vary from about 61 to 90 cm, which is in agreement with the lengths of the majority of the currency bars. Most of the larger blades have a discernible taper and in this they also agree with the currency bars. The shorter blades appear to be parallel-sided. These swords are all forged to a lozenge or very flat elliptical section.

Since many of the swords are in very bad condition or are rusted into their scabbards it has not been easy to obtain an accurate measurement of their weight and dimensions, but one sword from Llyn Cerrig Bach is unsheathed and in excellent condition. It has a total length of 89 cm and weighs 628 g. Surely the similarity between this and the currency bars can be no accident?

Unfortunately, few of the swords have been the subject of a metallurgical examination. But a sword from Isleham, Cambridgeshire, rusted into its richly decorated scabbard, has been the subject of examination by the British Museum.[145] Although the blade was in a very poor condition it contained an armourer's mark consisting of two circles, one above the other, in an oval or pelta-shaped field. A section taken across the blade suggests that it was made, like the Beckford currency bar, of three strips with a central core of a higher carbon content than the outer sections. The microhardness of the central strip and the cutting edges was in the range 370–450 HV, which suggests fast air cooling from 750 to 850°C. The metal near the central surfaces had a hardness of 320 HV. The composition is given in Table 78. This blade would seem to be one of the most effective of its period (50 BC–AD 50) and the armourer who left us his mark should have been proud of it.

The most comprehensive investigation is by McGrath,[95,96] who examined five fragmentary blades from Llyn Cerrig Bach, Anglesey. He was able to divide these into two categories: four out of five were placed in category 1. These were made by hot-forging heterogeneous pieces of wrought iron which had been carburized to various degrees, i.e. by piling, with the intention of making a fairly homogeneous blade of

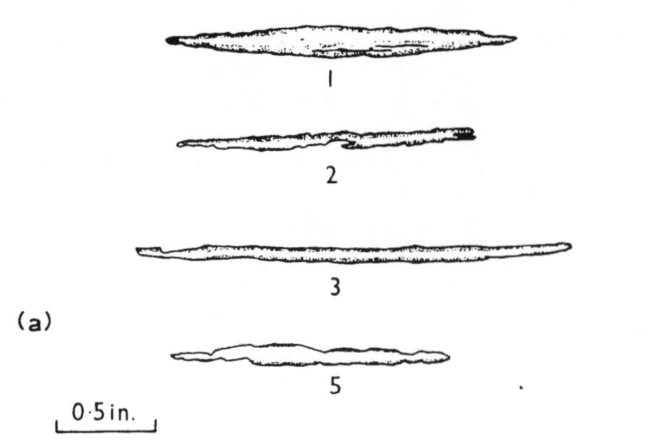

Cross-sections of the Category I blades. Shaded areas are carbon rich.

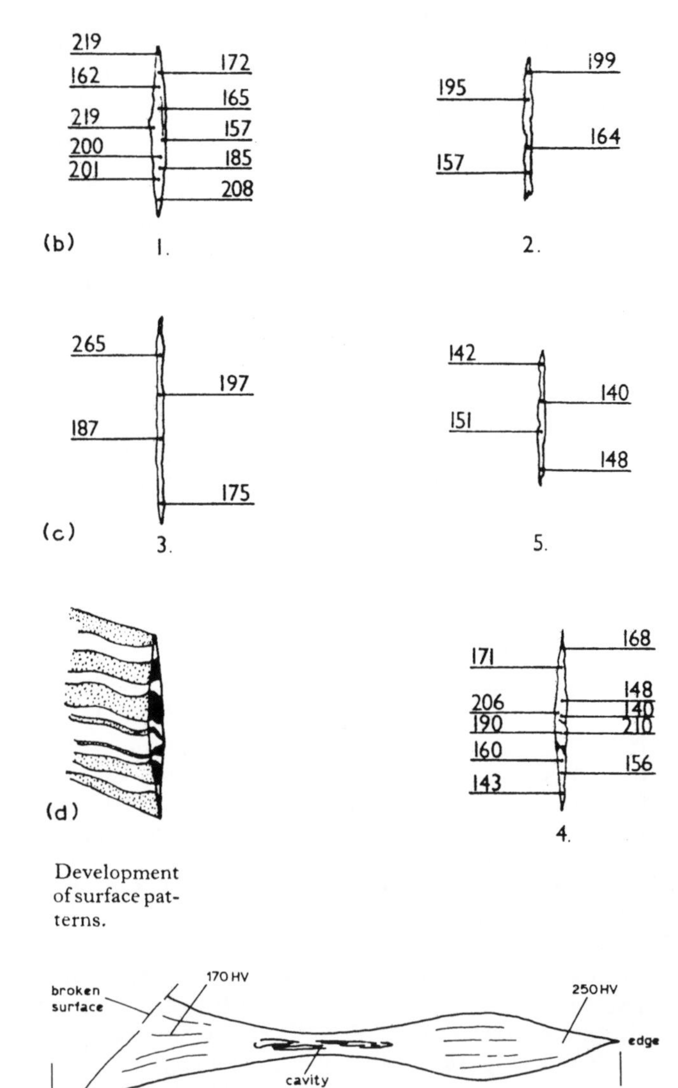

92 Sections through Iron Age swords showing structure and hardness (mostly after McGrath[95])

low-carbon steel. The sections of these blades together with hardnesses are shown in Fig. 92*a–d*. The hardnesses of this type vary from 140 to 265 HV, but there does seem to be a tendency for the highest hardnesses to occur at the cutting edge, as may be seen from the figures given. As there was no evidence of cold-working, the higher edge hardnesses must be due to higher carbon contents.

The category 2 blade was more like the pattern-welded blades of the Roman and post-Roman period (*see* Chapter 7). While the structure was coarsely piled to give about 14 layers, these alternated between high- and low-carbon content and were welded at right angles to the axis of the section (Fig. 92). This alternation of high- and low-carbon contents would show on the surface and besides being decorative would serve as a 'hallmark' of quality. Figure 92 shows that the hardness was in the same range as that of the other blades.

Another blade, from Waltham Abbey, believed to be of pre-Roman or Roman date, was examined and also found to consist of a piled structure containing 24–25 bands of 0–0·25% carbon steel of varying grain size. Again the average hardness was higher at the edge (250 HV) and lower at the centre (170 HV) (Fig. 92*e*). This blade was found to have a work-hardened cutting edge; the composition is given in Table 78. It was observed that there had been no carburizing or quenching.[97]

Other iron artifacts of the Early Iron Age

In contrast to the voluntary wearing of iron, as shown by the Spettisbury torcs, are the complex slave chains that were made and used in the immediate pre-Roman period. Such chains have been found at Llyn Cerrig Bach in Anglesey,[82] and on Belgic sites at Lords Bridge (Cambridge),[98] and elsewhere. A link of one of the chains from Llyn Cerrig Bach has been analysed (Table 78) and was proved to be of low-carbon wrought iron with 0·15% phosphorus. It was scarf-welded at the end, and the centre of the link was closed so that it rather resembles the modern stud-link chain. This operation minimizes twisting and entanglement of the chain in use.

During the Early Iron Age, iron razors gradually replaced those of bronze. A razor was found at Ewell but no microexamination has yet been done to determine whether it has been intentionally carburized.[99]

While fire-dogs were undoubtedly made and used during this period, perhaps one of the more ambitious pieces of ironwork is the anchor from Bulbury in Dorset,[100] which is a type in use from the 1st century BC in NW Gaul. It is 1·44 m long and was found with associated material dated to AD 1–40. These are no more than smithed wrought iron.

However, a chariot tyre from Llyn Cerrig Bach[82] consists of high-carbon steel (0·74–0·96%) with a very evenly distributed carbon content. The analysis is given in Table 78. This must be the earliest dated British find of a good quality steel. Cook[82] reports that 'there is no significant evidence of graduation through the thickness and therefore none to suggest superficial carburization of a wrought iron bar'. Nevertheless, the specimen is composite in character and evidence of layers of slightly differing carbon content was observed. It would appear, therefore, that a tyre measuring about 2·74 m in circumference and 3·8 × 2·5 cm in section had been made by welding carefully carburized strips of wrought iron together. The low phosphorus content (0·030%) would assist the even diffusion of carbon in the iron in contrast to the currency bars where the phosphorus is very high

Table 78 Hardness and composition of iron-based objects of the Early Iron Age (weight %)

Ref.	82 Llyn Cerrig Bach Chain 200 BC–AD 50	82 Tyre	35 Hengistbury Head Cast iron 1st cent. AD	97 Waltham Abbey, Herts. Sword	14 Odell, Beds.	101 Lough Mourne, Ireland Socketed axe 6th–3rd cent. BC	145 Isleham, Cambs. Sword (50 BC–AD 50)
C	0·07	0·74–0·96	3·49*	0–0·25		0–0·4	variable
Mn	tr.	<0·05	tr.	0·003 5		0·2	—
Si	—	<0·05	0·038	—		pres.	—
P	0·15	0·030	0·18	0·075		0·01	0·01
S	tr.	0·006	0·035	—		—	—
Cu	—	—	—	—		0·1	—
Ca						pres.	—
Co	—	—	—	0·030		—	0·01
Mg	—	—	—	—		pres.	—
Ni	—	—	—	0·125		0·01	0·03
Hardness (HV)				170–250		100–135	320–450

* = mostly combined (0·16% graphite); — = not sought; tr. = trace

(0·2–1·0%) and where the carbon graduation is very marked (*see* Tables 73 and 77).

These tyres are supreme examples of a smithing technique in which thin strips have been highly carburized and welded together into a long bar. The bar would then be cut and the pieces piled on top of each other, re-welded and forged out into a bar of the original length. This process would be repeated several times to give the homogeneous product that was required for heavy wear.

One of the artifacts that shows evidence of continuity with the Bronze Age is the socketed and looped axe. This was an attempt to shape a new metal into the style of the old. But the new metal could not be cast, so an object designed as a casting had now to be made as a forging. More than 21 of the looped axes are known from Britain. One from Lough Mourne has been examined by Scott,[101] who has shown how it was made from a single sheet of iron from which a flap projected; this was used to form the cutting edge. It appears to have been intentionally carburized on one surface so that when folded it consisted of ferrite and pearlite with 0·4%C in the centre of the section, surrounded by ferrite at the surfaces. The carburized area was exposed after sharpening the edge.

There were other types of socketed axes as shown by the unlooped axe from Feerwore, Co. Galway, in which the socket was formed by a simple butt-joint (unwelded) and which had a much thicker and stronger cutting edge. The iron sickle found with the bronzes in the transitional LBA–EIA site of Llyn Fawr in South Wales had been intentionally surface-carburized,[144] showing that this technique was known to the smiths of the time and had probably been introduced with the general technique of ironworking.

Knives are the next most important tools in the Iron Age and these could be expected to show a high level of metallurgical skill if they were to be of much use.

The site at Huckhoe,[66] has yielded an early form of broad-bladed knife (Fig. 93). The blade has been made from piled bloomery iron with a variable but low average carbon content (about 0·1–0·2%). The hardness of most of the section is 110–143 HV (1 kg), which suggests that the metal is low in phosphorus. One surface has been carburized to give a carbon content of about 0·8% and a hardness in the range 230–257 HV. The structure of the carburized layer is mainly coarse lamellar pearlite, but the ferrite + pearlite areas in the centre show a Widmanstätten distribution suggesting

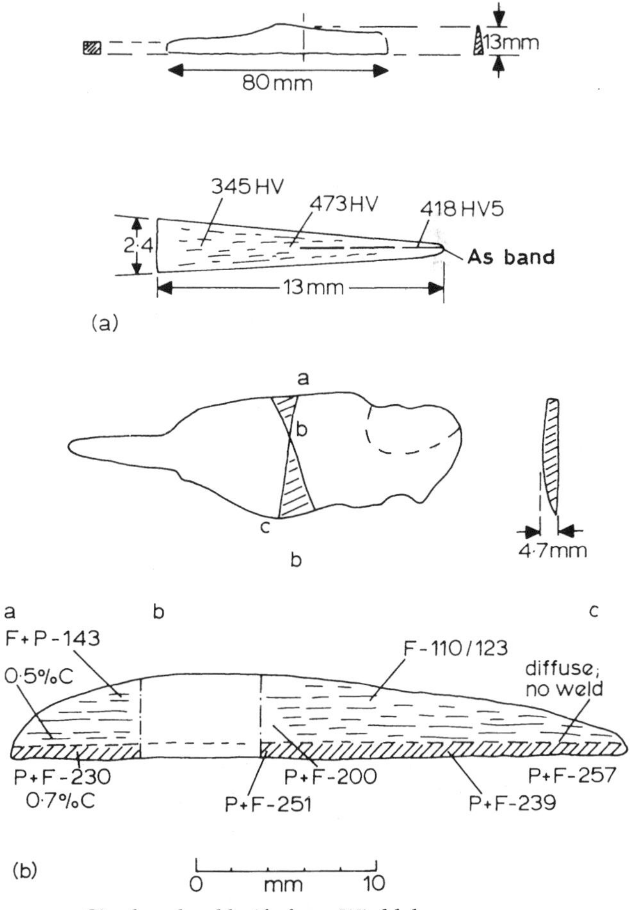

a Single-edged knife from Winklebury
b Double-edged knife from Huckhoe

93 Plans and sections of knife blades from Winklebury and Huckhoe, showing structure and hardness

that the blade has been cooled fairly rapidly in air after carburizing one surface at 900–1200°C, but that the cooling rate has been much slower below 750°C.

It seems that one surface was protected during carburizing by a clay envelope so that at least half the section would remain soft and ductile to support the harder and more brittle side which acted as the cutting edge. It is possible that the original intention was to harden the carburized surface by reheating the whole knife to a temperature of 750°C and quenching in water, but somehow this was omitted and the knife left in a carburized but unhardened condition.

Another knife comes from a 3rd–1st century BC level of the hillfort at Winklebury, Hants., where a currency bar has also been found. This was a narrow-bladed, single-edged knife (Fig. 93) and a section was cut near the tang end of the blade.[102] This had a fairly uniform carbon content and appeared to have been made by piling. The overall hardness varied from 345 to 418 HV (5 kg) and there was an arsenic enrichment band down the middle near the sharp end.[103] High-power examination showed a ferrite–martensite–troostite structure which clearly indicated heat treatment. Microhardness readings showed that the hardness of the martensitic regions was 500 HV (0·1 kg), the troostite 380 HV (0·1 kg), and the residual ferrite matrix about 193 HV (0·1 kg). It would appear that this was a medium-carbon steel which had been heated to between the upper and lower critical temperatures, i.e. between 700°C and 820°C, and slowly quenched. The carbon content was fairly uniform for an early steel but there were variations in the relative amounts of the three phases. This clearly shows a much improved technique over the Huckhoe knife and resulted in a very serviceable implement.

The Iron Age site at Gussage All Saints that was manufacturing chariot and harness gear also yielded moulds for bronze castings and a discarded steel bridle bit.[104] This latter is a very sophisticated piece of work, even from a site that shows evidence for lost-wax casting of bronze as well as bronze coating. Judging from what has been said earlier, the making of steel is not a great surprise, but such a combination of skilled crafts on one site is rare in Britain. The welded steel ring contained about 0·6% C and had a hardness of 221 HV (5 kg), which shows that it had not been quench-hardened. It was plated with a tin-bronze with about 10% Sn and a hardness of 90 HV (1 kg). The bronze plating had been applied by dipping, but first a tin coating had been applied. The plating had also been used as a 'brazing' medium to attach smaller rings to either side of the steel bit.

Other objects found on this site were a punch or drift and a piece of steel. Both of these had been heat treated. The first had been made of carburized piled iron with a variable carbon content which had been quench-hardened to give a hardness of 473–532 HV. The second had a hardness of 313–358 HV.

Both high- and low-phosphorus irons had been used and the site generally showed a high level of technology. No doubt full-scale excavation of other hillfort sites will yield more evidence to show that this level of metallurgical technology was normal in Iron Age Britain.

SUMMARY

We have seen how the Iron Age began, with a gradual transition from bronze, in which iron was at first used as a precious metal, as shown by its decorative role as an inlay to bronze and in the Spettisbury torcs. Its use for tools and weapons seems to have been delayed, except for occasional tools such as the knife from Huckhoe. But this picture may be distorted by the poor corrosion resistance of iron in the British environment, and the lack of work on knives and axes. Clearly the technique of carburizing was known at the beginning of the pre-Roman Iron Age, and quenching was practised towards the end of it.

A variety of types of furnace seem to have been in use before the Roman period, but it is possible that the slag-tapping shaft furnace was used in Britain well before the Roman period, perhaps as early as the 5th century BC. The other two – the bowl and the domed furnace – probably did not reach the slag-tapping stage until the Roman conquest.

NOTES AND REFERENCES

1. STRABO: Loeb edition, 1923, **2**, 255.
2. K. M. RICHARDSON and A. YOUNG: *Antiq. J.*, 1951, **31**, 132–48.
3 R. F. TYLECOTE: *Early Medieval Stud.*, 1973, **6**, 42–7.
3*a* R. F. TYLECOTE: *J. Hist. Metall. Soc.*, 1975, **9**,(2), 49–56.
4 R. F. TYLECOTE: *J. Iron Steel Inst.*, 1960, **194**, 451–8.
5 C. ROEDER: *Trans. LCAS*, 1901, **19**, 77–118.
6 R. E. M. WHEELER and T. V. WHEELER: 'The Lydney excavations', 92; 1932, Oxford.
7 R. C. C. CLAY: *Wilts. Archaeol. Nat. Hist. Mag.*, 1927, **43**, 59–93.
8 R. A. SMITH: *Proc. Soc. Antiq. Lond.*, (2nd ser.), 1905, **20**, 179–95.
9 R. E. M. HEDGES and C. J. SALTER: *Archaeometry*, 1979, **21**,(2), 161–75.
10 G. R. MORTON and J. WINGROVE: *J. Iron Steel Inst.*, 1969, **207**, 1557–64, 1972, **210**.
11 H.-G. BACHMANN: 'The identification of slags from archaeological sites', Occ. Publication No. 6, 1982, London, Inst. of Arch.
12 R. HAYNES: *J. Iron Steel Inst.*, 1956, **183**, 359–61.
13 M. E. CUNNINGTON: 'The Early Iron Age site at All Cannings Cross, Wilts.'; 1923, Devizes.
14 R. E. CLOUGH: 'The iron industry at the site of Odell, Beds.', (forthcoming).
15 G. J. WAINWRIGHT *et al.*: 'Gussage All Saints; A Dark Age settlement in Dorset', D. of E. Arch. Rep. No. 10, 1979, London, HMSO.
16 D. A. JACKSON *et al.*: *Britannia*, 1978, **9**, 115–242.
17 C. F. TEBBUTT and H. CLEERE: *Sussex Archaeol. Collect.*, 1973, **111**, 28–40.
18 R. F. TYLECOTE: *J. Hist. Metall. Soc.*, 1975, **9**,(1), 26–9.
19 R. F. TYLECOTE and E. OWLES: *Norf. Archaeol.*, 1960, **32**, 142–62.
20 J. PERCY: 'Metallurgy; Iron and Steel', 1864, 278–319.
21 J. H. MONEY: *Britannia*, 1977, **8**, 339–50.
22 W. GOWLAND: *Archaeologia*, 1899, **56**,(2), 267–322.
22*a* R. J. FORBES: 'Studies in ancient technology', Vols. 8–9; 1964, Leiden.
23 H. R. SCHUBERT: 'History of the British Iron and Steel Industry', 22; 1958, London.

24 H. F. CLEERE: *Antiq. J.*, 1972, **52**,(1), 8–23.
25 E. J. WYNNE and R. F. TYLECOTE: *J. Iron Steel Inst.*, 1958, **190**, 339–48.
26 W. GILLES: *Stahl Eisen*, 1960, **80**,(14), 943–8.
27 J. A. W. BUSCH: *Bull. Hist. Metall. Group*, 1972, **6**,(1), 28–32.
28 O. EVENSTAD (trans. N. L. Jensen): *Bull. Hist. Metall. Group*, 1968, **2**,(2), 61–5.
29 R. E. CLOUGH: 'Experiments in early iron smelting techniques', (forthcoming).
30 This can be seen in the museum at Wells. The photograph of it shown in Fig. 74 was kindly provided by Prof. M. J. O'Kelly.
31 A. FOX: *Trans. Devon Assoc.*, 1954, **86**, 21–62.
32 G. JOBEY: *Archaeol. Aelana* (4th ser.), 1962, **40**, 1–34.
33 A. KRUPKOWSKI and T. REYMAN, Metallurgical examination of iron bars found at Witowo and slag found at Igołomia. Sprawozdania Panstwowego Mus. Arch. Warzawa, 1953, **5**, 48–65.
34 R. F. TYLECOTE, A. E. WRAITH and J. N. AUSTIN: *J. Iron Steel Inst.*, 1971, **209**, 342–64.
35 J. P. BUSHE-FOX: 'Excavations at Hengistbury Head, Hampshire in 1911–12', Rep. No. 3, 1915, London, Soc. Antiq.
36 B. W. CUNLIFFE: 'Iron Age communities in Great Britain'; 1974, London.
37 O. G. S. CRAWFORD and R. E. M. WHEELER: *Archaeologia*, 1921, **71**, 133.
38 R. F. TYLECOTE: 'Metallurgical examination of material from the Bronze Age hoard from Gilmonby, Teesdale', (forthcoming).
39 R. HANWORTH and D. J. TOMALIN: 'Brooklands, Weybridge. The excavations of an Iron Age and Medieval site, 1964–65, 1970–71'; 1977, Res. Vol. of Surrey Arch. Soc. No. 4.
40 S. C. HAWKES: 'Excavations at Cow Down, Longbridge Deverill, Wilts.', (forthcoming).
41 E. M. CLIFFORD: 'Bagendon; a Belgic oppidum'; 1961, Cambridge.
42 G. BERSU: *Proc. Prehist. Soc.* (N.S.), 1940, **6**, 1–111.
43 C. I. FELL: *Archaeol. J.*, 1936, **93**, 57–100; T. C. George. VCH, Northants, **1**, 152.
44 E. T. LEEDS: *Archaeologia*, 1927, **76**, 205–37.
45 H. R. SCHUBERT: 'History of the British Iron and Steel Industry, 22; 1958, London.
46 A. M. APSIMON and E. GREENFIELD: *Proc. Prehist. Soc.*, 1972, **38**, 302–81.
47 H. O'N. HENCKEN: *Archaeologia*, 1933, **83**, 237–84.
48 A. FOX: *Trans. Devon Assoc.*, 1957, **89**, 18–77.
49 C. FOX: *Archaeol. Cambrensis*, 1927, **82**, 44–66.
50 H. O'NEILL: *Foundry Trade J.*, 1949, (Jun.), 16.
51 W. L. SCOTT: *Proc. Soc. Antiq. Scott.*, 1934, **68**, 200–23.
52 W. G. AITKEN: *Bull. Hist. Metall. Group*, 1971, **5**,(1), 15–23.
53 H. TAYLOR: *Proc. Speleol. Soc.*, 1922–3, **2**, 40.
54 A. BULLEID and H . ST. G. GRAY: 'Glastonbury Lake Village', Vol. I and II, 1911–17, Glastonbury Antiquarian Soc.
55 A. O. CURLE: *Proc. Soc. Antiq. Scott*, 1935–6, **70**, 153–69.
56 Analysis of a piece of peat charcoal from Devon; specimen kindly supplied by Diana Woolner.
57 D. A. SPRATT and R. INMAN: Excavated by Teesside Arch. Soc.; site No. 38 (Hayes) Roxby Moor. (Report forthcoming.)
58 R. F. TYLECOTE: *Bull. Hist. Metall. Group*, 1970, **4**,(2), 79.
58*a* R. H. HAYES: *J. Hist. Metall. Soc.*, 1978, **12**,(1), 18–26.
59 J. W. GILLES: *Stahl Eisen*, 1958, **78**, 1690–5.
60 J. GIBSON-HILL: *J. Hist. Metall. Soc.*, 1980, **14**,(1), 21–7.
61 H. M. SAVORY: *Archaeol. Cambrensis*, 1952–5, **104**, 14–51.
62 A. W. G. LOWTHER: *Surrey Archaeol. Collect.*, 1946–7, **50**, 9–46.
63 J. F. S. STONE: *Wilts. Archaeol. Nat. Hist. Mag.*, 1937, **47**, 466–89.
64 J. G. CALLANDER: *Proc. Soc. Antiq. Scotl.*, 1931–2, **66**, 42–66.
65 V. RAE and A. RAE: *Proc. Soc. Antiq. Scotl.*, 1952–3, **87**, 153.
66 G. JOBEY: *Archaeol. Aeliana.*, 1959, **37**, 217–78.
67 G. JOBEY: *Archaeol. Aeliana.*, 1968, **46**, 293–5.
68 D. L. C. WILSON: personal communication.
69 E. STRAKER: *Sussex Notes Queries*, 1937, **6**, 205.
70 H. F. CLEERE: *Sussex Archaeol. Collect.*, 1963, **101**, 48–53.
71 S. P. O'RIORDAIN: *Proc. R. Ir. Acad. (C)*, 1941, **47**, 77–150.
72 H. HENCKEN: *Proc. R. Ir. Acad. (C)*, 1950, **53**, 1–247.
73 M. J. O'KELLY: *J. Cork Hist. Archaeol. Soc.*, 1952, **57**, 18–40.
74 R. F. TYLECOTE: *Bull. Hist. Metall. Group*, 1970, **4**,(1), 3.
75 K. MUCKELROY: *Proc. Prehist. Soc.*, 1981, **47**, 275–98.
76 C. F. C. HAWKES: *Archaeol. J.*, 1940, **97**, 112–14.
77 N. H. HARBORD and D. A. SPRATT: *J. Hist. Metall. Soc.*, 1975, **9**,(1), 32–3.
78 T. MAY: *Iron Coal Trades Rev.*, 1905, **71**, 427 *et seq.*
79 H. E. BALCH: 'The great cave of Wookey Hole', 3 ed., 29; 1947, London.
80 J. A. SMYTHE: *Trans. Newcomen Soc.*, 1936–7, **37**, 197–203.
81 P. WHITAKER and T. H. WILLIAMS: *Bull. Hist. Metall. Group*, 1969, **3**,(2), 39–45.
82 C. FOX: 'A find of the Early Iron Age from Llyn Cerrig Bach, Anglesey'; 1946, National Museum of Wales, Cardiff.
83 S. A. HANDFORD: 'Caesar: the conquest of Gaul', 135; 1951, Harmondsworth.
84 E. W. HULME: *Antiquity*, 1933, **7**, 61–72.
85 D. ALLEN: *Proc. Prehist. Soc.*, 1967, **33**, 307–35.
86 C. FOX: *Antiquity*, 1940, **14**, 427–33.
87 R. A. SMITH: *Proc. Soc. Antiq. Lond.* (2nd ser.), 1905, **27**, 69–76.
88 *Proc. Soc. Antiq., Newcastle* (3rd ser.), 1915–16, **7**, 27.
89 W. BRITNELL: *Current Archaeol*, 1974, (45), 293–7.
90 W. HALDANE: *Bull. Hist. Metall. Group*, 1970, **4**,(2), 53–66.
91 J. MYERS: *J. Soc. Chem. Ind.*, 1922, **41**, 1331.
92 C. W. BREWER: *J. Hist. Metall. Soc.*, 1976, **10**,(1), 1–9.
93 R. V. RILEY: *Bull Hist. Metall. Group*, 1973, **7**,(2), 46–7.

93*a* L. V. GRINSELL: 'Archaeology of Wessex', 137; 1958, London.
93*b* J. BUCKMAN: *Archaeol. J.*, 1868, **25**, 46–59.
94 B. W. CUNLIFFE: 'Hengistbury Head'; 1978, London, Elek.
95 J. N. MCGRATH: *Bull. Board Cel. Stud.*, 1968, **22**, 418–25.
96 J. N. MCGRATH: *Bull. Hist. Metall. Group*, 1968, **2**,(2), 78–80.
97 J. LANG and A. R. WILLIAMS: *J. Archaeol. Sci.*, 1975, **2**, 199–207.
98 R. L. JESSUP: *Archaeol. J.*, 1932, **89**, 87–115.
99 A. W. G. LOWTHER: *Antiq. J.*, 1947, **27**, 82.
100 B. W. CUNLIFFE: *Antiq. J.*, 1972, **52**,(2), 293–308.
101 B. G. SCOTT: *Ir. Archaeol. Res. Forum*, 1974, **1**,(1), 9–24.
102 K. SMITH: *Proc. Prehist. Soc.*, 1977, **43**, 31–129.
103 R. F. TYLECOTE and R. THOMSEN: *Archaeometry*, 1973, **15**, 193–8.
104 M. G. SPRATLING and R. F. TYLECOTE: in 'Proc. of 16th Intern. Symp. on Archaeometry, Edinburgh, 1976', (ed. E. A. Slater and J. O. Tate), 268–92.
105 J. D. KENDALL: 'Iron ores of Great Britain and Ireland'; 1893, London.
106 G. S. SWEETING: *P. Geol. Assoc.*, 1930, **41**, 44.
107 S. TOMKIEFF: *P. Geol. Assoc.*, 1927, **38**, 518.
108 J. F. POTTER: *P. Geol. Assoc.*, 1977, **88**,(4), 229–41.
109 R. F. TYLECOTE: *J. Iron Steel Inst.*, 1959, **192**, 26–34.
110 T. MAY: 'The Roman fort at Templebrough, Rotherham', 1922.
111 V. B. PROUDFOOT: private communication.
112 D. A. JACKSON: *Northants. Archaeol.*, 1979, **14**, 31–7.
113 D. W. CROSSLEY: 'The Bewl Valley ironworks, Kent. *c.* AD 1300–1730', 1975, R. Arch. Inst. Monograph.
114 J. PICKIN: *J. Hist. Metall. Soc.*, 1983, **17**,(1), 4–11.
115 R. F. TYLECOTE: *J. Iron Steel Inst.*, 1966, **204**, 314–19.
116 R. F. TYLECOTE: *Bull. Hist. Metall. Group*, 1970, **4**,(1), 24–7.
117 R. F. TYLECOTE: *Norf. Archaeol.*, 1967, **34**, 187–214.
118 D. CROSSLEY: *Post Medieval Archaeol.*, 1975, **9**, 1–37.
119 R. F. TYLECOTE and J. CHERRY: *Trans. CWAAS* (N.S.), 1970, **70**, 69–109.
120 A. THOMPSON and E. HOLLAND: *Herts. Archaeol.*, 1974–6, **4**, 137–48.
121 S. M. LINSLEY and R. HETHERINGTON: *J. Hist. Metall. Soc.*, 1978, **12**,(1), 1–11.
122 N. MUTTON: *J. Hist. Metall. Soc.*, 1974, **8**,(2), 117.
123 B. NEUMANN: 'Die altesten Verfahren der Erzeugung technischen Eisens', 39; 1954, Berlin, Akademie Verlag.
124 J. E. STEAD: *J. Iron Steel Inst.*, 1915, (1), 140–98.
125 J. E. STEAD: *J. Iron Steel Inst.*, 1918, **97**, 388–409.
126 B. E. HOPKINS and H. R. TIPLER: *J. Iron Steel Inst.*, 1958, **188**, 218–37.
127 Cited by H. G. Baron and E. Costello in *Met. Reviews*, 1963, **8**, 416.
128 R. E. M. WHEELER: 'Maiden Castle, Dorset', Res. Rep. No. 12, Soc. Antiq., 1943, London.
129 H. R. SCHUBERT: 'History of the British iron and steel industry (450 BC–AD 1775), 21–6; 1957, London.
130 H. H. COGHLAN: 'Prehistoric iron', 43; 1956, Oxford.
131 W. J. WEDLAKE: 'Excavations at Camerton, Somerset', 1958, Camerton Excav. Club.
132 A. E. WILSON: *Sussex Archaeol. Collect.*, 1955, **93**, 59–77.
133 S. E. WINBOLT: *Sussex Archaeol. Collect.*, 1930, **71**, 223–36.
134 V. E. NASH-WILLIAMS: *Archaeol. Cambrensis*, 1939, **94**, 42–79.
135 S. E. WINBOLT: *Surrey Archaeol. Collect.*, 1932, **40**, 86.
136 H. E. KILBRIDE-JONES: *Proc. Soc. Antiq. Scotl.*, 1936–7, **71**, 401–5.
137 B. H. ST. J. O'NEILL: *Archaeol. Cambrensis*, 1942, **97**, 1–57.
138 N. R. THOMPSON: *Wilts. Archaeol. Nat. Hist. Mag.*, 1971, **66**, 58–75.
139 P. J. FOSTER *et al.*: *Northants. Archaeol.*, 1977, **12**, 55–95.
140 J. P. TOOMEY: *Brigantian*, 1976, 2–16.
141 C. F. TEBBUTT: *Bull. Wealden Iron Res. Group*, 1973, (5), 11–12.
141*a* C. F. TEBBUTT: *Surrey Archaeol. Collect.*, 1965, **41**, 79–92.
141*b* S. E. WINBOLT: *Sussex Archaeol. Collect.*, 1936, **77**, 245–9.
142 M. OLIVER: *Proc. Hants. Field Club*, 1978, **35**, 41–92.
143 G. SHAW SCOTT: *J. Iron Steel Inst.*, 1907, (3), 120–36.
144 J. P. NORTHOVER: 'The analysis of Welsh Bronze Age metalwork', Appendix to H. N. Savory, 'Guide Catalogue to the Bronze Age collections', 1980, National Museum of Wales.
145 I. M. STEAD *et al.*: *Proc. Cambs. Antiq. Soc.*, 1981, **71**, 61–74.
146 J. GOULD: *Trans. SSAHS*, 1969–70, **11**, 58–63.
147 L. L. KETTERINGHAM: *Surrey Archaeol. Soc.*, 1976, Guildford.
148 J. H. MONEY: *Sussex Medieval Archaeol.*, 1971, **15**, 86–111.
149 A film shown at the Symposium on Early Furnaces held at the British Museum in Oct. 1982.

7 The Roman Iron Age

As described in the previous chapter, the sources of iron ores in the British Isles are exceedingly numerous. There is no doubt that by the time the Romans arrived, native working of local ores was very widespread and given sufficient encouragement, it could be extended where necessary.

The first object was to obtain sufficient material for the military and naval establishments, where it was worked up by the smiths. The fact that a large number of smithing furnaces were found at Corstopitum (Corbridge) and little evidence of smelting in the form of slag suggests that military authorities did not bother to smelt their own material on these sites, but bought it from the natives.[1,2] How far this is true of the pacified areas further south is doubtful. The fact that large pieces of furnace slag, weighing as much as 36 kg, were found in the foundations of a villa at Great Weldon, Northants,[3] may be evidence of domestic iron smelting. The 2nd century site at Ashwicken, Norfolk, is certainly evidence of smelting on a large scale or over a long time, and is five kilometres from a villa at Gayton Thorpe.[4]

Mining was by the open cast method in the majority of cases. Only one case of deep mining for ore is recorded, and this is at Lydney in the Forest of Dean where a late 3rd century hut has sealed the opening to an adit which followed a band of ferruginous marl.[5] The ore was mainly obtained by working outcrops, as at Treborough and Luxborough in Somerset,[6] and by digging pits down to beds of nodular ore as at Ashwicken, Norfolk,[7] and in the Weald of Kent and Sussex.[8] There is no doubt that by Roman times other types of rich ore as well as nodular carbonates were being used on a large scale, particularly limonites. The method of smelting, in which a slag consisting of about 60% iron oxide was produced, made it absolutely necessary to smelt rich ores; the richness was judged by the iron plus manganese content after dressing and roasting.

The sites of Romano-British iron-working have been admirably summarized by Aiano[9] and others and, as they are so much more numerous than in the pre-Roman period, it will not be possible to discuss them all individually.

94 Remains of roasting hearth from Great Casterton, Rutland (after J. P. Gillam[11])

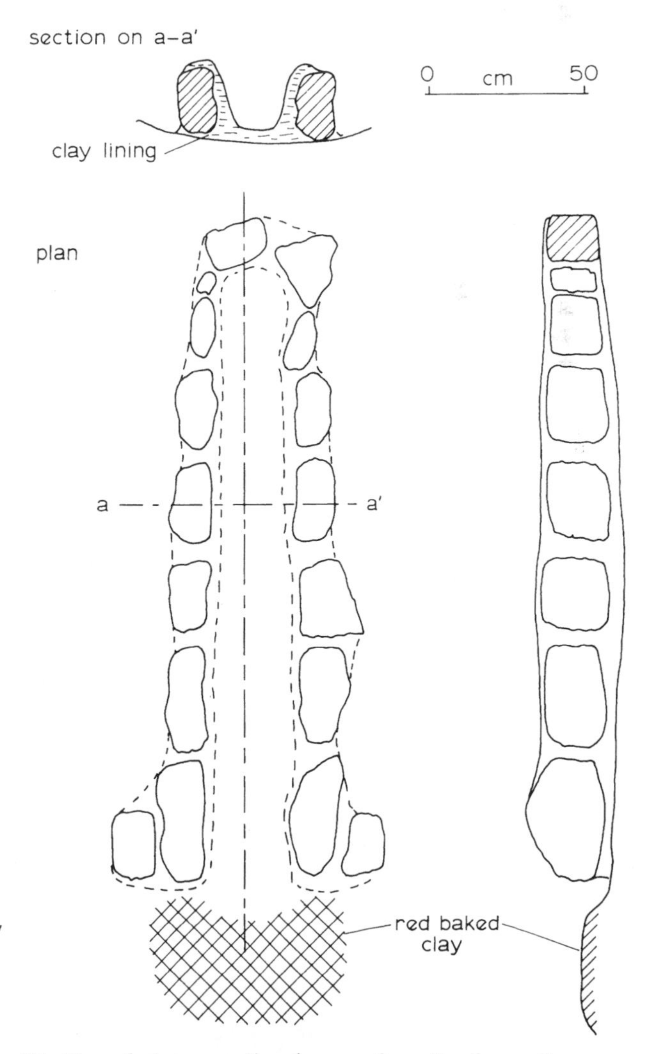

95 Trench-type roasting furnace from Bardown, Sussex (after H. F. Cleere[12])

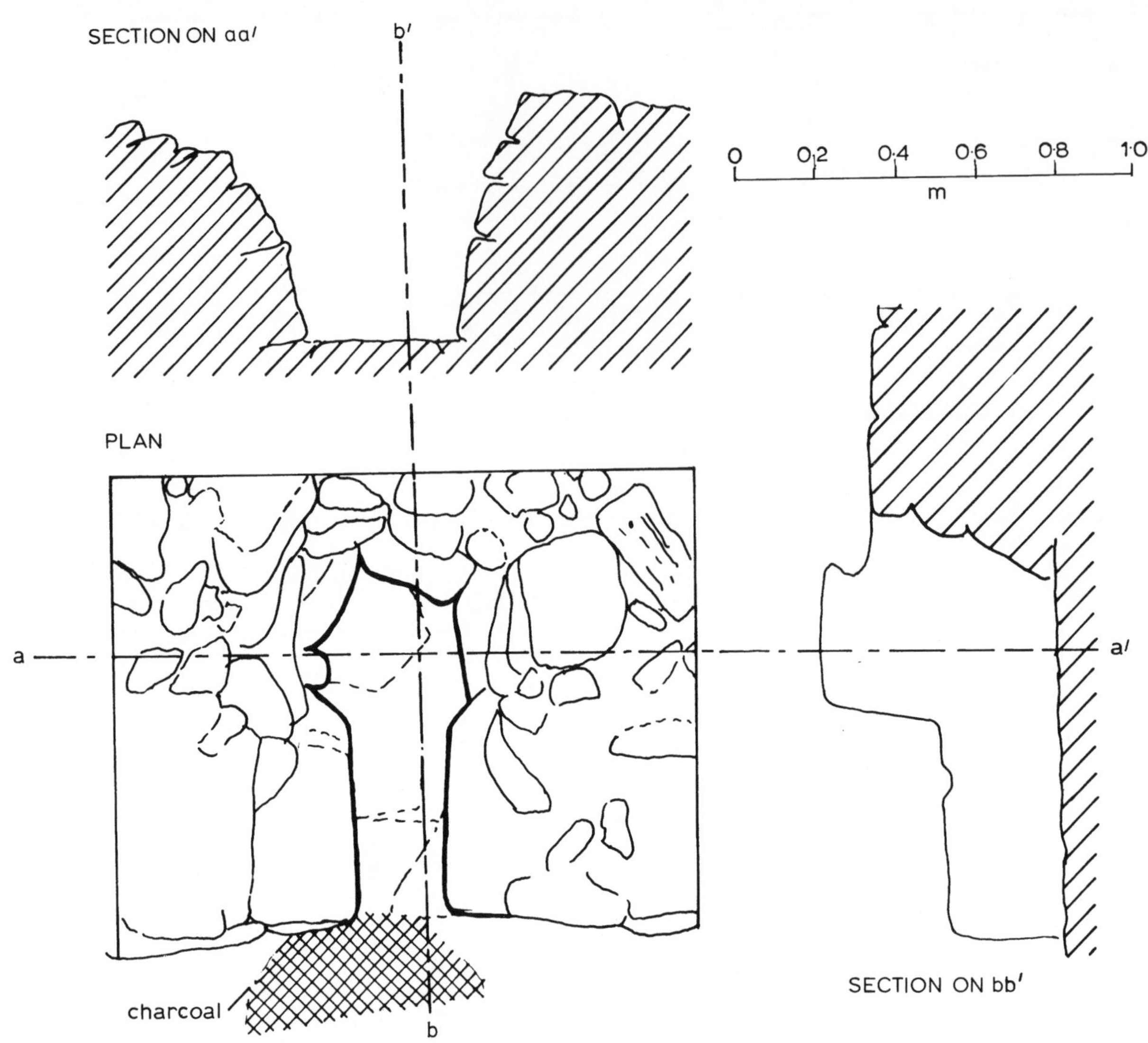

96 Roman 'keyhole' furnace from Cirencester (after Alan McWhirr[102])

ROASTING HEARTHS

Under this heading we have at least two well preserved hearths: those at Bedford Purlieus excavated by Dakin[10] and the one from Great Casterton[11] which is clearly of the same type. This latter (shown in Fig. 94) was previously thought to be a smelting furnace, but experimental work and the finding of similar hearths at Bedford Purlieus show conclusively that this description was wrong. These are too large for smelting, being as much as 2 m in diameter. They were made of yellow-red clay founded on stone and containing red silt, ash and charcoal and, in the case of Great Casterton, iron ore. However, this was not the only type of roaster used in this period.

At Bardown in the Weald, Cleere[12] found long trench-like furnaces which he identified as roasters and indeed used in his experiments at Horam.[13] These consist of stone-lined trenches dug in the soil, 2–2·5 m long and 0·30 m wide × 0·30 m deep. The stone lining was coated with clay, closed at one end and open at the other (Fig. 95). Cleere suggests that the furnace was charged with a 1:1 mixture of ore and charcoal. The charcoal seems unnecessary – wood would have been more efficient.

Why such long furnaces were used seems to be an enigma – although there are trench furnaces at Wakerley.[14] But several sites such as Cirencester[15] and Wilderspool[16] have produced what appear to be the bottoms of kilns more like those used for lime burning. We know from much later evidence that this type of lime kiln was very appropriate for calcining iron ore.

The lime-kiln type roasters are keyhole-shaped in plan (Fig. 96) and the furnace found at Corbridge[1] shown in Fig. 97 was almost certainly designed for lime burning as was that found at Weekley, Northants.[17] This more recent evidence makes it more likely that the furnaces found at Wilderspool, particularly the double one shown in Fig. 98, are roasting furnaces as their excavator, Thomas May, so described them.[16]

SMELTING FURNACES

The number of Roman iron-smelting furnaces is now quite considerable. Two main types can be distinguished: the shaft furnace and the developed bowl furnace. Both these are slag-tapping furnaces, and this

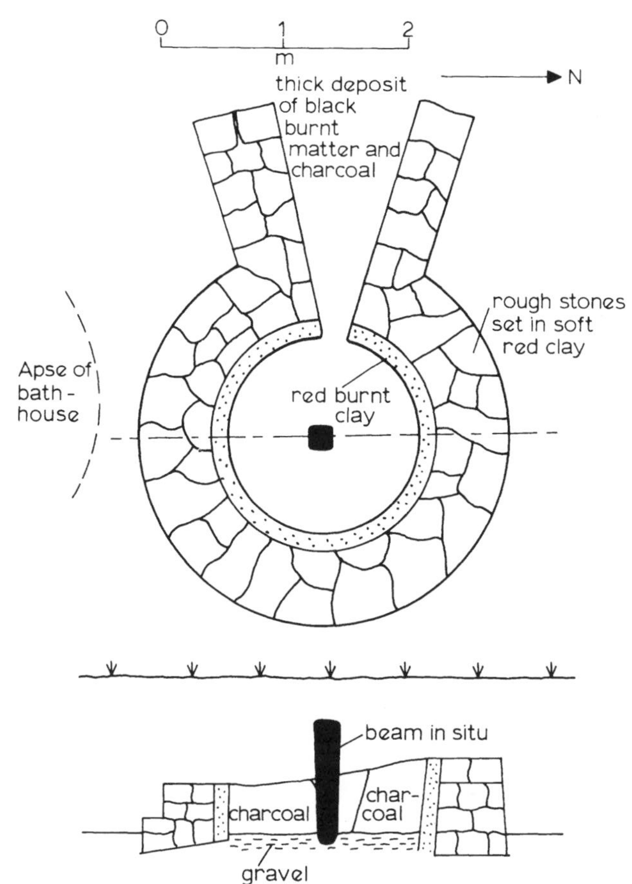

97 Lime-kiln re-used as a smithing furnace from Corbridge, Northumberland (3rd–4th century AD)

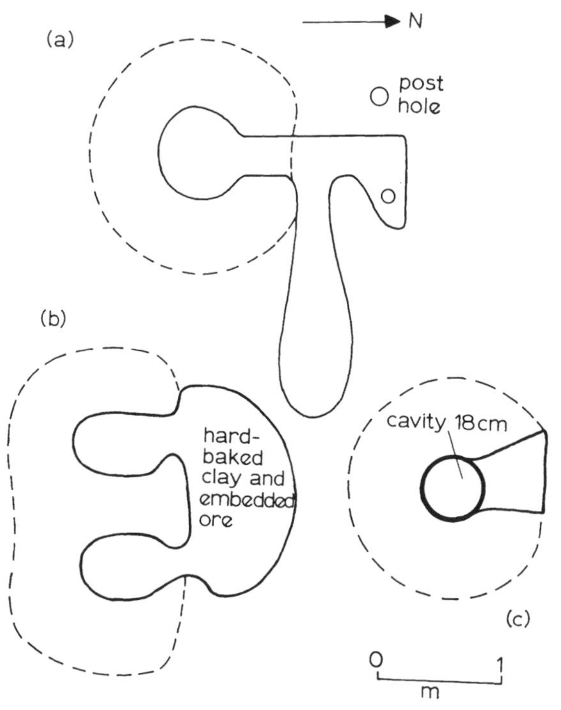

98 Furnaces from Wilderspool probably used as roasters (after May[16])

was undoubtedly a feature of the Roman period. At Ashwicken in Norfolk, cakes of slag weighing 18 kg were being produced in the shape of discs measuring 40 cm in diameter and 8 cm thick.[4]

Ashwicken would seem to be the iron-working site *par excellence*, and the one to produce the best shaft furnaces. Besides five clay-smelting furnaces, there

99 Remains of Romano-British shaft furnace from Pickworth, Lincs. (scale = 1 m (approx.))

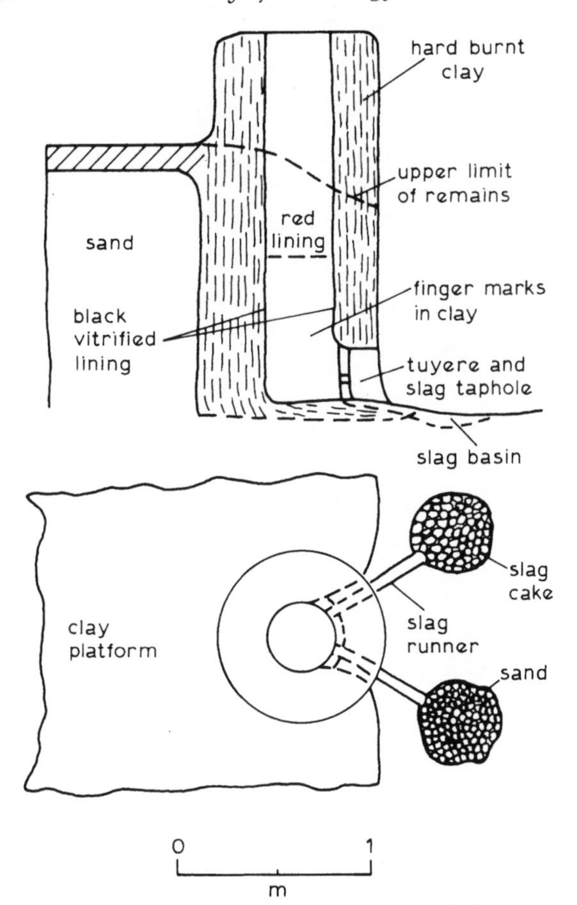

100 Reconstruction of Romano-British shaft furnace found at Ashwicken, Norfolk (2nd century AD)

were the remains of one or two bowl furnaces and some hammer scale. The smelting furnaces were shaft furnaces, and certainly blown by forced draught (Fig. 99). They were all 30 cm in diameter to within one centimetre, which suggests that they were used during a short period of time. If there had been a gap in the period of their use, it would be surprising if the new users adopted the old design to such close limits.

The remains of one furnace reached a height of 1·40 m, the walls of the shaft being vertical, except near the bottom where there was some enlargement. The slag was tapped, the bloom removed and the air blown in, all through one opening which formed an arch at the bottom measuring the full width of the furnace and which was about 30 cm high (Fig. 100). No tuyeres were found but a slag 'runner' was found in position in one of the furnaces. This showed that the slag had been tapped a short distance from the centre of the furnace into the sand at a slightly lower level. Unfortunately, it was very difficult to decide from the remains whether the furnaces had been worked singly or in groups of two or more. It would have been possible to have worked three together. It is believed that there are more units of this type in the Ashwicken area; when clay or ore ran out in the immediate vicinity of one set of furnaces the scene of operations would be moved to a new area a hundred metres or so away.

Since these furnaces were found another group was uncovered in a sand pit at Pickworth, Lincs.,[18, 18a] and more have been found at Holbeanwood[12] and Broadfield[19] in Sussex. The bases of a considerable number of furnaces in south-east Britain might well be of the same type.

In view of the large amount of evidence found by archaeological excavation, this type has been used as the basis for a series of more than 30 experimental smelts. As a result of these experiments it is believed that the original height of the furnaces at Ashwicken is probably no more than the 1·4 m found. This would be sufficient to obtain a good draught through a 1 m thick bed of charcoal and ore and to permit the smoke to be blown clear of the pit in which they were situated. The ore used in the furnace measured about 5–8 cm long by about 2–5 cm across, and the charcoal was of a similar size. The ore consisted entirely of nodules; these were first roasted to remove the water and then broken where necessary. They were obtained from measures in the Lower Greensand 3–13 m deep. The area of the site was capped with boulder clay, a point in common with another primitive iron-smelting site at Wormegay nearby. It is possible that the choice of sites was determined by the availability of clay as well as ore, since an enormous quantity of clay was used in the construction of the furnaces at Ashwicken.

Some of the linings of the furnaces show 30 cm thick layers of alternate red and blue fired clay. As is well known, to maintain clay in the blue state after firing, it has to be done in the absence of air, or under reducing conditions. It is suggested that the furnaces were constructed by using a tree-trunk as a pattern and plastering wet clay round it. This would be fired by a fire placed at the base of the clay-lined trunk and round it. After this layer had hardened, another layer of clay would be plastered on and in time hardened. Each layer would be about 2 cm thick and the outer layer of each layer would be burnt red while the inner half would be burnt blue, so giving the alternate red and blue layers found. By the time this lining had been completed the tree would have carbonized and shrunk, and could be removed easily. A similar method of construction had been suggested by Pleiner for some Polish shaft furnaces.[20] No definite sign of a roasting floor was found. The absence of unroasted ore in the immediate vicinity was very noticeable, which supports the theory that roasting was carried out some distance away.

Gowland[21] described and illustrated a furnace found by Artis at Castor, near Wansford.[22] Artis, however, thought that it might have been a Roman iron-smelting furnace for the production of cast iron, and shows iron slag issuing from the furnace (Fig. 101). After having seen the furnaces at Ashwicken, where the front wall has in most cases fallen away, one notices a marked similarity between these and the furnace from Castor. It is unfortunate that Artis's engraving gives no dimensions and the artist has added pig beds, no doubt to support the contention that the furnace was for the production of 'cast' iron.

At Holbeanwood and Broadfield the furnaces were about 30–65 cm in internal diameter and had either one or two slag-tapping channels (Fig. 79*b*, Chapter 6). Since the blowing mechanism would have to be situated in the front of the furnace in the region of the slag-tapping pit, the two-channel tapping pit gives room for placing the bellows between the two channels. In fact, according to Indian experience, there is no real problem as, when slag is about to be tapped,

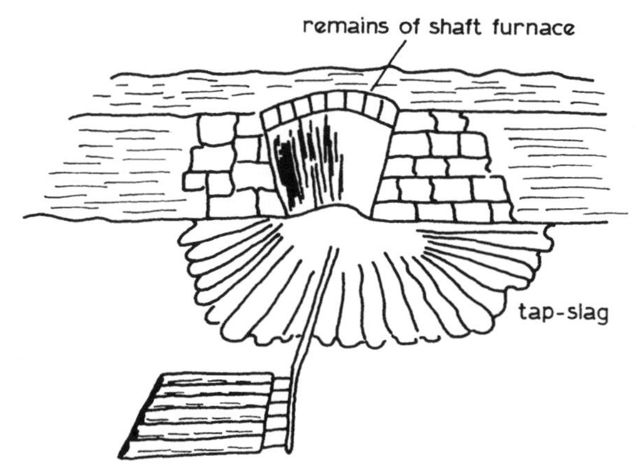

101 Shaft furnace found at Castor, Northants. (tracing of drawing in Artis's 'Durobrivae'[22])

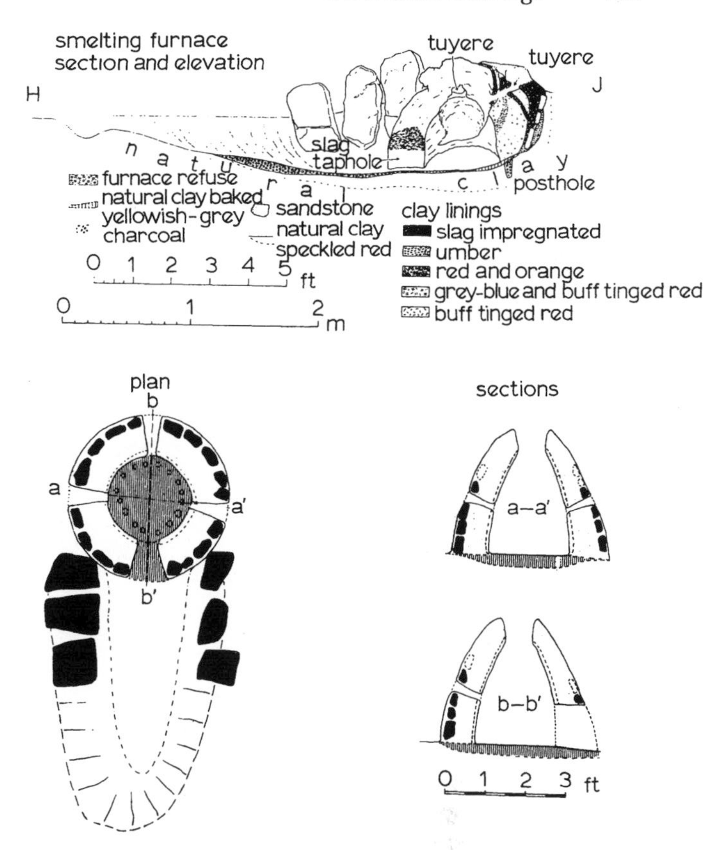

103 Section through a developed bloomery furnace from Minepit Wood with a reconstruction (both after Money[23])

the blower merely picks up his stool and bellows and retires to a discreet distance to return as soon as tapping is over. The Holbeanwood furnaces were 30–38 cm in diameter internally and at least one had a lintel stone placed over the single opening, which the Ashwicken furnaces lacked. The group of five furnaces found were very close together and it is unlikely that they were all worked at the same time.

The developed bowl furnace

The best example of this type found in Britain is that from Minepit Wood (alias Orznash) in Sussex, dated to the 1st–2nd century AD.[23] It consists of a 3 m long trench with a furnace at one end – a type that was to become popular in the medieval period. It is not unlike copper-smelting furnaces found in Israel, dated to the 12th century BC. Details of the Minepit Wood furnace are given in Fig. 103. Naturally we cannot be sure how high the bowl was and whether it would have been a shaft furnace situated at the end of its slag-tapping pit. The reconstruction shown gives a H/W of 1·4 at tuyere level. The taphole is too small for the withdrawal of the bloom and therefore it becomes a furnace from which the bloom is withdrawn through the top – a bowl or domed furnace. But a reconstruction could be produced with a higher shaft and a wider slag-tapping hole through which a bloom could be withdrawn, as in the Ashwicken type shaft furnaces. However, the incurving walls strongly suggest a domed or bowl furnace with H/W less than 1·5.

The bowl was constructed round a framework of 16 stakes. The lining consists mainly of clay with a stone backing, suggesting that a pit had been dug, lined with stones, a wattle, basket-like frame built at one end, and the intervening space filled with puddled clay. At least 3 tuyeres were inserted, 27 cm above the bottom. There may have been a fourth above the taphole. The height of the tuyeres would coincide with ground level. The furnace would have been charged with a mixture of roasted siderite and oak charcoal in alternate layers. The ore came from the Wadhurst clay which is exposed in the immediate vicinity.

We have the remains of other similar furnaces from the same area of Sussex, at Cow Park, Hartfield[24] and at Pippingford[25] (1st–2nd century AD) (Fig. 102). But the pits are not so long and the foundations resemble those of the Holbeanwood shaft furnaces with their stone lintels, rather than the one from Minepit Wood. Furthermore, there are no signs of tuyeres as noted for the Ashwicken furnaces. Some of the sections of the Cow Park furnaces show evidence of false bottoms – a characteristic of Meroitic (Sudanese) shaft furnaces, for example, where the slag flows over a sill at a level higher than the bottom of the furnace.

The furnaces at Wakerley, Northants[14] (2nd century BC–AD 400) mostly come into the category of shaft furnaces (Fig. 102), although some seem to be of the slag-pit type with sharply narrowing pits at ground level. This type was discussed in the previous chapter.

Other iron-smelting furnaces

We have discussed the two main types of furnace of the Roman period: the Ashwicken shaft furnace and the Minepit Wood developed bowl furnace. However, there are other examples which are difficult to classify.

At Ty Mawr[26] on Holyhead Island, Anglesey, a stone-lined cavity was found open at one side, like the furnace at Castor and most of those at Ashwicken. The cavity was 23 cm wide, about 30 cm deep and 40 cm high, and could well have been the remains of a shaft furnace from which the front wall had fallen away. Slag, charcoal and some pieces of limonite and other ferruginous material were found together with vitrified clay pipes which might have been tuyeres.

A series of bowl furnaces dated to the 1st and 2nd centuries was found at Aberllynfi in Powys.[27] Three out of the four found were excavated. These were all oval in shape, varying from 1 × 0·83 m to 0·68 × 0·45 m (Fig. 103); in depth they varied from 23 to 45 cm. In all cases the long axis was nearly on a north–south line.

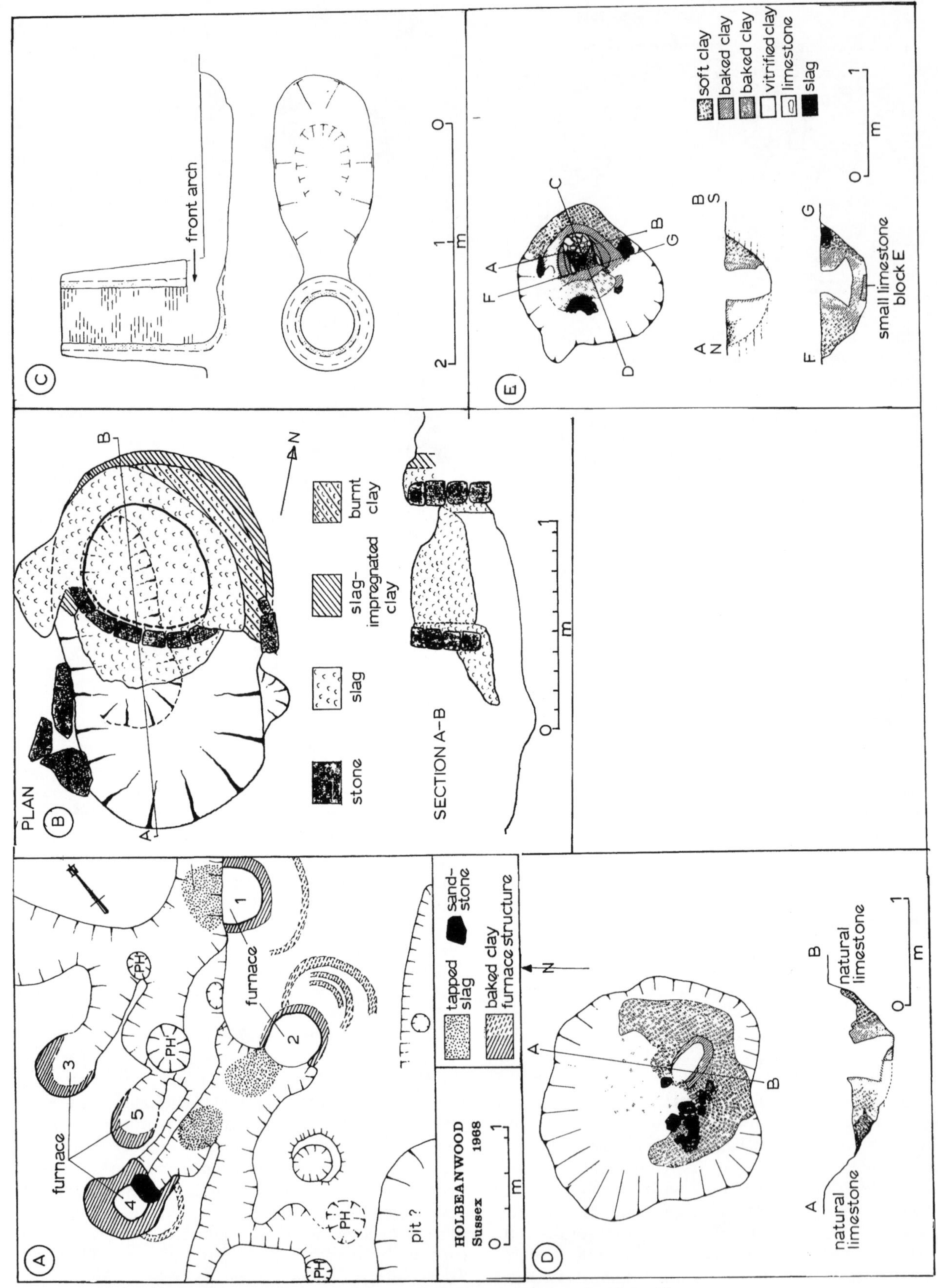

a Plan of group of shaft furnaces from Holbeanwood, Sussex (after H. F. Cleere[12])
b Plan and section of furnace at Pippingford, Hartfield, Sussex (after Tebbutt and Cleere[25])
c Reconstruction of type represented by *a* and *b* (after Gibson-Hill[19])
d and *e* Plan and sections of furnaces at Wakerley, Northants. (after D. A. Jackson[14])

102 Iron smelting furnaces of the Roman period

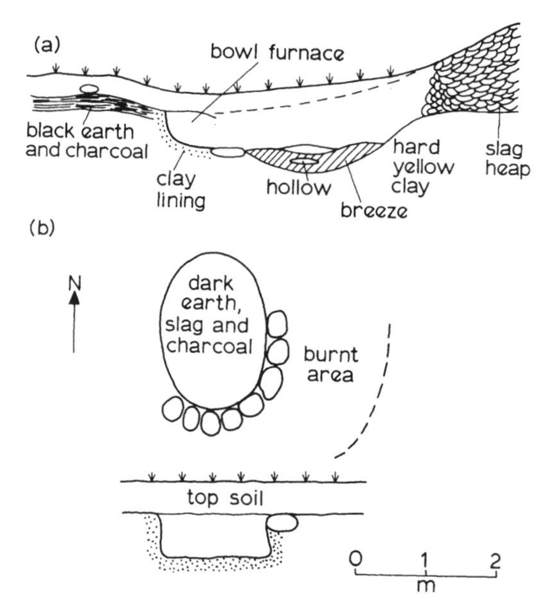

a Developed bowl furnace from Bryn Gefeiliau, Gwynedd (after Davies[28])
b Largest of three bowl furnaces from Aberllynfi, Powys (after Lloyd and Savory[27])

104 Welsh bloomery furnaces

These furnaces were straight-sided – a distinct departure from the smaller, pre-Roman bowl furnaces. One was in fact slightly undercut, but this may have been owing to some of the side coming away when cinder was being removed during their period of use. The fact that the slag had been tapped shows that these were smelting and not smithing furnaces; the excavator did not indicate how the slag was tapped.

A large section of clay lining was found in the remains of a developed bowl furnace at Bryn Gefeiliau, Gwynedd,[28] which showed it to have a depth of about 45 cm and a length of about 2 m with sloping sides (Fig. 104). This site, however, was not closely dated and could be as late as medieval. The slag contained about 18% zinc oxide and 48% FeO, but the quantity found shows that it was undoubtedly an iron-smelting site; the slag covered an area 30 m in diameter.

At Boho,[29] an Irish site probably used between the 1st and 5th centuries, two clay-lined bowl furnaces were found, 0·60 × 1–1·5 m long and 0·5 m deep. The clay lining showed signs of local heating to between 900 and 1100°C on one side only.

A 2nd century furnace in Constantine's Cave in East Fife[30] consisted of two low stone walls separated by an inner core of clay, producing a 'bowl' about 38 cm in diameter. Inside the bowl was found half a whinstone basin, which had been 38 cm in diameter and 10 cm thick. This basin lay on top of a deposit of partially slagged ore, clay and metal. The excavators were clearly under the impression that the basin was the bottom of the furnace, that it was sealed into the stone surround and surmounted by a clay superstructure. The presence of smelting debris below the basin suggests that the furnace was first worked without the basin, and that only later was the basin used to form the furnace base.

Several sites, for example Cantley near Doncaster[31] and Llantwit Major[32] in Glamorgan, have yielded remains that seem to indicate straight-sided cylindrical furnaces of unknown height. The fact that the internal diameter of these does not seem to be less than 0·76 m probably precludes their being shaft furnaces, since, judging by the evidence from Ashwicken, the shafts were of small internal diameter. However, a stone-clay wall 0·50 m high and more than 1·20 in inside diameter existed at Cantley, and a similar height of stone wall lined with 10 cm of clay at Llantwit. It seems here that we have the remains of bowl furnaces each with a permanent stone-clay wall and a semi-permanent lining. In section, these furnaces would be more like the straight-sided furnaces from Aberllynfi[27] (Fig. 104).

Straight-sided bowl furnaces dated to the 1st century have also been recorded outside the military part of the camp at Margidunum, Notts.[33] These appeared to have side gullies, but the evidence is not sufficient to allow the conclusion that these were smelting furnaces and not smithing hearths. Highly ferruginous slag was found, but no ore; the underlying clay had been burnt red, and the gully had a 15 cm diameter cylindrical opening which seems in many ways to resemble the smithing hearth found by May at Wilderspool[16] (Fig. 106 *below*).

The site at Din Lligwy[34] in Anglesey produced a number of 3rd–4th century hearths, some inside and at least one outside some huts. The hearth outside consisted of a clay-lined depression 30 to 45 cm in diameter surrounded by an area paved with stones 12 to 15 cm in diameter. The depression contained lumps of iron slag covered with blackish 'dust'. Some pieces of hematite and the products of forging were found in the huts.

There are a number of smelting sites which have not yielded furnaces. Among these is the smelting floor at Coed Newydd (Penrhos Lligwy) in Anglesey.[35] For its Roman dating we have only Gowland's authority, and it may be earlier or later. The floor is the result of the consolidation of slag, particles of ore, charcoal and coal dust, and like the medieval floor at High Bishopley (Durham)[36] it could be lifted in small slabs. On this floor were found a number of stones and two iron bars, 3·52 and 4·18 kg in weight respectively. If they consist only of one piece each, they give an idea of the size of bloom made. Their existence certainly points to a working-up process after smelting. Presumably the local coal was used for this operation and the charcoal from oak, birch and hazel, for smelting. The site was bounded by a low wall and measured 6·4 × 4 m.

The fort of Galava at Ambleside[37] produced some bog iron ore containing pieces of charcoal, and pieces of smelted iron. The close proximity of the ore and charcoal seem to be fair evidence of smelting, but no slag was reported.

Sharp[38] records that at King's Cliff, near Bulwick, Northamptonshire, remains of iron smelting were found consisting of charcoal, roasted material (possibly iron ore), ashes and slag, associated with Roman coins. This site is on the Northamptonshire ore-field, and there are reports of Roman period ore workings at Oundle, Laxton and other places in the county.[39, 40] Recent ore-mining in this area has revealed shaft furnaces and channel hearths at Brookfield Cottage near Bulwick.[41] The furnaces have deep foundations and

Table 79 Composition of Romano-British iron-smelting and smithing slags

Chapter Ref.	11		4			94		45	42		
	Great Casterton (Leics.)		Ashwicken (Norfolk)			Camerton (Somerset)		Corbridge (Northumb.)	Templebrough (Yorks.)		
%	Tap slag 1st cent.	Furnace slag 4th cent.	Tap slag 2nd cent.	Furnace slag 2nd cent.	Smelting slag 1st cent.	Slag 3rd cent.	3rd–4th building	From beam 3rd–4th cent.	Furnace slag 1 (2nd cent.)	Furnace slag 2 (2nd cent.)	Tap slag
FeO	55·5	64·4	62·1	54·5	46–62	62·2	55–61	58·0	—	—	51·8
Fe_2O_3	11·9	7·7	7·7	14·4	12–15	13·1	8–17·2	21·4	32·3	20·8	6·73
SiO_2	20·9	18·2	21·2	23·9	15–20	12·6	11–20	7·45	41·04	47·36	28·1
CaO	2·4	1·8	0·4	1·3	2·7–3·4	2·24	0·8–2·8	0·8	0·32	0·80	3·0
MgO	0·9	0·1	1·4	0·42	0·2–0·3	0·25	0·3–0·4	0·6	0·69	0·53	1·17
Al_2O_3	1·9	3·7	3·2	4·90	4·8–9·7	6·64	5·8–9·0	1·24	11·7	6·07	5·2
MnO	—	1·4	0·5	0·33	—	—	—	0·21	1·42	2·42	0·33
TiO_2	—	—	—	—	0·5–0·6	0·41	0·4–0·5	—	—	—	—
P_2O_5	—	0·1	1·72	0·31	0·12–0·27	0·59	0·3–0·7	0·174	2·27	0·69	0·43
S	—	0·02	—	—	0·11–0·55	0·15	0·08–0·16	0·170	0·04	tr.	0·06
H_2O	—	1·4	—	—	1·40–3·0	1·23	1·80–2·0	8·38	13·3	15·54	1·70
Alkali	—	—	—	—	—	—	—	0·278	—	—	—

Table 79 – *continued*

Chapter Ref.	16				96	97	14		55	55
	Wilderspool (Lancs.) 2nd cent.				Ariconium (Hereford)	Chichester (Sussex)	Wakerley (Northants.)		Northwich (Cheshire)	Manchester (Lancs.)
%	Black glassy slag			Vitrified lining	Slag runner Roman?	Vitrified lining?	Smelting slag		Smelting slag	Smithing slag
FeO	29·8	53·7	63·2	17·0	56·6	14·2	63·8	60·5	65·1	56·0
Fe_2O_3	23·8	10·0	8·6	1·3	23·9	5·3	13·0	—	—	—
SiO_2	39·7	28·0	21·2	80·4	<16·1	63·4	15·7	7·5	30·3	34·3
CaO }	2·7 (CaO + MgO)	1·65 (CaO + MgO)	1·4 (CaO + MgO)	—	1·25	2·36	1·1	2·4	1·0	0·19
MgO }				—	0·58	0·71	0·2	—	—	—
Al_2O_3	2·27	1·97	2·2	0·6	—	7·25	4·6	1·3	3·2	8·76
MnO	—	—	—	—	—	0·08	0·3	1·3	0·2	1·0
TiO_2	—	—	—	—	—	0·70	0·34	—	—	—
P_2O_5	0·3	0·73	0·43	0·64	0·40	1·27	—	0·37	—	—
S	—	—	—	—	0·95	0·01	—	0·08	—	—
H_2O	1·90	1·30	1·5	—	0·55	1·28	0·41	—	—	—
Alkali	—	—	—	—	—	—	—	—	—	—

appear to be only 15 cm in diameter, although the clay-lining is at least 10 cm thick. They closely resemble those from the nearby site of Wakerley.[14]

The Roman fort at Templebrough near Rotherham[42] produced ore and slag. The analyses (Table 79) allow little doubt that smelting was carried out.

Claxby, Lincs., even in recent times an ore-mining area, has produced evidence of smelting in Roman times. Roasted ore, iron slag and charcoal were found in association with Roman pottery.[43] Other sites in Lincolnshire where smelting has been carried out are at Bagmoor and Thealby, where Dudley found large lumps of slag – the contents of furnaces about 76 cm in diameter.[44]

Even when furnaces are not found, large lumps of furnace slag are frequently unearthed. These can be of enormous size, and since they comprise a semi-fused mass all in one piece, are clearly the contents of one furnace. The largest seen by the author is in the Castle Museum at Norwich, and must weigh several hundred kg. It measures about 0·75 m × 1 m and came from Hevingham in North Norfolk. These pieces of slag show how the size of the furnace has increased since the pre-Roman period. However, we cannot assume that these large lumps of slag are from successful smelts; they may be one result of failures, in which case it would be easier to build a new furnace than take them out.

SMITHING FURNACES

After the production of the raw bloom, further operations were required to cut up the bloom and remove the entrapped slag; finally, smithing hearths were necessary for the production of the objects required.

The size of Roman blooms is known from the fabricated beams found at Corbridge,[45] Chedworth,[46] and Catterick;[47] also from anvils found at Sutton Walls, Herefordshire[48] and Stanton Low, Bucks.,[49] which

105 Smithing hearth and anvil from the Catacomb of Domitilla, Rome (after W. H. Manning; courtesy Joint Museum of Antiquities, University of Newcastle upon Tyne)

may consist of a number of smelted blooms welded together. In the case of the Corbridge beam, about twenty blooms have been welded into a mass weighing 155 kg, giving a mean weight per bloom of 7 kg, which would make a cube with sides 10 cm long. We can assume that the original bloom removed from the furnace would be one-and-a-half to two times this size. On removal it would have adherent slag and charcoal hammered off it, and then be reheated in a smith's hearth for removal of entrapped slag. Until this had been done, the bloom would be rather brittle when cool, since solid slag has the characteristics of glass. The reheating operations would have to be carried out at a very high temperature to soften the slag sufficiently for beating out. For this reason smithing furnaces would generally be bellows-blown, bowl-type hearths.

The object of the smithing hearth is to provide a means of reheating a piece of iron so that it can be shaped efficiently. Metallurgically there are two basically different methods that can be used. The metal can be worked cold, whereupon it hardens and in due course becomes embrittled. To overcome this an annealing hearth is required which does not need to go higher than 700°C for iron. This technique was widely practised in the 19th century as we know from the general use of 'mower's' anvils in the field. It is still used in Africa today, but it must be backed up by smiths who have facilities for hot-working at temperatures of about 1 200°C. Most of the early work involved in shaping a tool is done hot; the heat treatment of steel requires temperatures in the range 700–950°C, so most of the early smithing hearths were capable of temperatures as high as 1 200°C and needed a forced, bellows-blown draught.

The minimum equipment for this purpose is a pile of burning charcoal with a tuyere inserted to raise the temperature locally. If this is to be done in the open a few stones are needed to prevent the wind from blowing away the light charcoal. But the most important piece of equipment may be a protection for the bellows. This can be provided by making a bowl hearth with the bellows placed above and to one side of the hearth, or by using a portable *essestein* which can be carried on a pony or donkey by an itinerant smith. This can be refractory stone or a plate of clay. In some cases the tuyere and hearth are integral, in others two or three separate stones can be made both to hold down the tuyere and to protect the bellows.

In the Roman period evidence is found of two types: a bowl or pit in the ground into which a tuyere was directed, and a low platform on which it was supposed that a fire was built and a tuyere inserted. Manning[50] has recently cast doubt on this rather facile explanation by calling attention to Roman illustrations of smithing hearths such as the one in the catacomb of Domitilla (Fig. 105). It is possible that a large number of different types were erected on the platforms.

The bulk of the Roman smithing hearths are no more than holes in the ground, i.e. bowl hearths in the true sense. They can be distinguished from smelting furnaces by the fact that they do not contain a heavily slagged lining as the walls have not been in contact with ore – merely charcoal. Many of these hearths are no more than horseshoe-shaped enclosures of stone and clay, urged with bellows from in front or behind. Tuyeres or pieces of tuyere are often found with them. The remains of many such hearths have been found at Corbridge and other military sites.

The most puzzling and yet most comprehensive collection of furnaces in any one site in Britain, was found at Wilderspool.[16] More recent work allows us to reassess the evidence unearthed by May. The main difficulty is that few remains exist to higher than about 30 cm. There are many furnaces or hearths with small permanent holes entering them upwards or sideways from the bottom. Since these holes would be easily stopped up by molten slag, the furnaces in which they were found cannot have been used for smelting. Since they must be ingoing air holes or tuyeres, their disposition points to smithing hearths of one type or another.

Also at Wilderspool[16] is found an example of the elongated or oval bowl hearth also used for smithing (Fig. 106*a*). Normally primitive smithing hearths

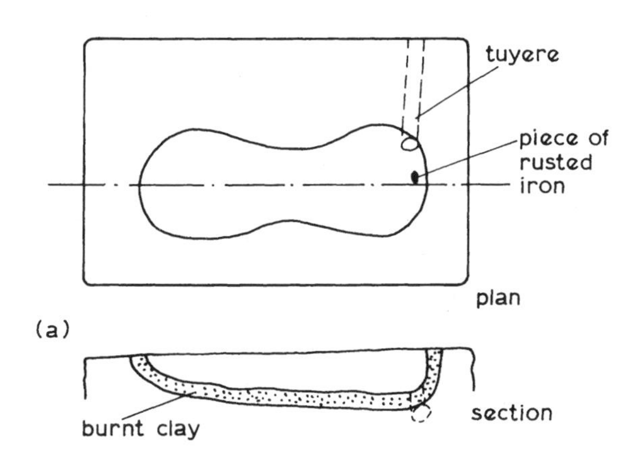

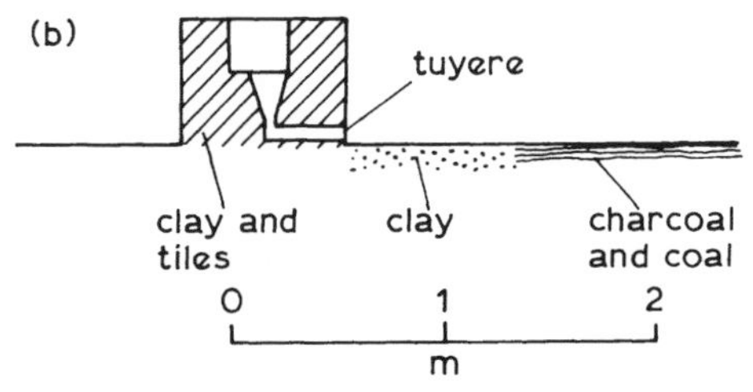

106 Smithing hearths from Wilderspool, Lancs. (after May[16])

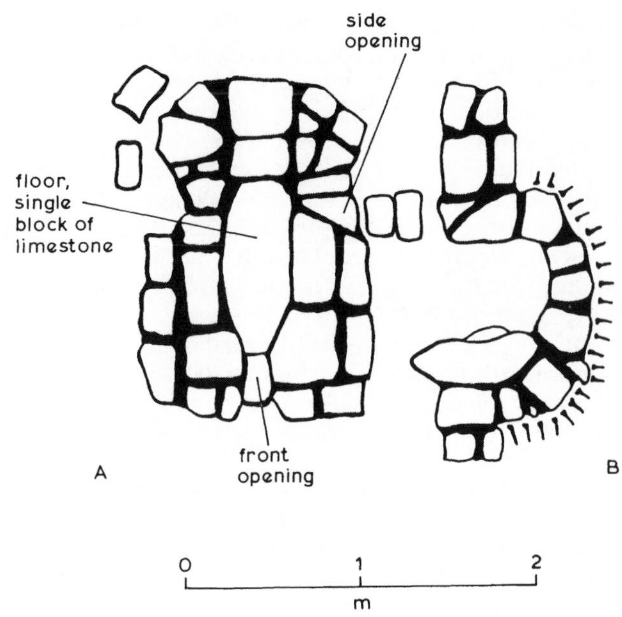

107 Smithing hearths from Tiddington, Warwicks. (3rd–4th century AD) (after May[16])

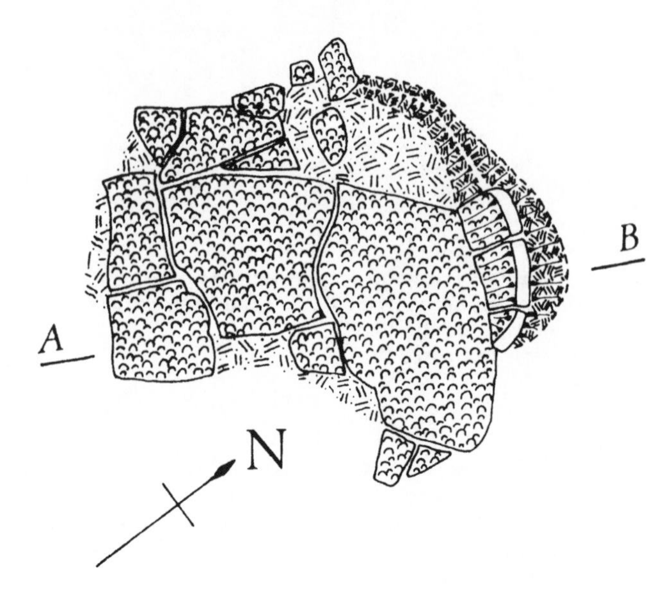

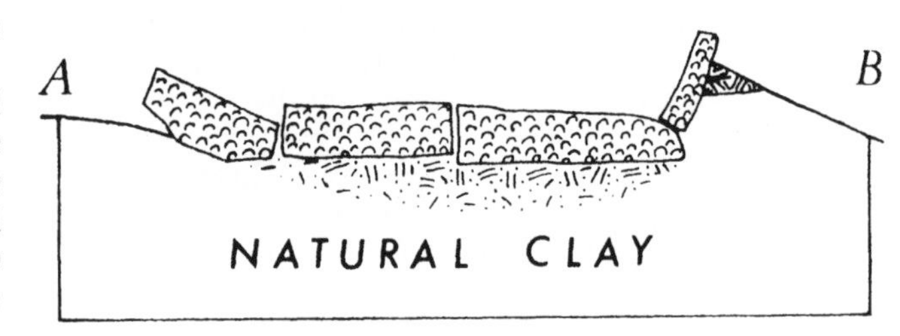

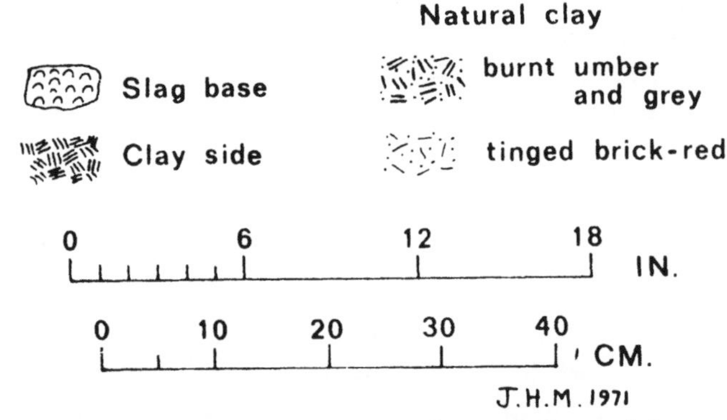

108 Smithing hearths from Minepit Wood, Sussex (after Money[23])

would be heated by the introduction of bellows where necessary, but this hearth has a figure-of-eight form. Normally for small articles, the right-hand end would be used alone with air supplied by bellows to the tuyere. For larger work the whole furnace would be needed, additional air being supplied to the left-hand end by supplementary bellows. Similar hearths have been found at Tiddington[51] (Fig. 107), Derby[52] (the Racecourse), and at Ware.[53]

Wilderspool also yielded a very interesting pot-shaped furnace, shown in Fig. 106. This was undoubtedly a smithing hearth since the bottom-situated tuyere would easily be closed by slag if used for smelting. It is very probable that originally this hearth was somewhat higher so that it would attain a considerable temperature, with air induced through the tuyere as in shaft furnaces. Alternatively, forced air from bellows could be supplied through the tuyere. It seems to have been mainly worked with charcoal, but quite a lot of coal was found in the vicinity.[54] Only one lump of heavy slag was found near the hearth, and the lining of the furnace itself was almost free of ferruginous material. This implies that forging hearths can be free of accretions or slag, and what little collects in them does not stick and can easily be cleaned out to give the ubiquitous smithing hearth bottoms (SHBs).

Sometimes we find tile-lined hearths or even clay-and-slag-lined hearths, as at Minepit Wood[23] (Fig. 108). It is quite possible that the two post holes found adjacent to the hearth supported an anvil of the type found at Hartfield.[24] A good example of the tiled hearth is furnace 15 from Manchester.[55] Hearths of this type would need some sort of protection for the bellows such as a clay plate, as shown in the reconstruction of the Huckhoe hearth (Fig. 86, Chapter 6), in fact an *essestein*.

An unusual furnace was found at Colsterworth, Lincolnshire,[56, 57] dated to the 2nd century AD. This is shown in Fig. 109; it consists of a rectangular red clay box, 60 × 90 × 58 cm high, with large openings back and front and 15 cm diameter holes in the middle of each side. It bears a marked resemblance to a smithing hearth seen by Buchanan at Magadi near Madras in about 1800.[58] The Indian hearth was blown by bellows through one of the side holes as shown in Fig. 109 and slag and ash were removed through a side hole at right angles. Charcoal was charged through another side hole and the bloom inserted for reheating through a hole in the top. In the Roman type the bloom was presumably put in and taken out through the same hole in one of the sides.

In Wales, at Rhostryfan,[59] Gwynedd, a rectangular 3rd century forge hut was found, containing two raised hearths in diagonally opposite corners. These hearths fitted the corners and were triangular in plan. A large amount of burnt clay and oak charcoal, some

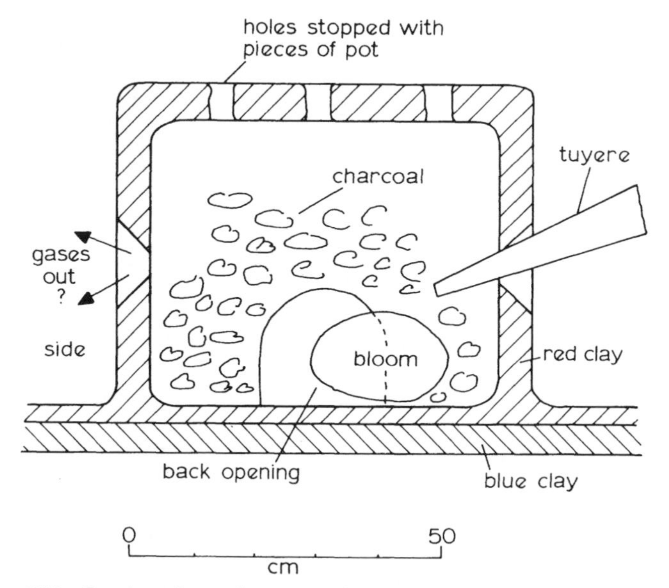

109 Section through centre of smithing hearth from Colsterworth, Lincs. (reconstruction based on Indian furnace described by Buchanan[58])

nails and slag were found near them. It appeared that this and another round hut were in fact smithies working up iron smelted outside.

Some sites such as the villa at Ilchester Mead[60] have rectangular stone platforms 1 m × 1·8 m which may well be the bases of the waist-level hearths described by Manning[50] and shown in Fig. 105.

Industrial and military sites such as Corbridge have provided numbers of small semicircular hearths like that shown in Fig. 107*b* from Tiddington. This and the oval-shaped bowl furnace were obviously the type preferred, unless the site was expected to have a long life, in which case more ambitious furnaces might be constructed.

At the fort of Benwell, Newcastle,[61] an area west of the granaries was covered by a thick, blackish deposit containing coal and a high proportion of hammer scale, which was believed to be the floor sweepings from a smithing workshop. There are many more sites of this type, but the most interesting is that at Corbridge[62] in which a 'furnace' was found with a beam in position (Fig. 97, *above*). This was originally designed as a lime kiln, and was later adapted to a smithing hearth. Presumably a fire was built around the 'stump' of the beam which was urged with bellows in order to take the temperature up to 1200°C. Meanwhile the next bloom would be heated in an adjacent part of the fire and when hot enough removed and hammered onto the stump. A similar technique was used in the construction of 19th century ships' stern frames.

IRON AND STEEL PRODUCTS

Beams and anvils

The last section referred to the making of iron bars by welding a number of small, smelted blooms to make anvils and beams for bath houses. Britain is by no means unique in having examples of these large fabrications. In Saalburg, Germany,[63] there are the remains of similar beams, and the column at Delhi (AD 319) and beams from the temple at Konarak, Orissa,[64] are examples of fabricated structural elements of this type. The British examples are listed in Table 80. Those from Chedworth and Catterick were all found near bath house stoke-holes. The one found at Corbridge *in situ* in the hearth was only half finished, comparing its shape with others. At Catterick Bridge (Cataractonium), one was found complete (Fig. 110) near the remains of the stoke-hole of the bath house hypocaust. It was much eroded in the centre by exposure to high temperatures. An incomplete one, only half the length of the other, had probably broken and been scrapped. The complete beams seem to have been about 2 m long. Two common features are that they have a square section, and are rather larger in the middle than

Table 80 Welded beams of the Roman period

Provenance	Date AD	Length m	Dimensions Section (cm) Ends	Centre	Weight, kg	Notes
Catterick I	2nd cent.	1·72	15 × 13	18 × 18	*c*. 250	Eroded in mid-span
Catterick II (half only)		0·89	15 × 16·5	18 × 18 (broken)	135	Examined and analysed
Corbridge (half only)	3rd–4th cent.	0·99	11 × 13	18 × 20 unfinished end	156	Examined and analysed
Chedworth I	2nd cent.	1·63	13 × 13	20 × 15	220	Complete
Chedworth II	2nd cent.	0·97	18 × 15	18 × 15	162	Halves only; do not fit together
Chedworth III	2nd cent.	0·99	18 × 15	18 × 15	116	
Leicester[99] (Jewry Wall)		(part) 0·90	12·1	—	*c*. 130	Splayed end
Chesters Fort (Northumb.)		(part) 0·25	12 × 8	—	*c*. 14	End fragment
Bayham Abbey[98]		0·356	10 × 7	—	13·6	End fragment
St. Albans (Herts.)[100]		0·43	13 × 15	—	—	End fragment

110 Beam found at Catterick (length 1·73 m) (photo courtesy of J. Wacher, University of Leicester)

at the ends. Both these features would have made them more suitable as beams. Their square section would have made them easier to place in stone-work, and the increase in size at the centre would have provided extra metal where the stress is greatest. The larger one has been cut and analysed.[47]

Iron has obvious advantages for stoke-hole supports as stone soon cracks under intense heat. Even iron beams will not last indefinitely, as shown by the pieces from Catterick and Chedworth, and the fact that the complete example from the former site was badly eroded in mid-span. In a drawing of the women's *caldarium* in the Stabian Baths at Pompeii, Mau and Kelsey[65] show a beam supporting the bronze water heater (*testudo*) above the furnace chamber. A recent examination of this beam shows it to be iron of 12 cm × 9 cm in section with an exposed length of 56 cm. Two other beams of similar section also span this furnace chamber and both have exposed lengths of 0·67 m. The finds of the remains of British beams have been discussed by Wacher.[66]

Anvils are not dissimilar to the ends of beams and their use is depicted in Fig. 111. Two examples come from Britain. One from Sutton Walls, Herefordshire,[48] weighs 50·5 kg, the other from Stanton Low, Buckinghamshire,[49] 23·2 kg (Table 81). Although the stratification in both cases was somewhat doubtful, there is no doubt that these are Roman in type. It is probable that they are made of a number of large pieces of metal at least 7 kg in weight. Both contain holes for heading nails or rivets, and are tapered at the bottom end so that they may be supported at working height in a baulk of timber or a tree-trunk as shown in Fig. 111.

The average carbon content of the Sutton Walls anvil was 0·5%; the structure was remarkably homogeneous.[67] If this carbon content was inten-

111 Roman-type anvil from Sutton Walls, Herefordshire (courtesy of the *British Welding Journal*)

tional, it certainly suited the intended application. Until we have more evidence, however, we must assume that it was fortuitous. The hardness was 190 HV; which was in keeping with its carbon content. The massive nature of the cementite within the pearlite showed that the centre had been heated for a long time at a temperature of about 700°C, presumably while the outside was raised to about 1 000–1 200°C for forging, and it is certain that it was made like the Corbridge beam.

Similar examples are known from Mainz[68] and Kreimbach.[69] The anvil from Mainz was approximately the same dimensions as that from Sutton Walls, but was designed to be placed on a flat surface. The other has a tapered base, and is in all respects very similar to the Sutton Walls anvil.

The anvil attached to the bloomery at Cow Park, Hartfield[24] is rather unusual in that it consisted of an iron plate 0·60 m long and 0·23 m wide by 4 cm thick resting on two tree-trunks 25 cm in diameter. This is dated to AD 50–135.

Blooms and billets

A number of individual blooms of iron have been found, and an examination of the Corbridge[45] and Catterick[47] beams shows the iron to be mostly a low-carbon iron (0·061–0·14% C) with some entrapped slag – in fact, a typical wrought iron. As is so often the case with these early irons, they show microstructural effects known as Neumann bands, which prove that very heavy hammering has been carried out when the temperature of the iron was below about 500°C.

Smelting experiments have shown that it is easier to produce low-carbon iron and that the yield is greater, as the slag separates more easily leaving a cleaner metal. Higher fuel/ore ratios will allow steel (i.e. iron with more than 0·5% C) to be produced, but the slag entrained in it is difficult to remove. It must also be remembered that the iron produced in the bloomery process is very heterogeneous and contains areas of high carbon content. These areas will be noticed during the smithing of the bloom and may well be picked out and saved for 'steeling'.

Steel was also produced, probably accidentally, by the almost complete carburization of quite large pieces. The small bloom weighing 305 g and containing 1·6% carbon, from Nanny's Croft, Arundel, Sussex, examined by Smyth[70] and dated to the 4th or 5th century, is probably an example of this. While the bloom from Arundel was steel and was probably kept for steeling edge tools, the larger bloom from Forewood had a low carbon content and was more likely a smelted product intended for general purposes such as nail making.

The piece from Cranbrook[71] and the bars from Strageath[72] were shaped into square-sectioned bars and were therefore *billets* rather than blooms. They were clearly semi-finished products, perhaps intended for the production of large nails or for military ironwork. The billet from Cranbrook is consistently high in carbon (Table 82) and had been quenched, no doubt to cool it after cutting off a piece with a cold chisel while hot.

The billet from Lower Slaughter[73] had only been flattened on one side and may have been intended as a bath house beam because of its size (Table 82).

In this section the odd pieces of cast iron and once molten high-carbon steel which are occasionally found on Roman sites might be considered. If the fuel/ore ratio is increased well above that necessary for low-carbon wrought iron in a furnace of the Ashwicken type, cast iron can be produced. It is clear from the pre-Roman example found at Hengistbury Head (*see* Chapter 6), and the Roman examples from Wilderspool[16] and Tiddington,[51] that this occasionally happened. There is no reason to think that it was intentional or that the Roman smiths learnt to make use of their mistakes, as in China where the perhaps accidental occurrence of cast iron was intensively exploited in the Han Dynasty (206 BC–AD 220). It would seem that because of its high sulphur content, the Wilderspool piece was the result of smelting with coal. But 0·49% S could have come from an ore which had not been sufficiently roasted (Table 83).

The two pieces of cast iron are mainly grey irons with large amounts of graphite; but more recently several white irons have turned up with lower carbon contents, more like steel. The pieces from Gestingthorpe[74] and Colchester[75] (Sheepen) are very similar

Table 81 Roman anvils

Provenance	Weight kg	Height cm	Breadth (cm) top	bottom	Notes
Sutton Walls, Hereford.	50·5	26·7	23 × 23	18·5 × 16·5	Ref. 67
Stanton Low, Bucks.	23·2	25·4	20 × 16	10 × 10	Ref. 49
Mainz, Germany	—	20·1	22 × 20	17 × 17	Ref. 68, flat bottom
Kreimbach, Pfalz, Germany	—	21·1	18 × 18	8·1 × 8·1	Ref. 69, flat bottom

Table 82 Roman blooms and billets

Provenance	Weight g	Size or shape cm length	cm breadth	Carbon % Min.	Carbon % Max.	Hardness HV Min.	Hardness HV Max.	Comments
Nanny's Croft,[70] Arundel, Sussex	305	'roundish'		—	1·6	480	770	4th–5th cent. AD bloom
Forewood Crowhurst,[70] Sussex	1 240	'roundish'		0	0·1	100	100	Bloom, Roman?
Cranbrook,[71] Kent	709	14	3·2 × 3·2 to 2·2 × 2·7	1·16	1·46	quenched		2nd cent. AD billet
Strageath 1,[72] Tayside	5 700	30	5·5	0·1	0·8	119	290	Flavian, 2nd cent. AD billet
Strageath 2, Tayside	6 900	30	5·5					
Strageath 3, Tayside	7 300	30	5·5					
Strageath 4, Tayside	7 400	30	5·5					
Lower Slaughter, Glos.[73]	10 910	*c.* 20	9·9 (approx.)	0	0·8	—	—	AD 250–400 billet or bloom

and consist of a plano-convex accretion of drops of solidified liquid metal of high carbon content (Table 83). The plate from Battle Abbey[76] was unstratified. It is 'burnt' (i.e. oxidized by exposure to high temperatures) for some time in an oxidizing environment and the structure consists of ferrite and graphite. This could have been a more recent cast iron plate for a bloomery, finery or fireback. But the other pieces are the accidental product of a smelting furnace. It must not be forgotten that the Japanese bloomery (Tatara) process was capable of making wrought iron, steel and cast iron all in one smelt; it is therefore not unreasonable to obtain some cast iron from the Roman bloomery process.

Innumerable smaller pieces of cast iron and pure iron have been found on many sites. But these have little significance, being the residue of smelting and off-cuts from smithing.

The composition of Roman iron

The composition of the majority of Roman irons, apart from proving that they are indeed wrought iron and not cast iron, seems unexciting. The ores, of course, contain varying proportions of sulphur, phosphorus, silica and manganese, besides oxides of iron. However, in the process used, as pointed out in Chapter 6, most of these elements are not so easily reduced as iron and do not go into the reduced iron but remain in the slag. The main exceptions to this are phosphorus and sulphur; the latter, however, can enter with the coal which, although found on many Roman sites, does not seem to have been used for smelting. There is abundant evidence that coal has been used for smithing and has perhaps been tried for smelting.[54] But there are other elements such as Cu, Ni, Co, which may tell us something about the ore deposit.

Representative analyses have been assembled in

Table 83 Composition of cast irons and cast steels

%	Wilderspool,[16] Lancs. 2nd cent.	Tiddington[51] (Warwick.) 570 g 3rd–4th cent.	Gestingthorpe[74] (Essex) 650 g Roman		Sheepen[75] Colchester (Essex) 1 000 g Roman	Battle Abbey[76] (Sussex) (unstrat.)	Braughing[84] (Herts.) 4th cent.
C (graphite)	3·0 }	3·21	2·26	2·49	—	1·12	} 3·3
C (combined)	0·23 }	0·31	—	—	3·13	—	
Mn	0·403	0·63	—	—	—	—	—
Si	1·05	1·92	3·44	—	5·86	1·64	2·62
P	0·76	0·77	0·62	0·90	0·72	0·57	0·26
S	0·49	0·05	0·13	0·17	0·18	0·61	0·043
Hardness, HV	—	—	235		210–224	130–162	110–119

Table 84 Composition of miscellaneous iron and steel objects

	Wrought irons									
Ref.	45 Corbridge (Northumb.)	16 Wilderspool (Lancs.)			80 Richborough		80 Folkestone	94 Camerton (Som.)		47 Catterick (Yorks.)
	Beam (mean)	Strip	Pin from smithing hearth	Nails	Nail	Nail	Nail	Rod	Lump	Beam (mean values)
	3rd–4th cent.		2nd cent.		1st cent.		2nd–4th cent.	3rd–4th cent.		
C	0·097	0·060	0·080	0·07	0·08	0·07	0·12	0·20	0·23	0·138
Mn	0·040	tr.	tr.	—	nil	nil	nil	nil	nil	nil
Si	0·046	tr.	0·060	nil	—	—	—	0·05	0·05	0·025
P	0·044	0·037	0·257	0·114	tr.	nil	0·001	0·052	0·066	0·077
S	0·025	0·027	0·031	0·013	0·046	nil	0·034	0·007	0·010	0·086
As	0·049	—	—	—	—	—	—	—	—	—
Cu	0·010	—	—	—	—	—	—	—	—	—
Ni	—	—	—	—	—	—	—	0·03	0·01	—
Cr	—	—	—	—	—	—	—	0·04	nil	—
Mo	—	—	—	—	—	—	—	—	—	—
V	—	—	—	—	—	—	—	—	—	—
N	—	—	—	—	—	—	—	—	—	—

Table 84 – ***continued***

	Wrought irons				Steels			
Ref.	92 Silchester	101 Brading (I. of W.)	49 Stanton Low (Bucks.)	91 Chesterholm	70 Nanny's Croft, Arundel (Sussex)	67 Sutton Walls (Hereford.)	71 Cranbrook (Kent)	73 Lower Slaughter (Glos.)
	Axe	Latch lifter	Coulter	Chisel	Bloom	Anvil*		
	4th cent.	2nd cent. or later	1st–4th cent.	2nd cent.	4th–5th cent.	1st–4th cent.	2nd cent. AD	250–400 AD
C	0·319	0·02	0·02	0–1·3	1·6	0·54	1·46	0–0·8
Mn	0·013	0·05	n.d.	nil	—	0·04	—	tr.
Si	tr.	0·11	0·01	0·038	—	—	1·20	tr.
P	0·061	0·30	—	0·016	—	0·11	0·025	0·085
S	0·007	0·015	—	0·011	—	0·035	0·03	0·007
As	0·071	—	—	—	—	—	—	—
Cu	0·015	—	0·01	—	—	—	—	—
Ni	0·033	—	0·08	nil	—	—	—	—
Cr	nil	—	n.d.	—	—	—	—	—
Mo	0·020	—	—	—	—	—	—	—
V	nil	—	—	—	—	—	—	—
N	0·0065	—	—	—	—	—	0·004	0·004

* Analysed in the Department of Metallurgy, University of Newcastle upon Tyne; n.d. = not detected; tr. = trace

Table 84. We see now, with the exception of two specimens from Wilderspool, that the phosphorus content of the steels and wrought irons is uniformly low.

It seems that during the Roman period there was a move away from the nodular carbonate ores of the pre-Roman Iron Age with their high phosphorus content, to a better class of ore such as limonite or hematite. The phosphorus irons are cold short, which means that they are brittle when bent cold, and though strong, are not attractive to those who understand well the technique of carburization. Even the cast irons from Wilderspool and Tiddington show relatively low phosphorus (0·76–0·77%) compared with the cast irons of today.

The uniformly low manganese content is very noticeable. This element, though present in the ores in many cases, is not reduced during the normal smelting process and finds its way almost exclusively into the slags (*see* Table 79). It is present in both the cast irons, showing that they have been made under abnormal conditions.

Sulphur is low, as would be expected where wood charcoal is used as fuel. Silicon behaves like manganese, and silica is not reduced except in the case of the cast irons. Considerable silicon is present in the slag inclusions contained in wrought iron (Fig. 72, Chapter 6). Where large amounts of silicon are found in wrought irons it is likely that it arises from the slag phase.

Wrought iron products

These range in size from nails to ship's anchors. In view of the importance of the nail in Roman times and

Table 85 Composition of nails from Inchtuthil, Scotland (after Angus *et al.*[77])

		Composition of heads, %				
Group	Length, cm	C	Si	S	P	Mn
A (i)	25–37	0·2–0·9	0·15	0·009	0·008	0·17
A (ii)	22–27	0·22–0·8	0·08	0·017	0·043	0·03
B	18–24	0·05–0·7	0·10	0·007	0·009	0·03
C	10–16	0·10–0·55	0·04	0·006	0·053	nil
D	7·2–10	0·05	0·08	0·01	0·035	0·03
E	3·8–7·0	0·06–0·35	0·05	0·003	0·16	nil

A (i) pyramidal heads; A (ii) flattened pyramidal heads

the finding of over 5 tonnes of nails at the fort of Inchtuthil in Scotland,[77] it seems worth discussing these in some detail.

The Inchtuthil hoard contained nearly 900 000 nails of various sizes. As the fort was built in AD 83 and evacuated soon after AD 87, this shows that the supply services must have been operating efficiently before the period of the *Notitia Dignitatum* in the 4th–5th centuries. From a metallurgical point of view there is nothing surprising about the composition or the structure of the nails. In many ways they resemble the construction of the iron cramps that key the stones of the Athenian Parthenon, i.e. they are made of forged, heterogeneous bloomery iron. But they were well made to exact dimensional specifications, and the larger nails contain rather more carbon than the shorter nails (*see* Table 85). This was intentional, since the larger nails would have needed to be stronger to withstand the increased driving force and it was probably achieved by selection of suitable blooms or parts of blooms.

There is no way of knowing where the iron was made; it was certainly made from low-phosphorus ores and the nickel content was also low. While such iron could have been obtained in Britain by this time, it would have been shipped by sea. All one can say is that its composition is typical of the Roman period. Since some of the nails show signs of a two-part die on the underside of the head, these at least were not headed in the nail hole of an anvil. Most of the nails had square-section shanks, but there were a few with a round section and a flattened end which must have been designed for a special purpose, perhaps for inserting into plaster.

While nearly all the nails from Inchtuthil must have come from a low-phosphorus ore, this was not always the case. Some hobnails from Cirencester had ferrite hardnesses of 169–159 HV which suggests a moderate phosphorus content.[78] Nails from Winterton Roman villa (not exceeding 11 cm in length) were all ferritic with hardnesses in the range 83–159 HV.[79] Post-Roman nails were invariably made with a high phosphorus content which conferred hardnesses of 200 HV in the annealed state, and over 300 HV at the junction of the heads and the shanks where work-hardening is likely to be most severe. As we see from the compositions given for wrought iron in Table 85, Roman iron was not usually of high phosphorus content and no attempt seems to have been made to select this type of iron which is ideal for nails.

Friend and Thorneycroft[80] examined some nails from Richborough and Folkestone. The carbon content was variable, but little else was present apart from slag (see Table 84). They also compared the corrosion resistance with modern mild steel (*see* Table 86), and found (as is now well known) that Roman wrought iron has a better resistance to corrosion than mild steel. The reason for this is not simple, as the investigations of Chilton and Evans have shown.[81] Wrought iron has two levels of segregation, which affects the corrosion resistance. One of these, microsegregation, stems from the solid-state reactions in the smelting or puddling furnace where the absence of a metallic liquid phase reduces the diffusion of impurities such as Ni, Cu, S, As, etc., some of which may be associated with the slag phase. The other, macrosegregation, stems from the piling or welding process and is the cause of the 'white' lines to which we have already referred in connection with arsenic enrichment (*see* Chapter 6[82]). Nickel enrichment in wrought iron may reach as high as 2·65% (from 0·2%), and this may modify the corrosion process as does the presence of sulphur in coal-smelted iron.

On the whole it would seem that early wrought irons were more resistant than modern mild steel, although this cannot always be said of the 19th century wrought irons with sulphur-associated slag inclusions. But much will depend on the initial stages of corrosion and the presence or absence of protective scale, as well as the ambient humidity.

Most specimens of early iron are badly rusted when found. A piece of iron found at Richborough and examined by Friend[80] consisted of a thick, earthy crust beneath which was a layer of scale, 3 mm thick, covering a core of iron 40 mm square. A further piece consisted of a bundle of strips, welded together with thick layers of 'impurity' between, which was 'surprisingly hard'. The 'impurity' was found to consist of 45·5% iron and 24·5% insoluble matter (mainly silica). If the iron was all oxidized to FeO, as is probable, the ratio of FeO to SiO_2 would be about 2:1; this would agree with the composition of fayalite $2FeO.SiO_2$ – a typical slag.

Generally, if the corrosion has not gone too far, the product is the magnetic oxide, magnetite (Fe_3O_4), and the object is identifiable by its magnetic nature, even if all the original metal has disappeared. However, it is possible for the corrosion to go further under certain conditions and convert the iron to hydrated oxide, FeO.OH (goethite), which is non-magnetic. Fortunately, since in most environments it takes more than 2 000 years to convert massive iron to hydrated ferric oxide, this stage is rarely reached in Roman objects

Table 86 Corrosion resistance of Roman nails compared with modern mild steel

	Richborough (2 months alternate wet and dry)		Folkestone (3 months alternate wet and dry)	
	Roman Iron	Mild Steel A	Roman Iron	Mild Steel B
Orig. wt. (g)	15·123	16·559	13·1274	13·0456
Loss (g)	0·243	0·419	0·3304	0·4674
Relative loss by corrosion	58	100	70·7	100

(After Friend and Thorneycroft[80])

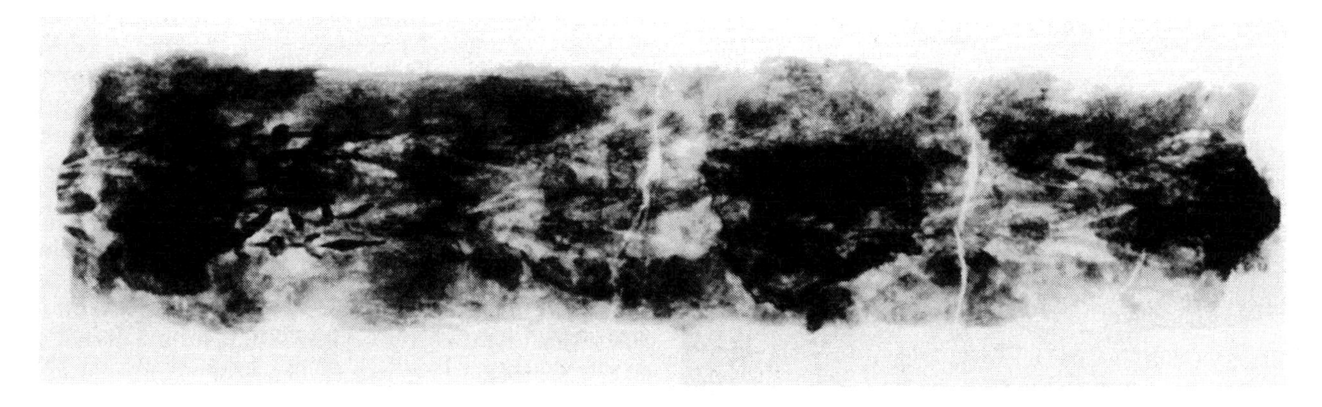

112 Radiograph of the hilt end of the blade of a pattern-welded sword from South Shields showing structure and bronze inlay (3rd century AD) (reproduced by permission of South Shields Corporation)

exceeding 1 cm in original thickness. However, one often wonders how many pieces of material identified by excavators and others as 'probably iron ore', are in reality the completely oxidized remains of iron tools.

The production and properties of swords

There were two types of Roman sword: the *gladius* (a short, stabbing sword), and the *spatha* (a long sword). Rather surprisingly, few of these swords have been metallurgically examined. A *gladius* from Bonn has been examined by Williams.[83] Another, probably a *spatha*, from Whittlesey, Cambs. (now in the Peterborough Museum), has been examined by Gilmour.[84] This example was found in a clay-pit, 2 to 3 m below modern ground level and associated with 2nd–4th century pottery. Its blade is 60 cm long.

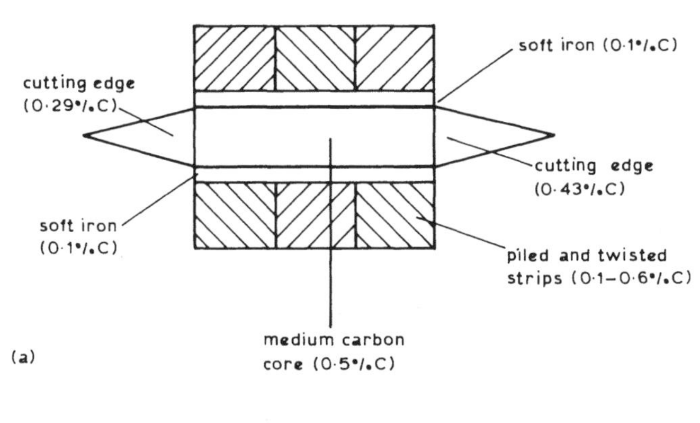

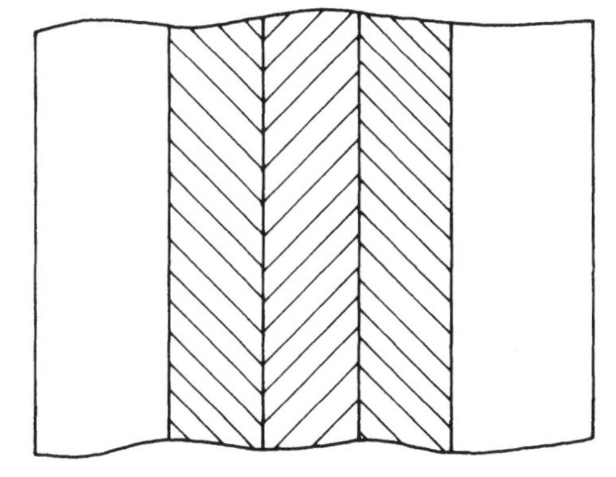

113 Structure of an 'N' type Nydam sword of the 3rd century AD: *a* Construction and composition (diagrammatic); *b* Plan of surface structure of finished blade

A section showed a well diffused structure of ferrite and pearlite with a higher carbon zone running through the centre, varying from 0·3% at the cutting edges to 0·25% at the centre. The carbon content of the centre decreased to about 0·1% at the surfaces. The hardness varied from a maximum of 150 to 120 HV. Slag inclusions marked the position of the welds between the centre and the outer portion of the blade, suggesting that it is basically of Type A structure (Fig. 116 *below*), i.e. composed of three welded strips but with the intention of obtaining maximum diffusion of carbon by holding for several hours at 750° under slightly reducing conditions.

In the pre-Roman Iron Age smiths were quite capable of making complex carburized swords, but not all of these were carburized or made of steel. In the Roman period we see the introduction of 'pattern welding', a method of sword-making which, although it can make use of carburizing, is mainly a complex forging process intermingling fibres of slag and metal.

Many weapons dating from the 3rd century to well into the Viking period have been made by this technique. The pattern probably started as a by-product of the method of manufacture but became a decorative feature – indeed a status symbol – and would serve as visual proof of the quality of the weapon.

A series of swords was found at South Shields, Durham, which showed this type of structure under radiographic examination. The most complete of these may be seen in the Museum at South Shields (*see* Fig. 112), and the whole series is closely dated to between AD 197 and 205.[85]

The first group of these swords to be recognized for what they were, was found in 1859 in a bog at Nydam in Denmark and dates from the first half of the 3rd century.[86, 87] Over a hundred swords were found in three boats, which also contained Roman coins, the latest dated to AD 217. The swords had been made from thin strips of iron carburized to a maximum carbon content of 0·6%. The phosphorus content varied between 0·16 and 0·21%. One of the swords examined was made of eleven pieces, the centre part consisting of a homogeneous core of medium-carbon steel. Layers of soft iron were welded to the core to assist the subsequent welding of three twisted and carburized strips to either side. The medium-carbon steel cutting edges were then welded to the narrow sides as shown in Fig. 113. The twisting of the thin strips was done in a number of different ways to give various

114 Inlay from one of the South Shields swords; Mars on one side; Eagle with palm branch between two standards on the other (courtesy South Shields Corporation)

patterns. Of the three swords examined no two were alike.

The need for such a complex method of construction stems from the difficulty, already referred to, of diffusing carbon into iron to make steel and the resulting brittleness of a high-carbon steel. The time required to carburize iron swords to a thickness of 3 mm would be of the order of four hours at 1100°C, and even then there would be a pronounced reduction in the carbon content towards the centre. The resulting blade would be very brittle and probably smash on impact. If carburization was not carried out to completion a fairly ductile core would be left in the centre.

These problems were fully appreciated by the smiths of this era, and the pattern-welding technique was devised to overcome the difficulty. They aimed at a mixture of high and low strength steels and slag to give good flexibility and reasonable strength. Such properties were only to be improved on by the 'damascening' process of the Middle Ages – a very similar process, resulting in a finer pattern and therefore even more intimate mixing of the high- and low-carbon areas.

Some idea of the time taken to make these weapons can be obtained from the work of Anstee,[88] who has experimentally reproduced the essential features of pattern-welded swords. The total time required was 75 hours, of which 28 were spent on the blade – which needed 128 separate welding heats followed by forging. The finished blade weighed 0·73 kg, of which 0·53 kg were removed by grinding. The hilt and scabbard were also made and are included in the total time taken.

Radiographs of recently made pattern-welded blades do not show as much detail as the Roman and medieval examples. This is because the structure shown in the radiographs is due to the penetration of the welds by the corrosion process, resulting in rust of lower density than that of the original metal. The patterns in the original and replicated swords are due to contrast on the surface between slag and metal on the one hand and between metal of differing carbon and alloy contents on the other. This could be caused by the type of microsegregation mentioned earlier. These patterns were originally developed by etching with naturally occurring acids, and by wear and cleaning processes.

In the smithing process the pattern is brought to the surface by twisting, forging and grinding. Greater skill would be required to get the pattern by means of carburization alone, since this would mean knowing exactly where to place the strips in the fire. However, the carburized pattern-welded sword would not have taken the early smith any longer to make than the slag-metal, pattern-welded sword that Anstee so skillfully produced.

One of the swords from South Shields has been decorated with a bronze inlay near the top of the blade (see Fig. 114). It is decorated on one side with the figure of Mars in full armour, and on the other side with an eagle bearing a palm branch. Thus in two respects these blades are the forerunners of the later Anglo-Saxon blades in which the arts of pattern-welding and inlay have been brought to full fruition. Other examples of Roman inlaying technique have recently come to light. A knife from Winchester[89] (? 4th century) was found to have a yellow, brass or bronze inlay, and a recent examination of iron plates from a Roman dagger scabbard found near Bridport in 1897 has revealed a gilded bronze inlay.[90]

The properties of edge tools

Although every Roman site produces a large number of iron objects, so far very few of these have been submitted to a metallurgical investigation. Even if they were, it would be expected that the vast majority of them would consist merely of wrought iron with no attempt at carburization or heat-treatment. For the great majority of purposes wrought iron was satisfactory and relatively cheap. However, for this very reason it is important that articles such as agricultural implements, tools, and knives (edge tools) which could be expected to show evidence of the higher levels of technique, should be subjected to examination. In the case of material from the Roman and later periods this has now been done, and in the last 20 years we have examined between 500 and 1000 edge tools in the British Isles.

Carbon-steel tools may be made in two basic ways: either by surface carburizing the finished tool or by welding onto the iron pieces of ready-made steel. By carburizing before the tool is fully fashioned it is

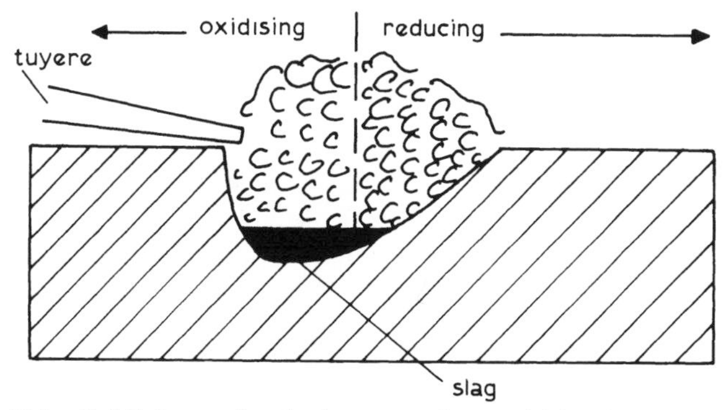

115 Oxidizing and reducing zones in a smithing hearth. Exact position of dividing line depends upon air rate and charcoal size. Position of SHB is usually in the hottest (oxidizing) zone below the tuyere as shown

possible to introduce carbon below the surface by repeated folding and welding.

It is not always easy to decide how the carbon was introduced. First, early iron was non-homogeneous. Certain portions of the charge would absorb more carbon than others in the upper parts of the hearths or furnaces, and if this part did not get too near the tuyere where conditions are conducive to the burning away of the carbon, it would form a high-carbon area in the bloom. This would certainly be the reason for the scatter of carbon contents in the Corbridge beam and the high carbon content in some small blooms of the Roman period. However, in the case of the Chesterholm chisel,[91] its very purpose would suggest to the smith the need for a high-carbon steel. It is possible that the smith could recognize the high-carbon blooms and select them, but it is more likely that he would try and increase the carbon content by placing the semi-forged tool in a reducing area of his hearth (Fig. 115). There would be a critical area where the conditions were reducing and the temperature sufficiently above 900°C to convert the iron into the austenitic state, in which condition it would absorb carbon most easily. The oxidizing zone near the tuyere would be hotter and perfectly acceptable for welding.

The Chesterholm chisel was 20 cm long by 20 mm thick. It was a typical mason's chisel, mushroomed at the top where it had been hit with the hammer or stone maul, and flattened at the end. Smythe[91] examined the edge in detail and found patches of the hard constituent, martensite, formed by quenching. But the carbon content of the edge was variable, so that the hardness varied from 579 HV down to 464. The constituents of the edge included work-hardened ferrite (almost pure iron hardened by hammering), martensite, martensite and troostite, ferrite and sorbite, and martensite and ferrite. This is evidence of the intention to harden by heating and quenching, but because of the poor carburization of the edge, this operation has not been uniformly successful. It is, however, proof of the smith's knowledge of the art. The bulk of the chisel consisted of a very low-carbon iron with a hardness of about 136 HV. As is so often the case with such low-carbon irons, the small amount of carbon existed as needles and grain boundary areas of cementite; but in some areas the carbon content was as high as 1·3%. There is no doubt that this chisel was a very serviceable tool and would have been even better if the smith had some means of judging the extent of the carburization of the edge.

On the other hand a 3rd–4th century tanged knife from Frilford, Oxfordshire, consisted of a piled wrought iron with a low carbon content.[92] The edge of the blade was so corroded that it was not possible to determine whether it had been carburized. However, it appears that domestic cutting tools were not always quench-hardened. A blade of a pair of shears from Silchester was examined by the author and found to consist of hammer-hardened, high-phosphorus ferrite with a hardness of 290 HV. Clearly, this was considered hard enough for domestic purposes.

A Roman axe, also from Silchester,[92] weighing 0·74 kg and dated to the 4th century, was made from carburized and piled iron. The analysis is given in Table 85, and the mean carbon content is 0·32%. Some areas have as much as 0·6%, others as little as 0·05%, and the whole blade has been heated for a long time in the range 650–700°C, as in the case of the Sutton Walls anvil. The hardness was 118–210 HV. Since no attempt has been made to harden the steel by quenching from a suitable temperature (about 850°C), it may be presumed that the carburization was unintentional.

However, an examination of many edge tools from a wide range of Romano-British sites shows that on most of them many of the edges have been carburized and not hardened. For example, the result of an examination of 17 edge tools from Gestingthorpe, Essex, is given in Table 87. Seven of these had a high enough carbon content to have benefited from heat-treatment, i.e. quench-hardening. Only one of these

Table 87 Structure and hardness of some edge tools from Gestingthorpe

Object	Structure	% Carbon	Hardness HV
Knife	Ferrite + phos.	nil	164
Knife	Ferrite + phos.	nil	153
Knife	Ferrite + phos.	nil	205
Knife	Ferrite + spher. pearlite	0·2	240
Knife	Ferrite + pearlite	0–0·4	(0·4% C) 168
Knife	Ferrite + spher. pearlite	0·15	153
Knife	Type A: Core Ferrite + pearlite	0·5–0·8	256
Knife	Ferrite + pearlite	0·6	223
Knife	Ferrite + pearlite	0·1	122
Knife	Ferrite + slag	nil	159
Set or chisel	Ferrite + martensite	0–0·6	440
Smith's chisel	Ferrite + pearlite	0·3–0·8	240 (CW)
Tanged chisel	Ferrite	nil	124
Smith's punch (side)	Sorbite	0·6	193
Smith's punch (side)	Ferrite + pearlite	0·1–0·2	290 (CW)
Paring chisel (edge)	Pearlite + ferrite	0·6	250

CW = Cold worked; phos = phosphorus; spher. = spheroidized

Table 88 Comparison of metallurgical levels achieved on various Romano-British sites

Site		Numbers of artifacts	
	Total	Carburized	Hardened
Gestingthorpe	31	11	2
Thistleton R.V.	13	3	0
Ware	12	5	2
Catsgore	11	6	1
Colchester	12	2	0
Brancaster	8	5	0
Wanborough	25	19	8

had been quench-hardened to give a hardness of 440 HV.

As we have seen, the presence of considerable phosphorus and work-hardening will give hardnesses up to 300 HV. It is clear that two types of metal were used in Roman tools: moderate- to high-phosphorus iron such as that from the Weald, and carburized iron such as that from the Forest of Dean. Both will give hardnesses in the range 150–250 HV, as forged.

While most sites conform to the situation at Gestingthorpe (Table 87), the site at Wanborough, Wilts., shows a slightly different picture.[93] Here, out of 25 edge tools examined, 19 had been carburized and of these 8 had been quench-hardened (Table 88).

It appears that the general level of smithing in the Roman period was low and that quench-hardening was not widely practised. But occasionally, as at Wanborough, a better technique prevailed (Table 89). It seems that here many tools were made by the local smith(s) who had an above-average standard of workmanship.

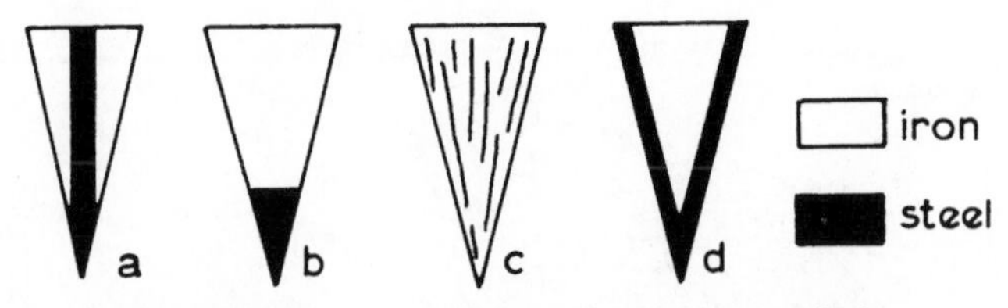

116 Types of blades in use from the Roman period onwards, showing methods of combining iron and steel

The 'steeling' of edge tools

The chisel from Chesterholm is an example of a method of local hardening by surface carburization. The principle of welding together, or piling, surface-carburized iron strips is another method of obtaining a steel. As far as the British Isles are concerned, these two methods were the only ones practised until well into medieval times.

By the Roman period we have many examples of the welding-on of steel to iron to make cutting edges for knives and tools. There were four methods in use for knives (Fig. 116):

(*a*) By having a layer of steel covered with two plates of iron in such a way that the steel projects at the thinned cutting edge. Sharpening such a knife will always give a steel edge to the blade.

(*b*) Welding-on a steel strip to the edge of a piece of iron, sometimes made by piling. The steel edge will wear away with sharpening until eventually there is no steel left and the knife becomes useless and is discarded.

(*c*) Using a piece of piled material which consists of

Table 89 Hardness and structure of some edge tools from Wanborough, Wilts.

No.	Object	Edge Structure	HV (max)	Back (matrix) Structure	HV	Comments
790281	Axe	P + F	269	F	156	Not hardened
790239	Cold chisel	F	153			
692356	Cold chisel	F	140			
771779A	Cold chisel	TM	800	F	182	Hardened
684280	Mortise chisel	F	129	F	224	Cold-worked head
800052	Socketed chisel	TM	414	F	210	Hardened
780169	Saw	F + P	245	F		Not hardened
435242	Saw	TM	470			Hardened
790274	Saw	F	185			
692337	Sickle	F + P	256			Not hardened
800060	Sickle	F + P	171			Not hardened
790723	Chopper	F + P	251			Not hardened
780856	Cleaver	F + P	158			Not hardened
707253	Knife blade	TM	720	F	148	Hardened
780163	Knife	F + P	240			Not hardened
684039	Knife	P + M	269			Hardened
790725	Knife	F + P	175			Not hardened
790296	Knife	F + P	185			Not hardened
790268	Knife	P + F	240			Not hardened
790726	Knife	F	133			
771180	Knife	TM	369			Hardened
771769	Spearhead	F + P	168			Not hardened
790277	Spearhead	F	105			
790279	Arrowhead	TM	390			Hardened
790887	File	TM	535	F + M	275	Hardened

F = Ferrite; P = Pearlite; TM = Tempered martensite; M = Martensite

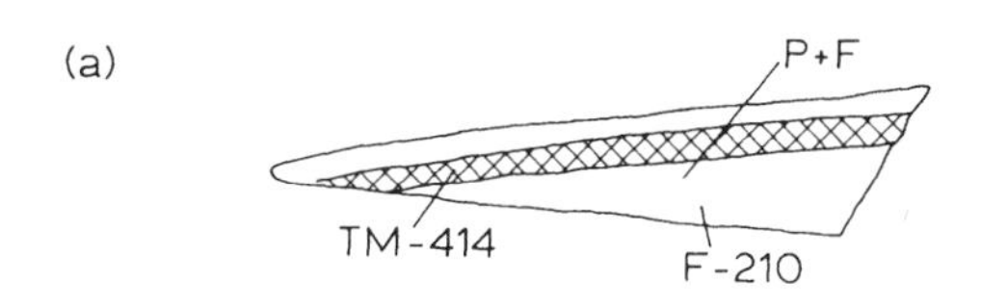

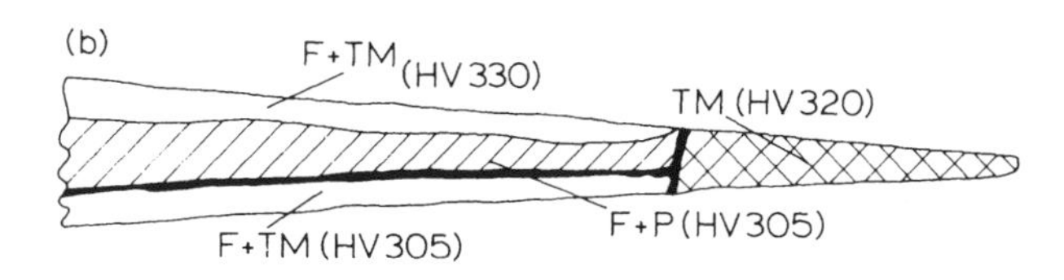

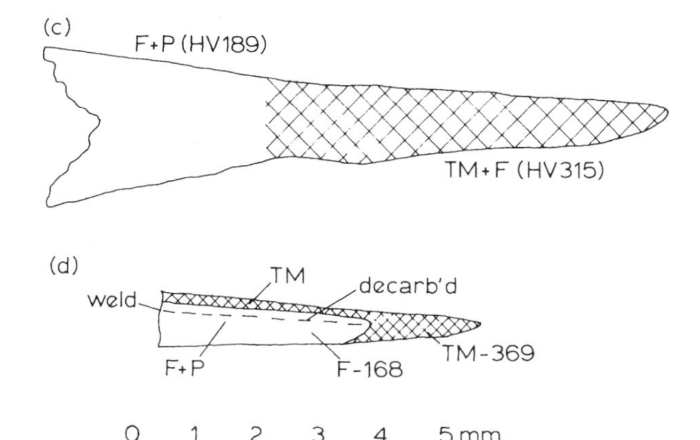

a Blade of a socketed chisel from Wanborough with edge hardened to 414 HV
b Edge of a gouge from Ware, Herts. The cutting edge has been welded-on to a piece of iron or steel, itself decarburized or carburized and welded as seen by the heavy dark lines
c Knife edge from Ware, Herts., made of a homogenous piece of steel; no weld
d Knife edge from Wanborough in which the edge has been steeled by wrapping a thin piece of steel sheet round the iron core. The bottom surface has been ground away

117 Sections through the blades of edge tools from Romano-British sites to illustrate the four basic types of steel-iron weld used

alternate layers of iron and steel. Given sufficient heat after welding, these layers may be homogenized so that the blade consists of a piece of homogenous steel.

(*d*) By having an iron core around which a piece of steel has been wrapped (i.e. the reverse of type *a*). This seems comparatively rare perhaps because it uses more steel than *a* in most cases and, more importantly, the lasting qualities of the tool are not much better than in the case of *b*.

Examples of these four techniques taken from several Roman sites are shown diagrammatically in Fig. 117.

These techniques were more or less international and knowledge of them must have spread widely during the Roman period, as we see evidence for all of them throughout Europe. Type *a* has a number of variations. If we call that shown in Fig. 116*a* the 'British' sandwich, we have examples of the open 'Danish' sandwich and some multi-decker sandwiches. But in all cases the welding was done late in the production of the tool and no attempt at homogenization was made.

Heat-treatment has been carried out in all cases on the tools shown in Fig. 117 giving hardnesses of 315–414 HV. It would appear that where the quenching temperature was adequate (which was not the case in *c*) tempering was allowed to occur by heat flow, back from the body of the tool. This resulted in hardnesses a good deal less than the 450–660 HV that we would expect today.

THE COMPOSITION OF SLAGS AND RESIDUES

Many more analyses have been carried out on slags than on metal. One reason for this is that slags survive in good condition with less chemical alteration than the metal itself. We do know, however, that some slags have weathered to some extent and show signs of hydrated iron oxides which were not there originally. Also, as mentioned elsewhere, we must be on our guard for the possibility of alkalis being leached out. Whereas this has little importance chemically, it will have the effect of raising the free-running temperature.

In Table 79, which gives the composition of tap slags and furnace slags of Roman date, only one determination of the alkali content has been made – in the slag found in the beam from Corbridge.[45] In this case it is unlikely that the alkali has been leached out as the slag was well surrounded by iron. The figure given, 0·278%, therefore seems a valid one.

The composition of Roman slags is very much the same as those from pre-Roman times (Table 76, Chapter 6) and medieval times (Table 91, Chapter 8). There are some conclusions we can reach, however. First, the lime content is uniformly low, in no case exceeding 3·7%. Since such low values could easily come from the ore we see that no attempt has been made to add additional lime as a 'flux'.

The phosphorus pentoxide content of the slags from undisputedly Roman sites, with two exceptions, is within the limits 0·1–0·9%. The exceptions (1·72 and 2·27%) come from Templebrough[42] and Ashwicken.[4] The latter seems to have used nodular ore exclusively and the former at least partly. The available evidence does not allow us to state that by the Roman period ores lower in phosphorus were being used, since we have too little pre-Roman material. However, the evidence of the iron produced does show that the ores themselves contained less phosphorus. We can therefore conclude that low-phosphorus ores have been sought and used or that some change in the process has altered the phosphorus partition coefficient (the ratio of phosphorus in the slag to that in the metal). At Camerton[94] metal and slag were analysed and we appear to have a metal/slag partition coefficient of the order of $\frac{0\cdot05}{0\cdot20} = c.\ 0\cdot25$ or $\frac{1}{4}$. Taking the further example of the Corbridge beam,[45] we find the coefficient is $\frac{0\cdot04}{0\cdot078} = c.\ 0\cdot5$. Therefore, with this and other evidence we can expect to find that the phosphorus content of the metal is about ½ to ¼ that of the slag.

It would appear that where there was no choice of ore, as at Ashwicken, high phosphorus ores were smelted, resulting in metal with high phosphorus content. Where there was a choice, as in most parts of the

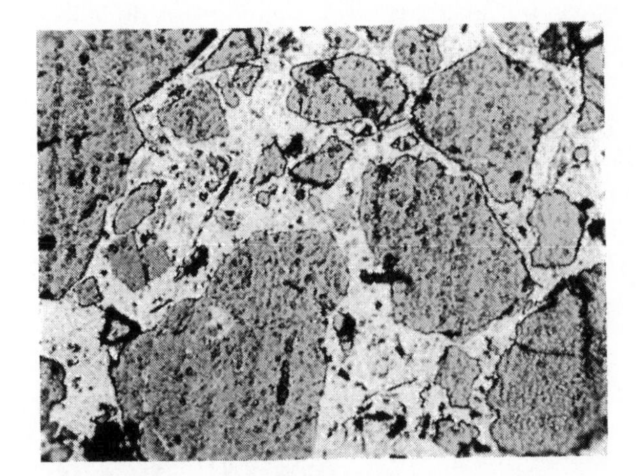

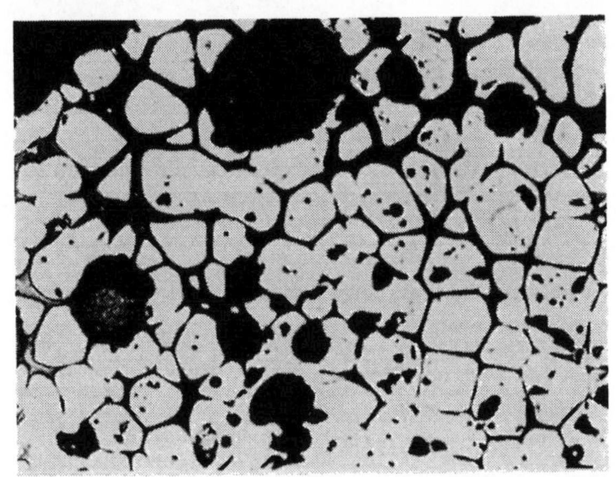

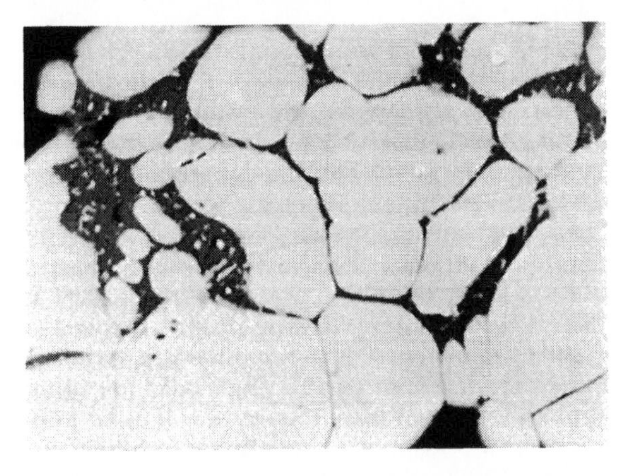

118 Hammer scales; *a* From Hucthoe, Northumberland: dark, hematite (Fe_2O_3); light, fayalite; *b* From Ashwicken, Norfolk: light, magnetite (Fe_3O_4); dark, fayalite; *c* Scale from modern wrought iron for comparison: light, wüstite (FeO); dark, fayalite

country, Roman period smelters seem to have been more discriminating than earlier people and to have selected the better ores. In all cases the true tap slags have high iron oxide contents of the order of 70%, and low silica contents of 10–20%. In some cases a good deal of lining has gone into the slag, and this tends to give an abnormally high silica value. The furnace slags should have undergone little change and therefore be very close in composition to that of the roasted ores. The presence of Fe_2O_3 shows that the slag contains a proportion of magnetic oxide of iron. This is a common characteristic of tap slags and the proportion of Fe_2O_3 varies between 0 and 25%. The main reason for this is oxidation of the wüstite in the slag by the air as it cools, showing that the sample included the surface layer.

Table 90 Composition of hammer scale from Roman sites

Ref. %	61 Benwell (Newcastle) 2nd cent.	4 Ashwicken (Norfolk) 2nd cent.	97 Chichester (Sussex)
Fe	7·7	—	0·11
Fe_3O_4	74·1	85·8	14·32
Fe_2O_3	8·5	—	48·48
SiO_2	1·45	9·94	22·68
CaO	1·80	1·20	0·31
Al_2O_3	6·47	0·90	3·18
MgO	—	0·30	0·19
P_2O_5	—	0·73	0·36

In the case of furnace slags, the amount of Fe_2O_3 and FeO may depend on the state of reduction reached. Slags which have spent a long time in the furnace under highly reducing conditions will have lower quantities of Fe_2O_3 and a large amount of FeO.

The other product which sometimes resembles slag is hammer scale. This material forms during heating under oxidizing conditions before and during the forging process. It consists, while adherent to the metal, of ferrous oxide (FeO) and magnetite (Fe_3O_4) in varying proportions. On removal from the metal during hammering or on quenching in water it may oxidize to a higher state, i.e. FeO may go to magnetite, and magnetite to ferric oxide (Fe_2O_3). Furthermore, changes may take place in these thin scales after deposition in the ground. These scales have a well-marked granular structure by which they can be identified, as it is quite distinct from forge cinder or smelting slag.

The photographs in Fig. 118 illustrate two examples that have come to the author's notice: (*b*) is a magnetic hammer scale from the 2nd century site at Ashwicken. It contains 85·8% Fe_3O_4, and the structure consists essentially of grains of magnetite cemented together with fayalite (Table 90). This was found loose in pockets in the ground and it is clear that it has not changed very much from the state in which it was deposited.

Sweepings from the floor of the forge at Benwell Fort, Newcastle, were examined and found to be mostly granular but uncemented pieces of magnetite (Table 90). A small amount was cemented and appeared very similar to that from Ashwicken. The silica in the fayalite cementing films in this material comes from the slag in the metal and possibly from sand used as flux in the smithing operation. Fig 118*c* shows the scale from a piece of modern wrought iron heated to 1 050°C for four hours. The structure is wüstite, approximating to FeO and fayalite in composition.

SUMMARY

The Roman period in England is characterized by large-scale smelting installations. The furnaces used were of two varieties: (*a*) shaft furnaces, and (*b*) developed bowl furnaces, much enlarged in comparison with the previous period and producing large quantities of tap slag and furnace slag. The bloom size has increased to something of the order of 7–9 kg. The ores seemed to have had lower phosphorus contents than those previously used, showing a discrimination which probably arose from the increased use of controlled carburization to obtain higher strength.

Steel was made by the selection of high-carbon parts of blooms or by carburizing wrought iron with low phosphorus content. Pieces of the latter were welded to common wrought iron, often with high phosphorus, to form the cutting edges of edge tools.

In the main, however, the majority of metallic artifacts consist of high- and low-phosphorus wrought iron without any attempt at carburization. The few treated in this way show some knowledge of the processes of carburization and quenching to obtain the maximum hardness, but a certain lack of control of the process is evident and quench-hardening is by no means common. The technique of pattern-welding, developing from the piled structure of the pre-Roman Iron Age, is now in existence, although not widely used. The few pieces of cast iron appear to be the result of accidents in smithing or smelting.

NOTES AND REFERENCES

1 R. H. FORSTER and W. H. KNOWLES (with contributions by F. Haverfield and P. Newbold): *Archaeol. Aeliana.*, 3rd ser., 1913, **9**, 230–80.
2 I. A. RICHMOND and E. BIRLEY: *Archaeol. Aeliana.*, 4th ser., 1940, **17**, 86–115.
3 D. J. SMITH: private communication.
4 R. F. TYLECOTE and E. OWLES: *Norf. Archaeol.*, 1960, **32** (3), 142–62.
5 R. E. M. and T. V. WHEELER: *Soc. Antiq. Lond.*, Report No. 9, 1939.
6 J. YATES: *Proc. Som. Archaeol. Soc.*, 1858, **8**, 1–34.
7 R. F. TYLECOTE: *Norf. Archaeol.*, 1967, **34** (2), 187–214.
8 G. SWIFT: *Wealden Iron*, 2nd ser., 1982 (2), 15–21.
9 A. R. AIANO: *J. Hist. Metall. Soc.*, 1977, **11** (2), 73–82.
10 G. F. DAKIN: *Bull. Hist. Metall. Group*, 1969, **2** (2), 66–7.
11 P. CORDER (ed.): *J. Rom. Stud.*, 1957, **47**, 212 (details kindly provided by J. P. Gillam).
12 H. CLEERE: *Sussex Archaeol. Soc.*, Occas. Paper No. 1, 1970, 23 pp.
13 H. CLEERE: *Britannia*, 1971, **2**, 203–17.
14 D. A. JACKSON *et al.*: *Britannia*, 1978, **9**, 115–242.
15 A. MCWHIRR *et al.*: 'Cirencester Excavations, II'; 1982, Cirencester, Cirencester Excavation Committee.
16 T. MAY: *Iron Coal Trades Rev.*, 1905 (Aug. 11), **71**, 427.
17 D. A. JACKSON: *Britannia*, 1973, **4**, 128–40.
18 R. F. TYLECOTE: *Bull. Hist. Metall. Group*, 1970, **4** (1), 24–7.
18*a* I. M. SMITH: *J. Rom. Stud.*, 1962, **52**, 173.
19 J. GIBSON-HILL: *J. Hist. Metall. Soc.*, 1980, **14** (1), 21–7.
20 R. PLEINER: *Tech. Civilis.*, 1956, **5** (4), 113–28.
21 W. GOWLAND: *Archaeologia*, 1901, **57** (2), 359–422.
22 E. T. ARTIS: 'The *Durobrivae* of Antoninus', 1828, London.
23 J. H. MONEY: *J. Hist. Metall. Soc.*, 1974, **8** (1), 1–19.
24 C. F. TEBBUTT: *Sussex Archaeol. Coll.*, 1980, **117**, 47–56.
25 C. F. TEBBUTT and H. CLEERE: *Sussex Archaeol. Coll.*, 1973, **111**, 27–40.
26 W. O. STANLEY: *Archaeol. J.*, 1870, **27**, 147–64.
27 J. C. LLOYD and H. N. SAVORY: *Brycheiniog*, 1958, **4**, 53–71.
28 O. DAVIES: *Archaeol. Cambrensis*, 1948, **100**, 90–3.
29 V. B. PROUDFOOT: *Ulster J. Archaeol.*, 1953, **16**, 41–57.
30 A. J. B. WACE and PROF. JEHU: *Proc. Soc. Antiq. Scotl.*, 1914–15, **49**, 233–55.
31 S. M. CREGEEN: *Yorks. Archaeol. J.*, 1956, **39**, 32–47.
32 V. E. NASH-WILLIAMS: *Archaeol. Cambrensis*, 1952–5, **102**, 89–163.
33 F. OSWALD: *Trans. Thoroton Soc.*, 1928, **31**, 66.
34 E. N. BAYNES: *Archaeol. Cambrensis*, 1908, **8**, 183–210.
35 E. N. BAYNES: *Archaeol. Cambrensis*, 1920, **20**, 91–8.
36 R. F. TYLECOTE: *J. Iron Steel Inst.*, 1959, **192**, 26–34.
37 H. S. COWPER: *Trans. CWAAS*, 1902, **2**, 31–7.
38 S. SHARP: *Archaeologia*, 1871, **43**, 118–30.
39 J. EVANS: *Archaeol. J.*, 1878, **35**, 263–74.
40 M. V. TAYLOR and R. G. COLLINGWOOD: *J. Rom. Stud.*, 1926, **16**, 223.
41 D. A. JACKSON: *Northampton Archaeol.*, No. 14, 1979, 31–7.
42 T. MAY: 'The Roman forts at Templebrough, near Rotherham'; 1922, Rotherham.
43 A. COOPER KEY: *The Mining J.*, 1896 (June 6), 734.
44 H. DUDLEY: 'Early days in north-west Lincolnshire'; 1949, Scunthorpe.
45 H. BELL: *J. Iron Steel Inst.*, 1912, **85**, 118–28.
46 C. BUCKMAN and R. W. HALL: 'Notes on the Roman Villa at Chedworth, Cirencester', 1872.
47 J. H. WRIGHT: *Bull. Hist. Metall. Group*, 1972, **6**, 24–7.
48 K. KENYON: *Archaeol. J.*, 1953, **110**, 1–87.
49 M. U. JONES: Records of Bucks., 1957–8; **16** (3), 198–215.
50 W. H. MANNING: 'Catalogue of Romano-British ironwork in the Museum of Antiquities, Newcastle upon Tyne'; 1977, Dept. Archaeol. Newcastle.
51 W. J. FIELDHOUSE *et al.*: 'A Romano-British industrial settlement near Tiddington, Stratford-on-Avon'; 1931, Birmingham.
52 J. DOOL: 'Excavations on the racecourse at Derby' (forthcoming).
53 C. PARTRIDGE: 'Excavation of Allen and Hanbury's site at Ware, Herts. (with a note on metallurgical material by R. F. Tylecote)' (forthcoming).
54 G. WEBSTER: *Antiq. J.*, 1955, **35**, 199–216.
55 G. D. B. JONES and S. GREALEY (eds): 'Roman Manchester'; 1974, Manchester.
56 I. C. HANNAH: *Antiq. J.*, 1932, **12**, 262–8.
57 'The Colsterworth iron furnace', Grantham Public Library and Museum, 10th annual report, 1931–2; 15.
58 FR. F. BUCHANAN: 'A journey from Madras . . .', 3 Vols.; 1807, London.
59 H. WILLIAMS: *Archaeol. Cambrensis*, 1923, **78**, 87–113; 291–302.

60 L. C. HAYWARD: (personal communication 25 Feb. 1970), J.R.S. 1963, **53**.
61 F. G. SIMPSON and I. A. RICHMOND: *Archaeol. Aeliana*, 1941 (4th ser.), **19**, 1–43.
62 R. H. FORSTER and W. H. KNOWLES: *Archaeol. Aeliana*, 1910 (3rd ser.), **6**, 205–72.
63 L. JACOBI: 'Das Römerkastell Saalburg, Homburg v.d.Höhe', 237; 1897.
64 H. G. GRAVES: *J. Iron Steel Inst.*, 1912, **85** (1), 187.
65 A. MAU and F. W. KELSEY: 'Pompeii: Its Life and Art'; 1904, London. (The beam was examined and measured by C. M. Daniels to whom I am much indebted.)
66 J. S. WACHER: *Britannia*, 1971, **2**, 200–2.
67 R. F. TYLECOTE: *Trans. Woolhope Nat. Field Club*, 1961, **37**, 56–61.
68 G. BEHRENS and E. BRENNER: *Mainzer Zeitschrift*, 1911, **6**, 114.
69 H. H. COGHLAN: 'Notes on prehistoric iron', 121.
70 J. A. SMYTHE: *Trans. Newcomen Soc.*, 1936–7, **17**, 197–203.
71 G. T. BROWN: *J. Iron Steel Inst.*, 1964, **202** (6), 502–4.
72 These were excavated by Prof. J. J. Wilkes of the Institute of Archaeology, London. I am indebted to Prof. Wilkes for permission to examine the bars and the soil in which they were found.
73 H. E. O'NEIL and G. T. BROWN: *Bull. Hist. Metall. Group*, 1966, **1** (7), 30–4.
74 R. F. TYLECOTE: in J. Draper: 'Excavations on the Roman site at Hill Farm, Gestingthorpe' (forthcoming).
75 D. T-D. CLARKE and M. R. DAVIES: 'Roman Colchester 1980' (2 ed.), 10; *Britannia*, 1979.
76 Excavations at Battle Abbey; report forthcoming.
77 N. S. ANGUS *et al.*: *J. Iron Steel Inst.*, 1962, **200**, 956–68.
78 A. MCWHIRR: (*see* Ref. 15).
79 R. GOODBURN: 'Winterton Roman villa' (forthcoming).
80 J. N. FRIEND and W. E. THORNEYCROFT: *J. Iron Steel Inst.*, 1925, **112**, 225–37.
81 J. P. CHILTON and U. R. EVANS: *J. Iron Steel Inst.*, 1955, **183**, 113–22.
82 R. F. TYLECOTE and R. THOMSEN: *Archaeometry*, 1973, **15**, 193–8.
83 A. R. WILLIAMS: *J. Archaeol. Sci.*, 1977, **4**, 77–87.
84 B. GILMOUR: personal communication.
85 I. A. RICHMOND: 'The Roman fort at South Shields; A Guide', 14 pp (no date).
86 E. SCHÜRMANN and H. SCHROER: *Archaeol. Eisenhüttenwesen*, 1959, **30**, 127–30.
87 E. SCHÜRMANN: *Archaeol. Eisenhüttenwesen*, 1959, **30** (March), 121–6.
88 J. W. ANSTEE and L. BIEK: *Medieval Archaeol.*, 1961, **5**, 71–93.
89 E. M. JOPE: *Antiq. J.*, 1946, **26**, 70–2.
90 K. J. BARTON: *Discovery*, 1960, **21**, 252.
91 C. E. PEARSON and J. A. SMYTHE: *Proc. Univ. Durham Philos. Soc.*, 1938, **9** (3), 141–5.
92 H. H. COGHLAN: 'Prehistoric iron', 189.
93 R. F. TYLECOTE: 'Metallurgical examination of iron work from Wanborough, Wilts.' (forthcoming).
94 W. J. WEDLAKE: 'Excavations at Camerton, Somerset'; 1958, Camerton Excavation Club.
95 R. F. TYLECOTE *et al.*: *J. Iron Steel Inst.*, 1971, **209**, 342–64.
96 G. H. JACK: 'Excavations on the site of Ariconium'; 1924, Hereford, Woolhope Nat. Field Club.
97 ALEC DOWN: 'Chichester excavations', Vol. I; 1974, Chichester. (H. Cleere: 'Notes on iron-making slags from Chichester, Sussex', unpublished.)
98 Bayham Abbey beam (report forthcoming).
99 Leicester. Jewry Wall beam (report forthcoming).
100 St. Albans beam (report forthcoming).
101 H. CLEERE: *Bull. Inst. Archaeol. Univ. Lond.*, 1958, (1), 55–74.
102 A. MCWHIRR: personal communication.

8 Iron in the early medieval period

This period may be defined as that started by Migration peoples from the East after the end of Roman rule and continuing up to the 15th century, when water power became widely used for mechanical hammers and the scale of operations increased greatly.

It is difficult to avoid comparing the level of technique used at the start of this period with that of the Roman period. This comparison may be justified, since it is almost impossible to imagine that the lessons taught during the Roman period could have failed to have a lasting effect. However, there is no sign in the period following the withdrawal of Roman power of any metallurgical techniques taught, although there is plenty of evidence for the carrying on of a pre-Roman Iron Age tradition of iron production. There are two possible explanations for this. One is that the knowledge of a more advanced smelting technique was not widely disseminated, and that when the Romans left, the old techniques were re-introduced by the Migration peoples. The other is that economic conditions did not favour the large-scale operations of the Roman period, since the military need no longer existed.

However, as far as the manipulation and heat treatment of iron and steel is concerned, the incoming peoples brought with them a vastly improved level of technique, which was widely disseminated.

ORE-PIT MINING

In this period there are the first signs of mining activity on a small scale with the aid of pits. These are not to be confused with the later, medieval bell-pits, which increase in diameter with depth but are roughly hemispherical with a depth of no more than 2–2·7 m. Unfortunately the pits found at West Runton, Norfolk, while associated with a Saxo-Norman smelting site 1 km away, were not securely dated.[1] However, although no pottery was found in the pits, there is little doubt that their use was associated with the smelting activities of the period AD 850–1150 (Thetford ware).

Ore-pits have also been excavated in the Weald[2] and found to be of similar shape and 2·5 m deep. Again no pottery was found, but some wood found in the pit at 2·3 m depth suggested that these pits had not been dug before AD 1600. Outlines of other pits could be seen in the sides of the brickworks at Sharpthorne, East Grinstead, in 1983. These have given a ^{14}C date of 1200 ± 50 AD.

Estimates were made of the yield of the West Runton pits. Each pit would have contained about 270 kg of ironstone nodules and required the excavation of 8 m of sand,[1] representing a yield of about 3%. This is similar to modern ironstone techniques where it was necessary to remove a considerable overburden, as in Northamptonshire.

ROASTING

There are now two types of ore-roasting hearth. The first is very similar to the Roman type found at Great Casterton[95] (*see* Fig. 94 Chapter 7); the second is a stone-built, horseshoe-shaped enclosure similar to those depicted by Agricola in 1540. Examples of the first type were found at Stamford,[3] Lyveden[4] and West Runton.[1] They are generally more shallow than the Roman examples and extend roughly 2–3 m. The linings consist of reddened clay impregnated with fine particles of ore and charcoal. Normally raw wood would be used as fuel and the temperature would not need to rise much above 600°C. They are usually situated close to the furnaces; those at West Runton and Stamford were only 2 m apart, but the furnace and roasting hearth at Stamford were not quite contemporary.

The best example of the horseshoe roaster is that from Minepit Wood, Rotherfield[5] (Fig. 119). It has been built upon the remains of the old furnace, which was saucer-shaped. This type was widely used for roasting copper ores in the 16th century, and is now known as a 'stall' roaster.

The ore would tend to get broken up during roasting if it was of the carbonate or limonite type, owing to the evolution of carbon dioxide and water. If it were hematite, then it might need to be thermally shattered by quenching in water, as in fire-setting. In either case there would be a need for final sizing by hammering on an anvil stone to give a consistent size, which would depend on the reducibility (porosity). A suitable size range would be 2–3 mm, and the fines would be discarded.

SMELTING FURNACES

The earliest comprehensive smelting site in this period was at Ramsbury,[6] Wilts., on the River Kennet. This was the seat of a bishop in the early 10th century and was in close contact with a number of ore sources in the Reading and Lower Bagshot beds, such as the one 5 km south in Savernake Forest. This interesting site showed a number of phases suggesting a continuous

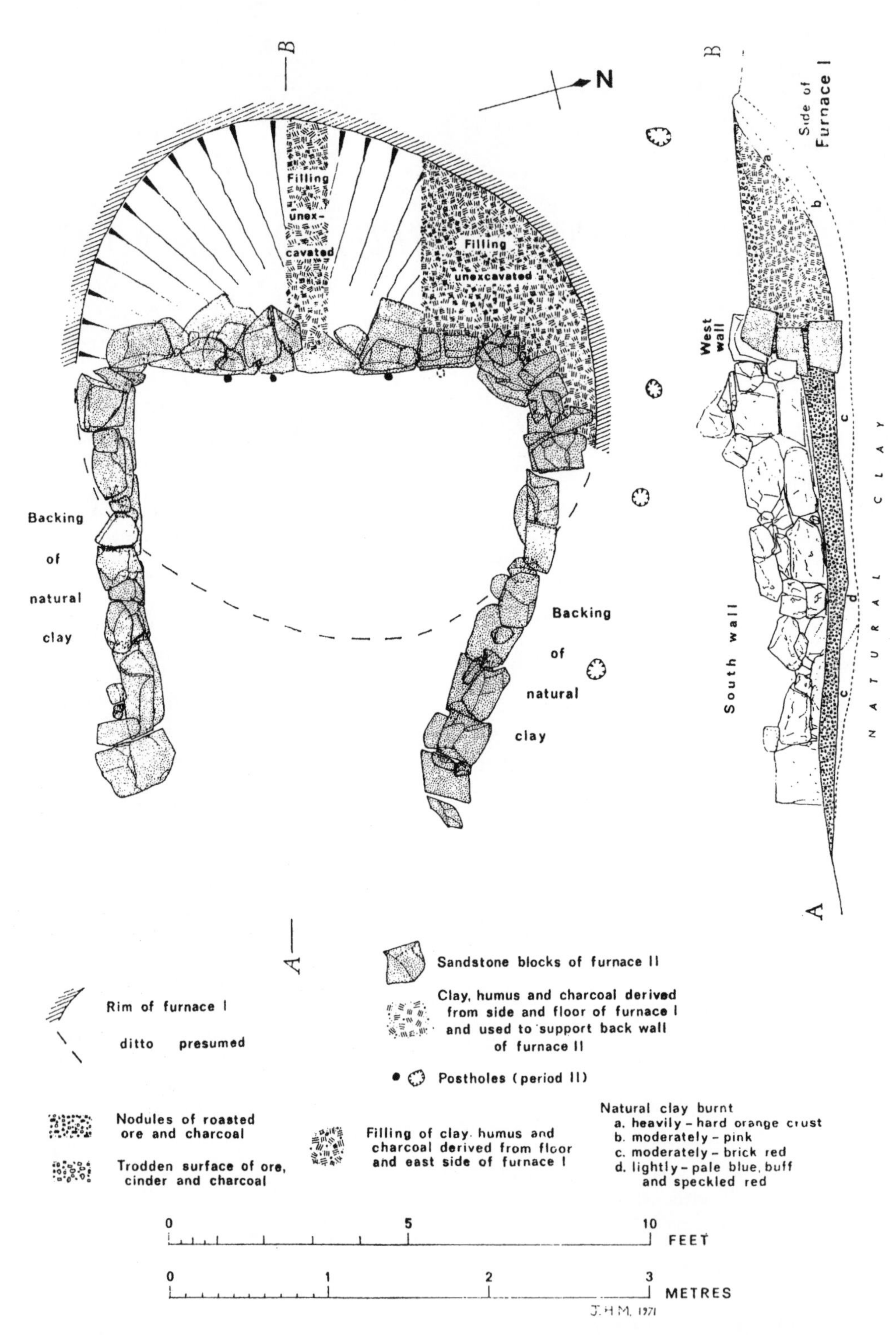

119 Stall roaster from Minepit Wood, Rotherfield (after J. H. Money;[5] courtesy Society of Medieval Archaeology)

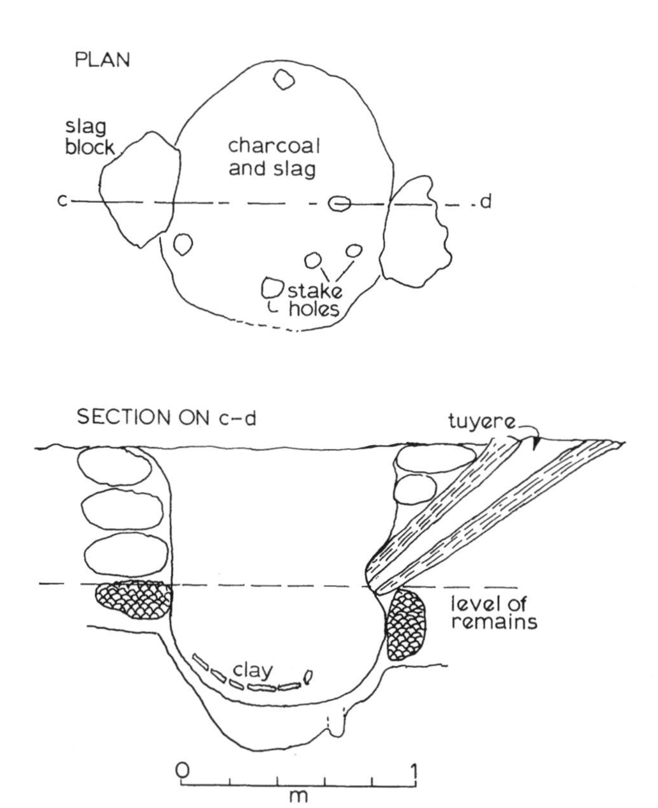

120 Reconstruction of iron-smelting bowl furnace from Ramsbury, Wilts., late Saxon period (based on J. Haslam[6])

development from a simple non-slag-tapping 'bowl' furnace to a developed slag-tapping furnace with unusual characteristics. If any one site shows that the new Migration people started with primitive types of furnace and developed independently, then this is it.

The ^{14}C dating covers the range AD 660–905 (six

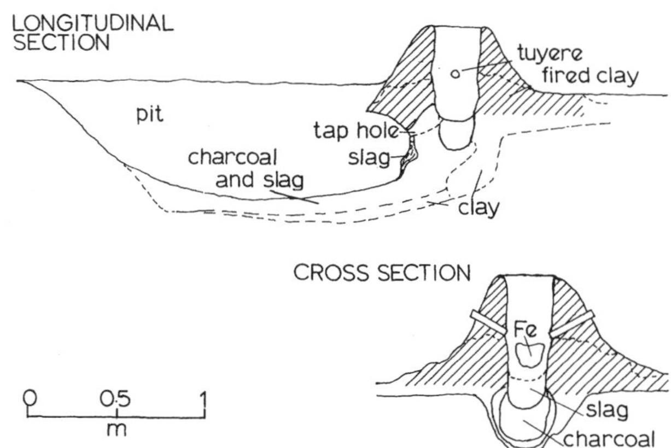

122 Reconstruction of developed bowl furnace from Ramsbury

dates). Taking all six dates as a group we get AD 820 ± 45, which agrees with the datable artifacts. The most primitive furnace is that shown in Fig. 120; from its size, this must have been an example of the manipulated bowl type. The furnace as excavated was 75 cm in diameter and contained a slag furnace bottom lying on charcoal. It had been constructed with the aid of a ring of stakes and had a refractory sand/clay lining (Fig. 121). In many ways this bears comparison with Chelms Combe and Round Pound, Dartmoor, both of the pre-Roman Iron Age (p. 133). There were three examples of this type, two with stake-holes in place, indicating an original diameter of 65–90 cm.

The fourth furnace had provision for slag-tapping (Figs. 121–122). The furnace shaft had been reduced to 25 cm in diameter, in keeping with the tendency seen at Broadfield and other sites. The taphole was quite small – perhaps only 15 cm – and remains of the last tap were still *in situ*. The taphole was too small for bloom

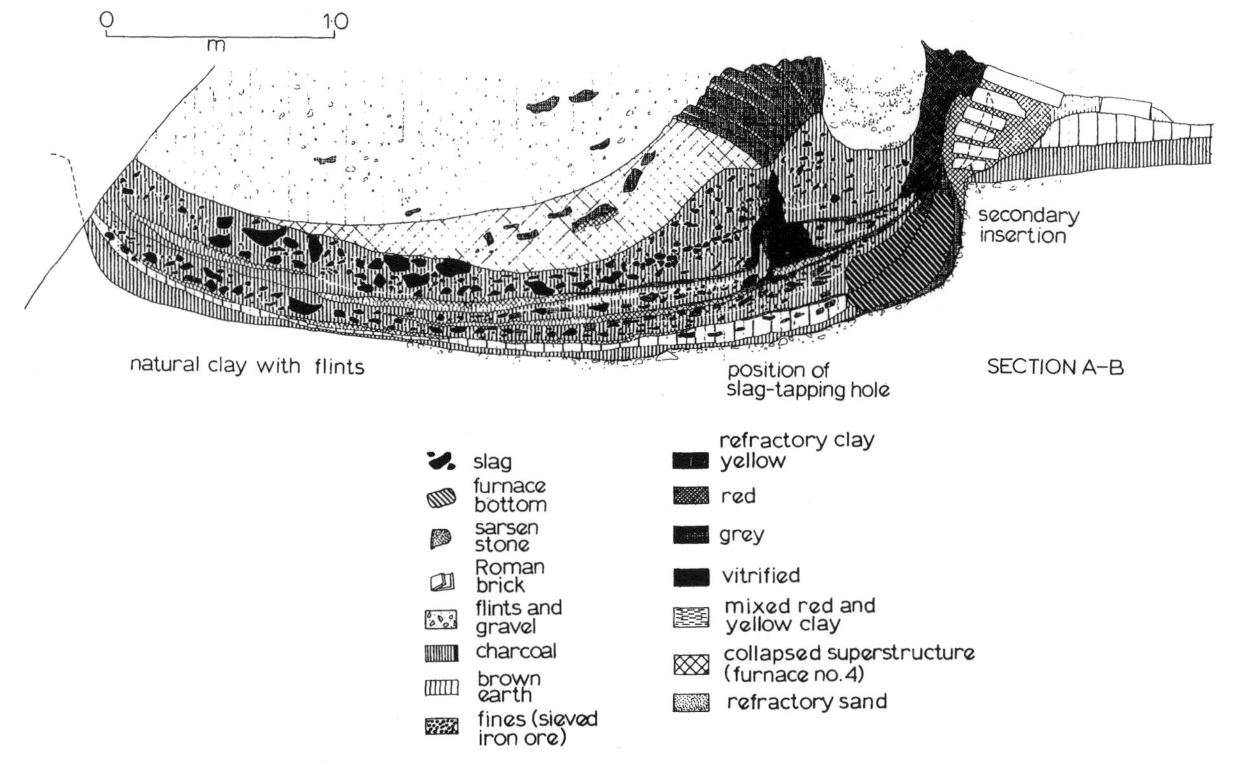

121 Developed bowl furnace from Ramsbury; longitudinal section (after J. Haslam;[6] courtesy Society of Medieval Archaeology)

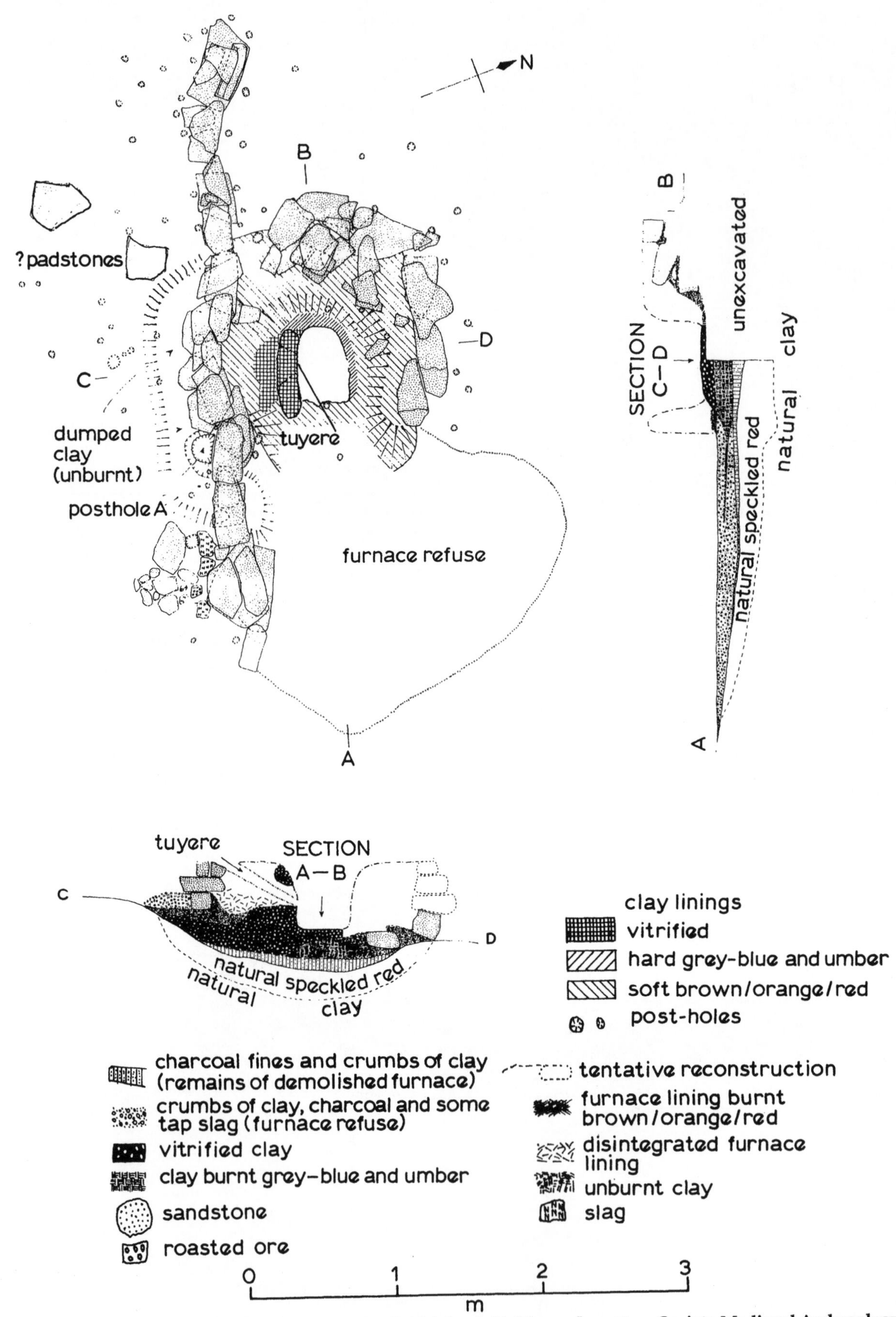

123 Plan and sections through bloomery at Rotherfield (after J. H. Money;[5] courtesy Society Medieval Archaeology)

124 Reconstruction of bloomery site at Rotherfield (after J. H. Money;[5] courtesy Society of Medieval Archaeology)

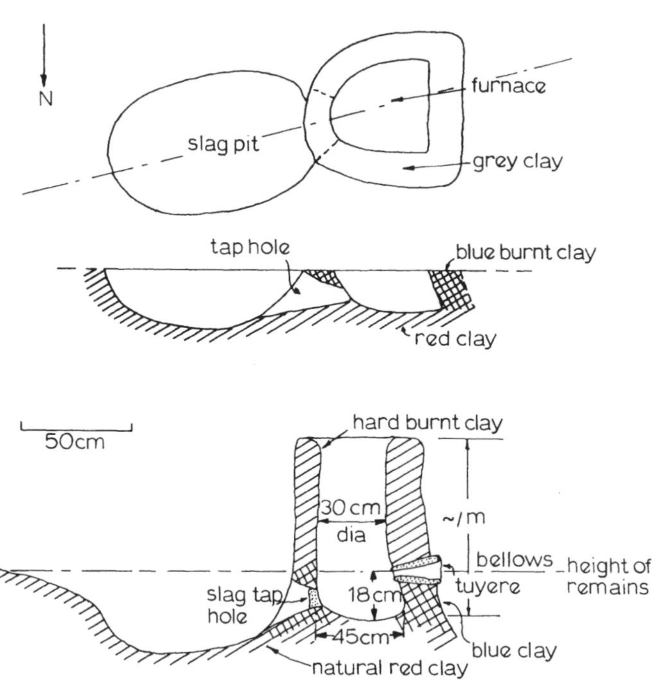

125 Reconstruction of slag-tapping, low-shaft furnace from Stamford, Lincs. (after Tylecote[10])

removal. The structure of the furnace shows progressive reduction of its cubic capacity by the addition of 13 'linings', and was clearly far removed from its original design, ending up more like a shaft furnace. Even so, during the final phases of operation the bloom must have been removed through the top, which would have restricted the height of the shaft to not much above the present height. Again, there were no signs of tuyeres, and it would seem that these were placed at the sides as shown in Fig. 122. In this case there would have to be much more shaft to hold the reducing charge, as shown. This would have to be at least 25 cm high and the working bottom would be a similar distance below the tuyeres.

This site shows a unique example of the thinking behind the development from the 'bowl' to the shaft, which is lacking on all the pre-Roman sites. Another, slightly different, solution to the problem may be seen in the domed furnace at High Bishopley, Durham[7] (*see* Fig. 128 *below*). In both cases the insulation inherent in the bowl is to a large extent preserved, but slag-tapping and therefore the production of a more metallic bloom is attained. It is doubtful if the bloom size exceeded a few kg.

Another 9th century Saxon bowl furnace, far less ambitious, has been found at Millbrook in Ashdown Forest in the Weald.[8] This had a clay-with-flint lining and a stake-hole ring diameter of 1 m. There was no provision for slag-tapping, but there were a number of small hearths around, which had been used for smithing.

One of the best post-conquest smelting sites is at Rotherfield, Sussex.[5] Here we have a complete and undisturbed site dated to the 14th–15th century, typical of the manually-powered sites such as Tudeley near Tonbridge, Kent, which is so far unidentified.[9] We have already discussed the roasting furnaces lying outside the forge building, but inside a stone-and-timber building was a bloomery furnace (Fig. 123). It would seem that the furnace was roofed but open all round (Fig. 124). The furnace was set beside a low wall which acted as a partition to protect the bellows from the dust and heat of the smelting furnace, and through which the tuyere was inserted (Fig. 123). Unusually, this tuyere terminated through a hole in a block of slag. The bowl hearth, probably of Catalan type, was tapped at right angles to the tuyere. No smithing furnaces were found, but it is possible that working-up was done in the smelting furnace. The furnace had a D shape in plan and in this respect resembles the Saxo-Norman furnace at Stamford[10] (Fig. 125), but it was tapped on the straight side of the 'D' and it would not have been as tall.

The enclosure shown in Fig. 124 housed the ore and charcoal (both covered), and the bellows would have been in the building behind the hearth. This was altogether a well organized establishment, of a standard well above that of the pre-conquest sites.

In contrast to the smithies most bloomery sites would have been away from settlements and near the ore-pits and forests. On the Moor of Rannoch in Tayside, Scotland[11,12] there are a large number of medieval, manually-powered bloomeries, all of much the same type, with a bowl hearth situated at one end of an oblong trench (Fig. 126).

The hearth proper is made of stones usually set on end and plastered with an inner lining of clay. The back is the hottest part and usually has slag adhering to it. No doubt the tuyere was placed in this region. It is not always easy to see how the slag could be removed from this type of furnace, although there is little doubt that it was tapped.

Similar furnaces have been found in monastic sites in North Yorkshire, at Glaisdale[13] and Baysdale.[14] The latter were tapped into a basin at a lower level. These were no more than 2 m long, but the same type with

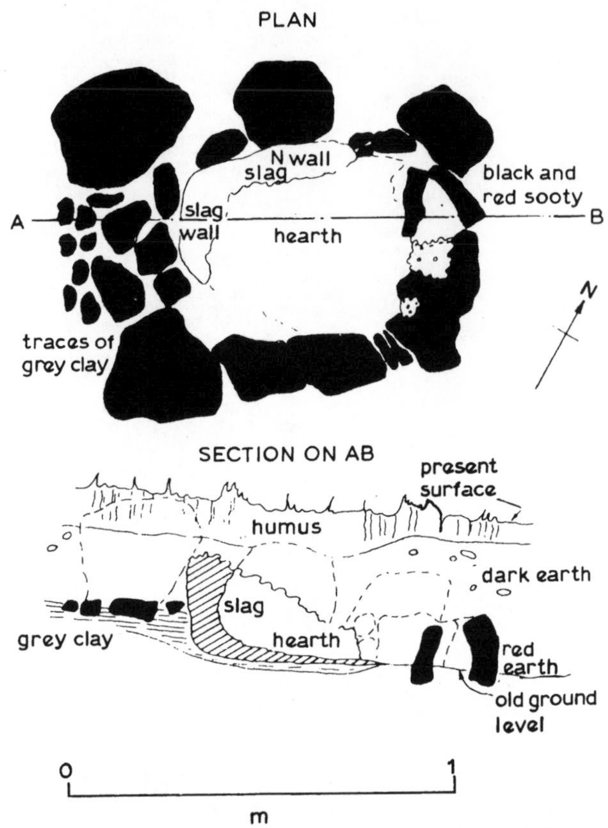

126 Early medieval bloomery furnace on Rannoch Moor, Perthshire (after Aitken[12])

much longer entrance channels leading to them have been found at Lyveden,[4] Northants., and at St. Neots, Cambs.[15] At Lyveden the furnace trench was 3·5 m long; at St. Neots it was 2·7 m long. In both cases a circular furnace had been inserted at one end. At Lyveden it was 0·8 m wide inside and was probably a shaft furnace.

Other oval or circular bloomery furnaces have been found at New Abbey, Dumfries,[16] and at Etchingwood, Sussex.[17] Unfortunately the extensively-excavated, Saxo-Norman site at West Runton[1] only yielded the foundations of a smelting furnace, but there is no doubt that slag was tapped from it.

Excavations at Pakenham, Suffolk[18] on a Pagan-Saxon site yielded iron objects and what appeared to be slag, in proximity to hearths measuring about 1·5 m square and 0·30 m deep. More recently slag has been found at Burrow Hill and other sites to the north of Ipswich.[19] There is also an unstratified furnace from Woodbridge,[20] and there is no doubt that smelting was carried out using the rich Tertiary ore deposits of the area (Burrow Hill, Table 68 Chapter 6).

At High Bishopley, in Weardale, Durham, the author excavated an iron-smelting site dated to the 12th or 13th century.[7] This site was about 310 m above sea level, well away from a stream, and was clearly chosen for its copious supplies of highly manganiferous bog iron ore. The fuel used was mainly oak charcoal, most of which seemed to have come from the small scrub oaks of the area. The site was located by the presence of tap-slag in the fields and farm roads covering an area several hundred metres in diameter.

Excavation of part of the site, which appeared to run along the 310 m contour for some hundred metres, produced two bowl hearths and a more ambitious furnace (Fig. 127). It would appear that the bowl hearths had had slag tapped from them. They were roughly 1 m in diameter and 0·2 m deep, and were filled with dark charcoal and slag. The area around was compressed into a highly ferruginous working floor measuring about 3 m × 3 m and 0·075 m thick. On the downward side of the hill was a charcoal dump, underneath which potsherds were found, also a slag heap containing tap slag, but no furnace slag or 'furnace bottoms'. The bowl hearth (no. 1 in Fig. 127) was the first to be used on the excavated part of the site, and slag was tapped towards hearth 2, which was made by enlargement of the slag-tapping pit when hearth 1 became disused.

The tapping of bowl-hearth slag is certainly an advance on the Early Iron Age sites. But furnace 3, found at High Bishopley, is a still further development. This was found to the west of hearth 2, and the slag had been left in place after tapping the furnace into the disused bowl hearth 2. The tapped slag was continuous with the cinder forming the bottom of the

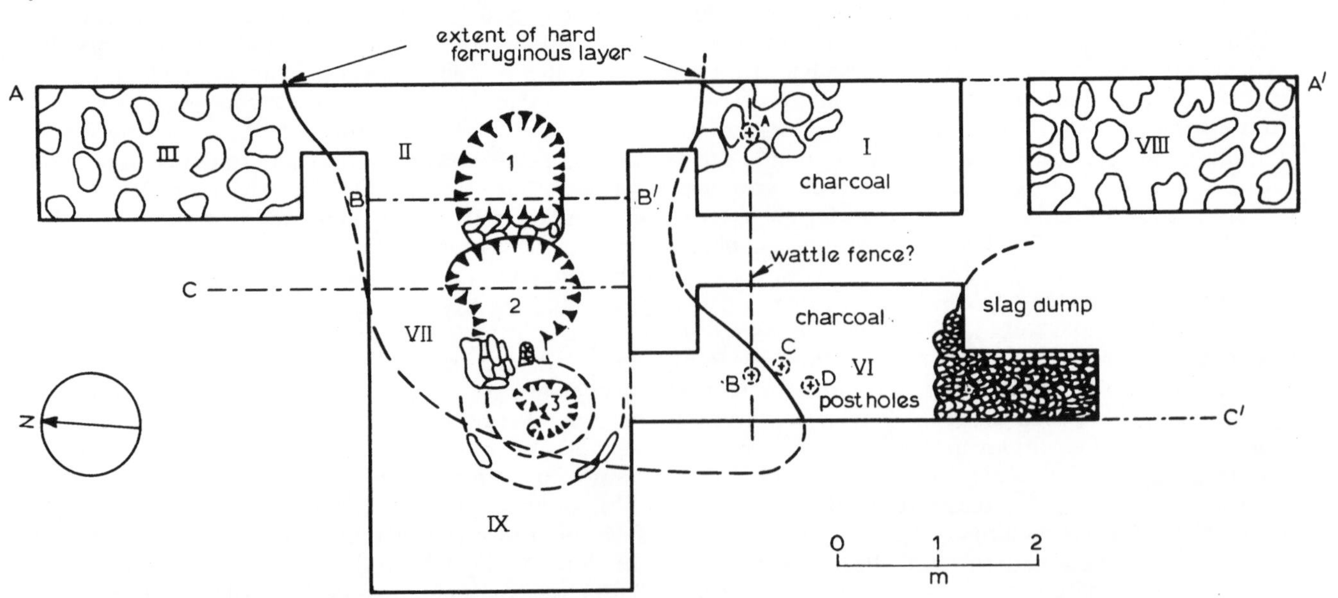

127 Early medieval smelting site at High Bishopley, Durham (after Tylecote[7])

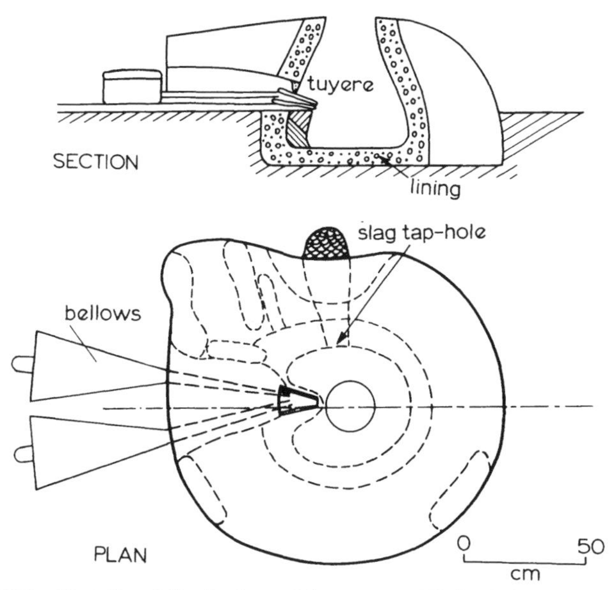

128 Details of final, domed furnace at High Bishopley, Durham (after Tylecote[7])

furnace. This furnace was kidney-shaped with a diameter of only 50 cm and was slightly re-entrant upwards. It was constructed of clay and sand, slag and stones, and had a very well-vitrified clay lining, which resisted the damaging effect of the cold chisel used to remove the hard slag inside. It was blown with the aid of bellows, which must have provided air through a tuyere placed on the clay platform (*see* Fig. 128). The bottom of this furnace was only 50 cm below the present surface level,[7] and because of the tendency for the inner diameter to get smaller nearer the surface, and the absence of any remains of a superstructure, it was concluded to be a low-domed furnace of an early type, blown with a forced draught.

The existence of a furnace of La Tène type at such a late date proves the continuation of a tradition through the Migration period in Britain. It would seem that this type, well known in the Siegerland in La Tène times, was used by Anglo-Saxon peoples in northern Germany and spread by them to the northern Pennines. This furnace could not have been used at the same time as hearths 1 and 2, but it may have been used at the same time as other bowl furnaces on the site. It had the one disadvantage of making the extraction of the bloom difficult. For this reason it gave way to the hearth-type furnaces used exclusively in Britain for the later medieval bloomeries.

Due to the coexistence of fuel and non-phosphorus-containing ore in the Lake District, this area was naturally exploited as soon as conditions here became sufficiently stable. In the Coniston area a number of bloomery sites have been excavated by Collingwood[21,22] and Cowper,[22,23] which seem to span the early–late medieval period. On Peel Island in Coniston Lake,[21] the remains of large D-shaped furnaces were found, but no slag. The pottery was Saxo-Norman in type, therefore this site can be dated to the earlier part of the period. The furnaces measured 3 × 2 m, and the presence of worked iron and charcoal with the absence of slag suggest that these were probably forge furnaces. We cannot, however, be certain that smelting was not carried out, as the slag might have been tipped into the lake. The island was only large enough to take the forge, and one wonders why an industry should have been established in such an awkward place.

The remains of a bloomery have been found on the northern end of Walney Island on a prehistoric site.[24] It is certain, however, that the bloomery is medieval and it probably worked ore carried from the opposite side of the channel. The amount of slag found suggests smelting rather than smithing, but no furnace structure was found. The sites at Eskmeals at the mouth of the Esk present a similar problem; the evidence would suggest a smelting area with ore and wood being brought across the river[25] (*see* Table 91).

Springs bloomery,[22,23] situated on a hillock on the west side of Coniston Lake, consisted of a mound 35 × 21 m, about half of which was slag 1·2 m deep. In the debris were found traces of four hearths consisting of stones packed with clay. One of these seems to have been a low-bloom hearth, tapped to one side, and blown with bellows. The slag found nearby certainly has the composition of a typical bloomery slag (*see* Table 91). Red hematite, charcoal, and a purple dross were also found; the latter, however, contained only a trace of iron and was thought to have come from a charcoal-making pit. It seems to be very similar to a dross found in a pit in a Roman-period barn at Great Weldon, Northants., and may have come from an elling hearth for making alkali.

The base of a 14th century hearth found at Torver, Lancs. has been illustrated by Fell[26] and Schubert.[92] It is clear that the smelting zone measured only 46 cm across; it was surrounded by a stone structure and was probably very like the one from Springs.

Similar furnaces were found at Knapping Tree and Harrison Coppice; these and many others in the Coniston area are supposed to date from the 13th to the 16th centuries, i.e. up to the dissolution of the monasteries. Altogether, Collingwood[21] estimates that there were thirty known sites in the Coniston area and many have been recently surveyed by Davies-Shiel (personal communication).

During the 16th century many protests were made about the depletion of the fuel resources of the area by the bloom-smithies. In 1546 many bloom-smithy leases were cancelled,[26] but in 1564 the local people were allowed to make iron for their own use. The ore resources were so plentiful that the lack of fuel merely led to the transport of the ore to other parts where ample fuel supplies still existed. Thus, the smelting of Lake District hematite was carried on for a while further south in the Rossendale area.

Kerr[27] reports having found the remains of such bloomeries at Miller Barn near New Church railway station, overlooking the Irwell Valley. Here were found stone furnace debris, burnt clay, slag and charcoal. Leading to this bloomery was a road 60 m long × 3·5 m wide that consisted entirely of slag 45 cm deep. It is said that another bloomery at Nook was worked as late as the 18th–19th century. Kerr believed that the ore came from Furness by water; it is possible that nodular ironstone was used, but unfortunately no ore was found. The majority of the bloomeries in this area ceased operations in the 16th century, and finally attention was turned to the fuel resources of the north of Scotland as a way of using the high quality ores of the Lake District.

Table 91 Composition of slags from the early medieval period (weight%)

Chapter Ref. %	98 Therfield, Herts. 12th cent.		3 Stamford, Lincs. Saxo-Norman		1 West Runton, Norfolk 850–1150 AD		5 Rotherfield, Sussex 14–15th cent.	8 Millbrook, Sussex 9th cent.	12 Lower Dall, Rannoch. 15th cent.	13*b* Wark Forest 14th cent.	99 Lindale, Cumbria pre-17th cent.	100 Odell, Bedford Saxon	94 Selside Beck, Coniston, Cumbria
FeO	68·0	53·0	65·9	52·7	58·4	56·8	45·5	30·0	55·17	(66·7)	52·3	—	46·55
Fe_2O_3	—	—	—	—	—	—	6·0	14·06	12·2	—	13·9	66·6	1·92
SiO_2	14·0	20·0	16·6	29·5	25·6	27·1	24·3	28·41	17·3	21·6	20·4	28·7	29·56
CaO	9·0	10·5	3·0	4·9	1·5	1·4	9·5	4·6	1·6	3·7	3·28	1·5	5·11
MgO	<1·7	<1·7	<0·1	—	0·2	0·8	1·7	—	0·7	1·8	0·73	0·5	0·96
Al_2O_3	<2·0	<2·0	4·6	7·4	9·1	10·2	6·4	11·05	1·5	5·7	—	2·2	7·08
TiO_2	<2·0	<2·0	0·3	—	—	—	—	—	—	—	—	—	0·27
MnO	—	—	<0·1	0·46	3·1	3·0	2·8	—	3·3	nil	—	—	0·21
P_2O_5	—	—	1·26	1·40	1·99	1·63	0·69	0·64	0·3	—	—	—	0·45
S	—	—	0·09	0·11	0·04	0·02	0·10	0·16	0·1	—	—	—	—
H_2O	—	—	} —	—	—	—	} —	—	0·7	—	2·71	—	—
CO_2	—	—	} 7·58	—	—	—	} n.d.	1·03	2·4	—	—	—	—
Na_2O	—	—	—	—	—	—	0·15	—	—	—	—	—	—
K_2O	—	—	—	—	—	—	0·72	—	—	—	—	1·0	—
C	—	—	—	—	—	—	—	—	—	—	2·30	—	—
Total	91·0	83·3	88·34	96·47	99·93	100·93	97·86	89·95	95·2	99·00	95·16		

Table 91 –*continued*

Chapter Ref. %	46 Dinas Powys, Glam. Slag	Slag	Furnace slag	95 Great Casterton, Leics. Medieval Tap-slag	96 York 9th–13th cent. Slag	Post-Roman Slag	97 Kirkstall Abbey, Yorks. 13th cent. Tap-slag	7 High Bishopley, Durham 13th cent. Tap-slag	Furnace slag	Slag	45 Downpatrick Co. Down 13th cent. Furnace slag	Slag	13 Glaisdale, N. Yorks. 12–13th cent. Tap-slag	25 Eskmeals, Cumbria Med. Slag
FeO	60·1	56·4	17·3	46·1	47·9	63·0	(63·7)	31·00	11·20	17·3	—	(57·3)	(67·2)	60·8
Fe_2O_3	14·0	12·0	72·1	3·2	5·72	5·7	—	26·4	40·3	22·1	60·23	—	—	6·1
SiO_2	18·7	25·7	4·24	26·2	27·4	17·8	20·8	20·3	22·7	48·9	24·0	18·9	22·8	28·8
CaO	1·4	2·24	0·72	7·0	2·6	2·1	--	2·7	2·6	1·3	—	—	2·3	0·73
MgO	0·73	0·35	0·20	1·1	0·54	0·54	—	1·9	<0·1	<0·1	—	—	0·1	0·39
Al_2O_3	3·44	2·16	1·14	9·5	3·43	2·05	1·83	2·4	3·4	3·7	4·47	6·5	2·1	1·23
TiO_2	—	—	—	0·45	—	—	—	—	—	—	0·29	0·42	0·1	0·1
MnO	—	—	—	0·7	nil	nil	—	13·1	8·8	3·5	1·1	1·2	0·3	0·21
P_2O_5	nil	nil	0·11	2·3	2·4	2·00	—	—	—	—	—	—	2·2	—
S	0·04	0·1	0·09	—	—	—	—	—	—	—	—	—	0·14	—
H_2O	—	—	—	} —	—	—	—	n.d.	6·0	1·2	—	—	—	0·50
CO_2	—	—	—	} 1·02	—	—	—	—	—	—	—	—	—	—
Na_2O														
K_2O														
C														

— = not sought; ( ) = by difference; n.d. = not detected

In Scotland there is evidence of a number of early bloomery sites. At Dun Cuier[28] on the island of Barra, Outer Hebrides, in the centre of a circular building was a hearth edged with chamfered stones, in places in a double row, all buried in deep layers of peat ash. In a bowl-shaped hollow lay a lump of iron slag with other large masses nearby; the slags were in various states of fusion. This site is dated to the early 7th century. Since there appears to be no evidence of iron ore or plentiful supplies of timber on Barra, it is supposed that ore and probably fuel were brought from the mainland. This may be so, although it is a far longer haul in this case than in the case of the islands off the Irish coast mentioned below. Until disproved by analysis one cannot neglect the possibility that bog iron ore and peat were used, as at Wiltrow in Shetland in an earlier period.[29]

On the neighbouring island of North Uist, at Garry Iochdrach,[30] lumps of iron slag and fragments of corroded iron were found together with a 4th century Roman coin. The Stack fort on Ugadale Point, Kintyre, yielded similar remains dated to before the 7th century.[31] A mote (AD 1200) at Bonnybridge, Central,[32] produced the usual high-iron slag. The excavators assumed that bog iron ore was used in this case, although Bonnybridge lies within one of the Scottish coalfields and Coal Measure ores such as the black band ore would have been available.

In Argyll, slag has been found at Gortanearn and other places between Fionn Port and Tarbert in Strathclyde.[33] As there is no local tradition of iron-working in this area, it would appear that it must have been early and, being on the coast, was probably supplied with Furness ore.

Macadam[34] and Fell[26] describe in great detail the rise of the Scottish iron industry which followed the great shortage of fuel in the Lake District. Macadam examined over eighty sites which all yielded slag containing a large amount of iron oxide, and little lime, and were therefore typical bloomery sites. In many of these bog ore was used (*see* analysis in Table 68 Chapter 6), but iron mines have been worked at Edderton (near Dornoch), Letterewe, Tomintoul, Abernethy, and in Grampian. Wood or peat charcoal was used for smelting.

One of the main sites investigated was near Gairloch, Ross, along the north shore of Loch Maree.[34] Here was found bog ore from South Erradale, and masses of slag 0·45–1 m square and 15–20 cm thick. At Letterewe was found the remains of a furnace made of Torridon sandstone and brick, and nodular ores that are not found in the vicinity and must have come from further south, possibly Fifeshire or England. There were also signs of ore from Furness. These had been brought in through Poolewe, whose harbour showed evidence of these imports. Later the furnace at Letterewe became a blast furnace and was used for casting guns, and cast iron was found near it containing 97% iron, 2% carbon and a trace of silicon. Limestone, containing 8·4% FeO, was found at Letterewe and some containing 5·3% FeO at Poolewe, which had probably been imported from the time when the blast furnaces were started. The blast furnaces in this area ceased work in about 1668, although local tradition says that they lasted longer. At Fasagh, near Kinlochewe, Macadam[34] found heaps of tap slag containing 66% Fe, and bog iron ore.

It would seem that most of the Scottish bloomery sites were worked up to the 16th–18th centuries in order to utilize local fuel and, latterly, Furness ore. Those with good fuel reserves would change over to blast furnace operation in the 18th century using Furness ore, until competition from areas near the coalfields using nodular and blackband ores, as at Carron, became too severe.

The presence of iron-smelting remains in almost all Irish sites of this period points to widespread and local iron-working in Ireland. It would seem that iron-working was a domestic industry and did not become centralized as it was in England at a slightly later date.

O'Kelly excavated a site at Ballyvourney, Co. Cork,[35] dated to between the 6th and 13th centuries AD but believed to belong to the earlier part of this period. Inside a round hut 6 m in diameter were crucibles, a piece of bronze, iron knives, ferrules and nails. The evidence of smelting was confined to 'pits' just inside the periphery of the house; these were about 30 to 60 cm in diameter and showed signs of intense burning.

From these pits came so-called furnace bottoms consisting of charcoal and slag which had formed below the bloom. Some of these were found outside the pits, but their shape showed that they had come from similar pits or furnaces. Most of these were 13–15 cm in diameter and about 7 cm thick, and it is clear that they had come from iron-smelting furnaces of a very primitive type with no provision for tapping slag. However, the quantity of these exceed those found elsewhere in Ireland, which shows that this must have been a smelting site of some importance or duration.

A number of clay discs with holes were also found. The diameters are about 13 cm, the diameter of the hole 2–5 cm and the thickness 2–8 cm. They are made of red clay and are slagged on one side. None of these had been found before on an Irish site. It would seem that these discs are part of a tuyere system, and this possibility has been discussed in Chapter 6. The fuel used was charcoal from ash, hazel, willow and poplar. No ore was found on the site, but it is most probable that bog ore was used. The composition of bog iron ore found in the vicinity is given in Table 68 Chapter 6; it shows a very high manganese content which should be reflected in the slag analyses.

At Ballyvourney there are furnaces of a type very like those of the pre-Roman Iron Age found by Lady Fox at Kestor, Devon.[36] It would appear that people untouched by Roman influence had been carrying on the pre-Roman tradition with no essential change in technique.

On Church Island, near Valencia, Co. Kerry, O'Kelly[37] has found the undoubted remains of iron-smelting in an early Christian monastic context. The astonishing thing is that the island was only 0·4 ha in size and possessed no ore or fuel; all that was required had to be brought from the mainland. A considerable quantity of slag and about 250 kg of charcoal were associated with a round timber hut. A shallow pit about 10 cm deep and 50 cm in diameter, containing charcoal and iron slag was found, which O'Kelly thought was the bottom part of a bowl furnace. Many other islands off the south-west Irish coast have yielded evidence of smelting, ore being obtained from the mainland.

At Townland Lissue in Antrim, Bersu[38] found iron slag and furnace debris together with a heavy iron bar about 15 cm long. Ardeloon[39] yielded a large lump of iron slag weighing about 2 kg. Ballycateen[40] yielded 14 kg of slag, some pieces being about 2 kg in weight. Hearths were found on the two latter sites, but there was no indication as to their purpose. Lagore Crannog,[41] a royal residence in the 7th–8th centuries AD, produced considerable metallurgical material of all kinds dating from the 7th to the 11th centuries. Slag existed in three forms: furnace bottoms, similar to Ballycateen and Ballyvourney, between 11 and 19 cm in diameter and 4 to 8 cm thick; smaller lumps containing stones and charcoal, probably cinder; and glassy, drop-like pieces. The ore used must have contained sulphur, and could not have been well roasted, as ferrous sulphide was present in some of the slags. The local ores consist of siliceous hematites from the Silurian rocks, bog iron ore and pyritic deposits. Apparently no identifiable pieces of ore were found on the site. A clay tuyere was found, 10 cm long and 8 cm in diameter with a 2 cm diameter hole (*see* Fig. 88 Chapter 6). The majority of the iron-working remains were dated to the 9th and 10th centuries and were probably connected with the Norse invasion.

The 7th century fort at Garranes, Co. Cork,[42] also yielded a furnace bottom, a piece of a clay tuyere, whose hole was reduced from 2 to 1 cm in diameter at the tip (*see* Fig. 88), and two pairs of tongs, one of which was 34 cm long. As on many other Irish sites, iron-working was carried out in conjunction with other metallurgical operations. A monastic site of the same period at Liathmore, Tipperary,[43] produced a furnace bottom and a fragment of iron.

Lough Faughan Crannog,[44] three miles south-west of Downpatrick, produced evidence very similar to that of Lagore. No analysis of the slag, ore, or the bloom has yet been made. At Downpatrick itself, excavations on the Cathedral Hill fort by Proudfoot,[45] produced extensive signs of iron-smelting, and the analysis of the slags and a piece of iron found are given in Tables 91 and 93. These remains date from about the 13th century.

Few Welsh sites of this period have been examined. Alcock has found slag and furnace bottoms on a 7th century site at Dinas Powys in Glamorgan.[46] Some pieces of ore also found consisted of the hydrated iron oxide, goethite (FeO(OH)), and had undoubtedly come from the Llanharry deposits nearby. The analyses given in Table 91 show slags low in phosphorus, which are in accord with the analysis for the ore deposit given in Table 67, Chapter 6. The furnace slags had undergone very little reduction, and the analysis is therefore almost the same as that of the ore used. The difference between the silica in the furnace slag and that in the furnace bottom would suggest a very high yield from this material. No furnaces were found, but the 'bottoms' and absence of tap slag suggest a very primitive technique. One small piece of heavily rusted iron was found with a carbon content of about 0·2–0·3%, which had been slowly cooled from about 1 100°C.

The early Welsh homesteads found on Gelligaer Common in the same county[47] and dated to the 13th and 14th centuries, also contained evidence of domestic iron-working. Iron slag was found embedded in the surface of a floor;[47] it was probably cooled in a furnace with a minimum diameter of about 20 cm near the base. On this site pieces of nodular ore were found that came from the Coal Measures.

HISTORICAL EVIDENCE FOR MINING AND SMELTING

The end of the 11th century is the beginning of the historical period of British metallurgy. Unfortunately, with few exceptions, the people who wrote about this subject were either accountants or historians, both ignorant of the principles of metallurgical processes, which often makes interpretation of the details given by them extremely difficult.

First let us consult the Domesday records.[48–51] We find that the distinction between smiths (*fabri*) and bloom-smithies (*ferraria*) is blurred. Also, we find references to *blooma, massae*, and *plubas ferri* or *plumbae ferri*; the latter presumably refer to iron. The word *plubas* has probably come from lead to mean 'pig', and has then been applied to iron to mean bloom. But since in the same record there are references to lead works as *plubaria* or *plumbariae*, one has to be careful in making the distinction.

The list of bloomeriės given in Table 92 is compiled from Domesday. If there was a reference to a *blooma* or a *plumbum ferri*, this has been accepted as indicating the existence of a bloom-smithy.

According to Domesday there was a considerable amount of activity in the south-west of the country. In Somerset, bloomeries existed at South Perrot, Seaborough, Lexworthy and Bickenhall. How far the choice of these sites was determined by water power requirements and how much by ore resources is uncertain. A bloomery was in operation in the Brendon Hills, where we know that there were ample ore resources. There were also bloomeries in the southern part of the county and the Mendips. While bloomeries were in operation in the Forest of Dean at Marcle, and Alwintune (Alvington near Lydney),[50] the largest recorded bloomery was at Pucklechurch near Bristol. This bloomery rendered 90 blooms per annum, probably 10% of its production, and must have been about the largest bloomery recorded in the land at the time. It was under the authority of the Abbey of Glastonbury and may have utilized the waters of the River Frome. Two types of ore were available: the nodular ores of the north Bristol coalfield and the brown hematites of the Lias, which have been worked at various places, such as Iron Acton, Frampton Cotterell and Pucklechurch itself. The deposit at Pucklechurch was probably worked out by the 13th century.[51] There are traces of old workings at Pucklechurch and Cold Ashton, and undated deposits of primitive smelting slag have been found at Iron Acton.

The Domesday evidence is patchy, since it only relates to land held by the King, not including Cornwall or Durham. There is no mention of iron-works in Kent, although according to the *Cartularium Saxonicum*,[52] as early as 689 AD an iron mine near Lyminge, Kent, was granted by Osuuyni (Oswy) King of Kent to Adrian, Abbot of St. Peter's, Canterbury. This shows that there must have been some interest in the local iron resources during this period.

Lyminge is on the North Downs, some 48 km east of the well known iron deposits of the Weald. The 'mine'

Table 92 Bloom-smithies mentioned in Domesday (*c*. AD 1080)

County (according to Domesday)	Site and river	No. of bloom-smithies	No. of blooms rendered
Somerset	Sudperit (South Perrot); Parrett or Axe	1	1
Somerset	Seveberge (Seaborough; R. Axe)	—	1*
Somerset	Lecheswrde (Lexworthy; streams from the Brendon Hills)	2	2
Somerset	Bichehall (Bickenhall; Fivehead. R. ?)	—	1*
Somerset	Not known; land owned by Baldwin of Exeter	1	2
Gloucester and Gwent	Stantune (Staunton, Glos.)	—	4
Gloucester	Pulcrecerce (Pucklechurch, nr Bristol. R. Frome)	6	90
Hereford (now in Glos.)	Alwintune (Forest of Dean; Alvington)	—	20
Hereford	Merchelai (Much Marcle?)	—	50
Cheshire (now Clwyd)	Atiscross (Croes Ati, Flint; R. Cloith)	(Iron mines only?)	—
Cheshire, Clwyd and Lancs.	Aldedeford (Aldford-on-Dee, Cheshire)	—	8
Yorks. and N. Lancs.	—	6	—

* = from every free man

probably consisted of ore-pits going through the edge of the chalk to nodular ore in the Greensand below. This makes it more probable that the ore used on the Roman site at Springhead (also near the North Downs) was from the same strata, and not the Wealden which would have been a considerable distance away.

Several iron mines were being worked near Flint, in the Hundred of Atiscross[48] (Croes Ati), where there were also lead works in the Roman period. Lead and iron go together here as they do in Weardale, where one part of a vein may be worked for lead while at the same time another may be worked for iron. The works at Aldford-on-Dee were probably supplied with ore from the Coal Measures of the south Denbighshire [Clwyd] coalfield, about 24 km away.

The Domesday account relates to the period *c*. 1086 and, as can be seen, there is already a considerable concentration of iron-smelting in the river valleys. At Staunton and Aldford the reference to mills permits no doubt that water power was already in use by this time, almost certainly for driving the bellows, since it is believed that mechanical hammers did not come in until the end of the 14th century. This is in contrast with High Bishopley, which is well away from the available water in Weardale, suggesting that water power was only being used in the more developed south of the country (*see* Chapter 9).

In areas where water power is scarce, for example in the Northamptonshire plateau, there is no mention of bloom-smithies but frequent mention of forges. It is difficult to believe that these were not working up local material, and we can only assume that the individual output of the local bloomers was too small to get the attention of the king's accountants.

In the north, in spite of political troubles, iron-smelting was being carried on. In about 1150, an iron mine at Rookhope in Weardale was granted by Bishop Pudsey.[53] Between 1211 and 1213 the sale of iron bars from Durham figures several times in the Pipe Rolls,[54] iron being sent as far afield as Portsmouth and Ireland. By 1307 Bishop Bek was receiving an income from coal mines in the Chester-le-Street area, which probably also provided iron ore.[55] In 1368 the lease of a bloomery at Gordon and Evenwood is mentioned.[56]

There is some historical evidence from Northumberland.[57] In the 14th century, Tyndale (probably the Manor of Wark on the North Tyne) had a very high output of iron. The figures were as follows:

1333	294 blooms
1350	204 blooms
1351	143 blooms
1353	26 blooms

There is little doubt that the decline in production is mainly associated with the Black Death, which reached this area in 1350. It would seem that the normal annual output up to the Black Death was about 4 tons, and that the bloom size in this period was about 20 kg. A recent survey of the area revealed some enormous slag heaps immediately to the west of Birtley village, in which sherds were found dated to the 14th century. The heaps contain typical tap slag, and it is very probable that excavation on the plateau above and immediately adjacent to them will show the remains of furnaces. These heaps contain many hundred of tons of slag, and are certainly much greater than those in either Teesdale or Weardale. It would seem that the Black Death marked the decline of the manually worked bloomeries of this area.

There is an account roll for this period relating to a manual or foot-powered bloomery at Tudeley, Kent,[9] working the Wealden deposits. The accounts relate to two main periods: 1329–34 and 1350–4, i.e. before and after the Black Death. The site has not been located, but Tudeley is near Tonbridge. It originally consisted of a wattle-and-daub hut and a hearth. In the early period the output was about 200 blooms per year, the weight of the bloom probably being of the order of 20 kg. The ore was first roasted, and only one hearth was used, so presumably the same hearth was used to extract the slag. This part of the process was later (at Byrkeknott[7]) to be done at the bloomery in a separate hearth known as a 'stringhearth'.

The main charge in working such a bloomery in its earliest phase was the cost of conversion of wood to charcoal, which accounted for 50% of the running cost, in spite of the fact that the timber itself came from the owner's estate. Wages and salaries accounted for only a third, but out of this the main expense was the wages of the blowers. The cost of manual or foot-blowing would therefore be about 20–30% of the total. With the change to power-blowing, which was soon to come

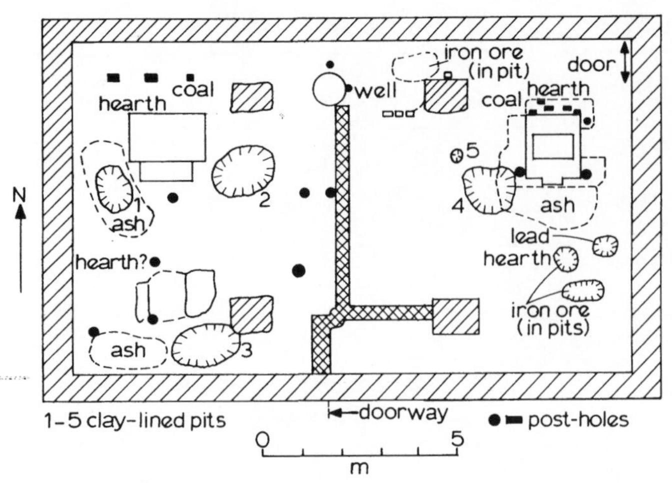

129 Simplified plan of the forge at Waltham Abbey (after Huggins[58])

130 Low-level smithing hearth of the 14th century (from *Romance of Alexander*, AD 1340, Bodleian Library, MS 264 f 84T) . (Drawn from the original by L. L. Ketteringham by kind permission of the Trustees of the Bodleian Library, Oxford and reproduced through the courtesy of the Council for British Archaeology.)

elsewhere, much of this would be saved, but the chief gain was the production of a bigger bloom and consequently greater fuel efficiency.

It is fairly certain that during this period the standard type of hearth in use in Britain was the final development of the bowl hearth. The shaft furnace of the Roman period was to give way in Western Europe to the bowl hearth and the Catalan hearth, while in Central Europe and Scandinavia it developed into the Stückofen, a high bloomery furnace. This was later destined to become a blast furnace for the production of liquid iron due to the prolonged contact of charcoal and ore and the opportunity given for carburization.

131 Waist-level smithing hearth (from British Museum Sloane MS 3983 f 5r; drawing based on the original by L. L. Ketteringham and reproduced with the kind permission of herself and the Council for British Archaeology)

SMITHIES AND SMITHING HEARTHS

The Roman smithy hearth shown in Fig. 105 (Chapter 7) was a foretaste of what was to come eventually to Britain. It was not situated out in the open or even under some primitive shelter but in a substantial building. Waltham Abbey, Essex has provided such a building with at least two hearths.[58] This monastic forge was a flint-and-stone walled, 3-bay aisled building which is datable principally to the 12th century, although it had 10th and 11th century phases (Fig. 129).

Of course, not all smithies had waist-high hearths; many hearths were still no more than pits in the ground as at Goltho[85] and Tresmorn. (There was clearly an advantage in working for a monastery.) At Alstead, in Surrey, we see the stages of development from the pit to the slightly raised hearth depicted in the Bodleian MS of AD 1340 (Fig. 130), to the waist level hearths of the Sloane and other MS of the same period (Fig. 131).

The forge at Waltham Abbey had two hearths on stone or brick bases, but there is no way of knowing their exact height. A hood or canopy would take away the fumes, and a manually worked bellows placed either to one side or behind would supply the draught through a tuyere; in front would be an anvil. The position of these is indicated on the plan. Finally the smith would need a water bosh for cooling the iron or the tools; this could be directly in front or to one side. Perhaps it was in a pit in the ground. The forge also had a hearth for melting lead (and probably other non-ferrous metals), as well as an iron ore pit, which suggests that iron was occasionally smelted in the smithy. This is not unusual; a Slovak smith once stated (in 1965) that he was able to smelt iron if necessary.

In case it is thought that a monastic smithy was a special case where capital was unusually easy to obtain, let us now look at the smithies of the Manor of Alstead.[59] The iron-smithing period has three phases, all in the period AD 1250–1405. The first period produced small bowl hearths 25 cm in diameter and 28 cm deep. Some slag adhered to the bottom and the sides, which indicates that it might have been used for smelting. Other hearths contained soft grey ash but were free of slag, so could only have been used for smithing or non-ferrous melting.

By about 1270, layers of clean clay and charcoal were laid over the old iron-working area, which was neatly squared off with a flint and clay bank. This provided a 3·5 m square enclosure, possibly containing a tile kiln. But during the period 1270–1340 renewed iron-working activity occurred, resulting in an adjacent mound containing over 2 tonnes of iron-working debris.

Between 1395 and 1405 a hearth was placed over the remains of the tile kiln and its adjacent puddling tank to give a low platform about 2·1 × 2·7 m and 0·53 m high. In front was an anvil base. The whole structure was probably covered by a tiled hood, suggesting the reconstruction shown in Fig. 132. This is believed to be the equivalent of the waist-level hearth shown in Fig. 131. Such a hearth is unique in 14th century Britain.

This forge lay against the back wall of a building 6 m square, made by extending the original walls. The details of the smithing hearth, which is made of sandstone slabs, are seen in Fig. 132. It is believed that the fire was placed in *a* and the bellows and tuyere would have been to the left as shown here, rather than behind as shown in the MS of 1340 AD (Fig. 131). There would be a chimney behind and a working surface, *b*; *k* would support the anvil. Smith's hearths in the 19th and early 20th centuries were not unlike this, and one is preserved in the White Swan Hotel at Heddon-on-the-Wall, Northumberland.

ANGLO-SAXON AND EARLY MEDIEVAL ARTIFACTS

Again, the starting point of any smithing operation leading to an artifact would have been the bloom. While many smelting hearths have forging hearths associated with them, we are not certain whether these were used for consolidating the bloom and expelling some of the slag (i.e. as stringhearths), or for general smithing operations. Since the latter were normally performed in the settlements, it is more likely that they were used as stringhearths.

After smelting, the bloom would be cut up, first with an axe and later with a chisel to give smaller pieces or 'gads'. In Ireland the townland of Dernaglug and Drumaa in Co. Fermanagh has yielded three split blooms.[60] These were found at a depth of 50 cm, weigh about 5·45 kg and are about 20 cm long. After smelting they have been partially cleaved with an axe producing a **V**-shaped notch with an angle varying from 25° to 35°. They were found buried in a bog and two were interlocked. One of these was analysed, and the composition is given in Table 93. It is clear that these have been cleft when hot, probably to determine their iron content and malleability, or in order to make them easier for working-up by the smith. The analysis shows that they consist of a pure, low-carbon iron with some patches of higher carbon at the outer edges and in the junction of the 'V'. The phosphorus and sulphur contents are both low and indicate a low-phosphorus ore and the use of charcoal. The splitting of blooms to test their purity was common in Iceland in medieval times. Percy[61] shows the splitting of a bloom in his plate depicting the 'Osmund' process in 18th century Sweden. It is probable that the Fermanagh blooms date from the medieval period, since the area was almost devoid of woodland by the 18th century. No remains of smelting furnaces were found.

At Carrigmuirish in Co. Cork was found a small oval bloom (Fig. 133) dated to AD 500–1000, weighing 3·34 kg.[62] This was typical bloomery iron with the carbon varying from 0 to 0·7% (average 0·3%), and with a hardness varying from 92 to 195 HV. It had been partly consolidated but still had considerable porosity as shown by its density of 4·12 g cm^{-3} (solid iron = 7·8). The slag content was normal for wrought iron, i.e. about 10% (analysis is given in Table 93).

The Downpatrick 'bloom' is very much smaller but is dated to the 13th century.[63] It only weighed 640 g but like the one from Carrigmuirish was very porous, and it contained 7·5% slag. All this suggests that it is a piece cut from an unconsolidated bloom. It consisted entirely of ferrite (and slag) and had a hardness of 137 HV, (*see* Table 93).

Billets, osmunds and gads

The bloom was forged into standard-sized pieces for sale. Although so-called 'currency bars' of axe-shape

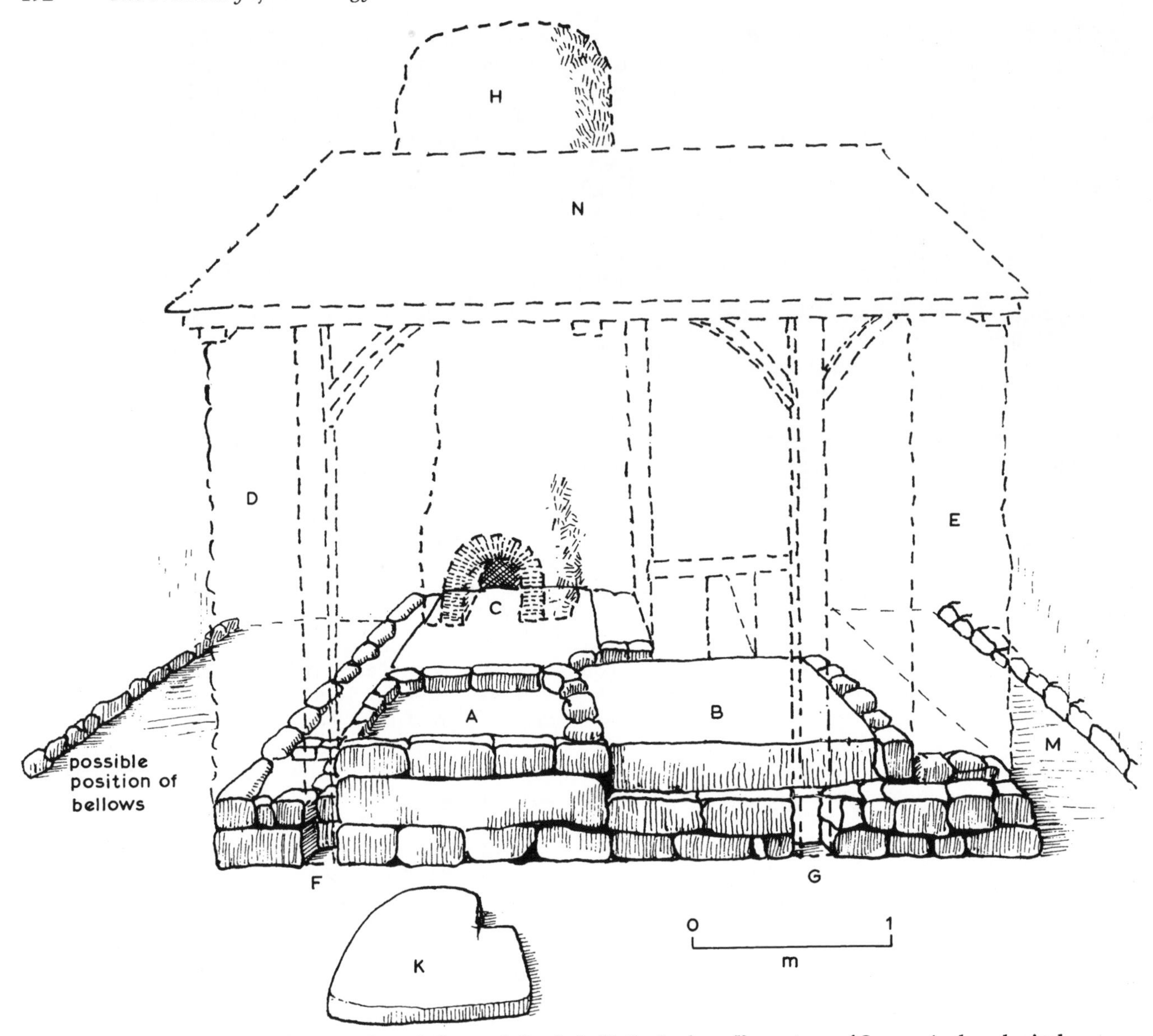

132 Reconstruction of the forge at Alstead, Surrey (after L. L. Ketteringham;[59] courtesy of Surrey Archaeological Society)

continued to be used in Scandinavia and Slav Central Europe, there is no indication of such a bar in Britain. But in medieval times there are references to 'gads' and 'osmunds'. The osmund was a pure iron bar imported from Scandinavia in barrels, so it must have been fairly small although neither the Scandinavians nor the British have any precise idea of its size or shape. We do know that the osmund decreased in weight from some unknown value in the 13th century to 300 g in the 16th.[64] There is a finished bar from medieval Winchester[65] weighing 1·28 kg, which resembles in shape the bars found on the Viking site of Hedeby[66] weighing 58–1 450 g.

133 Iron bloom from Carrigmuirish, Co. Cork; 8 cm wide; 3.6 kg when complete

Carburizing, nitriding and phosphiding

The process of carburizing followed the principles laid down in Chapter 6, but was much more widely and intelligently used. When steel was needed it was usually made separately and welded to the artifact where required. However, there is written evidence for other steel-making and hardening processes which we will examine.

In the story of Weyland, he files the sword blades into fine particles and feeds them to 'tame birds'.[67] Does this introduce nitrogen or change the phos-

Table 93 Composition of iron objects of the early medieval period (weight %)

Ref.	86 Axe R. Kennet, Reading 6–7th cent.	86 Axe Stratford, Essex 9th cent.	96 Piece York Saxo-medieval	86 Arrowhead Woodeaton	63 Bloom Downpatrick 13th cent.	60 Bloom Fermanagh Med.	77 Sword Donnybrook, Dublin Viking centre	Edge	62 Bloom* Carrig-muirish, Co. Cork E. Christ.
C	0·23	0·049	0	0·05	0·08	0·05–0·047	0·2	0·3–0·4	0·3
Si	n.d.	0·04	—	tr.	0·16	0·01	0	0	—
Mn	0·01	0·01	—	tr.	0·02	0·01	0·1–1·0	0·1	0·01
S	0·008	0·011	—	—	0·038	0·014–0·019	0·01	tr.	0·065
P	0·13	0·445	0·17	—	0·061	0·081–0·063	0·02	tr.	0·053
As	0·049	0·042	—	0·008	—	—	n.d.	n.d.	—
N_2	0·005	0·0085	—	—	—	—	—	—	—
Cu	0·01	0·01	—	0·01	—	0·01	—	—	—
Ni	0·02	0·07	—	0·004	—	0·01	n.d.	0·2–0·3	—
Ca	—	—	—	—	—	0·005	—	—	—
Mg	—	0·005	—	0·003	—	—	—	—	—
Al	—	0·04	—	0·01	—	0·01	<0·01	n.d.	—
Cr	n.d.	nil	—	nil	—	0·01	—	—	—
Mo	n.d.	nil	—	nil	—	0·02	—	—	—
V	tr.	nil	—	nil	—	0·02	—	—	—
Co	—	—	—	—	—	—	n.d.	n.d.	—
Slag					7·25				

*Analysis kindly provided by Dr K. C. Barraclough; — = not sought; n.d. = not detected; tr. = trace

phorus content? The answer seems to be that there is no change in the phosphorus content, which can often be high when required, by ore selection. It was also unlikely that nitrogen would be absorbed in the low-temperature regime of a bird's gut. However, when the filings were mixed with fresh chicken manure and heated to 1 000°C in a simple forge for 30 min, the result was quite different: the nitrogen content had been increased from 0·019 to 0·13% in weight.

It is possible to phosphide in much the same way by using a phosphate mixture such as bone ash and charcoal with sand, i.e. under reducing conditions. If the iron is heated in such a mixture for 2 h at 1 230°C in an iron box, a layer containing about 1·7% P and 0·1 mm thick can be obtained. Two of these layers will give a 'white band' of retained ferrite when welded.[68] So it is possible to get layers of high phosphorus, arsenic, or nitrogen content as well as the more normal carbon by cementation and heating at high temperatures in various media.

Of course, the presence of nitrogen in early iron is very common and can be seen as nitride needles in the ferrite. But this amount of nitrogen does not contribute much to the hardness of the ferrite.

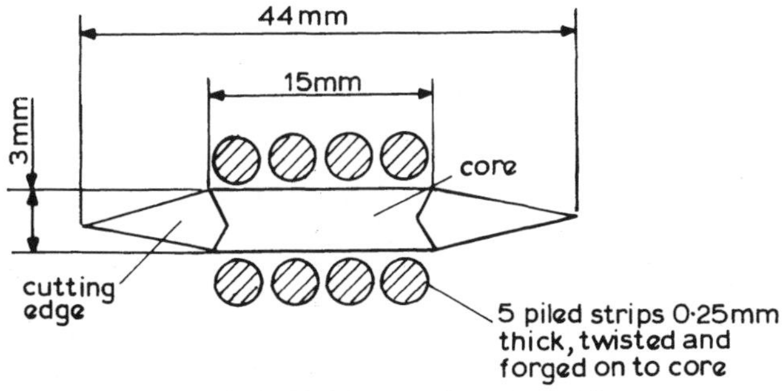

134 Structure of Ely Fields sword; not to scale (after Maryon[69])

Swords

The pattern-welding technique, the principles of which were discussed in chapter 7, had now got well into its stride and been widely adopted. The earliest of the pattern-welded swords to be examined was found at Ely Fields Farm, near Ely, Cambridgeshire, and dates from the Pagan-Saxon period between the 5th and 7th centuries.[69] This sword has a good deal in common with the Roman period Nydam-type swords. It consists of a central core, on each side of which were welded four bundles each of five strips only 0·25 mm thick (*see* Fig. 134). The bundles were twisted at intervals of 28 mm, thus giving alternate straight and twisted sections; there were twenty-two twists in a length of 67 cm. In all, forty-three separate layers were present in this sword, which had a total length of 86 cm, a maximum width of 4·5 cm and a thickness of 3 mm. The story of the battle of Swanfirth against Snorri and his folk shows that good and bad swords were used concurrently. Steinthor found that 'the fair-wrought sword bit not whenas it smote armour, and oft he must straighten it under his foot'.[70] Radiographic examination can distinguish between pattern-welded and non-pattern-welded swords, but we do not know how many of the latter type were in use during this period. Since the pattern-welded swords were known from Roman times, and the Anglo-Saxon examples must have been similarly made, we can assume that the non-pattern-welded sword found in the River Witham in 1848,[71] and dated to the 10th century, was superior to the pattern-welded swords.

The majority of early swords were pattern-welded

Table 94 Composition of Merovingian swords and scramasaxes (after Salin[73])

Artifact	Identification	Composition %				
		C	Si	Mn	S	P
Swords (Pattern-welded)	M. 7 Blade	0·12	—	0·01	0·02	0·21
	M. 10 Blade	0·09	—	0·05	0·02	0·30
	M. 11 Blade	0·08	—	nil	0·02	0·16
	M. 11 Edge	0·2	—	nil	0·02	0·14
	Luneville; blade	0·01–0·05	—	nil	0·03	0·18
	Luneville; edge	0·02–0·03	—	—	—	—
Scramasaxes (5th–7th century)	No. 21 Blade	0·46	0·30	nil	0·0033	0·041
	No. 18 Blade	0·11	0·06	nil	0·003	0·17
	No. 14 Blade	0·11	0·061	nil	0·01	0·29
	No. 17 Blade	0·22	0·075	0·008	0·0016	0·088
	Puxieux Blade	0·18	0·11	0·03	0·01	0·09

and resemble those from Ely Fields and the River Witham (1954).[72] All had a central gutter or fuller, the distinguishing feature between the La Tène (Early Iron Age) swords and those of the Migration period.

The composition of the French pattern-welded swords (Table 94) shows a generally lower carbon content but a higher phosphorus content than the scramasaxes (sword-knives) which have also been examined.[73] The mean carbon content was 0·08% and the phosphorus 0·2%.

The edges of the swords fall into three groups: (1) uncarburized, 0·1–0·2% C; (2) carburized, 0·4–0·6% C; (3) carburized and quenched. These show three well-defined techniques, but it is probable that they were used concurrently rather than successively, and merely represent differences in the skill of the smiths.

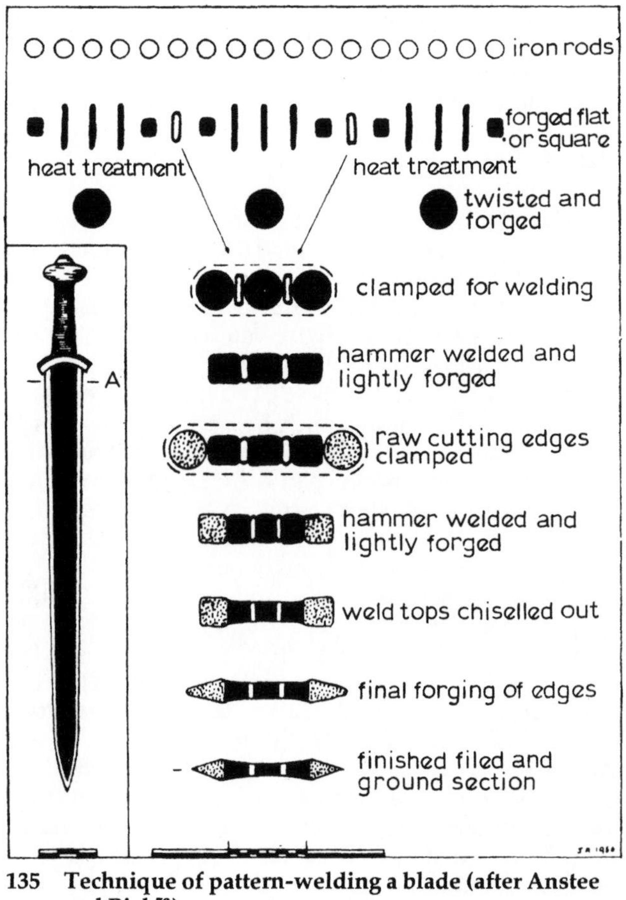

135 Technique of pattern-welding a blade (after Anstee and Biek[72])

It is clear from the low carbon and phosphorus contents that these swords could probably be bent in use, although they would have been superior to the Early Iron Age weapons. This would depend a good deal on the degree of carburization obtained in the heating and forging processes. Oakeshott[74] reports that three Norwegian pattern-welded swords found in 1889 had carbon contents of 0·414, 0·401 and 0·52%, considerably higher than those examined by Salin.[73]

It would seem that the pattern-welding technique was commonplace among Anglo-Saxon smiths. In a cemetery at Finglesham in Kent,[75] one of the women's graves contained a so-called weaving sword. These were pattern-welded, sword-shaped bars of iron about 4·6 cm wide and 51 cm long. They had tangs at both ends and might well have been made from cut-down swords. They could have been used for beating the weft but may have been merely badges of office. In Viking times they were replaced by bone.

The subject of pattern-welding has inspired a comprehensive investigation of the techniques used.[72] These are shown in diagrammatic form in Fig. 135. Apart from the cutting edges, the strips and rods may all be pure wrought iron of low carbon content, but the forging and entrapment of scale and slag does seem to confer an increase in the mechanical properties of the core. This seems to be due to a more intimate mixing of iron, slag and oxide scale. Attempts to carburize over short periods (30 min) were, not unexpectedly, a failure. But there were some signs that treatment in a mixture of organic material and pigeon droppings was having a beneficial effect by hardening up to about 20 HV on the original material. Since this was not on account of carburizing it must have been connected with the absorption of nitrogen or phosphorus.

The 9th century sword from the Palace of Westminster which was pattern-welded seems to be a rather ineffective weapon.[72] The pattern-welded areas down the centre consisted of 0·2% carbon iron together with fine and coarse grained ferrite and small globules of slag. The hardness varied from 186–188 HV. The edges consisted of fine-grained mild steel with a ferrite + pearlite structure but, surprisingly, only had a hardness of 136–145 HV. It would seem that the pattern-welded areas had a higher average phosphorus content.

During the 9th century the pattern-welded sword

Table 95 Some Anglo-Saxon, Viking and medieval swords with inscriptions

	Provenance	Date AD	Pattern-welded	Inscription and remarks	
1	Ely Fields (Cambridge)	500–600	Yes	No inscription on blade	69
2	Sutton Hoo (Suffolk)	*c.* 700	Yes	No inscription on blade	102
3	Rainham (Essex)	Anglo-Saxon	Yes	Fragments of 2	104
4	R. Witham (Lincoln) (1954)	9th cent.	Yes	'N' pattern-weld; silver bands on hilt with niello	105
5	R. Witham (Lincoln) (1848)	9th–10th cent.	No	+ LEUTLRII; Sulphur on reverse; both iron inlay on blade; hilt decorated with copper and brass	74
6	Ballinderry (Co. Westmeath) (1928)	9th–10th cent.	—	HILTIPREHT on guard and another name on blade	74
7	Kilmainham (Dublin)	9th–10th cent.	No	ULFBERHT on blade in iron inlay; HARTOFLR on hilt	74
8	R. Nene, Wisbech (Cambs.)	900–1100	—	ULFBEHRT (misspelt); INLEFRII on other side; (Probably misspelling of INGELRII); all iron inlay on blade	106
9	R. Thames (Windsor)	10th cent.	Yes	Possibly reused blade from earlier period (B.M. No. 1929, 2–6, 1)	B.M.
10	R. Thames (Temple)	10th cent.	—	INGELRII; iron inlay on blade	106
11	R. Thames (Shifford, Berks.)	800–1100	—	ULFBERHT; iron inlay on blade; (at Reading)	74
12	Abingdon (Oxon.)	9th cent.	—	Hilt only; silver and niello	107
13	London (?)	11th cent.	—	INGEL . . . and a carrocium; iron inlay on blade	74
14	Farnham (Norfolk)	1171	—	SESBENEDICTAS; white metal on blade in 1/8in letters	74
15	Shaftesbury (Dorset)	1130–1170	—	GICELINMEFECIT in small iron letters on blade; zoomorphic heads on guard	74
16	R. Witham (Lincoln)	1250–1300	—	+ HDOXCHDRCHDXORVI + on centre of blade in brass	74
17	R. Trent (Cawood Castle, Yorks.)	1250–1320	—	NED ———	74
18	Canwick (Lincs.)	Late 13th cent.	—	ANTANANTANANTAN in silver letters 6 mm high along blade	
19	Tower (London)	*c.* 1300	—	Wolf mark, inlaid with white or yellow metal	74
20	Thames (London) (1959)	13th cent.	—	Wolf mark, inlaid with white or yellow metal	74
21	Thames (Westminster Bridge)	*c.* 1325	—	Maker's punch mark on hilt *and* blade	74

was in decline.[76] A sword found at Donnybrook[77] was of Viking-Medieval type and was inscribed with a now indistinct iron inlay which could be ULFBERHT (*see below*). It was not pattern-welded but was made by piling and carburizing. The core was made of fine-grained ferrite with spheroidal pearlite (*see* Table 93 *above* for composition); the edge had a higher carbon content and had been quench-hardened to give a hardness of 520–550 HV – altogether a very efficient weapon.

An ULFBERHT sword from Norway, dated to the 10th century had a carbon content of 0·75% and also was not pattern-welded.[74] This could have been made by straightforward piling without twisting, and the surface finally carburized, and the structure would not be determinable by radiographic examination.

A Frankish 9th–10th century sword was found on Canwick Common, Lincoln.[78] This blade had been made by piling or folding carburized iron to give laminae about 2 mm thick. Neither the carburizing nor the heat treatment had been sufficient to give complete homogenization, so the structure varied from high-carbon martensite to a lower-carbon structure throughout the thickness. The hardness varied from 306–630 HV.

The swords of this period tended to increase in both size and weight. Compared with the pre-Roman Iron Age sword of about 0·7 kg, the Anglo-Saxon swords weighed about 1·10–1·40 kg, and 14th century swords 1·80–2·30 kg. After this the lighter sword regained its popularity. The lengths varied a good deal, some reaching as long as 100 cm, but the majority were between 76 and 90 cm. The more interesting swords embracing the whole period are listed in Table 95.

The scramasax

The other weapon, differing from the sword in that it is single-edged, is the scramasax, seax or sword-knife. Belonging to the earlier part of our period, these blades have areas of pattern welding, like some of the medieval knives.

Unfortunately a full examination has not been made on any of the British examples. But Salin[73] has examined those found in France, and since it would be expected that the technique used and the composition of the material would not be very different from that of the English blades, it is useful to summarize his conclusions. They were all made from low to medium carbon iron, some showing a 'piled' structure. The edges were highly carburized, as shown in Fig. 136, but in many cases they were not quenched and so maximum hardness had not been attained. The nitro-

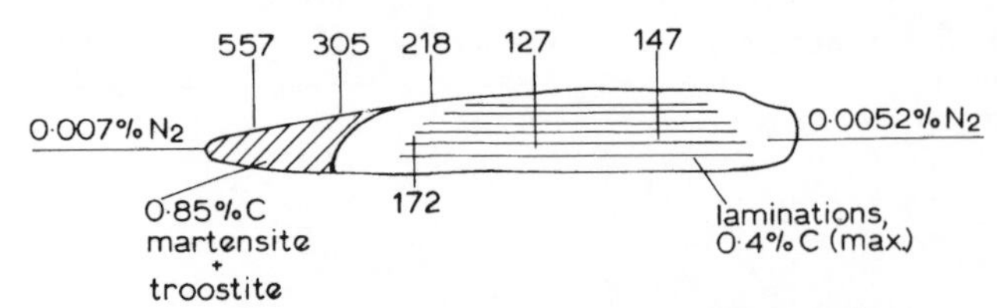

136 Section of piled and steeled scramasax blade from France giving hardness; 5 cm wide (after Salin[73])

gen content of the carburized edge was higher than that of the rest of the blade, showing that it had been carburized in a nitrogenous organic medium.

The compositions are given in Table 94. The phosphorus content is variable but tends to be higher than in Roman times, the average for the five scramasaxes being 0·13%. The average carbon content is 0·22%. It would appear that the scramasax from the Dorset County Museum had been made with a steel edge (now rusted) and a pattern-welded back which served as decoration, rather like the inlay often applied to knives and swords.[79] The structure of the remains consisted of alternate layers of high-phosphorus ferrite and 0·3% carbon steel. The hardness varied from 200 to 260 HV, the higher figures being given by the ferrite. But it is likely that the blade had been quenched in order to harden the edge, which may contain as much as 0·85% C (Fig. 136).

The scramasax from Northolt Manor, Middlesex was 60 cm long including the tang which was 24 cm long.[80] This corresponds with the proportions of the one from Purton, Wilts. It is dated to the 7th–8th century and along the back of each face are a number of grooves, some of which contain brass inlay near the hilt in a zigzag design.

It would seem that scramasaxes were made by scarf welding-on a steel edge, as shown in the French example (Fig. 136).

Spearheads

Spearheads often show noticeable 'pattern-welding' effects under radiography but prove disappointing after sectioning. Some edges are steeled, and the application of the steel envelope may be done in such a way as to give a pattern-welded appearance. The Saxon spearhead from Clifton-on-Trent was no more than heterogeneous bloomery iron varying from pure ferrite in the blade, with a hardness of 114 HV, to pearlite and ferrite (in the socket!) with a hardness of 240 HV.[79]

A comprehensive study was carried out by Swanton,[81] who concluded that the blade consisted of a core concealed between two layers of pattern-welded material, with high-carbon edges welded-on (Fig. 137). In no case was there any heat treatment. The

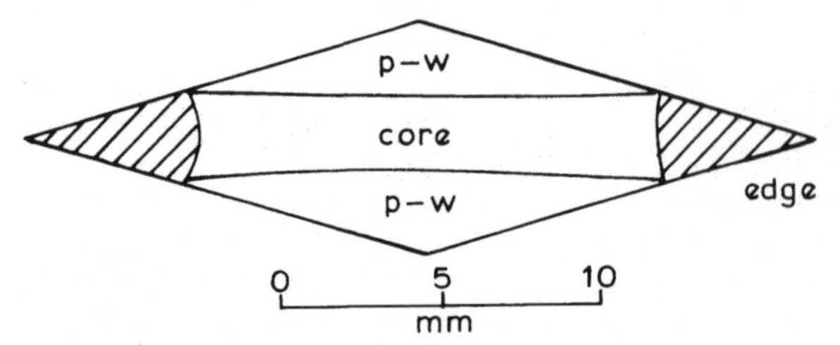

137 Section of Anglo-Saxon spearhead showing construction (after Swanton[81])

Table 96 Hardness and structure of spearheads (mostly after Swanton[81])

Site	Hardness HV			
	Core	Pattern-welded area	Edges	Socket
Clifton-on-Trent	114	—	—	240
Kentmere	112	—	142–211	—
Barrington (Cambs.)	116	106–118	122	183
Ely (Cambs.)	119	114–119	184	114
East Anglia	208	183–203	312–262	245
Cambridge	106	114–129	—	120
Kennet (Reading)	97	—	219	—
Newbury I	—	None	165 (F)	—
Newbury II	133	None	160 (F + P)	—

carbon content of the edges was of the order of 0·6%, and the hardness of the various component parts is shown in Table 96. Clearly, the pattern-welded part does not contribute anything to the strength, but is merely meant to look decorative. It would seem that the edges are appreciably harder than the core, but heat treatment has not been attempted.

Many Viking spearheads have also been found, and one of these from Kentmere, Westmorland [Cumbria],[82] was found to have a 'herringbone' pattern. This example was 34 cm long and had a split socket. The maximum width of the blade was 3·6 cm, and wavy lines suggestive of pattern-welding were visible on the blade. Another, from Lough Gur, had been inlaid with gold.[108] A Viking period spearhead was found in the River Kennet near Reading.[83] This had been made by piling twelve thin strips, which do not appear to have been more than lightly carburized in their central regions, since their hardness was only 97 HV. The edges of the blade had a hardness of 219 HV, and a carbon content of 0·45%. This appears to have been obtained by final carburizing, and no attempt has been made to harden it by heat treatment. However, a pattern has been introduced, which may have been made by 'cold stippling'. The details are summarized in Table 96.

Arrowheads

Arrowheads were also steeled, presumably for the better penetration of mail. Ring-mail using brass or

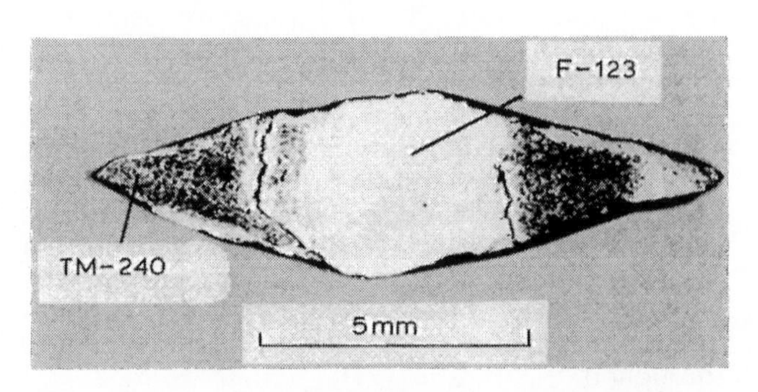

138 Microstructure and hardness of medieval arrowhead from Lambourn, Berks. (after Coghlan and Tylecote[101])

Table 97 Hardness and structure of medieval arrowheads

Site/area and No.	Type/remarks	Structure and hardness HV Core	Barbs or cutting edge	Ref.
Newbury; S 466(1)	Barbed and socketed	123 (F)	240 (M)	C & T[101]
Newbury; S 466(2)	Barbed and socketed	(F)	—	C & T
Newbury; S 468(3)	15th cent.	110 (F)	—	C & T
Newbury; S 470(4)	Point only	165–178 (F)	—	C & T
Newbury 1924(5)	Hunting type; barbed and socketed	185 (F)	185–(F)	C & T
Hadleigh, Essex	Hunting type	—	245 (P + F)	Brewer[84]
Winchester (BS3433)	13th cent. socketed; Sandwich, Type *a*	276 (P)	210 (F + P) (Outer layer)	Biddle[65]

F = Ferrite; P = Pearlite; M = Martensite

iron had been introduced during the Roman period and this was soon followed by scale armour – *lorica segmentata*. In Saxo-Norman times iron mail was more common, as we see on the battlefield at Hastings (Bayeux tapestry). It would appear that in order to kill, an arrow need not penetrate the mail, but the heavier iron or steel arrow could do mortal damage when mail was worn. However, steel arrows could not penetrate the later (AD 1400) plate armour, which was often hammer-hardened or carburized.[84]

There were at least two varieties of arrowhead: the hunting type with wide, thin barbs, and the more compact, military type. As we see from the arrowheads listed in Table 97, the barbs are on the whole of higher carbon content than the core, but only in one case has any attempt been made to harden them by quenching. This is shown in Fig. 138.

Knives

It is reasonable to regard knives as showing the highest level of metallurgical technology of the time. They must be light and sharp, and any deficiencies cannot be made up by using other physical properties such as weight. It is in this field that one sees the greatest improvement over the standards of the Roman period.

Over 71 knives from 7 sites dated from the immediate post-Roman period (Poundbury, Dorset) to the 14th century (Goltho, Lincs.), have been examined as part of an extensive programme of work on the quality of early edge tools. Out of this sample, more than half have edge hardnesses exceeding 300 HV, and of these 50% have hardnesses in the range 500–900 HV, which would signify a blade up to modern standards of non-stainless knives.

We will consider two examples: first, a type *b* knife from Winchester dated to the 13th century, made of a piece of steel folded over and welded twice[65] (Fig. 139). In turn this was welded to a pattern-welded back consisting of laminations of ferrite and pearlite, which could have been polished and etched to show decorative, wavy patterns on the surface.

The cutting edge consisted of tempered martensite with a hardness of 575 HV, which declined to 386 HV at the join with the pattern-welded back. This had a hardness ranging from 201 to 225 HV. Altogether this was a very efficient and beautiful knife although, when excavated, it was little more than a rusty piece of scrap iron, without a tang and hardly recognizable as a knife (Fig. 140).

Another knife came from the 14th–15th century deserted medieval village of Goltho in Lincolnshire (Fig. 141).[85] This knife was of type *a* construction with a piece of steel inserted between piled wrought iron. In this figure the wrought iron is black and the steel light. The steel was again tempered martensite with a hardness of 557 HV while the back, which consisted of 14 layers of ferrite, had a hardness of 151 HV. The ferrite that normally provides a soft protection for the edge has been well ground away exposing the steel for about half the cutting edge.

Tools and constructional material

Two axes have been examined on behalf of the Pitt Rivers Museum, Oxford.[86] One of these, a Danish battle-axe, comes from the River Lea near Stratford,

139 Pattern-welded and steeled knife from Winchester (after Tylecote[103])

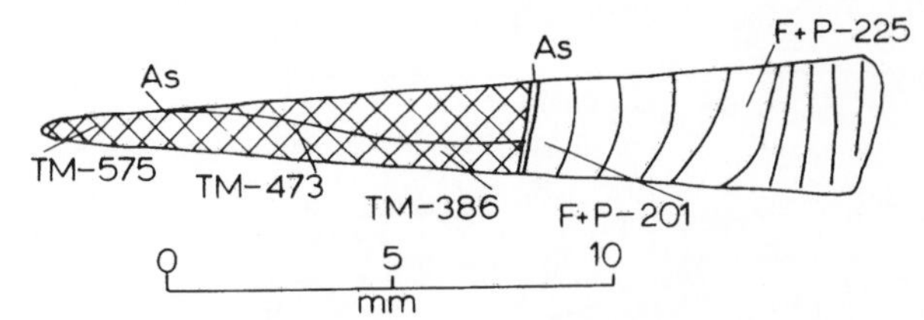

140 Structure of Winchester knife (after Biddle[65])

Essex, and is dated to the 9th century. Like the spear-heads this has been made from piled pieces of low-carbon wrought iron, the composition of which is given in Table 93 *above*. The grain flow is parallel to the cutting edge, and the socket has been made by lapping over a tongue and welding it to one side of the blade. The hardness, of 160–170 HV, is almost entirely owing to the high phosphorus content. The cutting edge is a piece of welded-on steel and has been quenched, resulting in a hardness of 350–450 HV. The structure is that of martensite, showing that the cutting edge has been very efficiently quenched. The final hardness, however, is low for a martensite and shows that not much more than about 0·25% carbon was introduced.

The other axe, very similar to a modern axehead, was found in the River Kennet near Reading and dates from the 6th–7th century. This also has a piled structure, but in this case the smith has been comparatively successful in introducing carbon, so that the average carbon content of the heavy section near the socket was 0·23%[86] (Table 93). Unfortunately the edge was almost carbon-free (0·04%), but owing to cold-hammering and the presence of 0·13% phosphorus, had a hardness of between 154 and 165 HV. The axe had been cooled from a forging temperature of over 1 000°C but not quenched, and had then been reheated to about 700°C and finally hammered on the edge of the blade. No attempt had therefore been made to heat treat the edge. In this sense, therefore, it would not be as efficient an implement as the battle-axe from Stratford.

A Viking-period chisel from Crayke, near Easingwold, Yorks. was found to consist of a steel of varying carbon content.[87] Near the head the composition was 0·9%; elsewhere it varied from 0·4–1·2%, with a hardness varying from 238 to 283 HV. The blade of the tool had been quenched, resulting in a hardness of 657–879 HV. This is one of the most satisfactory pieces of early smithing and heat treatment so far found, but is not uncommon in this period. The tool had been finally heated to about 1 000°C and only the blade quenched

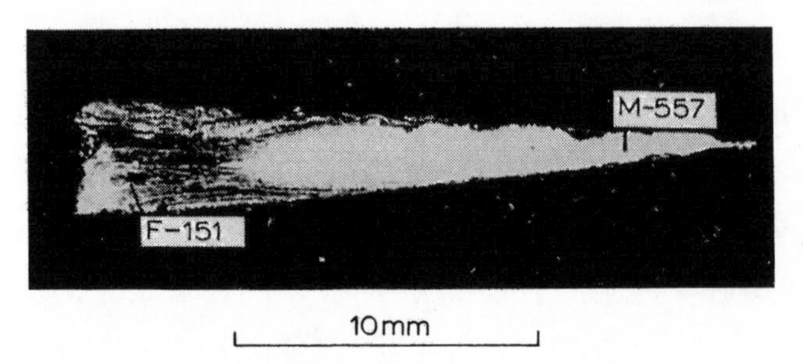

141 Structure of knife from Goltho, Lincs., with hardnesses (after G. Beresford;[85] courtesy Royal Archaeological Institute)

in water. Smythe does not report that the structure was laminated (or piled), but it would appear to be the only method for giving such an even distribution of carbon. It would seem that here is displayed the even carburization of the pre-Roman chariot tyre, coupled with a good knowledge of hardening by quenching.

Steel was occasionally used in constructional items to resist wear. The vertically-shafted Saxon water-mill from Tamworth, Warwickshire, is one example[88] (*see* Fig. 142, Chapter 9). The lower bearing originally consisted of a rectangular piece of steel, 6 × 8 cm across by 2·5 cm thick, with a V-shaped hollow forged into the centre of one surface. This piece was made of piled wrought iron with a considerable slag content which had been unevenly carburized. It would seem that the hardness of the original bearing surfaces was in the range 169–221 HV5, and that fairly rapid wear occurred so that the bearing was reversed to give new wearing surfaces. These had hardnesses in the range 150–258 HV. But the whole bearing had been quenched at some stage from a high temperature (>1 000°C) to give structures varying from martensite to ferrite and pearlite. The hardness varied from 150–874 HV5. Clearly, this object presented the smith with a problem in quality control. He knew what he had to do but had no way of checking whether he had succeeded until the object had been in service.

Inlay and inlaying techniques

A feature of Anglo-Saxon and Viking weapons is the technique of inlay. This too is merely a development of a technique utilized earlier in the South Shields sword (*see* Chapter 7). Basically there are now two types: inlay using a non-ferrous metal such as silver or copper, and a special iron inlaying technique first seen in 10th century swords, such as that from the Thames near the Temple,[89] and in the sword from Wisbech, Cambridgeshire. The non-ferrous inlay was probably applied by cutting a groove and cold-hammering in the non-ferrous metal wire. The adherence would be mainly due to mechanical interlocking, which is implicit in the technique described by Evison[90] and Theophilus.[91] If the wire and iron surfaces were clean there might be some degree of pressure-welding. This point could only be settled by a microexamination. The 1954 Witham sword has silver bands on the hilt inlaid with niello, and the earlier (1848) Witham sword hilt has been decorated by hammering copper or brass foil onto the roughened surface of the iron. In the case of the scramasaxes (Table 94 *above*), the blades have been inlaid with lettering composed of copper or silver, in the manner of the inlay of the South Shields sword.

The technique of iron inlay seems to have been introduced in the 10th century. The blades of the 1848 Witham sword and the Thames (Temple) sword are good examples of this technique, which mainly used letters about 2·5 cm high. According to Maryon,[71] the iron inlay on the 1848 Witham sword was applied by first making a deep chisel mark to locate the position of the inlay. The sword was then heated to a white heat and short pieces of iron rod, probably cold, were laid in position and hammered. All the limbs of the letter E, in the inscription + LEUTLRI⅃, were made from separate pieces of metal. He could find no evidence of undercutting to key the inlay. The letters were revealed by light etching, when the pure iron letter showed well against the steel background.

Maryon's interpretation of the technique is mainly based on radiographs and leaves certain metallurgical questions unanswered. First, if the sword was preheated to a welding temperature, how was oxidation prevented? The applied lettering would be cold but as soon as it was laid on, the hot metal would rapidly heat up to a similar temperature, and no difficulty would be experienced in welding it to a clean, oxide-free surface. The process would seem to need very careful control to carry out the heating of the area to be welded under non-oxidizing conditions. Given this, there would seem to be no difficulty in obtaining a true smith's weld between the blade and the lettering. However, the manipulation of short lengths of wire, some less than 2 cm long, would need a very delicate technique.

We now have many examples of the iron inlaying technique from Britain and many more from abroad, principally from the Scandinavian countries. The non-ferrous type was widely used for all purposes and it is possible that by the 8th century the English smiths excelled in inlaid work of this type. It is thought that England, and Northumbria in particular, was renowned for the production of such well made and decorated knives and weapons.[92,93] In the year AD 758, twenty knives were sent as a present from the abbot of Monkwearmouth and Jarrow to an English bishop of Mainz. Since richly decorated cutlery and swords were being made in the latter area in the 4th century, Schubert[93] concludes that the English articles must have been of a very high standard indeed, and that the English had in a space of several centuries surpassed the Rhineland in this technique.

SLAGS

The composition and constitution of medieval bloomery slags have been fully discussed by Morton and Wingrove,[94] and it is not necessary to discuss them in great detail here but merely to point out the differences that they show compared with the previous period.

It is clear from the analyses shown in Table 91 (*above*) that by the medieval period the silica content was higher and the iron content was lower, on average. In a sample of 27 slags taken at random, the silica content had risen from 20% to 27%. This was accompanied by an increase not in the free-running temperature but in the anorthite content ($CaO.Al_2O_3.2SiO_2$), as shown by Morton and Wingrove. This phase usually forms a glass and arises from the use of ores containing appreciable amounts of lime and alumina. This usually means rather poor ores as compared with the best of the Roman period, but ores with amounts of CaO and Al_2O_3 so balanced as to give anorthite–$2FeO.SiO_2$ eutectics with free-running temperatures in the range 1 100°–1 300°C. Occasionally the alumina is high compared with the lime and the compound $FeO.Al_2O_3$ (hercynite) is formed. The existence of such slags with their lower iron contents seems to show a careful selection of suitable but comparatively low-grade ores. The slags listed in Table 91 (*above*) show a range of 5–10% of both CaO and Al_2O_3, always with sufficient silica to form anorthite.

Some of the slags have very high phosphorus contents which would indicate a high-phosphorus iron. The mean content of P_2O_5 in the Roman slags is 0·76% while that for the early medieval slags is 1·21%. This is not denying that high-phosphorus ores were used in the Roman period, but such use was a good deal commoner in the medieval period. No doubt the higher strength (cold-shortness) conferred by certain ores had been noticed and metal from such ores had found its way into general wrought iron products such as nails, while the need for a high purity iron for steelmaking was increasingly met by imports through the Baltic and elsewhere.

SUMMARY

The early medieval period shows a big contrast between the level of extractive metallurgy that seems to have been developing from a pre-Roman level and that of the physical metallurgy of the artifacts themselves. The use of steel for cutting edges is now widespread and its heat treatment reaches a very high level in a large number of the artifacts examined. Clearly, the Migration peoples brought with them a high level of smithing technology.

The size of the bloomery industry increased markedly during our period to support the developing need for tools in agriculture, building, and weapons of war. This meant the opening up of an increasing number of smaller deposits of iron ores, many of which produced high-phosphorus iron unsuitable for steel-making. This deficiency was rectified by the increasing imports of steel and good quality iron through the Baltic.

NOTES AND REFERENCES

1 R. F. TYLECOTE: *Norf. Archaeol.* 1967, **34**, 187–214.
2 G. SWIFT: *Wealden Iron*, 1982, (2), 15–19.
3 A. BURCHARD *et al*: in 'Excavations in Stamford, Lincs. 1963–69', (ed. C. Mahany *et al*.), 133–44; 1982, No. 9, Soc. Medieval Archaeol. Monograph.
4 JOHN STEANE: *J. Northampton Museum*, 1975, (12), 4–22.
5 J. H. MONEY: *Medieval Archaeol.* 1971, **15**, 86–111.
6 J. HASLAM *et al.*: *Medieval Archaeol.*, 1980, **24**, 1–68.
7 R. F. TYLECOTE: *J. Iron Steel Inst.*, 1959, **192**, 26–34.
8 C. F. TEBBUTT: *Sussex Archaeol. Collect.* 1982, **120**, 19–35.
9 M. S. GIUSEPPI: *Archaeologia*, 1912, **64**, 145–64.
10 R. F. TYLECOTE: *Bull. Hist. Metall. Group*, 1970, **4**, (1), 24–7.
11 W. G. AITKEN: *Proc. Soc. Antiq. Scotl.*, 1969–70, **102**, 188–204.
12 W. G. AITKEN: *Bull. Hist. Metall. Group*, 1971 (1), **5**, 15–23.
13*a* G. STAINTHORPE: *Ryedale Hist.*, 1966, (April), **2**, 47–49.
13*b* R. F. TYLECOTE: 'Medieval bloomeries in the North of England'. (ed. W. U. Guyan), in 'Vita pro Ferro – Festschrift to R. Durrer', 115–34; 1965, Schaffhausen.
14 D. M. WILSON and D. G. HURST: *Medieval Archaeol.* 1965, **9**, 218.
15 C. F. TEBBUTT and G. T. RUDD: *Medieval Archaeol.* 1966, **10**, 158–60.
16 J. WILLIAMS: *Trans. DGNHAS*, 1967, **44**, 126–32.

17 P. ARCHIBALD and C. F. TEBBUTT: *Sussex Archaeol. Collect.*, 1975, **115**, 190–1.
18 B. J. W. BROWN *et al.*: *Proc. Suff. Inst. Archaeol.*, 1954, **26**, 189–207.
19 D. WILSON and J. G. HURST: *Medieval Archaeol.* 1981, **25**, 164–5.
20 M. L. WEIGHT: *J. Hist. Metall. Soc.*, 1981, **13**, (2), 107–9.
21 W. G. COLLINGWOOD: *Trans. HSLC*, 1901, **53**, 1–22.
22 H. S. COWPER: *Archaeol. J.*, 1898, **55**, 88–105.
23 H. S. COWPER and W. G. COLLINGWOOD: *Trans. CWAAS*, 1898, **15**, 211–28.
24 F. BARNES: *Trans. CWAAS*, 1958, **55**, 1–16.
25 J. CHERRY: *Trans. CWAAS*, 1966, **66**, 46–56.
26 A. FELL: 'The Early Iron Industry of Furness and District', 1908, Ulverston.
27. J. KERR: *Trans. HSLC*, 1872, **24**, 57–70.
28 A. YOUNG: *Proc. Soc. Antiq. Scotl.*, 1955–6, **89**, 290–327.
29 A. O. CURLE: *Proc. Soc. Antiq. Scotl.*, 1935–6, **70**, 153–69.
30 E. BEVERIDGE: *Proc. Soc. Antiq. Scotl.*, 1931–2, **66**, 32.
31 H. FAIRHURST: *Proc. Soc. Antiq. Scotl.*, 1954–6, **88**, 15–21.
32 S. SMITH: *Proc. Soc. Antiq. Scotl.*, 1933–4, **68**, 59–68.
33 A. GRAHAM: *Proc. Soc. Antiq. Scotl.*, 1918–19, **53**, 112–13.
34 W. I. MACADAM: *Proc. Soc. Antiq. Scotl.*, 1886–7, **21**, 89–104.
35 M. J. O'KELLY: *J. Cork Hist. Archaeol. Soc.*, 1952, **57**, 18–40.
36 A. FOX: *Trans. Devon Assoc.*, 1954, **86**, 21–62.
37 M. J. O'KELLY: *Proc. R. Ir. Acad.*, (C), 1958, **59**, 57–136.
38 G. BERSU: *Ulster J. Archaeol.*, (3rd ser.), 1947, **10**, 30–58.
39 E. RYNN: *J. R. Soc. Antiq. Irel.*, 1956, **86**, 203–14.
40 S. P. O'RIORDAIN and P. J. HARTNETT: *Proc. R. Ir. Acad.*, (C), 1943, **49**, 1–43.
41 H. HENCKEN: *Proc. R. Ir. Acad.*, (C), 1950, **53**, 1–247.
42 S. P. O'RIORDAIN: *Proc. R. Ir. Acad.*, (C), 1941, **47**, 77–150.
43 H. G. LEASK and R. A. S. MACALISTER: *Proc. R. Ir. Acad.*, (C), 1945, **51**, 1–14.
44 A. E. P. COLLINS: *Ulster J. Archaeol.*, (3rd ser.), 1955, **18**, 45–82.
45 V. B. PROUDFOOT: *Ulster J. Archaeol.*, 1954, **17**, 97–102.
46 L. ALCOCK: 'Dinas Powys'; 1963, Board of Celtic Studies, Cardiff, Wales U.P.
47 A. FOX: *Archaeol. Cambrensis*, 1939, **94**, 163–99.
48 J. S. MOORE (ed. and trans.): 'Domesday Book'; 1982, Chichester.
49 H. ELLIS: 'A General Introduction to Domesday Book'; 1833, London.
50 I. COHEN: *Trans. WNFC*, 1954, **34**, 161–77. (I am indebted to Ian Standing for details of iron-working in the Forest of Dean.)
51 *Victoria County Hist.*, Gloucester. **2**, 235.
52 W. DE GRAY BIRCH: 'Cartularium Saxonicum', **1**, 107; 1885, London.
53 'The Boldon Book and other documents', (trans. W. Greenwell), **25**, App. p. XLVI; 1852, Durham, Surtees Society.
54 *Ibid.*, App. pp. XII, XVIII, XXII.
55 *Ibid.*, App. pp. XXIV, XXVIII.
56 W. HUTCHINSON: 'The History and Antiquities of the County Palatine of Durham', **3**, 418; 1823, Durham.
57 T. ROGERS: 'A history of agriculture and prices', **2**, 465, 467, 472; 1866, Oxford.
58 R. M. and P. J. HUGGINS. *Essex Archaeol. Hist.*, 1973, **5**, 127–84.
59 L. L. KETTERINGHAM: 'Alsted; Excavation of a 13th–14th century iron works in Netherne Wood, Surrey'; 1976, Guildford, Surrey Archaeol. Soc.
60 E. E. EVANS: *Ulster J. Archaeol.*, 1948, **11**, 58–64.
61 J. PERCY: *Iron and Steel*, 1864, 321.
62 The bloom was kindly provided for examination by the late Prof. M. J. O'Kelly. The analysis was arranged through the kind offices of Dr K. C. Barraclough.
63 R. F. TYLECOTE: *J. Hist. Metall. Soc.*, 1977, **11** (2), 83. (The bloom with details of its excavation was kindly provided by Prof. V. B. Proudfoot.)
64*a* E. THOLANDER: *J. Hist. Metall. Soc.*, 1975, **9** (2), 68–9.
64*b* E. THOLANDER: 'Osmondsjarnet i ny belysning', 9–10; 1973, Stockholm, Bergsmannen.
65 M. BIDDLE (ed.): *Winchester Studies*, **7**, (2), (forthcoming).
66 R. THOMSEN: in 'Berichte über die Ausgrabungen in Haithabu', Bericht 5, 1971, 9–29.
67*a* D. WILLEN *et al.*: *J. Hist. Metall. Soc.*, 1976, **10**, (2), 84–6.
67*b* G. MARTIN and V. FOLEY: *J. Hist. Metall. Soc.*, 1979, **13**, (1), 38–9.
68 R. F. TYLECOTE and R. THOMSEN: *Archaeometry*, 1973, **15**, 193–8.
69 H. MARYON: *Proc. Cambs. Antiq. Soc.*, 1948, **41**, 70–6.
70 W. MORRIS and E. MAGNUSSEN (transl.): 'The Story of the Ere-Dwellers', 120; 1892, London.
71 H. MARYON: *Antiq. J.*, 1950, **30**, 175–9.
72 J. W. ANSTEE and L. BIEK: *Medieval Archaeol.*, 1961, **5**, 71–93.
73 E. SALIN: 'La civilisation mérovingienne d'après les sculptures, les textes et le laboratoire–3 ème partie: les techniques', 331; 1957, Paris.
74 R. E. OAKESHOTT: 'The Archaeology of Weapons', 143; 1960, London.
75 S. E. CHADWICK: *Medieval Archaeol.*, 1958, **2**, 1–71.
76 H. R. ELLIS DAVIDSON: 'The sword in Anglo-Saxon England,' 1962, Oxford.
77 R. A. HALL *et al.*: *Medieval Archaeol.*, 1978, **22**, 64–83.
78 J. LANG and A. R. WILLIAMS: *J. Archaeol. Sci.*, 1975, **2**, 199–207.
79 C. W. BREWER: *J. Hist. Metall. Soc.*, 1976, **10**, (1), 1–10.
80 V. EVISON: in J. G. Hurst, *Medieval Archaeol.*, 1961, **5**, 211–299.
81 M. SWANTON: 'The spear in Anglo-Saxon times', Ph.D. thesis, 1966, Univ. Durham.
82 C. FELL: *Trans. CWAAS*, 1956, **56**, 67–9.
83 A. E. P. COLLINS and H. H. BEENY: *Man*, 1950, **50**, 124–5.

84 C. W. BREWER: *J. Hist. Metall. Soc.*, 1981, **15**, (1), 1–8.
85 G. BERESFORD: 'The Medieval Clay-land village', 85; 1975, Soc. Medieval Archaeol., Monograph No. 6.
86 H. H. COGHLAN: 'Notes on prehistoric and early iron in the Old World', Occasional Papers on Technology No. 8, 190–191; 1956, Pitt Rivers Museum, OUP.
87 T. SHEPPARD: 'Metal objects made by the Vikings'; 1941, Hull Mus. Publ. No. 212.
88 E. M. TRENT: *J. Hist. Metall. Soc.*, 1975, **9**, (1), 19–25.
89 V. I. EVISON: *Trans. CWAAS*, 1968, **68**, 31–4.
90 V. I. EVISON: *Antiq. J.*, 1955, **35**, 20–45.
91 J. G. HAWTHORNE and C. S. SMITH (trans. and eds.): 'On divers arts; the treatise of Theophilus'; 1963, Chicago.
92 H. R. SCHUBERT: 'History of the British Iron and Steel Industry from *c* 450 BC to AD 1775'; 1957, London.
93 H. R. SCHUBERT: *J. Iron Steel Inst.*, 1949, **163**, 8.
94 G. R. MORTON and J. WINGROVE: *J. Iron Steel Inst.*, 1972, **210**, 478–88.
95 P. CORDER: 'The Roman town and villa at Great Casterton, Rutland, 1951–53'; 1954, Univ. Nottingham.
96 I. M. STEAD: *Yorks. Archaeol. J.*, 1958, **39**, 515–38.
97 L. ALCOCK and D. E. OWEN: *Proc. Thoresby Soc.*, 1955, **43**, 62.
98 M. BIDDLE: *J. Br. Archaeol. Assoc.*, 1964, **27**, 53–91.
99 J. W. JACKSON: *Trans. CWAAS*, 1914, **14**, 256–61.
100 R. E. CLOUGH: personal communication.
101 H. H. COGHLAN and R. F. TYLECOTE: *J. Hist. Metall. Soc.*, 1978, **12**, (1), 12–17.
102 P. HUNTER BLAIR: 'An introduction to Anglo-Saxon England', 283; 1959, Cambridge. ('The Sutton Hoo ship burial; a provisional guide', T. Brit. Mus., 1957, describes the hilt of the sword in detail but does not state that the blade is pattern-welded.)
103 R. F. TYLECOTE: *Journées de Paléométallurgie*, Université de Technologie de Compiègne, 1983, (22–23 Feb.), 155–67.
104 V. I. EVISON: *Archaeologia*, 1955, **96**, 159–95.
105 A. GREEN and H. R. SINGLETON: 'City of Sheffield Museum, Annual Report, 1955–56', 9 and Pl. 2*b*.
106 R. E. OAKSHOTT: *Antiq. J.*, 1951, **31**, 69–71.
107 D. B. HARDEN: *Oxon.*, 1952–3, **17–18**, 261–2.
108 S. P. O'RIORDAIN: *Proc. R. Ir. Acad.*, (C), 1948–50, **52**, 39–111.

9 Water-powered bloomeries: the end of an era

It is appropriate to open this chapter with a discussion on the introduction of water power. At the time of the Domesday survey it is estimated that Britain had over 5 000 water mills for the milling of grain,[1] and it is likely that these were of the 'Norse' type with a vertical shaft and a horizontal wheel like that at Tamworth (Fig. 142). These can still be seen at Millbridge in Orkney, and in Shetland, and are more suited for small flows. Furthermore these are simpler to make as they do not require gearing like the Vitruvian vertical-wheeled mill. Wulff[2] states that in the Near East this type used a wooden jet with a bore of 11·4 cm and 7·6 m head, running at 164 rpm to give about 10 hp (7·5 kw). Clearly this amount of water is not always available; then the jet may be reduced to 7·5 cm and the rate to 151 rpm to give 4·5 hp. The bottom bearing is like the one shown for the Tamworth mill and could be adjusted to raise the stone by means of a lever. A pair of millstones (1·52 m in diameter) requires 4 hp (3 kw); most 19th century mills had at least 2 pairs and required 8 hp, which was normally obtained from a Vitruvian-type water wheel.

While the Norse-type mill is difficult to apply to the working of a hammer it can be adapted to working bellows, as the Chinese have shown[3,4] (Fig. 143). We do not know whether any of the early Norse-type mills were adapted in this way, but since the bellows would be the most used part of an early iron mill of the Domesday period it is quite possible that the Domesday iron mills were using this type of bellows' drive and that the hammer was manual. It is estimated that a mill like the Tamworth mill could produce 1 to 2 hp,[5] which is more than sufficient for bloomery bellows.

The earliest reliable allusion to the existence of a corn mill in England occurs in AD 762 in a charter granted to the owners of a monastic mill in Kent.[1] This is almost certainly of the Vitruvian type as this was being introduced by the monastic orders from the Continent.

One of the earliest dated and excavated sites to produce a mill is Chingley, Kent,[6] where a race dated to the 13th century has been found. Unfortunately this was not accompanied by much slag and is therefore more likely to have powered a cornmill or fulling mill with a Vitruvian wheel. There were at least 63 fulling mills in use in England in the 14th century[7] which worked, in principle, very like a powered hammer; there is little doubt that water-powered hammers were known as early as fulling mills.

The Wakefield court rolls have produced the earliest reference to a mechanical hammer in Britain.[8] It is now certain that the word 'oliver', which was once considered to be a mechanical hammer – not necessarily water-driven (Fig. 144) – refers to the whole mill in the same way as 'hammer' refers to Sussex forges at a later period. Such an oliver existed in the Calder valley in 1349/50 and required underwood for its working. It is argued that since the oliver was sited near Wortley, Halifax, which is a hilly area with numerous streams near or on the Coal Measures, it was an ideal place for a water-powered bloomery forge with a hammer.

The power required for bellows is less than 1 hp (0·75 kw), but 5 hp is required for a hammer; thus, while human power can be used for bellows, even for long periods, it is not a very practical proposition for a mechanical hammer used for breaking up large blooms (50 kg upwards). However, the blooms of this period were small – not exceeding 20 kg – and therefore available water power was more likely to be used for bellows.

We have some data of the power required for bellows driven by Vitruvian-type wheels on the 17th century site of Rockley Smithies. Here were found two overshot wheels 30–36 cm wide by 3 m in diameter, driving bellows which were calculated by Dr Strange (in Crossley[9]) to require 30 to 45 l/sec to provide about 1 hp (0·75 kw) each. This agrees with similar estimates made in regard to the late bloomery at Muncaster Head.[10] As far as hammers and fulling mills are concerned about 8 hp (6 kw) is needed, i.e. 8 times the power required for one set of bellows.

By 1408, when the bloomery at Byrkeknott in Durham was built,[11,12] there is no doubt that the main user of power was the hammer, since we know that the bellows were often, if not always, worked manually. Perhaps they were worked in the manner shown by the woman in Percy's drawing from Swedenborg.[13,14]

It is clear that the trend towards the application of water power to bloomery hammers was underway by 1352 at Creskeld.[11] Of 63 fulling mills in use in England in the 14th century, 30 were in the North (West Yorks.

142 Norse–Arab type water mill at Tamworth (after Rahtz[5])

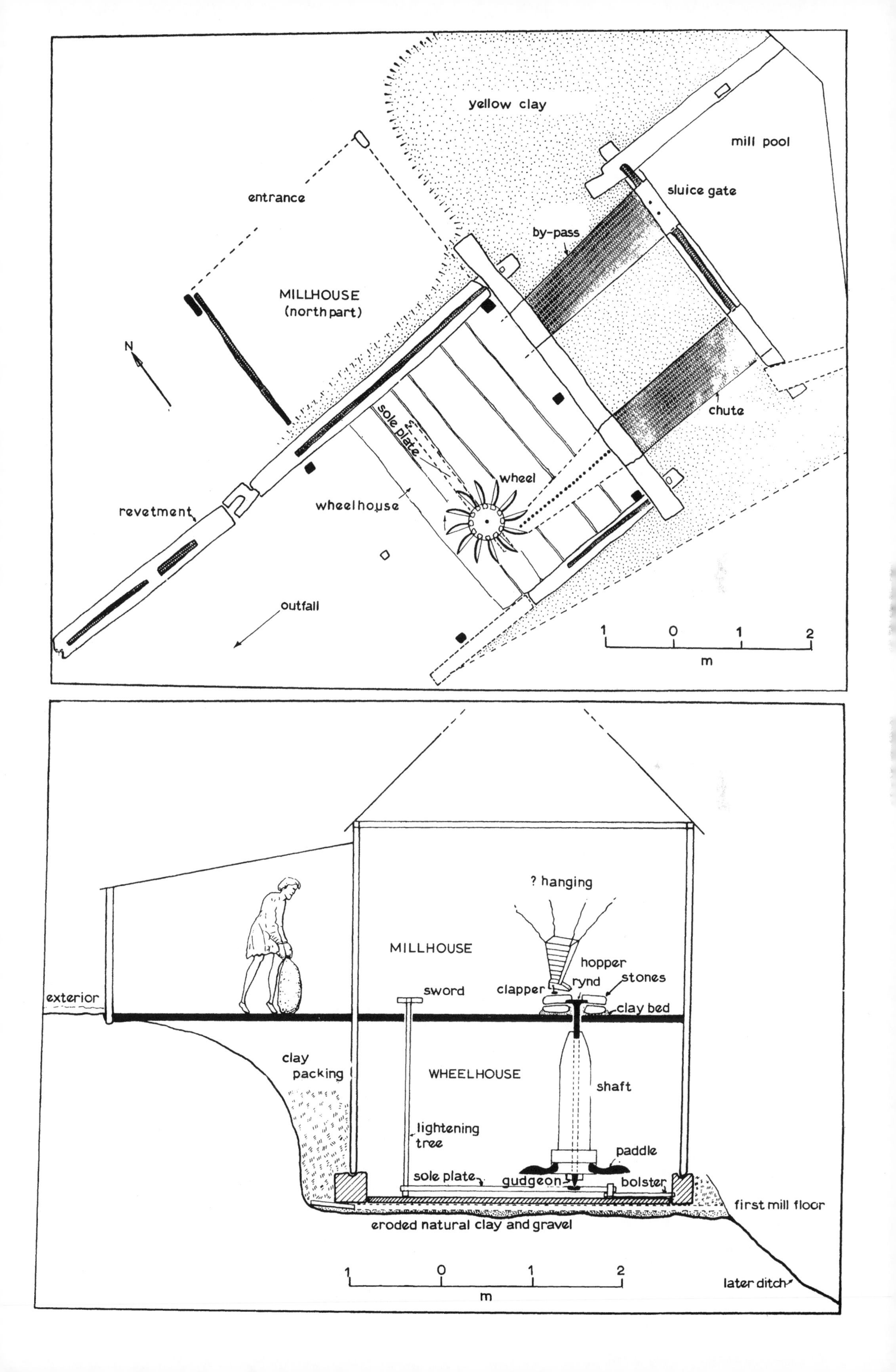

yellow clay
mill pool
sluice gate
entrance
by-pass
MILLHOUSE
(north part)
N
chute
sole plate
wheel
wheelhouse
revetment
outfall
1
0
1
2
m
? hanging
MILLHOUSE
hopper
rynd
stones
clapper
sword
clay bed
exterior
clay
packing
WHEELHOUSE
shaft
lightening
tree
paddle
sole plate
gudgeon
bolster
first mill floor
eroded natural clay and gravel
1
0
1
2
m
later ditch

143 Chinese vertically-shafted, mill-driving bellows; about 1313 AD (after Wertime;[3] courtesy *Technology and Culture*). It is more likely that the rotary drive would be geared down rather than up, as these wheels normally run too fast for bellows.

and the Lake District) where water power was most easily available.[7]

The available head for an overshot wheel at Byrkeknott was 15 ft (5 m).[15] The 100 m long leat had a cross section of 1·2 × 1·2 m and should be able to carry water at a rate of 0·5 m/sec (0·68 m³/sec).[16,17] The so-called mill pond is no more than a head race and would only hold water at this rate for about 17 min, but such a flow applied to a 3 m diameter, well constructed, overshot wheel would provide a power of 24·5 hp (18 kw). If we assume that such an early wheel is not up to modern standards we can divide this by two and get at least 10 hp (7·5 kw). If we assume the wheel is 1 m wide, it would consume about 0·21 m³/sec to produce this power. So the full capacity of the leat would not be needed most of the time; 8–10 hp would be sufficient for one hammer and two hearths.

144 Treadle hammer (after Schubert;[25] courtesy Routledge and Kegan Paul)

IRON ORE MINING

Apart from the brief reference to the iron 'mine' at Lyminge in Kent in AD 689, and the exploratory drift of the Roman period at Lydney, it is only in the later medieval period that substantial evidence is found for the deep mining of iron ore.[18] At Evenwood in Durham ore is found beneath the shales in the Coal Measures. In 1959 open-cast coal excavations were being carried out in this area, and one of these revealed the filled-in 'bell-pits' from which iron ore was extracted in medieval times.[15] Today it seems astounding that at a depth of about 3·7 metres a 1·2 m thick coal seam was penetrated without attempting to work it; such pits were continued down to 7 m, at which level the deposits of nodular iron occur. Some of the coal was back-filled into the disused pit, showing the disdain with which medieval people viewed this now valuable material. It was useless for iron-smelting on account of its sulphur content, and presumably the houses of the time were ill-adapted for its burning.

Similar workings were found near Sedgeley on the Worcestershire–Staffordshire border in 1890.[19,20] No signs of these were visible on the surface as they were covered with a level, 38 cm thick layer of topsoil. Under the topsoil were found 1·5 m diameter openings to bell-pits which went down 4·5–6 m, increasing in diameter to 3·7 m at the bottom. The ironstone lay immediately below a thick coal seam and, as at Evenwood, no real attempt had been made to cut into the coal; again the coal accidentally removed was found in the back-filled pits. Here the iron ore workings had penetrated as far as possible from the conical shaft – a distance of 2·5–3 m – and timbering would have been necessary to penetrate any further. Similar pits to these have been made into the seam of Tankersley ironstone near Huddersfield[21] and near Barnsley, Yorks. (Fig. 145).

ROASTING

While roasted ore was found at Harthope Mill,[15] no roasting hearth was located; this is more likely to have been near the woods and coal pits. At Rockley,[9] however, a well burnt area was found belonging to the earliest period (16th century), clearly used for roasting. This measured 2 × 2 m and contained bloomery slag, red gritty dust (fines), charcoal dust and clay.

EXCAVATED SMELTING AND SMITHING SITES

Tudeley, Kent,[22] is the best documented site of the pre-water power period, and has given us considerable insight into conditions of iron-working just before and after the Black Death. The resulting shortage of labour which followed undoubtedly assisted the move to the sources of water power which had already begun in the period covered by the previous chapter.

This process was continuing in the north with its greater water power potential. In Weardale in 1408, the Bishop of Durham, Langley, decided to enter the iron-smelting business on his own account instead of farming out his property for the purpose. He set up a

145 Aerial photograph of bell-pits (after M. Beresford;[39] courtesy CUP; Crown copyright reserved)

bloom-smithy near the Bedburn and appointed John Dalton as ironmaster.[11]

The record of the building of this bloom-smithy and the accounts for a year's working have survived. Unfortunately, the exact position of the site is unknown, as the name Byrkeknott, given to it by the bishop, is now no longer in existence.[12] Nevertheless, certain names mentioned in Lapsley's transcription of the account roll and the description of the site suggested to the present author that it might be identified with a now disused cornmill known as Harthope Mill.[15] Excavation below the floor of the ruined mill revealed a ferruginous working floor containing iron ore, charcoal, slag, and potsherds, datable to the 14th–15th centuries. There is no doubt that this mill with its leat and head race was the scene of iron-smelting operations involving water power in the 14th or 15th century.

The construction of the leat, which is about 100 m long, took four or five people thirty days. A forge house was constructed by a carpenter, so it must have been timber-framed, probably with wattle-and-daub walls. A bloom hearth and a string hearth were made, both with the assistance of a bloomer from another bloomery. The string hearth would be used for working and cutting up the bloom after smelting. The carpenter constructed a water-wheel, which could have had a diameter of about 5 m, this being the head of water available at Harthope. The wheel had sufficient power to blow the bellows for the two hearths and a hammer for most of the year, but the record mentions several times that the wife of one of the workers had to help with the bloom-hearth bellows. It has been suggested that this means the bellows were manually operated as at Tudeley, and that the power was used for a hammer. However, there is mention of a 'great hammer' and considering the power available it is possible that this was water-driven. However, it is also possible that the water supply was not always sufficient and the bellows sometimes had to be manually operated.

The total quantity of ore used in the year (170 tons) is given under one item in the account roll, suggesting that it was all delivered at the beginning of the financial year which ran from Michaelmas to Michaelmas (November 1408–November 1409). The ore came from the mines of Sir Radulphus Eure in the Evenwood district where there was a bloomery in 1368.[23] The ore found at Harthope Mill is of the nodular type, and could well have come from the bell-pits revealed by open-cast mining at Rowntree Farm in 1959.

In the Byrkeknott account roll there is the passage: *'sindres a dicto campo de Hopyland . . . pro ferro novo ibidem cum ejusdem temperando'*. This seems to refer to the use of slags made in an earlier period to modify the iron then being made. All the slag analyses show that the iron contents of bloomery slags often reach 50% and have therefore been made from good ores according to modern standards. But the process worked at Byrkeknott was no different from that of previous periods and there is no substantial difference in the iron content, as can be seen by comparing the analyses of the

Table 98 Input, output and yield at Byrkeknott bloomery AD 1408–1409

Total ore supplied for one year	170 tons
Iron produced in one year	24·3 tons
Weekly output	545 kg
Size of bloom	87 kg
Overall yield; kg iron/kg ore	14·3%
* Actual yield; kg iron/kg iron in ore	33·0%
Charcoal consumption (2 hearths per week)	12·9 tons
Charcoal rate; per kg of iron smelted	12·0 kg
Total weight of slag produced per year (ash content of charcoal = 3%)	155 tons

* Based on analysis of ore from Harthope Mill

slags from Harthope Mill with those of other periods (*see* Tables 91 and 100).

In one respect the analyses of ore and slag from Harthope Mill do not agree; this is in regard to manganese. The ore which was found on the site only contained small amounts of manganese (0·88%), which is a characteristic of nodular ores in general. Yet the slags contained from 2·44–6·73% MnO, which suggests that material from another source was added to the smelting charge. In a field some 250 m away from this site was found some slag datable to the 12th century, and therefore to the same period as High Bishopley.[24] During this period bog iron ore was used, which has an MnO_2 content of 15·5% and which yielded slags of similar manganese content. It is suggested that such slags were found somewhere in the Hoppyland area, near Harthope Mill, and used in the process. It is not expected that the result would modify the iron in any way, since under normal conditions manganese is not reduced in the bloomery process and therefore would not be taken into the iron itself. But it would slightly improve the yield of iron and might have a good effect on the free-running characteristics of the slag. The yield could be improved by taking low silica slags with about 20% SiO_2 and raising the smelting temperature from the minimum of 1 150°C to about 1 250°C so that the slags are capable of absorbing more silica, thus raising the total to 30–35% SiO_2. The manganese in the ore would replace iron in the slag, so releasing more of the available iron.

Table 99 Bloomery yields

Date AD	Provenance	Weight of iron/ Weight of ore kg of iron/ 100 kg ore	Prob. weight of iron/ Weight of iron in ore; kg of iron/kg of Fe in ore	Ref.
1240	Coupland (Cumb.)	17·5	29·0	25
1409	Byrkeknott (Durham)	14·3	33·0	15
1531	Llantrisant (Glam.)	12·5	19·0	25
1540	Rievaulx (Yorks.)	12·0	27·0	25
16th cent.	Low Moor (Yorks.)	7·0	24·0	35, 36
1775	N. Lancs.	33·0	55·0	37
1841	Pyrenees	33·0	72·0	13

On the assumption that Harthope Mill can be equated with Byrkeknott, the author has calculated the yield for this particular bloomery, and the details are given in Table 98. With the exception of the slag produced and the iron content of the ore, the data in the table are from the account roll itself. Comparison with other bloomeries of the period shows that the yield of 33% of the iron in the ore is a good one (Table 99).

Discussion of bloomery yield requires an awareness of certain facts that are not always brought out in published figures. Because the actual iron content of the ore is often not known, the figures for yield from primitive bloomeries are based on yield of iron per kg of ore. It is presumed, since the ore has been roasted near the mine and not at the bloomery, that we can regard this yield as being from the roasted ore and not the raw ore as mined. But this is by no means certain. The figures given in the first column of Table 99 are taken from the records and it is assumed that they relate to roasted ore. Those in the second column are based on the probable or actual iron contents of the ores used. We see, therefore, that even with the best ores in this country, the real yield was only about 30%.

Since we are now entering a historically well recorded period, which has been discussed by Schubert in his work on the history of the iron and steel industry,[25] there is less need for the excavator. But many metallurgical aspects are not discussed in the records and it is possible for some of these problems to be settled by excavation of selected sites, and examination of the metal, slags and ores found on them.

Many of the old bloomery sites were converted into finery forges after the arrival of the blast furnace around AD 1500. Examples are: Stony Hazel in Cumbria and Woolbridge, Mayfield and Brookland in the Weald,[26] but we are lucky to have two excavated sites which ended their activities as water-powered bloomeries.

Rockley Smithies is one of the best bloomery sites so far excavated as it gives evidence of three hearths but, apparently, no powered hammer. The water was supplied by a pond fed from a leat, and water was fed to overshot wheels, as shown diagrammatically in Fig. 146. The wheels were narrow and only capable of powering bellows for the bloomery, string and chafery hearths. The bellows' rooms were found in all cases, but very little of the leather and wood of the bellows themselves. The bloomery hearth existed to a height of 65 cm, but it is not clear where the tuyere entered (Fig. 147). The slag would have been tapped off in the passage leading from the north-west of the hearth. The working platform of the hearth was 60 cm above the slag-tapping level, which indicates the minimum effective height of the hearth. This should be compared with the 1 m of the 1840 Catalan hearth as shown by Percy.[13]

Beside the bloomery hearth was a water butt let into the floor of the platform and part of an old millstone which could have been used as an anvil while the bloom was being rid of loosely adherent slag and charcoal. The other two hearths were let into the floor but

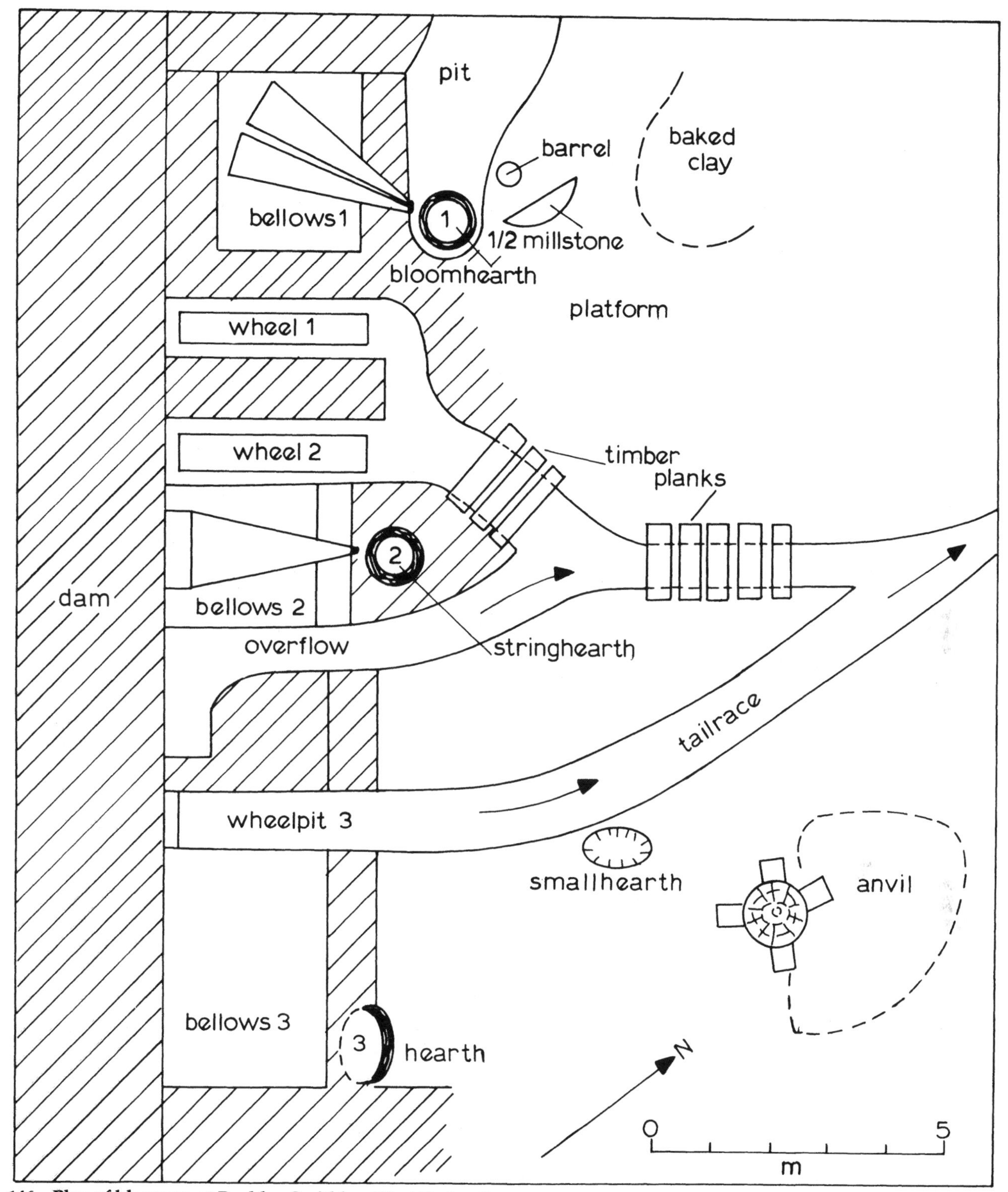

146 Plan of bloomery at Rockley Smithies, West Yorks. (based on Crossley and Ashurst[9])

had no arrangements for slag-tapping. It is for this reason that it was decided that they were string hearths or chaferies.

An anvil base was found surrounded by hammer scale, but it was not possible to connect this with a race and so have it water-powered. It is possible that the hammer was of the mechanical type well known to chain makers in the 18th century (Fig. 144 *above*), but this is very unlikely in such a well ordered works. It is also possible that another and larger wheel was sited in the unexcavated area to the south of the site or, perhaps, operated from the south wheel-pit.

Muncaster Head in Cumbria was probably the last bloomery to be established in the British Isles. It was built on a virgin site in 1636 and lasted until about 1720, by which time the blast-furnace with its attendant finery forge had been introduced into the area. The site showed evidence of a large race-and-wheel pit but little sign of either a hearth or a hammer. There was a large quantity of very high-grade hematite ore and furnace bottoms of slag together with the remains of

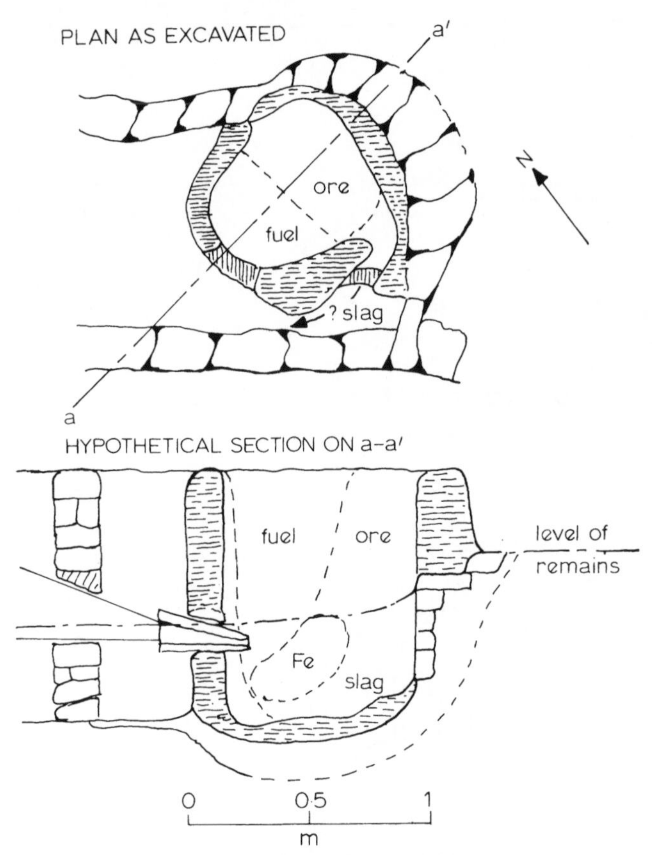

147 Bloomery furnace at Rockley Smithies (after Crossley and Ashurst,[9] with a possible reconstruction by the author)

one of the blooms. On the basis of these remains and details of the Catalan hearth given by Percy,[13] the reconstruction shown in Fig. 148 is offered.

The original agreement of 1636 referred to a hammer, anvil, and three bloomery hearths with bellows, and there is no doubt that this site had a hammer of the type shown. It is estimated that the power required and available was about 10 hp (7.5 kw).

The ore found was very rich (*see* Table 68, Chapter 6), and from the slag composition (Table 100) it is certain that some ore with a higher phosphorus content, or charcoal with a high bark content was being used. If it was the ore, this would have been from the West Cumberland Coal Measures and, not being so rich, would help to 'flux' the high-grade local hematite ore.

The furnace bottoms weighing as much as 22 kg were a surprise at this late date, as the Catalan process suggests that all the slag was tapped. It appears that in this case the hearths were deep and not all the slag was able to leave during tapping. The height of these bottoms suggests that the tuyere was at least 30 cm above the hearth bottom. This is very different from Rockley and seems to suggest that the principle of the slag-pit furnace may still have been in fashion in 17th century Cumbria. History does not relate what happened to the slag in the bloomery furnaces of Evenstad[27] and Busch,[28] but it is certain that it was not tapped (*see* Fig. 149).

Amongst the metal-working remains found in the Abbey at Kirkstall near Leeds[29] were lumps of tap slag

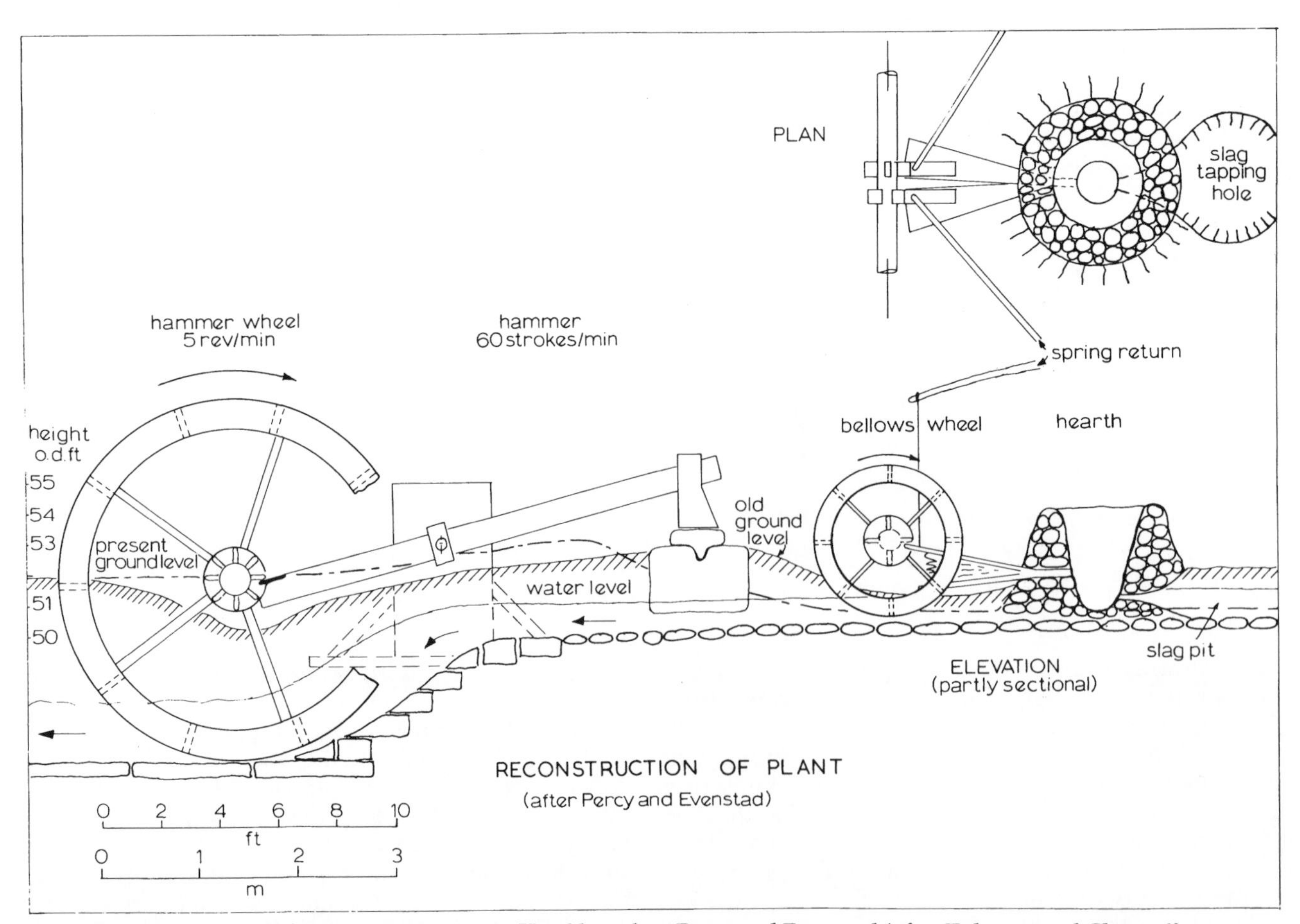

148 Reconstruction of bloomery at Muncaster Head based on Percy and Evenstad (after Tylecote and Cherry;[10] courtesy Cumberland and Westmorland Archaeological Society)

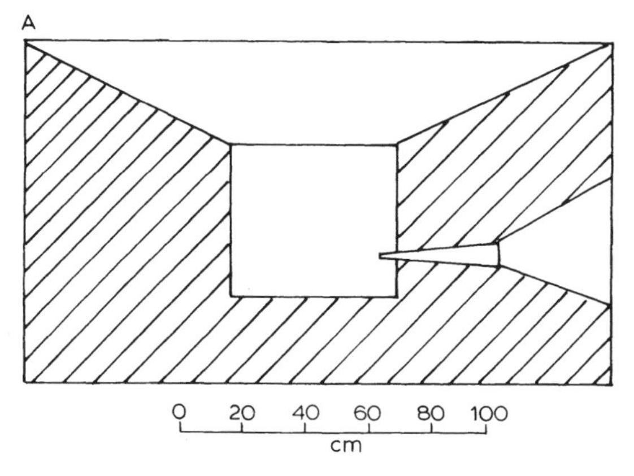

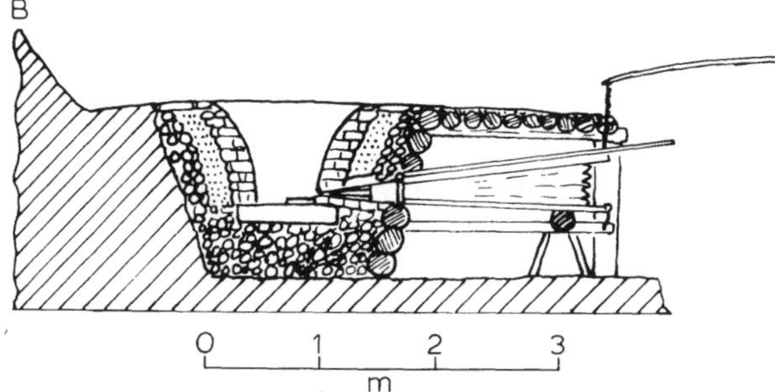

149 Two Scandinavian bloomery furnaces (A, after Busch[28] and B, after Evenstad[27])

which had the composition given in Table 100. It is recorded that land was granted for making a mill race from the River Aire prior to AD 1220, which makes this site one of the earliest definitely known to use water power. The ore undoubtedly came from the nodular ore beds in the vicinity.

Another water-powered bloomery site is that at Bourne Pool Aldridge, Staffs.[30] Documentary evidence equates this with the 'iron mill' of Sir Simon Montford who was executed for treason in 1495. The excavator believes that the wheel was used for a hammer rather than bellows. The mill used the ore from the Walsall–Rushall area (composition in Table 68, Chapter 6) and roasted it before smelting. The slag was a relatively high alumina slag (shown in Table 100), and the phosphorus content of the iron would be in the range 0·1–0·2%.

Historically recorded details of the bloomery process

Before we close our discussion on bloomeries it would be useful to look at the Catalan hearth of 1840 as described by Percy[13] and the Scandinavian bloomery as depicted by Evenstad[27] and Busch.[28]

The Catalan hearth was 104 cm high and 100 cm wide at the top, narrowing to 60 cm at the bottom. As we have said, the ore was charged down one side and the fuel down the other (tuyere) side. It is possible that the Rockley hearth was of this type, but it is difficult to see how we could get a front-wall height exceeding 70 cm unless the hearth was partly above the upper floor level (as shown in Fig. 147 *above*). This would be quite possible.

At Rockley the ores would be of the normal Coal Measure quality and might give a yield of 30% with much tap slag. Busch[28] makes no provision for slag-tapping, but a richer ore might explain the relative absence of slag in this case. However, Evenstad[27] puts great emphasis on 'blending', and it is quite clear that the bloomery process cannot be worked on rich ores alone as the slag volume is not then great enough. This was found in our experiments with shaft furnaces; with rich ore a very porous and loosely sintered bloom was formed which was difficult to consolidate.[31] So here we have a reason for the addition of a leaner ore to the rich Eskdale ore at Muncaster Head.

The working of the Scandinavian bloomery (manipulative type) is best seen in the description given by Busch,[28] which is basically the same but a little clearer than that given by Evenstad.[27] Since one has to stand above the furnace to manipulate it, the position of the hearth at Rockley was ideal. Once a good fire is going in the hearth, either by charring wood or by the addition of ready-made charcoal, finely-divided, blended ore is sprinkled over the whole surface of the hearth. The blast can be regulated by control of the water supply to the 1·5 m diameter bellows (breast) wheel running at 7–11 rpm. Ore charging was repeated at 3 min intervals and the whole packed down with a rake.

Table 100 Composition of slag from water-powered bloomeries (% by weight)

Ref.	10 Muncaster Head 17th cent. Tap slag	Furnace bottom	34 Rushall (Staffs.)	34 Bourne Pool (Staffs.) 15th cent.	34 Nuns Well (Staffs.)	15 Harthope Mill (Durham) 15th cent.	34 Burbage (Sheffield) <1300	34 Rockley (W. Yorks.) 17th cent.	38 Kirkstall (Leeds) 13th cent.
FeO	62·3	34·6	40·10	35·40	47·5	38·96	46·36	37·20	(63·7)
Fe_2O_3	—	32·7	3–13	9·0	6·90	4·98	1·32	2·40	—
SiO_2	23·9	15·7	31·10	23·44	24·20	29·44	30·20	26·90	20·8
CaO	4·5	4·7	6·40	7·00	1·10	5·46	2·79	7·00	
Al_2O_3	5·5	4·0	10·29	8·64	14·50	9·68	10·01	13·70	1·83
MnO	0·1	<1·0	1·03	2·14	2·50	2·44	1·95	2·09	
MgO	0·7	<1·0	6·50	6·16	0·30	4·90	3·24	1·80	
TiO_2	0·2	<0·5	—	—	—	0·40	0·03	—	
K_2O	—	0·74	—	—	—	—	} 2·10	—	
Na_2O	—	0·15	—	—	—	—		—	
S	0·01	0·08	0·03	0·01	—	0·14	—	—	
P_2O_5	0·65	0·29	0·40	1·0	3·20	0·36	1·18	—	
H_2O						1·20			
CO_2						0·33			(C) 7·8

— = not sought; ( ) = includes any Fe_2O_3 if present

Table 101 Comparison of the efficiency of the bloomery process with a Swedish charcoal blast furnace (after Busch[28] and Percy[13])

	Scandinavian bloomery 1851 (after Busch)	Catalan bloomery 1840 (after Percy)	Blast furnace Kalmar, 1851 (after Busch)
Ore, blended (kg)	32·2	510	8 270
Wood and* charcoal (kg)	41·0	460	5 200
Bloom (kg)	9·74	—	—
Cast iron (kg)	—	—	1 862
Finished wrought iron (kg)	6·70	152 (165)	1 241†
Time (h)	3	6	24
Slag (kg)	—	189·5	—
Ratio fuel:ore	1·27	0·9–1·09	0·63
'Coke' rate; Fuel:iron (kg)	6·12	2·78	4·35
Fuel (kg/h)	13·6	76	2·17
Yield % Fe:ore	21	33	15·0‡
Yield % Fe/Fe in ore	42§	72	30

*50% Fe in ore. Assumes packed density of dry wood and charcoal is 200 kg/m³.
†Based on 2/3 recovery in the finery.
‡Busch claims that the bloomery is more efficient than the BF in yield of iron from ore; this seems to be borne out by the results, but is very unusual (see Table 104).
§Assumes 10% weight of iron as fuel for conversion.

Then a hole was pierced through the middle of the charcoal (and ore?) pile and some stove ash (wood ash) was added. Small charcoal from just above the tuyere was pushed into the centre and replaced with a larger piece to encourage the deflection of the blast towards the opposite wall. Then the ore from the sides was repeatedly pushed towards the central hole. More charcoal was placed above the tuyere and more ash was needed. The slag could be seen to be flowing freely (away from the tuyere area).

This process sounds very complicated. It would seem that the ore is spread on the surface in order to be reduced by the upward-flowing gases. Then, when reduction is complete, it is pushed into the central hole where it is quickly separated into reduced metal and slag which moves downwards – the slag flowing to the periphery of the bowl, while the iron agglomerates in front of the tuyere from which it must be removed before it burns. The parameters of this process according to Busch are given in Table 101 and compared with the blast furnace process. As a producer of ductile (wrought) iron, the latter does not show any advantage over the bloomery process.

The Catalan process as described by Percy[13] seems to be far simpler and more logical, with its division of fuel and ore. In terms of iron production there is not much difference between the two bloomery techniques, but the system of making charcoal *in situ* in the furnace in the Scandinavian process could be expected to be more thermally economic.

The advantages of the direct process led to its continuing use in North America and Russia;[32] in both cases a hot blast was used. The Russian furnace was an improved version of the Stucköfen based on a design by Husgafel.

LATE BLOOMERY IRON AND ITS COMPOSITION

There is little iron that can be related to this period. The piece of bloom adhering to the slag furnace bottom from Muncaster Head had the low phosphorus content that is associated with hematite ores (Table 102). These figures are compared with others from late bloomeries, such as the Catalan hearth and the hot blast bloomeries of the late 19th century. It is unlikely that any of the Mn and Si are in the metal itself – both are present in the residual slag inclusions in the bloom.

THE COMPOSITION OF LATE BLOOMERY SLAGS

The composition of the slags from water-powered bloomeries is given in Table 100 (*above*). As far as the silica content is concerned there is no change from the previous period, but the big difference is in the

Table 102 Composition of metal from late bloomeries

Ref. %	10 Muncaster Head *c.* 1650	13 Catalan hearth 1840	32 North American hot-blast bloomery, 1884	32 Russian hot-blast Stuckofen, 1886
C	0·02–0·2	tr.–0·42	0·1–0·3	0·07
Mn	nil–0·01	0·025–0·531*	0	—
Si	0·01–0·33*	0·020–0·037	0·018	0·08
S	0·005–0·017	—	tr.	0·08
P	0·059–0·097	—	0·015	0·29–0·85
Cr	0·01	—	—	—
Ni	0·01–0·02	—	—	—
Slag	—	4·5–6·32	0·25–0·50	—

* = from slag?; tr. = trace

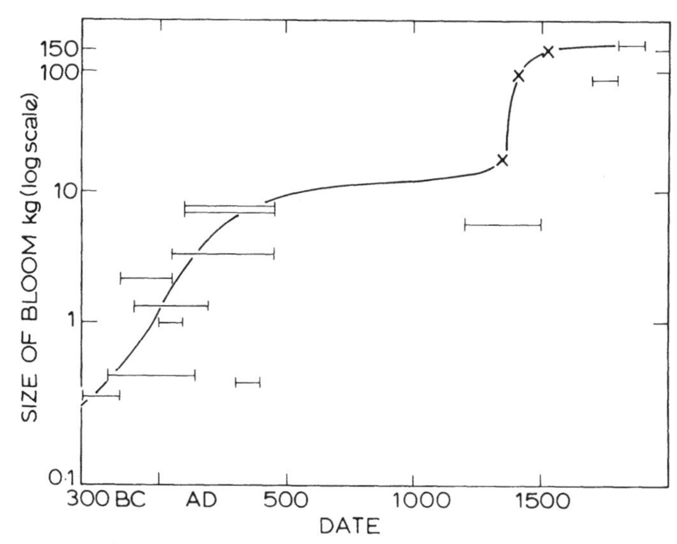

150 Increase of bloom weight with time

alumina content, which reflects the use of Coal Measure carbonate ores found in shale beds, and the Clay Ironstones of the Weald. As can be seen from the analysis of the Northumbrian Coal Measure ores (Table 68, Chapter 6), alumina need not be high in the nodule itself. However, if such nodules are not encouraged to weather and lose their adherent shale, some of this may be taken up by the charge. This could be beneficial, as the system $FeO–SiO_2–Al_2O_3$ at 15% Al_2O_3 has a free-running temperature in the range 1 088–1 205°C, depending on the SiO_2 content.[33] Higher contents of Al_2O_3 give rise to a new phase – hercynite ($FeO.Al_2O_3$) as shown by Morton and Wingrove[34] and others.[31].

Some of the Jurassic ores, such as those from Northants., also contain high alumina; but when the lime is also high, anorthite will be formed. The need to increase the lime content to form anorthite may be the reason for the apparent addition of lime at Bourne Pool,[30] although normally there is sufficient lime in the fuel ash to provide all that is needed for this purpose.

The phosphorus content of the Coal Measure ores is moderate, and the 0·1–0·2% that is likely to be present in iron from this source is not likely to have much effect on strength, although it will slow down carbon penetration during carburizing (*see* Table 73, Chapter 6).

CONCLUSION

Since this is the last chapter discussing the bloomery process, which for nearly 2 000 years was the only process used for the production of iron in this country, it is useful to conclude with a brief note showing its progress.

Fig. 150 shows the growth in the size of the bloom produced. The detailed figures on which this curve is based are shown in Table 103. There is a steady growth in the size of bloom produced, from the 200 g of the pre-Roman period up to the 15 kg or so of the mid-14th century. Only then is a sudden increase apparent, which appears to coincide with the introduction of the water-driven hammer.

There is too little bloom material to decide the average composition of the metal produced, and the figures given in Table 102 cannot be taken as in any way representative. But from the author's personal experience with metallographic examination of nails and other artifacts that do not include edge tools, the average level of phosphorus must have been high. This is in marked contrast to the steel of the edge tools, which has been selected because of its low phosphorus content. The high phosphorus level is borne out by the slag analyses given in Table 100 (*above*). Normally the phosphorus content is about one quarter of that present in the slag.

The slags from the water-powered bloomeries are not very different from other medieval slags. The silica content is about 30% and there is a nice balance between the lime and the alumina contents, as can be expected from the use of Coal Measure nodular ore. The use of such ores and the lower iron in the slag produced from them was instrumental in raising the yield.

NOTES AND REFERENCES

1 M. T. HODGEN: *Antiquity*, 1939, **13**, 261–77.
2 H. E. WULFF: 'The traditional crafts of Persia'; 1966, London, MIT Press.
3 T. A. WERTIME: *Tech. Cult.*, 1964, **5**, 391–7.

Table 103 Weights of individual iron blooms

Provenance	Date	Weight (kg)	Bellows power
Swallowcliffe Down (Wilts.)	300–150 BC	0·23	Hand or foot
Ewell (Surrey)	200 BC–AD 150	0·34	Hand or foot
Wookey Hole (Som.)	150 BC–AD 50	2·05	Hand or foot
Hengistbury Head (Hants.)	1st cent. AD(?)	0·91	Hand or foot
Forewood, Crowhurst Park (Sussex)	EIA–Roman	1·25	Hand or foot
Nanny's Croft (Sussex)	Roman	0·30	Hand or foot
Corbridge (Northumb.)	3rd–4th cent. AD	6·8	Hand or foot
Catterick (N. Yorks.)	Roman	7·7	Hand or foot
Coed Newydd (Anglesey)	Roman	3·6*	Hand or foot
Carrigmuirish (N. Ireland)	Medieval	3·6	Hand or foot
Fermanagh (N. Ireland)	AD 500–1000	5·5	Hand or foot
Tudeley (Kent)	*c.* AD 1350	20	Hand or foot
Byrkeknott (Durham)	1409	89	Water
Rievaulx (Yorks.)	1541	131	Water
Pyrenees	18th cent.	80	Water
Pyrenees	19th cent.	150	Water

*Possibly welded; EIA = Early Iron Age

4 N. BARNARD: 'Bronze casting and bronze alloys in ancient China'; 1961, Canberra.
5 P. A. RAHTZ: in: 'Medieval Industry', Research Report No. 40, (ed. D. W. Crossley), 1–15; 1981, London, CBA.
6 D. W. CROSSLEY: 'The Bewl Valley ironworks, Kent. *c.* 1300–1730 AD, R. Archaeol. Inst. Monograph, 1975.
7 P. N. WILSON: *Water Power*, 1952, **4**, 308–13.
8 H. JEWELL *et al.*: *J. Hist. Metall. Soc.*, 1981, **15**, (1), 39–40.
9 D. CROSSLEY and D. ASHURST: *Proc. Medieval Archaeol.*, 1968, **2**, 10–54,
10 R. F. TYLECOTE and J. CHERRY: *Trans. CWAAS*, 1970, (N.S.), **70**, 69–109.
11 G. T. LAPSLEY: *Econ. Hist. Rev.*, 1899, **14**, 509–21.
12 R. A. MOTT: *J. Iron Steel Inst.*, 1961, **198**, 149–61. (The original document has an inkblot over the first letter of the name which was believed to be 'Byrkeknott' by Lapsley; although Mott prefers to read this as 'Kyrkeknott', we prefer to keep Lapsley's original reading.)
13 J. PERCY: 'Iron and steel', 254–321; 1864, London.
14 E. SWEDENBORG: 'Regnum subterraneum sive minerale de ferro (De Ferro)'; 1734, Dresden et Leipsig.
15 R. F. TYLECOTE: *J. Iron Steel Inst.*, 1960, **194**, 451–8.
16 D. MCGUIGAN: 'Small scale water power'; 1978, Dorchester.
17 F. L. LONGLAND: 'Field engineering'; 1948, Dar es Salaam.
18 W. B. CRUMP: *Proc. Thoresby Soc.*, 1954, **12**, 300–8,
19 L. MEACHEM: *Trans. Fed. Inst. Min. Eng.*, 1894, **6**, 554–7.
20 T. E. LONES: *Trans. Newcomen Soc.*, 1930–2, **11–12**, 42–54.
21 D. A. WRAY: 'The mining industry in the Huddersfield district', 6; 1929, Tolson Memorial Museum Handbook.
22 M. S. GUISEPPI: *Archaeologia*, 1913, **64**, 145–64.
23 W. HUTCHINSON: 'The History and Antiquities of the County Palatine of Durham', **3**, 418; 1823, Durham.
24 R. F. TYLECOTE: *J. Iron Steel Inst.*, 1959, **192**, 26–34.
25 H. R. SCHUBERT: 'History of the British iron and steel industry from 450 BC to AD 1775'; 1957, London.
26 I am indebted to D. W. Crossley and C. F. Tebbutt for this information.
27 O. EVENSTAD (trans. N. L. Jensen): *Bull. Hist. Metall. Group*, 1968, **2**, (2), 61–5.
28 J. A. W. BUSCH: *Bull. Hist. Metall. Group*, 1972, **6**, (1), 28–32.
29 L. ALCOCK and D. E. OWEN: *Proc. Thoresby Soc.*, 1955, **45**, 62.
30 J. GOULD: *Trans. SSAHS*, 1969–70, **11**, 58–63.
31 R. F. TYLECOTE *et al.*: *J. Iron Steel Inst.*, 1971, **205**, 342–63.
32 H. M. HOWE: *Eng. Min. J. (Suppl).*, 1887 (Mar.), 167–316, (269–74).
33 E. M. LEVIN *et al.*: 'Phase diagrams for ceramists'; 1964, Ohio, Columbus.
34 G. R. MORTON and J. WINGROVE: *J. Iron Steel Inst.*, 1972, **210**, 478–88.
35 J. KERR: *Trans. HSLC*, 1872, **24**, 57–70.
36 L. C. MIALL: *Yorks. Archaeol. J.*, 1870, **1**, 110–15.
37 A. FELL: 'The early iron industry of Furness and district'; 1908, Ulverston.
38 L. ALCOCK and D. E. OWEN: *Proc. Thoresby Soc.*, 1955, **43**, 51–66.
39 M. W. BERESFORD and J. K. S. St. JOSEPH: 'Medieval England; An aerial survey', 234; 1958, Cambridge.

10 The charcoal blast furnace and the finery

The blast furnace represents the final development of the shaft furnace for the production of molten cast iron. Its use in Britain dates from the beginning of the historical period (*c.* AD 1500) and is therefore well documented. Since the early history of the blast furnace in Britain is very fully covered by H. R. Schubert[1] and the technical aspects considered elsewhere by myself,[2] it is not intended to discuss it in detail here.

The limitation of the bloomery process lay mainly in its poor yield. The yield from the best ores rarely exceeded 55% of the iron in the ore during the use of bloomeries in Britain, and many were operating with yields of only about 20%. The remaining iron was lost in the slag, and the use of a proportion of earlier slag (as at Byrkeknott) did little to improve the yield in the Middle Ages.

The ultimate yield of 72% was achieved by the Catalan hearth in the 19th century. The survival of this process in the Pyrenees and other less industrialized parts of Europe was probably due to this large improvement in yield, the low labour costs, and the relatively good fuel consumption of 2·8 kg of charcoal per kg of bar iron produced.[3] It must be remembered that the bloomery was capable of doing in a single process that which later required two stages, since the iron produced by the blast furnace required a conversion process before it could be wrought. This usually meant reheating the cast iron from the blast furnace, and therefore it required almost double the fuel consumption of the bloomery process. Admittedly the bloomery process was often followed by the further process of refining in the string hearth, but this was also needed for breaking down and removing the slag from the large blooms produced by the conversion process. After the advent of the blast furnace, the breaking down process was done in the 'chafery', while the conversion process was carried out in the 'finery'.

The old bloomery process would only work with charcoal; when charcoal became expensive it was natural to look for a process which would use the more widely available fuel – coal. For the same reason that prevented its use in the bloomery (i.e. its sulphur content), coal was not permissible in the early blast furnace or the finery. However, it could be used in the chafery. It was not until the 18th century that iron-blast furnaces could be fuelled with coal or coke, and even then many conversion forges found difficulty in converting pig iron made in this way into malleable iron. Thus the only advantage that the early blast furnace had over the contemporary bloom hearth was that it produced liquid metal which could be cast into the final shape required.

THE INTRODUCTION OF THE BLAST FURNACE TO BRITAIN

The blast furnace was introduced from abroad because it had evolved from a type of bloomery furnace which had not been used in Britain since the Roman period. The shaft furnace of the Ashwicken type died out in Britain, but in Central Europe and Scandinavia it persisted in spite of its disadvantages. By the 18th century the Scandinavian bloomery furnace[4] had a height of 2–3 m while the *Stückofen* or high bloomery furnace of Austria had reached a height of 4·9 m by the 19th century.[5] The height reached by this type when the blast furnace was introduced into Britain is not known, but it cannot have been less than 2 m. A shaft with a height of 2 m or more would enable the ore to be in contact for a much longer time with the charcoal and the reducing gas, carbon monoxide, coming up the shaft. Some of the iron oxide in the ore would be converted to iron carbide (Fe_3C), which would then be oxidized further down in the area of the tuyere. Normally, this oxidation was more or less complete so that a solid bloom was formed, and only by accident would cast iron be obtained. When the accident occurred, the smelter would make sure that the next charge had a rather smaller proportion of charcoal so that a solid bloom was produced. It is not known who first turned this accident to good account and decided that there was a real advantage in tapping the iron from the furnace in a molten condition. The brittleness of cast iron in thin sections would be a disadvantage but the value of casting would be obvious, and was soon put to use in making guns. Since the art of using cast iron was known from at least the 4th century BC in China,[6] it may be supposed that the closer contact with the East in medieval times led to a realization that cast iron could be put to good use. The knowledge of how to produce it must have been available in the western hemisphere, since its accidental production in the high bloomery furnaces must have been a fairly common occurrence. Once its value was made obvious, conversion of the Stückofen type furnaces to cast iron production would have been rapid.

Table 104 Early blast furnace yields (Schubert[10])

Furnace	Date AD	Kg Fe/100 kg ore as mined	Probable Fe/100 kg Fe in ore: cast	Probable Fe/100 kg Fe in ore: wrought
Newbridge (Hartfield) Sussex	1548	12	27	18
Weald	1674	33	74	49
Frith, Sussex	1648	41	92	61
Heathfield, Sussex	1738	27	60	40
Hales, Worcs.	*c*. 1700	29	65	43

As far as Britain is concerned, the first blast furnace definitely known to be in existence was at Newbridge, in the parish of Hartfield, Sussex, at the end of 1496.[7] This appears to have received the encouragement of the Crown, for in 1496 Henry VII commissioned Henry Fyner to engage in the production of iron ordnance for the war against Scotland. The name Fyner is interesting, since fining is the term used for the process of converting cast iron to the wrought iron in use at this time, and it suggests that the process of fining, hence iron casting, must have been carried on for a short time at least before 1496.

According to Schubert, the first gun in English history definitely known to be of cast iron was manufactured at Newbridge in 1509.[8] We know that many of those responsible for the production of iron ordnance had French, Dutch (or German) and Belgian names, and there is now considerable evidence to show that the technique was introduced from the Continent.[9]

Typical figures for the yield of the blast furnace process, which have been taken from Schubert,[10] are given in Table 104. It is clear that by the middle of the 17th century it was possible to achieve a true yield of cast iron of 92%, which may be compared with a bloomery yield of wrought iron of 55% at about the same period. The figure of 92% was not universal, but it is probable that the variation is owing mainly to the working of different types of ores.

A truer comparison is between 'fined' (i.e. wrought) iron and bloomery iron. The last column in Table 104 is based on a yield of 2 parts wrought from 3 parts cast iron, which seems to be a typical conversion factor for the finery. As will be seen, this figure is not so favourable to the blast furnace, but is of course countered by a far more favourable 'coke' rate, i.e. tonnes of fuel per tonne of iron produced.

EARLY BRITISH BLAST FURNACES

Quite a number of British, 16th century, charcoal-blast furnace sites have now been examined or excavated and there is a large amount of data from this period. Many of these sites continued in use until the 17th century in the south-east, by which time charcoal was becoming expensive and the smelting of iron moved to the west and the north of the British Isles. Most of the Wealden furnaces were concerned with gun manufacture and are associated with gun-casting pits. In one case there is a reject gun and part of its associated boring apparatus;[11] in another case there is a complete boring-bar with steel-edged cutting tools.[12]

The iron ores were mainly of the carbonate type as may be seen from Table 105, and for the easiest smelting they needed roasting. There are at least two examples of roasted ores: one from the 18th century Wealden site at Pippingford[11] and the other from the 17th century Northumberland site at Allensford.[13] It is not known how the roasting was done in the Weald at this time, but the site at Allensford has produced a very nice roasting kiln, not unlike a lime kiln (Fig. 151).

Perhaps the most typical and complete plan of a 16th century blast furnace is the one from Chingley, Kent[14] (Fig. 152). This shows the foundations of the square stone casing with openings at the top and right sides supported by the stone 'pillar'. The wooden supports

Table 105 Iron ores used in early British blast furnace practice

Ref. / %	18 Coed Ithel, Gwent 17th cent.	11 Pippingford, Sussex 18th cent. Roasted	11 Hartfield, Sussex (deposit)	14 Panningridge, Sussex 16th cent.	13 Allensford, Northumb. 17th cent. Roasted	— Melbourne, Derbys.* 18th cent.
FeO	—	—	45·5	28·9	35·74	—
Fe_2O_3	87·5	64·0	—	19·4	14·44	68·00
SiO_2	0·5	17·9	8·9	15·3	8·51	13·9
Al_2O_3	0·1	5·8	3·35	3·47	8·73	7·1
CaO	0·1	0·8	6·20	1·87	4·74	4·1
MgO	0·1	1·3	1·84	1·56	0·99	5·4
TiO_2	0·1	—	—	0·38	—	0·1
MnO	0·1	2·8	1·28	1·31	0·85	1·0
P_2O_5	0·08	0·7	0·37	0·45	—	1·40
S	0·01	0·03	0·09	0·14	—	0·01
H_2O	} 10·07	—	4·1	} 22·5	—	—
CO_2		1·60	25·4		(32·7)	—
Na_2O	—	—	—	—	2·29	—
K_2O	—	—	—	—	1·00	—
Total	99·06	94·23	97·03	95·28	100·0	101·00

— = not determined; ( ) = by difference

*Analysed by the United Steel Co. Ltd through the good offices of Dr H. F. Cleere

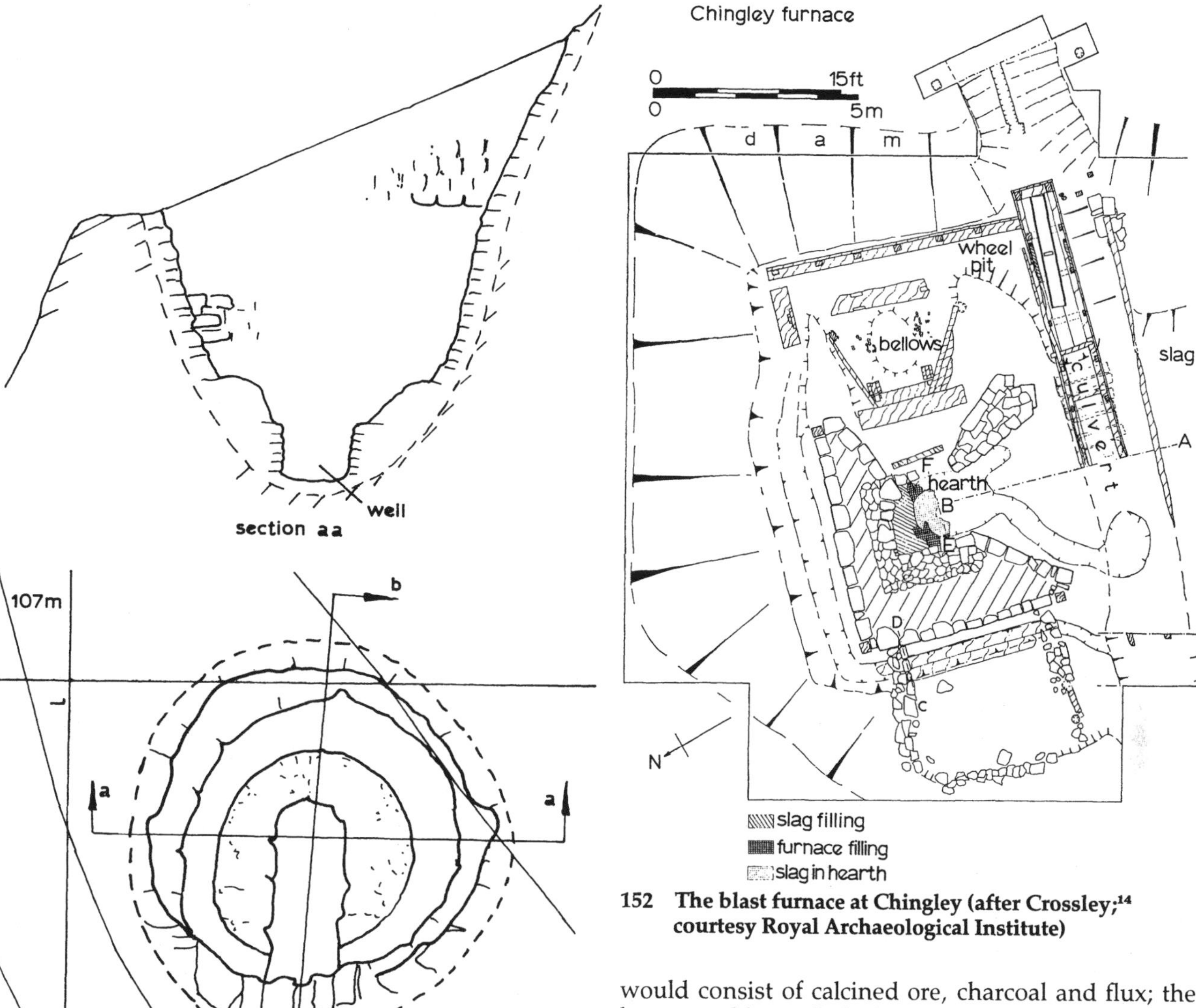

151 The roasting kiln at Allensford (after Linsley and Hetherington[13])

152 The blast furnace at Chingley (after Crossley;[14] courtesy Royal Archaeological Institute)

of the bellows are clearly visible at the top (east side), and the tapping arch on the south side led across the culverted tail race. The remains of a 0·30 m wide wheel were still in position in the wheel pit.

At this time there must have been at least two types of furnace 'profile' or internal shape. One is the shallow bosh with separate crucible mentioned by Swedenborg[15] as being used at Lamberhurst, Kent. The first example of this type was found at Allensford (Fig. 153) and is believed to exist at Wheelbirks,[16] also in Northumberland, and possibly at Ashburton in Devon.[17] The other type is the bosh-cum-crucible in which both components form a cone with an angle of about 70–80°. This is typified by the furnace at Coed Ithel where the cone joins a pyramidal 'inwall' or top portion of the shaft[18] (Fig. 154).

The height of these furnaces would be about 6 m and they would be charged from a 'high-line' reached by a road or ramp which, in the case of the furnace at Chingley, would be on the north side. The charge would consist of calcined ore, charcoal and flux; the latter would usually be limestone or, as in the Weald, ferruginous limestone.

METAL, ORES AND FLUXES

Cast iron was a new material with new uses, and the need for a malleable iron still persisted. It seems that the process of converting cast iron to wrought iron, carried on in the finery and chafery, was introduced from the start, as a finery and hammer are referred to in 1518 at Parrock, Sussex.[19] Therefore it must be concluded that the indirect process with its three stages of blast furnace, finery and chafery was introduced from the Continent as a whole. It would be logical to make use of the bloom hearths to convert the iron.[20] In most cases, probably for reasons of water power, the blast furnace was not located near the finery, but the finery and chafery were usually grouped together in the 'forge'.

The indirect process, with its three stages, gives rise to a number of products. First there is cast iron, which in many cases was run directly from the furnaces into suitable moulds. Cast iron occurs in two forms, depending on its silicon content. The silicon content is decided by the conditions of production – higher temperatures favouring the production of high silicon or 'grey' cast iron, and lower temperatures 'white' cast iron. Of the three specimens of pre-medieval cast iron found in Britain, the earliest, from Hengistbury, was white with low silicon and the other two, from Roman

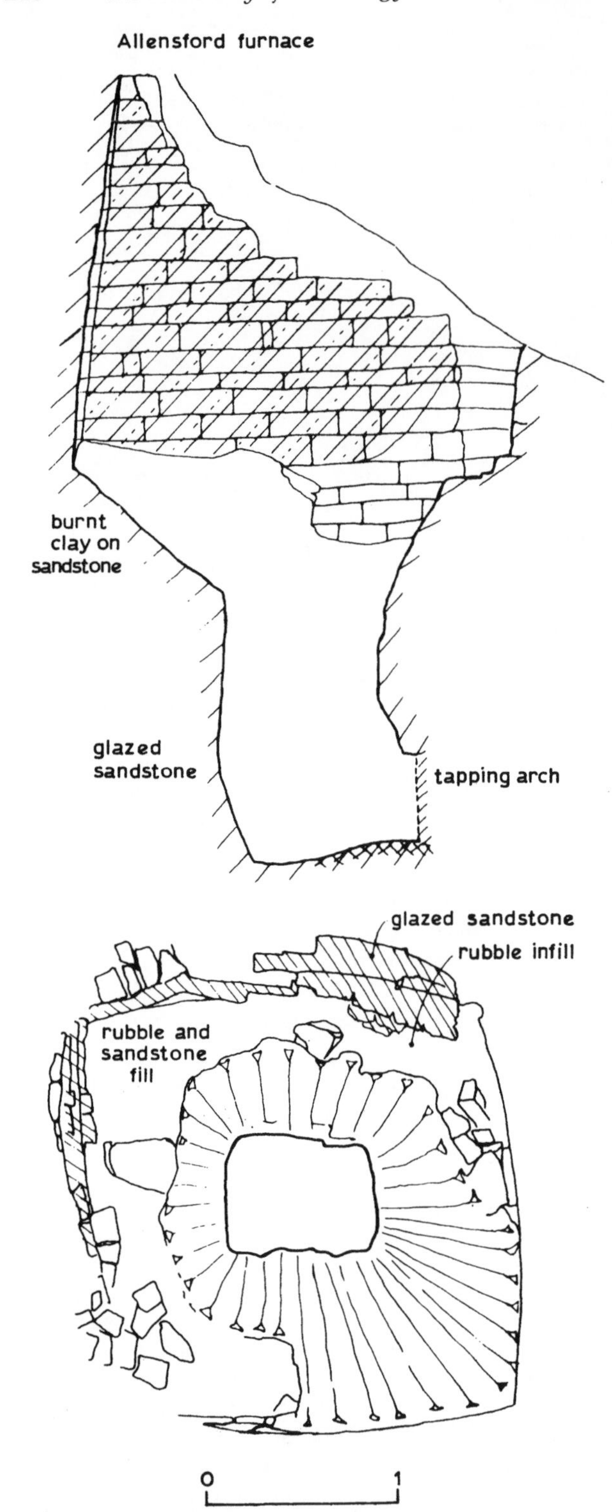

153 Plan and section of the blast furnace at Allensford (after Linsley and Hetherington[13])

levels, grey. The specimens illustrated by Schubert[21] and dated to the 16th and 17th centuries are all grey.

Cast iron is a brittle, non-malleable material which cannot be forged; white iron is rather more brittle than grey iron, therefore grey iron is generally aimed at for final products. Grey cast iron consists of areas of pure carbon in the form of graphite, which accounts for most of the 3·5% C that it contains, and areas of fairly pure iron or steel. It is the presence of the weak graphite that accounts for its brittleness and low tensile strength. White cast iron, on the other hand, is devoid of graphite and owes its brittleness to the presence of large amounts of iron carbide (Fe_3C) known as cementite.

As has been shown, the bloomery process produced slags containing a considerable amount of iron oxide, the remainder being mainly silica. The modern blast furnace, which uses lime in its charge to keep iron out of the slag, produces a slag of very different composition (*see* Table 106) with considerable lime and silica, and almost no iron. But we are told very little about the charge put into the earliest blast furnaces, and whether or not it contained lime.

It is becoming quite clear that one of the principal parts of the charge of the earliest blast furnaces was bloomery slag which, with its 50% or so of iron, was as rich as many ores. If this was smelted with sufficient lime or chalk at a high enough temperature, all the iron could be recovered and the slag would be a calcium silicate devoid of iron.

Straker[22] found at Dry Hill Camp, near Lingfield, Surrey an unusually light-coloured slag, which he thought might have come from a 16th century blast furnace at Scarlets in the valley below.[23] The composition showed it to be a lime-containing slag from the blast furnace. Together with this was found bloomery slag and Cyrenae limestone which is ferruginous. The 18th century Sloane MS refers to 'bottom', a sort of limestone which is found below the ore about 5–9 m below the surface.[24] If Straker is right in connecting his slag from Dry Hill with the 16th century furnace, this is confirmation of the use of limestone as a flux at this time.

There is also a reference to the use of shells or 'greys' (i.e. shelly ironstone) at Robertsbridge, Sussex, in the 18th century.[25] However, in the 17th century a mixture of half ore and half old bloomery cinder was being used in the Forest of Dean.[25] In the same period old bloomery slag was being exported from this area to Ireland.[26] 'Hammer slag' from the finery was added to the charge at Rievaulx in 1605.[27] In Lancashire in the 18th century 5–10% of limestone by volume was being added to the charge.[27] Furthermore, the clay ironstones are high in alumina and could be self-fluxing. High alumina slags seem to be a feature of the 19th century. The mixing of ores in order to obtain self-fluxing characteristics was well known in the late 17th century.[27] Macadam[28] mentions that the early Scottish blast furnace slags contained 0·5–14% iron, but an analysis of a slag from the 17th century furnace at Red Smiddy near Poolewe, Ross showed 20% iron, 14% lime and more than 50% silica. It is clear that it was not always possible to attain conditions which would allow the addition of enough flux to keep all the iron out of the slag.

In spite of the iron content of some of the blast furnace slags (Table 106) the coming of the blast furnace brought about a real improvement, since the weight of slag for each kg of metal produced was very much less.

THE FINERY AND THE CONVERSION FORGE

The second stage of the indirect process – only needed when it was required to produce wrought iron – was the finery. The principle here is the oxidation of the carbon present in cast iron, be it white or grey, by

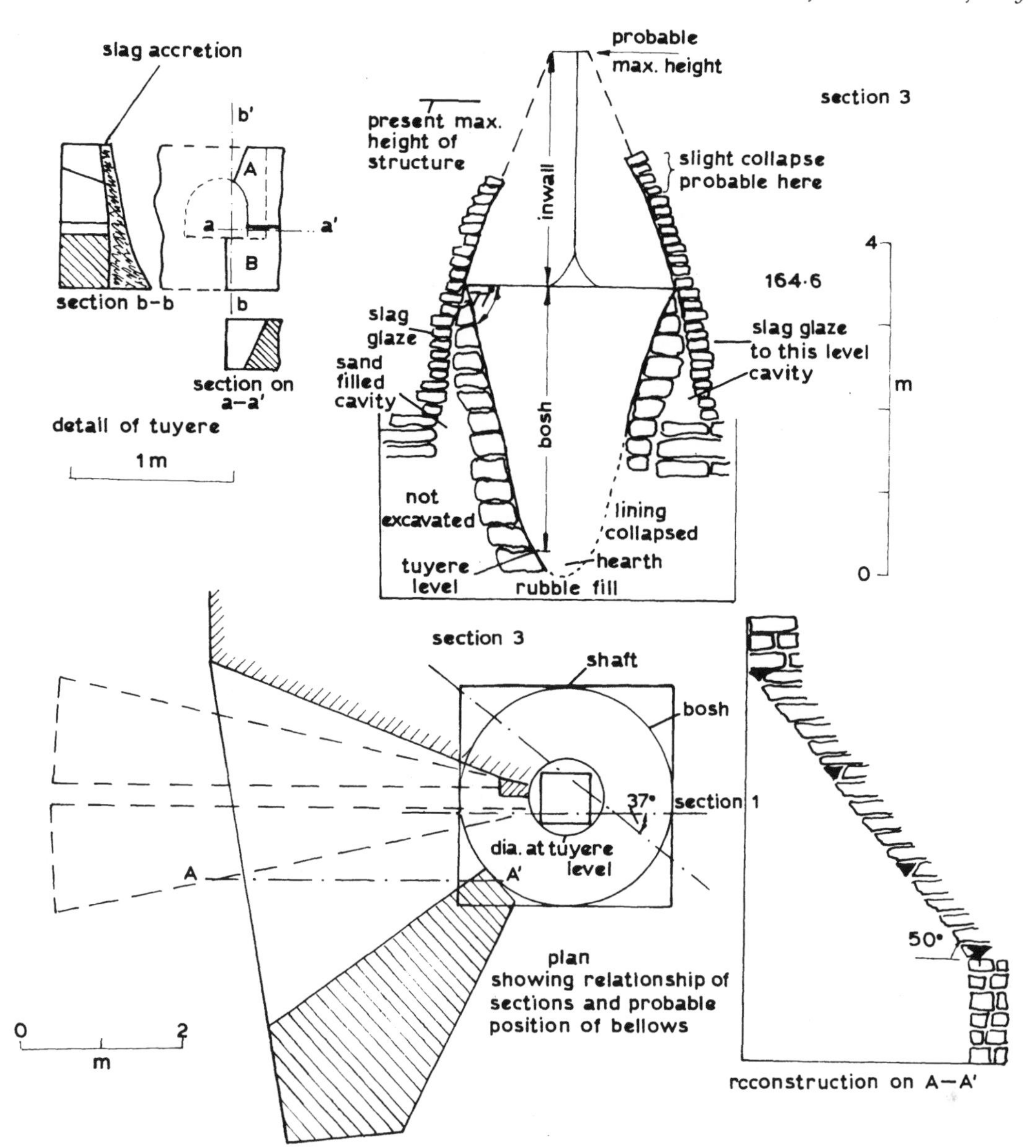

154 Plan and section of the blast furnace at Coed Ithel (after Tylecote[18])

Table 106 Composition of charcoal blast furnace slags

Ref.	14 Panningridge, Sussex	38 Angidy Valley, Gwent	18 Coed Ithel, Gwent	39 Michael's Wood, Glos.	11 Pippingford, Sussex	13 Allensford, Northumb.	40 Shifnal, Salop.
%	16th cent.	18th cent.	17th cent.		16th cent.	17th cent.	
FeO	nil	2·17	4·75	1·49	7·30	9·71	5·0
Fe_2O_3	0·05	0·33	—	—	—	0·71	0·4
SiO_2	45·5	48·90	(62·8)	58·83	52·7	48·25	51·8
Al_2O_3	19·2	8·31	7·3	8·67	13·4	20·33	18·2
MnO	4·85	0·46	0·4	0·46	3·4	1·38	2·3
CaO	20·6	30·69	15·9	17·66	16·8	6·81	13·5
MgO	9·6	4·97	8·4	9·14	2·6	2·65	5·0
P_2O_5	tr.	—	0·13	—	0·1	—	0·1
S	0·12	0·10	0·01	—	0·14	—	—
TiO_2	—	0·40	0·30	—	1·10	0·77	0·77
K_2O	—	2·50	—	3·61	2·20	3·09	—
C	—	0·10	—	—	—	—	—
Na_2O	—	—	—	—	—	1·79	—
Total	99·95	98·93	100·0	99·86	99·74	99·11	97·00

( ) = by difference; — = not determined; tr. = trace

Table 106 – *continued*

Ref.	41 Cannock, Staffs.	43 Rievaulx, Yorks.	44 Sharpley Pool, Worcs.	45 Rockley Yorks.	46 Charlcot, Salop.	43 Melbourne, Derbys.	47 Duddon, Cumbria
%	16–17th cent.	16–18th cent.	17th cent.	17–18th cent.	18th cent.	18th cent.	18–19th cent.
FeO	4·37	—	2·70	—	—	—	2·60
Fe_2O_3	—	—	nil	—	—	2·57	—
SiO_2	49·66	45·3	49·30	45·9	52·5	41·6	56·4
Al_2O_3	23·16	22·48	11·4	19·07	20·17	22·73	12·4
MnO	3·29	1·17	0·84	2·95	1·86	3·01	9·8
CaO	11·92	22·8	22·8	18·4	17·0	14·12	14·6
MgO	7·16	3·69	12·0	9·19	4·57	14·19	3·6
P_2O_5	0·07	0·055	tr.	0·50	—	0·023	—
S	0·1	0·22	tr.	0·16	0·01	0·227	—
TiO_2	—	—	—	—	—	—	—
K_2O	—	—	2·0	—	—	—	—
C	—	—	—	—	—	—	—
Na_2O	—	—	—	—	—	—	—
Total	99·73	95·72	101·04	96·17	96·11	98·47	99·40

means of air from a tuyere. The furnace was essentially similar to the bloomery hearth, but almost always square or rectangular in plan (Fig. 155). In the early days this operation would have been carried out in the then (in many cases) redundant bloomery hearths. In the Walloon process, which seems to have been the forerunner of the method used in Britain, the pig from the blast furnace was melted in front of the tuyere and in so doing a good deal of the carbon was oxidized away (Fig. 156). As has been explained in earlier chapters, there is no difficulty in obtaining a high temperature under oxidizing conditions in the smith's or bloomery hearths. The end of the cast iron bar would be brought to fusion point, and drips would run off, oxidizing rapidly as they fell towards the bottom of the hearth. By the time they reached the bloom forming at the bottom they would be solid, since the removal of carbon raises the melting point of iron.

155 Section through a Walloon finery hearth (after Maréchal[52])

There would be some molten slag covering the bloom through which the solid particles of decarburized iron would fall. The cast iron pig was fed into the tuyere zone progressively, so turning it into bloomery iron, with impurities forming a fluid slag that could be tapped off at intervals. The impurities would be mainly phosphorus, manganese and silicon, which, with some oxidized iron, would form a slag very similar to the normal bloomery slag (*see* Table 107). Some slag would be entrapped in the bloom and most of this would be removed during the next stage in the chafery.

The Walloon finery used two hearths – a finery and a chafery. However, there are systems such as the German finery process which do both fining and reheating for forging in one hearth; although slower, this is more economical in fuel and man-power.[29]

There are several sites which were undoubtedly used for fining in the British Isles. The best preserved of these is probably at Stony Hazel[30] in Cumbria, where there appears to be only one hearth; this means that in its last phase, at any rate, it worked as a German hearth.[31] The plan and sections of this hearth are shown in Fig. 157; it has a pig hole like that shown for the Walloon hearth in Fig. 155. This was not a normal feature of the German hearth, where the cast iron was added in smaller pieces, but it could indicate that this

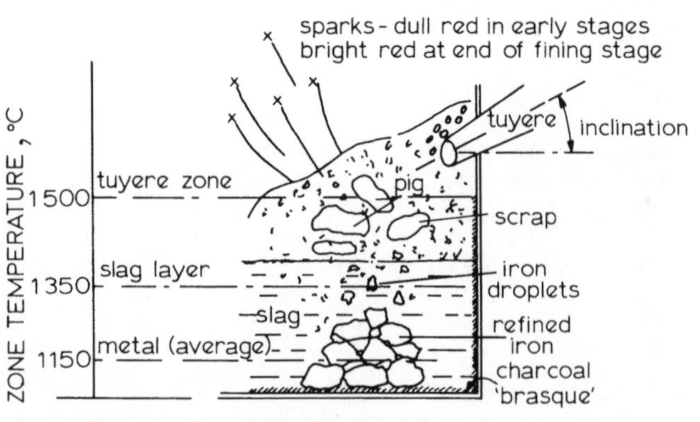

156 Diagram showing oxidation of cast iron in a finery (after Morton[48])

Table 107 Composition of finery and chafery slags

Ref. %	48 Finery Sparke Forge, Cumbria	49 Sowley Pond, Hants.	48 Chafery Nibthwaite, Cumbria	Sparke Forge, Cumbria	Penny Bridge, Cumbria	50 Puddling furnace 'Tap cinder'
FeO	33·60	57·3	65·53	33·10	29·30	39·8–66·3
Fe_2O_3	50·40	—	9·43	27·60	31·40	8·3–23·8
SiO_2	8·16	23·9	16·16	21·32	11·04	7·71–15·3
Al_2O_3	4·65	5·6	4·30	2·52	4·30	tr.–2·84
MnO	—	3·4	—	—	—	0·57–6·17
CaO	2·60	7·7	2·10	5·68	4·34	0·28–4·70
MgO	0·54	0·8	1·20	1·77	3·14	0·24–0·42
P_2O_5	0·57	—	0·53	0·25	0·12	1·34–8·07
S	—	—	0·47	—	3·69	0·17–2·56

hearth had been used previously as part of a Walloon finery and that owing to a shortage of fuel or water supply or a change in working conditions its output was reduced and single-hearth working was possible.

It is almost certain that this site was originally a bloomery, and of course the slag would have gone to the blast furnace when this was introduced. No doubt most of the Walloon slag went this way too, until only the large slag blocks, known locally as 'mossers', were left.

A more conventional two-hearth refinery, dated to the 16th–17th century, was found at Ardingly in Sussex.[32] This had two water channels – one supplying power for both sets of bellows and the other power for a hammer. A suggested layout based on the existence of two races, two hearths and the base of an anvil is given in Fig. 158. This site was converted into a fulling mill in the 18th century using the larger, southern channel for power.

Another Sussex site, at Chingley[14] (in its 16th–17th century phase) had a different layout with the hammer and the chafery bellows provided with water via one leat and the finery proper from the other (Fig. 159). Considering the closeness of the anvil base and the wheel, it is most likely that the hammer was of the side belly-helve type with the hammer shaft and the wheel shaft parallel to each other[33] (Fig. 160).

Hammer heads of cast iron of the type shown in this

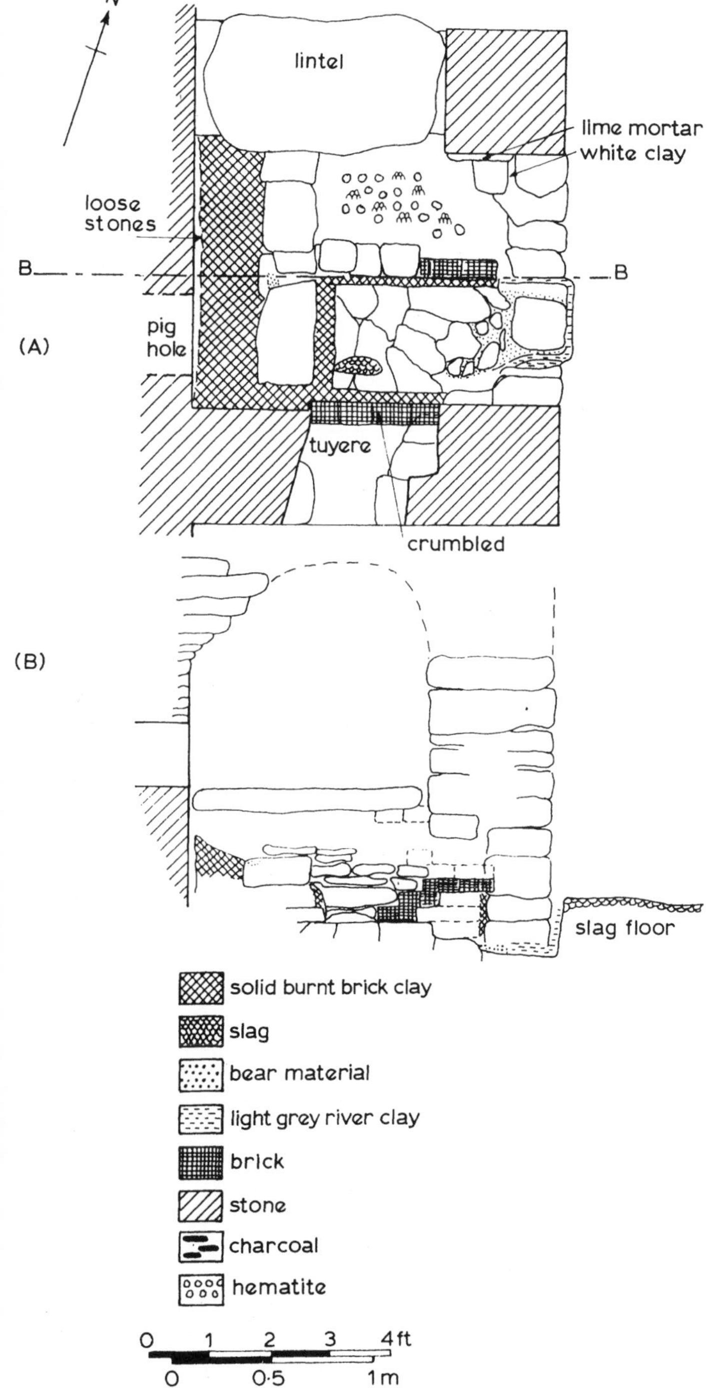

157 **Plan of Walloon-German finery hearth at Stony Hazel, Cumbria (after Davies-Shiel[30])**

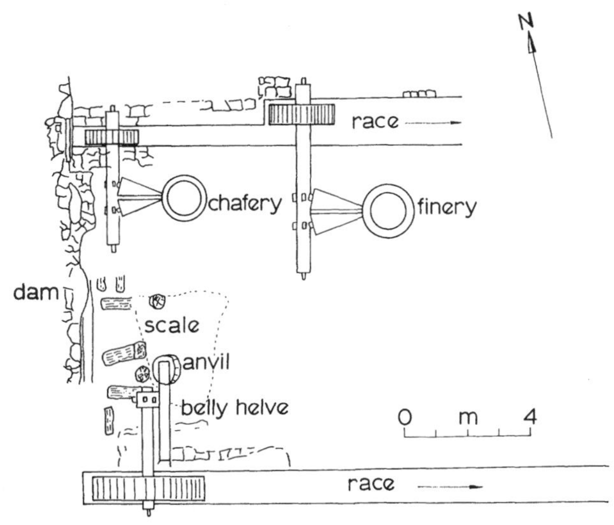

158 **Possible reconstruction of finery forge at Ardingly, Sussex (based on Bedwin[32])**

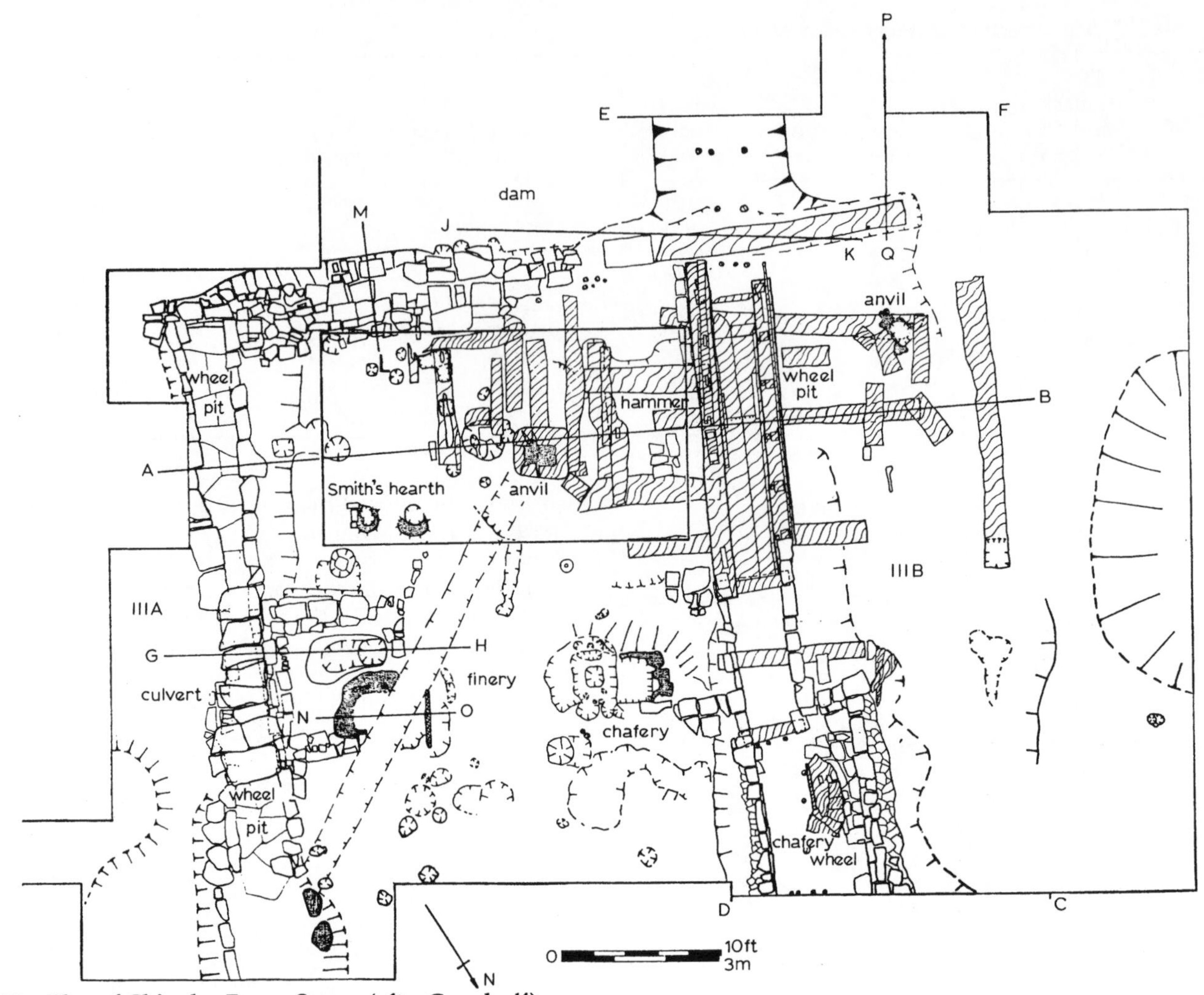

159 Plan of Chingley Forge, Sussex (after Crossley[14])

figure are well known from the Weald,[34] and half of one was found in the middle of the finery floor at Stony Hazel.

The stabilization of the anvil was a problem, and Rockley,[35] Chingley and Ardingly all show evidence of this. The anvil at Ardingly was enclosed in an iron casing which had 'wings' or cells to distribute the load over a wide area.

In the later phases of bloomeries and fineries the hearths were lined with cast iron. No iron was found at Stony Hazel, but a slab of cast iron 130 × 60 cm was found at Ardingly.

The chafery served essentially the same purpose as the string hearth. The bloom, which weighed about 50 kg, was reheated here and cut up into small pieces, in the process of which much of the entrapped slag was removed. Little or no slag was formed in the process, but a good deal of hammer scale resulted, looking no different from the Roman product shown in Fig. 118*b* (Chapter 7). Both the finery slags and the hammer scale can be returned to the blast furnace where convenient. The fuel of the chafery was built up to form a 'hollow fire', usually of coal; the high sulphur content of some chafery slags is evidence of this (Table 107). The metal to be heated was inserted into the hollow where it was heated by radiation.

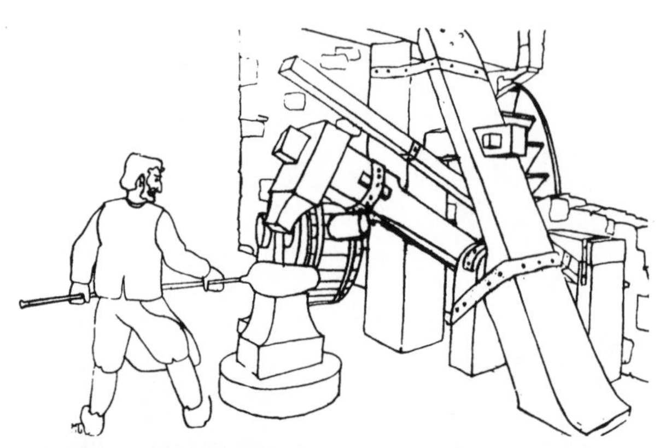

160 Side belly helve hammer

During the late 18th and 19th centuries, coal-smelted iron was converted to wrought iron by the puddling process. The puddling furnace produced tap cinder very similar physically and chemically to that produced by the bloomery. This can be distinguished from bloomery slag by its sulphur content (*see* Table 107). This varies between 0·2 and 3%, while that from charcoal bloomeries rarely contains more than 0·1%. Even so, it is unwise to attempt to date or type slag from these processes without taking into account its associated material. Puddling furnace slag was often resmelted in blast furnaces, consigned to tip-heaps, or made into roads, where it is usually to be found associated with fire-bricks, pottery, and other more closely datable remains.

Finery or refining slags can have a 'ham-bone' shape

Table 108 Composition of early British cast irons

Ref. %	11 Pippingford, Sussex 16th cent. Runner	44 Sharpley Pool 17th cent. Pig	36 Little Aston, Staffs. Pig	18 Coed Ithel, Gwent 17th cent. Runner	14 Panningridge, Sussex 16th cent. Runner	47 Duddon, Cumbria Pig	43 Maryport, Cumbria Pig (coke)	43 Barepot, Cumbria Coke pig	43 Clifton Cumbria Coke pig
C (graph.)	2·57	3·90	3·37	3·89	4·11	4·30	2·12	1·56	2·73
C (cem.)							0·48	1·24	0·48
Si	0·56	0·49	0·28	0·58	0·94	0·65	3·26	2·10	1·54
Mn	0·86	0·05	0·91	0·08	0·56	0·10	0·40	2·45	0·71
S	0·11	0·068	0·081	0·043	0·16	0·023	0·12	0·15	0·16
P	0·57	0·31	1·01	0·39	0·66	0·125	0·06	0·36	0·37
Hardness HV	298			239					
Structure	mottled	—	—	grey	—	—	grey	mottled	grey

owing to the manner of tapping. Little Aston forge has produced some well recorded and researched examples,[36,37] and the analysis with its high sulphur content shows that it arises from a coal-using process – probably a finery converting coal-smelted cast iron or a coal-fired chafery.

As for the metal, the cast iron used in the finery would normally be white iron with a low silica content (Table 108). Grey iron with high silicon is normally a product of the coke-blast furnace; but with slow cooling and hard driving (a good blast), mottled or grey iron can be produced by a charcoal-blast furnace, as may be seen from the structures given in some of the examples listed in Table 108.

The whole object of the finery is to produce a wrought iron of similar composition and structure to that produced in a bloomery. This is clear from the figures shown in Table 109.

The other products are hammer scale, which is no different from that of previous periods and, occasionally, swarf from such processes as gun-boring. Those examined have rusted badly, but their characteristic structure was still visible under the microscope.[12]

SUMMARY

The main object of this chapter is not to discuss the processes, which are dealt with in much greater detail elsewhere, but to recognize the products and to differentiate them from similar ones of other processes.

Luckily the blast-furnace slags are easily distinguished from other slags, owing to their higher contents of lime and alumina and their glassy character (*see* Table 106). They are often green and blue and sometimes black, which is owing in all cases to residual iron. The purpose of the addition of lime as flux is to decrease the iron content of the slag, but it is not always possible to reduce this to zero because of the problem of achieving a high enough blowing rate to raise the temperature in order to melt slags with a high lime content.

The composition of finery and chafery slags is similar to the bloomery slags (cf. Tables 107 and 100). The large lumps or 'mossers' are, however, physically different from the bloomery tap slags although not from the furnace bottoms such as were found at Muncaster Head bloomery. At Chingley forge, slag was found with ring-porous wood replicated in slag within it. This wood, most likely beech or birch, is more likely to be found in a finery than a blast furnace, but of course could have arisen from an earlier bloomery.

It has been shown how the blast furnace developed and why. Since most of the domestic market required wrought iron and not cast iron, outside periods of war, most of the output of the blast furnace was taken to the finery. The Weald, because of its proximity to ports in south-east England, supplied most of the arms requirements in the early periods, but as charcoal became more expensive in the south, armaments were supplied by furnaces in more distant areas.

The conversion forges were also heavy users of fuel, but it is more likely that the shortage of water led to their eventual removal elsewhere. In the Lake District and Scotland it still made economic sense to install

Table 109 Analyses of wrought iron

Ref.	Morton[48]		Ward[51]	Chilton[53]	
%	Nibthwaite Charcoal 19th cent. Bolt	Penny Bridge Fined iron Axle band	Yorkshire Puddled iron (Shingled ball) 20th cent.	Swedish Charcoal iron 20th cent.	British Coal-puddled iron 20th cent.
C	0·028	0·035	0·08	0·034	0·15
Mn	0·13	0·09	0·10 (max)	0·145	0·02
Si	0·23	0·14	0·10	0·010	0·13
S	0·024	0·12	0·05	0·015	0·01
P	0·31	0·209	0·04	0·032	0·151
Ni	—	—	—	0·030	0·19
Cu	—	—	—	0·010	0·15

bloomeries as late as the 17th century rather than import the indirect process of blast furnace and finery with its high capital requirements.

Finally, with the improvement in the technical economy of the blast furnace, the bloomery was rendered obsolete and the indirect process became general. By this time coal smelting had been introduced, and the fuel situation was no longer a problem.

NOTES AND REFERENCES

1 H. R. SCHUBERT: 'History of the British iron and steel industry from *c.* 450 BC to AD 1775'; 1957, London.
2 R. F. TYLECOTE: 'A history of metallurgy'; 1976, London.
3 J. PERCY: 'Metallurgy; Iron and steel', 310; 1864, London.
4 IBID., 321.
5 IBID., 327.
6 J. NEEDHAM: 'The development of iron and steel technology in China', 47; 1958, London.
7 H. R. SCHUBERT: op. cit.; 162–3.
8 IBID., 164.
9 B. G. AWTY: *Econ. Hist. Rev.*, 1981, **34**, (4), 524–39.
10 H. R. SCHUBERT: op. cit.; 244–5.
11 D. W. CROSSLEY: *Post-Medieval Archaeol.*, 1975, **9**, 1–37.
12 D. S. BUTLER and C. F. TEBBUTT: *Post-Medieval Archaeol.*, 1975, **9**, 38–41.
13 S. M. LINSLEY and R. HETHERINGTON: *J. Hist. Metall. Soc.*, 1978, **12**, (1), 1–11.
14 D. W. CROSSLEY: 'The Bewl Valley ironworks, Kent. *c.* 1300–1730'; 1975, R. Archaeol. Inst. Monograph.
15 E. SWEDENBORG: 'Regnum subterraneum sive minerale de ferro'; 1734, Dresden and Leipzig.
16 D. RICHARDSON: *Trans. NHSN*, 1904, (NS), **1**, (2), 283–7. (This has been recently re-surveyed by S. M. Linsley; report forthcoming.)
17 I am indebted to T. A. P. Greeves for showing me this furnace, which was reported in the *Trans. Devon Assoc.*, 1924, **56**, 94–6 by J. S. Amery during his presidential address.
18 R. F. TYLECOTE: *J. Iron Steel Inst.*, 1966, **204**, 314–19.
19 E. STRAKER: 'Wealden Iron'; 1931, London. (Repr. Newton Abbott, 1969.) This subject has been updated in a new book edited by D. W. Crossley and H. F. Cleere: 'Ironworks of the Weald', 1985, Leicester University Press.
20 This seems to have been the case at Brookland and Woolbridge (Mayfield Parish), Sussex. I am indebted to D. W. Crossley and C. F. Tebbutt for this information.
21 H. R. SCHUBERT: op. cit., Appendix 8.
22 E. STRAKER: op. cit., 98, 224.
23 D. W. CROSSLEY: *Post-Medieval Archaeol.*, 1979, **13**, 239–49.
24 V. C. H. SUSSEX: Sloane MS 4020, Fol. 189 (cited by Straker in 'Wealden Iron', 103).
25 H. R. SCHUBERT: op. cit., 229.
26 G. BOATE: 'Ireland's natural history', 138; 1652, London.
27 H. R. SCHUBERT: op. cit., 232–3.
28 W. I. MACADAM: *Proc. Soc. Antiq. Scotl.*, 1886–7, **21**, 89–131.
29 A. DEN OUDEN: *J. Hist. Metall. Soc.*, 1981, **15**, (2), 63–87; and 1982, **16**, (1), 29–32.
30 M. DAVIES-SHIEL: *Bull. Hist. Metall. Group*, 1970, **4**, (1), 24–32.
31 A. DEN OUDEN: *J. Hist. Metall. Soc.*, 1982, **16**, (2), 76–80.
32 O. BEDWIN: *Post-Medieval Archaeol.*, 1976, **10**, 34–64.
33 The drawing shown in Fig. 160 is taken from Straker (Ref. 19) and drawn by J. Lewis, FSA. A hammer of this type can be seen in working order at Wortley Top forge near Sheffield. (*See also J. Hist. Metall. Soc.*, 1970, **4**, (2), 83.)
34 T. HONYWOOD: *Sussex Archaeol. Collect.*, 1866, **18**, 195.
35 D. W. CROSSLEY and D. ASHURST: *Post-Medieval Archaeol.*, 1968, **2**, 10–54.
36 J. GOULD: *Trans. SSAHS*, 1969–70, **11**, 58–63.
37 T. TURNER: *J. Iron Steel Inst.*, 1912, **85**, (1), 203–12.
38 J. PICKIN: *J. Hist. Metall. Soc.*, 1983, **17**, (1), 4–11.
39 A. DUNN and AMINA CHATWIN: personal communication.
40 N. MUTTON: *J. Hist. Metall. Soc.*, 1974, **8**, (2), 117.
41 G. R. MORTON: *Iron and Steel*, 1966, (Dec.), 563–6.
42 G. R. MORTON: *Trans. SSAHS*, 1964–5, **6**, 21–38.
43 ANON: *Bull. Hist. Metall. Group*, 1964, **1**, (3), 15.
44 M. M. HALLETT and G. R. MORTON: *Bull. Hist. Metall. Group*, 1966, (Jan.), **1**, (6), 41–3.
45 G. R. MORTON: *Bull. Hist. Metall. Group*, 1966, **1**, (6), 53–7.
46 N. MUTTON: personal communication.
47 G. R. MORTON: *J. Iron Steel Inst.*, 1962, **200**, 444–52.
48 G. R. MORTON: *Metallurgist*, 1963, (Sept.), **2**, 259–68.
49 E. STRAKER: op. cit., 96.
50 J. PERCY: op. cit., 668.
51 H. D. WARD: *J. Iron Steel Inst.*, 1972, **210**, (6), 396–405.
52 J. R. MARÉCHAL: *Tech. Civil.*, 1955, **4**, 129.
53 J. P. CHILTON and U. R. EVANS: *J. Iron Steel Inst.*, 1955, **183**, 113–22.

11 Fuels and their ashes

The earlier version of this book relegated this subject to an appendix consisting of one table giving analyses of fuel ashes. This may have surprised some readers who remember that John Percy started his famous work on metallurgy with this subject. All pyrotechnical processes – and this includes most metallurgical processes – require fuel, and the subject is treated at length in any metallurgical textbook.

In early times, using more primitive processes, the chemistry of fuel appears to have been of little significance. But its solid product, i.e. ash, entered into both the metal produced and the slag. In primitive smelting, which was not economic in the use of fuel, the contribution from this source could be quite large and must be taken into consideration.

Metallurgical fuels may be divided into long-flame fuels such as wood and bituminous coals, and short-flame fuels such as charcoal and coke. Metallurgical operations that require the highest reducing conditions use charcoal and coke, since wood contains too much moisture and oxygen, but certain types of coal can be used in blast furnaces to good effect.

Where conditions may be oxidizing such as in roasting and in puddling furnaces, long-flame fuels are necessary or advantageous. Since most coals and cokes contain appreciable amounts of sulphur they cannot be used with iron except under special conditions, or where the slag is capable of absorbing sulphides, or where the iron is relatively solid and does not have large surface areas exposed to sulphur-containing gases. These conditions usually apply in chaferies and smith's hearths. In most cases, then, wood is charred, either in pits or clamps. Peat is also charred when it is used by smiths, although in its dried state it has been widely used in blast furnaces and in lead-ore hearths. It would appear that wood has been charred (in pits) since the EBA[1] and probably much earlier, once the special properties of charcoal were appreciated.

WOOD

Wood is not a homogenous product; it consists of heartwood, sapwood, and bark. Most analyses apply to heartwood, but it is well known that bark has a higher concentration of ash and phosphates. Furthermore, bark helps to retain the moisture in the wood, so the moisture content of wood including the bark will be appreciably greater than wood which has been debarked and allowed to dry. Debarked wood can lose 30% of its original weight in four months. Naturally the sapwood and young stems and branches contain more moisture than the older trunks, and leaves have a higher ash content.

Since bark was valuable (e.g. for tanning), good husbandry demanded that it be separated and sold as such. At Cannock it was recommended that the wood should not be cut till May when the sap is ascending, as this loosens the bark and makes it come off easily,[2] and allows the wood to dry faster. Bark has a higher content of phosphorus, according to Percy,[3] so whether bark is left on or not will determine the composition of the charcoal, to some extent.

As has been seen, dried wood can be used as a fuel or coaled within the smelting furnace, thus conserving the heat of coaling. But most metallurgical charcoal was coaled in pits or heaps (meilers).

Raw wood has been found uncharred in iron-smelting slags from Weldon, Northants. in the Roman period.[4] This probably results from the preheating of the furnace before smelting. Once slag formation had started it would prevent the complete burning of residual wood at the bottom.

Where raw wood *per se* is advantageous is in the smelting of lead, where it is necessary to expose the lead ore to oxidizing conditions in the top of an open fire or 'bole'. Brushwood with its interlocking branches assists in preventing the rapid fall of the lead ore to the bottom where it meets the reducing conditions.

There are accounts for 1303/4 referring to the use of wood for lead-smelting near Baxenden in Rossendale, Lancs.[5] The wood (often called white coal), was dried in a kiln. It is possible that such kilns have been mistaken for 'elling hearths' in which potash lye was made from brushwood.[6] Raistrick[7] believes that several kilns at Grasswood near Grassington, among older hut circles of the Iron Age, are medieval wood-drying kilns. They consist of deep, bowl-shaped hollows about 4–5 m in diameter at the top, and 2–2·5 m deep, lined with carefully built drystone walling, and paved at the bottom. On one side is a narrow passage going down to the floor level; where the kilns are built into the side of the hill, these passages go in from the low side on the level.

Numerous reports of the early medieval period refer to brushwood, but Hughes[8] has found in the Blackett

Table 110 Composition of fuel ash

%	Cow dung English	Fuel ash Iona	Fuel ash Hawk's Hill	Vitrified Indian ash (Wandalli)	Mixed wood Burnt	Charcoal Vitrified	Beech charcoal ash	Oak charcoal ash	Ash leaf ash	Peat ash	Seaweed ash (washed)	Coal ash Pulverised coal	Fly ash	Bone ash
Ref.	37	—	37	12	9	9	37	38	3	3	38	35	36	*
SiO_2	56·31	(81·7)	80	53·10	55·20	68·70	8·0	(55·30)	5·50	3–30	(67·1)	45·6	65·3	
Fe_2O_3	} 17·57 (Fe_2O_3, Al_2O_3, MnO_2)	15·8	tr.	17·27	2·06	2·56	tr.	0·56	0·1	10–20	0·22	14·2	3·4	
Al_2O_3		nil	tr.	4·51	3·30	4·10	0·92	—	tr.	0–5	4·80	26·7	25·2	
MnO_2		—	—	—	0·15	0·19	31·62	1·83	0·1	0–1	0·08	—	—	
CaO	17·56	2·5	4·2	10·89	14·40	17·80	24·96	22·54	49·5	24–30	4·80	5·5	3·1	
MgO	0·35	—	0·55	3·89	1·26	1·56	7·64	3·56	10·6	1–7	8·15	2·3	0·89	
Na_2O	1·68	—	1·65	2·54	0·46	0·56 }	} 22·0 (Na_2O + K_2O)	1·31	1·2*	0–2	15·3	} 0·06 (Na_2O + K_2O)	1·6	
K_2O	1·34	—	7·20	1·11	5·63	7·00 }		14·9	19·0*	0–1	4·39		0·89	
SO_3	0·61	—	—	—	—	—	—	—	2·0	5–10	—	1·3	—	
Cl^-	0·07	—	—	—	—	—	—	—	—	—	—	0·04	—	
P_2O_5	2·46	—	5·06	1·17	—	—	4·85	—	7·07	—	—	—	—	
CO_2	nil	—	—	5·30	16·46	nil	—	—	—	—	—	—	—	
$CaCO_3$														9
$Ca_3(PO_4)_2$														87
CaF_2														3
$Mg_3(PO_4)_2$														1·7

( ) = by difference; — = not sought; + = includes other compounds of these metals *From Kingzett's Chemical Encyclopedia, 1945

Matfen MSS. a report of a trial in which coal was used instead of brushwood. This took place in 1701, at Brickburn, Northumberland, and it was concluded that although the yield of lead from the ore was 2·2% less when coal was substituted, the lower price of coal gave it a financial advantage.

Ash from the burning of wood is usually a blue-grey powder. A much higher temperature than that required to burn it is needed to cause it to vitrify and form a fuel ash slag (>1 000°C). This causes the alkali carbonates to lose their CO_2,[9] and is the main reason for the difference between column 5 and 6 in Table 110.

DUNG AND DUNG CAKES

One of the earliest domestic fuels was dung. When the fibre content is low it can be increased and the heat content improved by using the dung as an adhesive to make cakes of chaff. These are put out to dry and then stacked in piles. This process can be seen in a number of countries in the Near East today and was still used in late 17th century England, according to Celia Fiennes.[10]

The remains of enormous cinder mounds consisting of vitrified dung have been reported in India.[11] According to Allchin,[12] these are the result of ritual burning of cattle compounds in which the dung and straw have been mixed by trampling and then sun-dried. This could easily catch fire and result in the vitrified ash, analysed in Table 110. Burning haystacks are another source of a similar vitrified product, often taking in some of the earth beneath them.

There is no difficulty in getting temperatures high enough to cause vitrification (>1100°C) by such means, but the vitrification of smaller amounts would require either an induced draught, as in the stokehole of a pottery kiln, or the use of bellows.

The burning of bracken and brushwood in heaps or 'kilns' has long been used as a source of alkali for soap;[6] Celia Fiennes witnessed this process at Cannock.[10] The ashes were ground up and made into balls, then sold in the country markets or sent to London for use in washing and scouring. Such material is used in Afghanistan today.

While all organic ash contains alkali in the form of sodium and potassium compounds, dung naturally has a high phosphorus content, which can only be equalled by bark or leaf ash (Table 110). Generally this is the feature which distinguishes dung from wood ash, but it should be clearly stated that there is almost always an area of overlap in the composition of different fuel ashes and their vitrified products.

PEAT

The significant difference between peat-ash and charcoal-ash is the low alkali and high sulphur content of peat-ash. Its fluxing action will not be so marked, and there is likely to be trouble with sulphur contamination when it is used for iron-smelting. It is a pity that no blooms were found in the alleged smeltery at Wiltrow in Shetland.[13]

According to Percy,[3] copper has been profitably extracted from peat near Moel-Hafod Owen in North Wales. It is not likely that this material would be sought after by primitive people as a copper ore, but it does show one method by which a small copper content could be transferred to iron during smelting.

According to Woolner[14, 15] and Crawford,[16] charred

peat is good for smithing and is so used by Scottish Highland smiths.[16] It is prepared rather like wood charcoal, by digging a trench 2·4 m long × 1 m wide × 1 m deep. This was filled with 12 cartloads of dry black peats. The peat was lit at both ends and in the centre and when glowing, covered with turves. This took two days. The result was light, blue-grey, and very brittle. It was not as hot as coal but cleaner, and gave good results.

Dartmoor peat charcoal was used in the tin-blowing houses of Devon and Cornwall. The remains of charring heaps could be seen in the northern parts of Dartmoor in 1965.[14, 15]

The analysis of peat ash given by Percy[3] suggests a high sulphur content (Table 110). For this reason a test was carried out on some Dartmoor peat charcoal sent by Mrs Woolner, and the sulphur content was found to be only 0·16%. This is very much less than the average sulphur content of coke, but rather more than the average for wood charcoal. However, in view of the sulphur content of most iron ores, it is unlikely to make a significant contribution to the overall sulphur availability in iron-smelting. In non-ferrous smelting much higher sulphur contents can be tolerated until the final stage.

Peat was often used in lead-ore hearths, and is referred to in the manuscript prepared by James Mulcaster in 1780.[17] Presumably it was well dried, either in the open or in kilns.

In the 18th century, peat and mixed peat-ore briquettes were used in conjunction with charcoal in the iron-blast furnace at Leighton in North Lancashire [now Cumbria].[18] This was an attempt to reduce the charcoal content of the charge, but the peat tended to burn away too rapidly. The reason for mixing the ore and the peat in the form of a briquette was to get the ore reduced by intimate contact before the peat was consumed. Small amounts of clay had to be used as a binder.

CHARCOAL

Generally, the earliest metallurgical fuel used was charcoal, which would have been plentiful at the beginning of the Bronze Age in this country. There seems to have been no selection of charcoal in pre-Roman Britain. The strength of the material was evidently not considered important. A 4th century BC site in the Chilterns on which iron-working remains were found yielded charcoal from birch, willow, ash, sycamore, and hazel.[19] Another iron smelting site dated 1–120 AD in Sussex yielded alder, hazel, ash and oak.[20]

Since oak was the primary component of native woodland over most of the country in the Roman period, the preponderance of oak in the charcoal is not unexpected. Charcoal at Ashwicken was 80% oak, the rest being made up from elm, ash, hazel, maple and birch.[21] The preference for oak was obvious in the 13th century site at High Bishopley;[22] by the 15th century, birch was mainly used in this area.[23]

The only important property required of a metallurgical charcoal is that it should be strong enough to avoid crumbling under pressure in the furnace. Since primitive furnaces are always low, almost any charcoal will satisfy this requirement. However, to avoid too much loss as dust, it is important that it should not crumble during cartage. This is one of the reasons why hard charcoals from oak, beech and birch are preferred to those from conifers.

During smelting the ash from charcoal will be incorporated in the slag, and may therefore have a significant effect on the free-running temperature. As can be seen from Table 110, wood charcoal contains appreciable alkali usually in the form of potassium, which has a marked effect in this respect. The ash content of charcoal is usually in the range of 2–4%, but because of the inefficiency of the smelting process, considerable amounts of fuel ash will be incorporated into the slag.

Very little is known about the making of charcoal in the early periods. Charring hearths ('pit-steads') are known from bloomery sites in the Lake District and elsewhere.[24] Charring pits are shown by Biringuccio in the 16th century.[25] However, it would seem that there is some dating from the EBA on an East Anglian site! The trenches found on the Roman iron-smelting site at Wakerley may well be examples of these.[26]

A trench would be dug, the wood placed in it and covered with turf or soil except at the ends. The wood would be ignited at one end and would burn with the aid of a controlled draught along the trench, as in the making of peat charcoal described above.

As can be seen from Table 110, wood charcoal is relatively low in silica but high in lime and potassium. The iron and manganese contents depend very much on the soil in which the wood is grown. Wood charcoal is almost unique in having a higher potash than sodium content. Seaweed ash (kelp) is exactly the opposite.

Much has been written on the coaling of wood in more recent times,[27] and Percy[3] should be referred to for a good description of the methods used in the 19th century.

At present, wood is largely coaled in kilns or retorts designed to save the heat of combustion for the next charge and, in many cases, for recovering the wood-tar by-products.

COAL AND COKE

Coal has been reported from Roman times onwards. The industrial settlements at Tiddington, Camerton and Wilderspool all used coal, and Webster[28] and Cunnington[29] have summarized the evidence relating to the use of coal for industrial and domestic purposes in Roman times. There is no doubt that where coal outcrops, early peoples used it long before the Roman period. The deeper mining of coal, however, probably dates only from medieval times.

The use of coal for iron-smithing purposes is accepted; what is not accepted, however, is the use of coal for iron-smelting. Coal has a much higher sulphur content than charcoal,[30] and can introduce sufficient sulphur into iron while it is being smelted to embrittle it. This does not happen in the case of massive iron during smithing operations, but the embrittlement conferred on the iron by sulphur during smelting cannot be removed in smithing.

The sulphur content of coal varies from 0·39–1·23%.[3] This is mostly in the form of iron pyrites (FeS_2). On coking, half the sulphur is removed, but since there is a loss of weight due to the distillation of the volatile constituents, which amounts to about 33% by weight,

Table 111 Organically vitrified building material: composition (weight %)

% Ref.	Burnt building Huntingdon 31	Vitrified fort Finavon, Scotland Country rock 32	 Intercryst glass	Burnt building Chan Chan, Peru 34	Vitrified fort France 33
SiO_2	rest	48·5	65·4	62·3	70·3
Fe_2O_3	~4–8	13·3	2·41	6·10	3·1
Al_2O_3	~10	15·9	20·3	15·0	16·9
MnO_2	0·04–0·1	—	—	—	—
CaO	~0·7	5·14	0·84	4·35	} 2·4
MgO	~0·3–0·7	8·43	1·81	2·92	}
K_2O	~7·0	1·36	5·36	2·37	} 9·3
Na_2O	~3–8	2·59	4·03	5·10	}
S	—	—	—	0·04	—
P_2O_5	<0·2	0·1	0·5	0·80	—
TiO_2	—	1·03	0·58	0·88	—
LOI	—	—	—	—	1·0
Total				99·86	103·0

— = not determined

the sulphur content of the coke is not very much less than that of coal.

However, the difference in sulphur content between different seams of coal is very great, and it is possible by careful roasting of ore to remove the sulphur from this source and the choice of a low-sulphur coked coal, to get an iron with an acceptably low sulphur content. This is clearly what Abraham Darby did in 1750 when he made a pig iron that could be converted to a wrought iron of passable quality. This was helped by the use of the two-stage process, since it is possible to remove some of the sulphur with the slag arising from the blast furnace process, which is not possible in the case of the older single or direct process.

As can be seen from Table 110, coal ash tends to have a high alumina content and a low alkali content as compared with wood. It can be differentiated from peat by its higher alumina content and lower lime content. On the whole, total vitrification is not easy under primitive conditions and the structure of the coal-shale can usually be recognized somewhere in the burnt product.

BONE

Small dried bones burn very intensively, as many will know. This is owing to the organic content of bones, which varies a great deal with source and size. No doubt the small bones were saved for use as fuel while the larger bones might have been burnt for their ash, which is mostly the refractory calcium phosphate (Table 110). This material formed the basis of the cupellation hearths used at Hengistbury Head and other Roman sites.

VITRIFIED CLAY AND STONE

There is one other vitrified product that ought to be mentioned here: the 'slag' that results from the burning of buildings reinforced with organic materials such as wood, thatch, wattle, etc. The mixture of clay daub with wood can produce a vitrified material that is sometimes difficult to distinguish from a metallurgical slag[31] (Table 111).

The main source of such material in this country is from 'vitrified forts', i.e. timber-laced stone forts of the Early Iron Age.[32, 33] The alkaline ash from the burning timbers and the high temperatures produced in the burning vertical columns have been sufficient to slag the stone into a conglomerate (analyses in Table 111). For many years the slaggy material at Chan Chan in Peru[34] was thought to have a metallurgical origin. Now it is known to be the result of firing daub and a material rather like thatch. Half-timbered mud-brick buildings at Kultepe-Kanesh in Central Turkey dated to the 18th century BC have been burnt so as to give a material resembling overburnt brick.

NOTES AND REFERENCES

1 This seems to be the case at the EBA–MBA site of Mildenhall (MNL), Suffolk, according to information kindly provided by E. Martin of the Suffolk Archaeol. Unit, Bury St. Edmunds.

2 A. C. JONES and C. J. HARRISON: *Econ. Hist. Rev.*, 1978, **83**, 795–810.

3 J. PERCY: 'Metallurgy; Fuels and Refractories'; 1875, London.

4 The slag found in a hearth at the Roman villa of Great Weldon, Northants. contained the imprint of charcoal, which is not unusual, and also pieces of decayed wood.

5 J. KERR: *Trans. HSLC*, 1872, **24**, 57–70.

6 M. DAVIES-SHIEL: *CWAAS*, 1972, **72**, 85–111.

7 A. RAISTRICK and B. JENNINGS: 'A history of lead mining in the Pennines' 81; 1965, London.

8 Blackett Matfen MSS, 'Ledger of rents 1685–91', Newcastle upon Tyne, University Library.

9 H. A. GHAZNAVI: 'Trace element partitioning in early copper smelting'; MSc. dissertation, 1976, University of Newcastle upon Tyne.

10 C. MORRIS (ed.): 'The Journeys of Celia Fiennes', 161, 166; 1947, London.

11 F. E. ZEUNER: *ULI Archaeol. Bull.*, 1959, **2**, 37–44.

12 F. R. ALLCHIN: 'Neolithic cattle keepers of South India: A study of Deccan ash mounds'; 1963, CUP.

13 A. O. CURLE: *Proc. Soc. Antiq. Scotl.*, 1935–6, **70**, 153–69.

14 D. WOOLNER: *Devon Cornwall Notes*, 1965.
15 D. H. and A. H. WOOLNER: *Ind. Archaeol. Rev.*, 1966, **3**, (4), 270–1.
16 I. A. CRAWFORD: *Scottish Stud.*, 1964, **8**, (1), 108.
17 J. MULCASTER: *Bull. Hist. Metall. Group*, 1971, **5**, (2), 45–62.
18 G. R. MORTON: *Iron and Steel*, 1965, (Aug.), 421–4.
19 K. M. RICHARDSON and A. YOUNG: *Archaeol. J.*, 1951, **31**, 132–48.
20 S. E. WINBOLT: *Sussex Archaeol. Collect.*, 1930, **71**, 223–36.
21 R. F. TYLECOTE and E. OWLES: *Norf. Archaeol.*, 1960, **32**, (3), 142–62.
22 R. F. TYLECOTE: *J. Iron Steel Inst.*, 1959, **192**, 26–34.
23 R. F. TYLECOTE: *J. Iron Steel Inst.*, 1960, **194**, 451–8.
24 J. GOULD: *Trans. SSAHS*, 1973–4, **15**, 40–2.
25 C. S. SMITH and M. T. GNUDI (eds.): 'Pirotechnia of V. Biringuccio', 178; 1943, New York.
26 D. A. JACKSON *et al.*: *Britannia*, 1978, **9**, 115–242.
27 L. REYNOLDS: 'The manufacture of wood charcoal in Great Britain'; Forestry Commission: Forest Record No. 19, (revised 1961).
28 G. A. WEBSTER: *Antiq. J.*, 1955, **35**, 199–216.
29 M. E. CUNNINGTON: *Antiquity*, 1933, **7**, 89–90.
30 H. F. ADAMS *et al.*: *Geol. Mag.*, 1965, **102**, (6), 469–73.
31 D. W. BROWN: personal communication, 6 Mar. 1968.
32 E. YOUNGBLOOD *et al.*: *J. Archaeol. Sci.*, 1978, **5**, 99–121.
33 R. A. DAUBRÉE: *Rev. Archéol.*, 1881, **41**, 18–28.
34 H. LECHTMAN and M. E. MOSELEY: *Nawpa Pacha*, 1972–4, **10–12**, 135–70.
35 B. BUTTERWORTH: *Trans. Br. Ceram. Soc.*, 1954, **53**, (4), 293–313.
36 G. L. FISHER *et al.*: *Science*, 1976, (May), 553–5.
37 R. T. EVANS and R. F. TYLECOTE: *Bull. Hist. Metall. Group*, 1967, **1**, (9), 22–3.
38 D. C. SANDERSON and J. R. HUNTER: *Sci. Archaeol.*, 1981, **23**, 27–30.

Appendixes

Table A Analyses of 'Pure' Copper Artifacts from the Copper and Early Bronze Ages

Ref.	Artifact	Provenance	As	Sb	Ag	Pb	Ni	Sn	Fe	Others
1	Flat axe	Ireland	0·77	nil	0·12	0·07	tr.	0·05	0·14	
2	Flat axe (?)	Ireland	tr.	nil	0·35			0·17	0·2	
3	Thick-butted axe	Ireland	0·28	0·26	0·2					
3, 4	Flat axe	Ireland	0·20	0·36	0·3					
3	Thick-butted axe	Ireland	0·01	0·006	0·03					(Probably native)
3	Bevel axe	Ireland	0·05	0·08	0·07					
3	Flat axe	Trillick, Tyrone	0·01	0·15	0·15					
3	Flat axe	Co. Cork	n.d.	0·05	0·005					
3	Bevel axe	Eriswell, Suffolk	0·31	0·09	0·1	0·22				Au, 0·58
5	Flat axe	Irish	—	—	0·15			0·03		
5	Flat axe	Cork	0·18	0·10 (+ Sn)						
5	Flat axe	Londonderry	0·76	tr.	0·25					
5	Flat axe	Irish	0·78	0·61	0·17			0·05		
5	Flat axe	Irish	0·13	0·83 (+ Sn)	0·07					
5	Flat axe	Irish	0·28				0·12			
5	Flat axe	Galway	0·76	0·79 (+ Sn)						Zn, 0·44
6	Flat axe	Irish	0·80	0·18						
6	Flat axe	Co. Down	0·05	0·04	0·2	0·05		0·04		
6	Flat axe	Cork	n.d.	0·05	0·005				0·01	
3	Thick-butted axe	Irish	0·31	0·32	0·14					
3	Thin-butted axe	Irish (N.)	0·50	0·65	0·07	0·02				P, 0·006
3	Bevel axe	Icklingham, Suffolk	0·50	0·20	0·10					P, 0·010
3	Bevel axe	Newry, Co. Down	0·48	0·84	0·20					Zn, 0·02
3	Thick-butted axe	Ballypatrick, Antrim	0·88	0·35	0·20					
3	Thick-butted axe	Irish (N.)	0·50	0·40	0·05	0·20				
3	Thick-butted axe	Armoy, Antrim	0·35	0·30	0·10					
3	Thick-butted axe	Co. Limerick	0·57	0·51	0·20					
3, 4	Halberd	Bellanamellard, Co. Fermanagh	0·2	—	0·07			<0·07		
3	Halberd	Irish	0·20	0·18	0·2					
3	Halberd	Falkland, Fife	0·55	0·38	0·27					
3	Halberd	Irish	0·005	0·01	0·18					
3	Halberd	R. Bann, Londonderry	0·05	0·11	0·05		0·26			
7	Halberd	Irish	nil	nil	0·26			0·22	0·04	
7	Halberd	Antrim	0·18	0·14	nil			0·31	nil	
7	Halberd	Galway	nil	nil	nil			0·22	0·17	
7	Halberd	Cork	0·37	0·27	nil			0·30	nil	
3	Knife	Mere, Wilts.	0·17	0·04	0·3					
3	Knife	Dorchester, Oxon.	0·10	0·17	0·14		0·20			

Table B Analyses of 'Impure' Copper Artifacts from the Copper and Early Bronze Ages

Ref.	Artifact	Provenance	As	Sb	Ag	Pb	Ni	Sn	Fe	Others
3, 4	Thick-butted axe	Irish	2·3	0·21	0·3					
3	Thick-butted axe	L. Gur, Limerick	2·1	0·34	0·28	0·3				
3	Thick-butted axe	Castletown Roche, Cork	2·0	0·93	0·28	0·2				
3	Thick-butted axe	Castletown Roche, Cork	2·1	0·48	0·33					
3	Thick-butted axe	Irish	0·9	0·35	0·10					
3	Thick-butted axe	Irish	0·7	0·25	0·10					
3	Thick-butted axe	Killarney, Eire	0·7	0·60	0·06					
3	Flat axe	Irish	1·38	0·34	0·20					
3	Flat axe	Irish	1·92	0·23	0·20					
3, 4	Flat axe	Irish	2·50	0·57	0·20					Se, 0·03
3, 6	Thick-butted axe	Irish	2·30	0·30	0·30					
3, 4	Thick-butted axe	Ireland (N.)	1·92	0·007	0·005	0·11		0·4	0·5	
7	Flat axe	Irish	1·54	nil	0·25			0·18		
2	Axe (?)	Irish	2·17	nil	0·21	0·06		0·08		
5	Flat axe	Waterford	tr.	0·05 (+Sn)	nil	2·74	0·21	—	0·25	
5	Flat axe	Irish	nil	nil	0·06	nil	nil	1·09	0·08	
5	Flat axe	Tyrone	1·56	0·51 (+Sn)	0·25	0·17	nil		0·10	
5	Flat axe	Irish	1·86	0·27 (+Sn)	0·11	0·17	nil		nil	
5	Flat axe	Irish	1·04	nil	0·16	tr.	nil	0·12	0·17	
6	Flat axe	Co. Cork	≥1·0	≥1·0						
6	Flat axe	Co. Cork	tr.	0·3	3·0		0·04			
3	Thick-butted axe	Co. Mayo	1·30	0·25	0·15					P, 0·012
3	Thick-butted axe	Irish	1·67	0·58	0·30					
3	Flat axe	Irish (?)	1·50	0·14	0·30					Zn, 0·1
3	Flat axe	Irish	2·06	0·62	0·30	0·35				
3	Thin-butted axe	Irish	1·78	0·57	0·30	0·19				
8	Flat axe	Irish	1·5		0·10	tr.	tr.	<0·05		Si, 0·1
4	Halberd	Cavan, Eire	1·30	0·005		0·30				
3, 4	Halberd	Faversham, Kent	4·60	0·2			0·17			
4	Halberd	Stoke Ferry, Norfolk	2·1	0·3	0·2					
3, 4	Halberd	Letterkenny	2·2	0·2	0·1					
3	Halberd	Tullamore, Offaly	2·23	0·81	0·57					
3	Halberd	Irish	2·54	0·86	0·35					
3	Halberd	Irish	2·90	0·89	0·48					
3	Halberd	Irish	2·64	0·82	0·30					
3	Halberd	Irish	2·35	1·5	0·28					
3	Halberd	Irish	2·02	1·03	0·44	0·2				
3	Halberd	Irish	2·50	0·56	0·44					
3	Halberd	Irish	2·39	1·07	0·55					
3	Halberd	R. Suck, Galway-Rosc.	2·57	0·66	0·43					
3	Halberd	R. Suck Galway-Rosc.	1·97	0·63	0·24					
3	Halberd	L. Derrycassan, Cavan	0·90	0·52	0·28					
3	Halberd	Irish	2·42	0·44	0·30					
3	Halberd	L. Gur, Limerick	1·72	0·52	0·23					
3	Halberd	Cotton, Co. Down	3·02	0·52	0·34					
3	Halberd	Cotton, Co. Down	2·68	0·60	0·20					
3	Halberd	R. Shannon	2·89	0·41	0·14					
3	Halberd	Clonloghan, Clare	1·23	0·24	0·13					
3	Halberd	Irish	2·78	0·48	0·14					
3	Halberd	Ballyhaise, Co. Cavan	0·55	0·60	0·14					

Table B –*continued*

Ref.	Artifact	Provenance	As	Sb	Ag	Pb	Ni	Sn	Fe	Others
3	Halberd	Tullyvallan, Armagh	1·35	0·58	0·30					
3	Halberd	Irish	1·90	0·60	0·30					
3	Halberd	Irish	1·93	0·91	0·27					
3	Halberd	Co. Cavan	3·99	0·37	0·16					
3	Halberd	Co. Offaly	2·55	0·55	0·29					
3	Halberd	Tullamore	1·14	0·53	0·07					
3	Halberd	Ballybogy, Antrim	2·14	0·19	0·13					
3	Halberd	Irish	2·61	0·91	0·26					
3	Halberd	Keeloge, R. Shannon	1·68	0·58	0·18					
3	Halberd	Irish	2·10	0·75	0·20					
3	Halberd	Irish	2·66	0·30	0·20					
3	Halberd	Portara Ford, L. Erne	1·71	0·36	0·20					
3, 4	Halberd	Irish	2·00	0·50	0·44					Se, 0·03
3, 4	Halberd	Ballina, Co. Mayo	1·70	0·2	0·3					Zn, 0·02
3, 4	Halberd	Co. Meath	2·51	0·33	0·18					
3, 4	Halberd	Ballybogy, Antrim	1·80	0·10	0·14					
3, 4	Halberd	Maena Fen, Cambs.	0·30	0·10	0·01			0·5		Au, 1·3
3, 4	Halberd	Falkland, Fife	2·1	0·70	0·41					
3, 6	Halberd (?)	Argyll	4·80	0·48	0·30				0·3	Se, 0·03
3	Knife	Winterslow, Wilts.	2·0	0·43	0·15					
3, 4	Knife	Faversham, Kent	7·0	0·05	0·02		0·1			

Table C Analyses of Bronze Age Bronzes

Ref.	Artifact	Provenance	Sn	Pb	Fe	As	Sb	Ni	Ag	Others
	Early-Middle Bronze Age									
3	Thin-butted axe	Cobbinshaw, Edinburgh	8·9	0·1		0·3	0·26		0·1	
3, 4	Flat axe	Irish	12·0			0·01	0·20		0·2	Bi, 0·0086
3, 6	Dec. axe	Irish, Co. Carlow	5–10			0·2	0·3		0·2	
3, 4	Dec. axe	Irish	11·4			0·3	0·003		0·1	
3	Dec. axe	Bushmills, Antrim	9·1	0·40		0·3	0·007		0·01	
3, 6	Thin-butted axe	Cornwall	9·6	8·2		0·3	0·3		0·30	Bi, 0·12, Zn, 2·2
2	Flat axe	Yaxley Fen, Cambs.	8·99	tr.	0·16	0·68	0·15	0·11	0·09	
2	Flat axe	Butterwick, Yorks.	10·74	tr.	0·10	0·56	0·15	0·06	0·16	
2	Flat axe	Cambs.	11·04	tr.	1·22	tr.	nil	0·11	tr.	
2	Flat axe	Plumpton Plain, Sussex	11·34	tr.	0·14	0·78	0·31	0·37	0·17	Mn, tr.
2	Flat axe	Near York	11·73	1·21	0·10	nil	nil	0·14	tr.	
2	Flat axe	Irish	12.52	tr.	0·19	0·68	0·26	nil	0·21	
10	Flat axe	Rhymie, Aberdeen	~8·0	n.d.	tr.	0·50				
11	Flat axe	Asterton, Salop.	14·1	0·04	0·07	tr.			0·03	
3	Flanged axe	Icklingham, Suffolk	11·1			0·05	0·004	0·35	0·02	
12	Wing-flanged axe	Birtley, Durham	17·29		tr.			tr.		
3	Halberd	Limavady, Londonderry	10			1·01	0·13		0·28	
3	Halberd	Co. Cork	10·78			1·5	0·30		0·2	
3	Halberd	Breaghwy, Co. Mayo	10·8			0·68	0·29		0·4	

Table C–*continued*

Ref.	Artifact	Provenance	Sn	Pb	Fe	As	Sb	Ni	Ag	Others
13	Knife,	Todmorden,	16·4	—	0·012	0·26	—	0·28	0·005	Bi, 0·0083
	Awl	Lancs.	9.3	0·23	0·0086	1·12	0·19	0·25	0·15	Bi, 0·0083
3	Halberd	Irish	11·2			0·39	0·03		0·28	
3	Halberd	Moylough, Sligo	11·8			0·05	0·05		0·03	
3	Halberd	Rock Forest, Tipperary	9·5			n.d.	0·05		0·02	
3	Halberd	Co. Cavan	9·1	0·3		1·3	0·005		0·07	
3	Knife	Oakley Down, Dorset	6·7			0·15	0·24		0·2	
3	Knife	Eynsham, Oxon.	9·06			1·10	0·05	0·04	0·05	
14	Knife	Penmaenmawr	~10·0	0·005	0·1	0·05	0·01		0·001	
	Middle-Late Bronze Age									
6	Palstave	Irish	12	0·6				0·4		
6	Palstave	Surrey	6	0·4				0·4		
6	Palstave	Kent	9	0·8				0·4		
6	Palstave	Irish	15	2·0				0·4		
6	Winged palstave	Irish	13					<0·05		
11	Trunnion axe	Meole Brace, Salop.	14·2							
11	Looped palstave	Meole Brace, Salop.	13·0							
11	Looped palstave	Meole Brace, Salop.	12·7							
11	Palstave	Asterton, Salop.	14·2	0·02	0·04	tr.			0·03	
15	Palstave	Southall, London	18·25	nil	tr.					
15	Palstave	Fife	18·31							
6	Socketed axe	Carlow	6	2·0				0·1		
6	Socketed axe	Ireland	10	2·0				0·05		
17	Socketed axe	Heathery Burn, Durham	8·06	24·3	1·0	tr.				S, 0·18
18	Leaf s. sword	Ballycroghan, Co. Down	9–10	5–10						
19	Palstave	Roundhay, Leeds	13·15	0·24	0·11	—	—	tr.		
18	Leaf s. sword	Ballycroghan, Co. Down	9·3	13						
20	Sword	Irish	5·2	8·4	3·0					
21	Sword	Whittingham, Northumb.	~9·0							
22	'Awl'	Latch Farm, Wilts.	28·0*	4·2		tr.	tr.	1·2		
16	Trumpet	'Dowris', N. Ireland	10·9	9·1						
20	Cauldron	Scotland	7·2	8·5	0·03					
16	Cauldron	Irish	9·46	1·66	tr.					
16	Shield	Ely, Cambs.	11·72					0·4		
23	Anvil	Flax Bourton,	20	little	little	tr.	—	little	—	Zn, tr.
23	Palstave	Som.	13							
8	Socketed axe (Breton type)	Vale of Menteith (?)	1·46	43·3	0·12	0·09	0·06	0·02		S, 0·22
9	Socketed axe	Allendale, Northumb.	9·4	0·68						

* Tin probably increased by selective dissolution

References

1 W. GOWLAND: *Archaeologia*, 1899, **56**, 267–322.
2 W. GOWLAND: *J. Inst. Met.*, 1912, **7**, 23–49.
3 H. H. COGHLAN and H. J. CASE: *Proc. Prehist. Soc.*, 1957, **23**, 91–123.
4 H. J. CASE: *Man*, 1954, **54**, 18–27.
5 G. COFFEY: *J.R. Anthropol. Inst.*, 1901, **31**, 265–79.
6 H. H. COGHLAN: *Man*, 1953, **53**, 97–101.
7 G. COFFEY: 'The Bronze Age in Ireland'; 1913, Dublin.
8 S. S. MITCHELL and M. E. CRICHTON MITCHELL: *Proc. Soc. Antiq. Scotl.*, 1934–6, **69**, 424–30.
9 ANON: *Proc. Soc. Antiq. Newcastle*, 1928, (4th ser.), **3**, 215.

10 E. VOCE: *Man*, 1951, **51**, 6.
11 L. F. CHITTY: *Antiq. J.*, 1925, **5**, 409–14.
12 A. RAISTRICK and J. A. SMYTHE: *Proc. Univ. Durham Philos. Soc.*, 1931–7, **9**, (2), 47–54.
13 Information kindly supplied by D. Britton.
14 W. E. GRIFFITHS: *Proc. Prehist. Soc.*, 1960, **26**, 303–39.
15 C. G. READ: *Proc. Soc. Antiq. Lond.*, (ser. 2), 1895–7, **16**, 328–34.
16 J. EVANS: 'Ancient bronze implements of Great Britain and Ireland'; 1881, London.
17 W. GREENWELL: *Archaeologia*, 1894, **54**, (1), 88–114.
18 E. M. JOPE: *Ulster J. Archaeol.*, 1953, **16**, 37–40.
19 E. K. CLARK: *Proc. Soc. Antiq. Lond.*, 1905, **20**, 258.
20 W. GOWLAND: *Archaeologia*, 1901, **57**, 359–422.
21 J. D. COWEN and H. MARYON: *Archaeol. Aeliana*, 1935, **12**, 280–309.
22 C. M. PIGGOTT: *Proc. Prehist. Soc.*, 1938, **4**, 169–87.
23 R. M. ORGAN: *Proc. Speleol. Soc.*, 1955–6, **7**, 184–6.

NOTE ON ANALYSES OF PREHISTORIC AND MEDIEVAL METALS

Since the earlier version of this work ('Metallurgy in Archaeology') was published there has been an enormous increase in the number of analyses, which has made it impossible to attempt to list the individual analyses beyond those more isolated ones on copper and bronze given in Tables A–C.

The reason for this increase is due mainly to the new techniques, which have made the analysis of small samples a good deal easier, and partly to the desire of many museums to know more about their collections.

One of the first major programmes was by Jolowicz and Blin-Stoyle on Middle and Late Bronze Age bronzes, reported in a supplement to *Archaeometry* in 1959, Part II. This was also summarized by M. A. Brown in *Proc. Prehist. Soc.* (1959, **25**, 188–208). In 1979 M. J. Hughes re-analysed some of the lead values given in this work after discrepancies were noted in a standardization exercise; he found that many of the higher values were seriously in error, owing to the use of a faulty method for the analysis of lead. Some of the problems of high lead analyses were further discussed and interpreted in 1982 by Hughes, Northover and Staniaszek (*Oxford Journal of Archaeology*, **1**, (3), 359–63). M. A. Brown has added some further figures for swords in her paper on 'Swords and sequence in the British Bronze Age' (*Archaeologia*, 1982, **107**, 1–42).

One of the foremost authorities to add to our analytical programme was H. H. Coghlan who published 128 analyses of British and Irish Copper and Bronze Age artifacts with D. Britton and I. M. Allen in a monograph in the Pitt Rivers Museum series of occasional papers (No. 10, 1970, Oxford). This was preceded by 125 analyses of Irish coppers and bronzes in an occasional paper (No. 17) published by the Royal Anthropological Institute in 1963 under the title 'Ores and Metals'. This paper overlapped with one published with H. J. Case (*Proc. Prehist. Soc.*, 1957, **23**, 91–123), which also included a number of British copper analyses. A further group of analyses from the Cork and Dublin Museums deriving, like the previous groups, from a project initiated by the former Ancient Mining and Metallurgy Committee of the Royal Anthropological Institute, was listed in *J. Hist. Metall. Soc.* (1979, **13**, (2), 90–105). Coghlan continued to analyse the artifacts in the Newbury Museum, and the results on 32 axes and palstaves appeared in 1970 (Borough Museum, Newbury). Thirty artifacts from the Yattendon hoard were published, with partial analyses, from the same museum in the same year.

At Oxford, D. Britton published a programme of analyses (*Archaeometry*, 1961, **4**, 40–52) comprising data from 107 artifacts relating to the Wessex culture and Arreton axe industry. This work was extended in *Proc. Prehist. Soc.* (1963, **29**, 258–325) with the publication of additional analyses and a summary of results from other sources making a total of 167.

Another worker in Oxford is J. P. Northover, who has been using the electron microprobe technique. He carried out the analyses for a study of the metalwork of the Welsh Bronze Age organized by the Board of Celtic Studies of the University of Wales. The conclusions from nearly 600 analyses are summarized in an appendix to the Guide Catalogue to the Bronze Age collections in the National Museum of Wales, published by H. N. Savory in 1980. The full publication of these results is currently in press (C. B. Burgess and J. P. Northover, 'Bronze Age Metalwork in Wales', University of Wales Press, 2 vols.). A similar publication of some 400 new and existing Bronze Age analyses from south-west England is also in preparation by Northover and Pearce. Another 280 analyses from LBA Wilburton-type hoards have been published in the *Oxford Journal of Archaeology* (1982, **1**, (1), 69–109). Metallurgical appendixes, with numerous analyses, have been provided for the Prähistorische Bronzefunde Series published by Beck in München on the rapiers of the British Isles and the swords of Great Britain.

The British Museum, like many other museums, has been steadily analysing its collections, and reports by Dr P. T. Craddock and his colleagues have been appearing at intervals although, naturally, a large amount of this material remains unpublished. Much has appeared, mainly in graphic form, and a good deal of it relates to artifacts from abroad. One example is 'The composition of the copper alloys used by the Greek, Etruscan and Roman civilizations', which has been appearing in parts in *J. Archaeol. Sci.* since 1976. Craddock has published over 500 analyses in his paper on 'Deliberate alloying in the Atlantic Bronze Age' in the *Proceedings of the 5th Atlantic Colloquium* (Dublin, 1979, 369–85), and many more in graphic form on microfiche. The Museum's work on the Roxby hoard (18 analyses) was published with R. J. Harrison (*Antiq. J.*, 1979, **59**, (2), 231–44).

Some museums abroad have quite considerable collections of material of British origin. One example is the Royal Ontario Museum of Canada, which has published 184 analyses by P. T. Craddock in its catalogue of British and Irish prehistoric bronzes in 1980 under the authorship of F. Pryor. In 1969, J. V. S. Megaw published 9 analyses of British and Irish axes in the

Nicholson Museum, Sydney (*Proc. Prehist. Soc.*, 1969, **35**, 358–64).

A large series of analyses of Copper and Early Bronze Age material in the National Museum of Antiquities of Scotland was carried out by Junghans and his colleagues (SAM 1968, **2**, (3)). With other analyses by this team for other Scottish museums (SAM 1968, **2**, (3); SAM 1974, **2**, (4)), J. M. Coles has contributed an important paper on Scottish EBA metalwork (*Proc. Soc. Antiq. Scotl.*, 1968–9, **101**, 1–110), presenting over 200 analyses. SAM, **2**, (3) and **2**, (4) contain between them some 1 000 analyses of Irish artifacts from the Copper, Early and Middle Bronze Ages.

Analyses of gold

The largest corpus of analyses on this subject is that assembled by J. Taylor, largely from the work of Hartman (SAM, **3**) in 'Bronze Age goldwork of the British Isles' (1980, CUP). This lists the analyses of over 180 artifacts and 15 native metals. Gold analysis is being continued at the British Museum Research Laboratory, which has provided analyses for a number of hoards (e.g. 'Capel Isaf, Llandeilo', published by H. N. Savory; *Archaeol. Atlantica*, 1977, **2**, 37–53; 'Craig-yr-Wolf, Llanarmon-yn-Ial' by H. S. Green, *Antiq. J.*, forthcoming).

Analyses of Iron Age bronze

There are few recent analyses of Iron Age bronze. A number of artifacts from the hillfort at Danebury have been examined by Dr Northover along with a few sword and dagger scabbards from the River Thames. Most of these will be published in 'Danebury, the first ten years' by B. W. Cunliffe (ed.) (CBA, in press). Some Iron Age mirrors have been analysed at the British Museum Research Laboratory, as has the Iron Age bronze from the Breiddin hillfort.

Roman and medieval metals

Some analyses of Roman pewter and bronzes were carried out by J. A. Smythe and are reported in chapters 2 and 4 of this book. More recent analyses on lead have been carried out by A. Wyttenbach and P. A. Schubiger (*Archaeometry*, 1973, **15**, (2), 199–207). J. Carlson has been concentrating on medieval pewter and has published the results on 15 British artifacts (*Archaeometry*, 1977, **19**, 147–55), and on over 300 artifacts from Britain and America (Winterthur Portfolio No. 12, 1977, 65–85. *See also* A. M. Pollard, *J. Hist. Metall. Soc.*, 1983, **17**, (2), 83–90).

There are approximately 50 analyses of gold, silver and copper alloys from Sutton Hoo and related material in the three Sutton Hoo reports. (Vol. 3, East and Bruce-Mitford (ed.), is now being published by the British Museum.)

There are also the analyses of Hughes and Hall (*J. Archaeol. Sci.*, 1979), predominantly on Romano-British silver.

Dr R. Brownsword and E. E. H. Pitt have been concentrating on post-medieval brasses and bronzes, and 33 analyses have appeared in *Metall. Mater. Sci.* (1981, **13**, (4), 184–5) and 31 in *J. Hist. Metall. Soc.* (1983, **17**, (1), 44–9).

The largest group of analyses of artifacts of the EIA and later is that of coins, and the principal workers in this field have been L. Cope, and W. A. Oddy, who have published their results in the various numismatic journals. A useful source on this subject is the monograph edited by E. T. Hall and D. M. Metcalf entitled 'Methods of chemical and metallurgical investigations of ancient coinage' (*R. Numismatic Soc. Spec. Proc.*, Publ. No. 8, 1972, London).

From the above examples it may be appreciated that an enormous amount of analytical work has been carried out since 'Metallurgy in Archaeology' appeared in 1962. This work is continuing on an increasing scale; the above notes are by no means complete but merely a selection of results now available.

Table D Crucibles

Table ref.	Provenance	Date	Shape	Size mm O. Dia. or side*	Height	Capacity cm^3	Type	Fig. No., Museum Register No., etc.
1	Dainton, Devon	LBA	O + 3 legs	270 × 170	70		J2	Fig. 50, 25
2	Mucking, Essex	LBA	O (Oval) ?	? × 120	50		J1	Sherd; date dubious
3	Breiddin, Powys	LBA	O	75 × 100	33		J1	Fig. 50, 24
4	Rathgall, Wicklow	LBA	O	95 × 80	35		J1	Fig. 50, 23
5	Jarlshof, Shetland	LBA	Δ	150	76	—	—	Fragment, 'Boat-shaped' *Proc. Soc. Antiq. Scotl.*, **68**, 276
6	Dalkey, Eire	LBA	O	63	—	—	—	Sherds only
7	Long Wittenham, Berks.	EIA	Δ	40	38	—	A2	
8	Winterbourne Monkton Down, Wilts.	EIA (?)	Δ	53	50	*c.* 20	A2	Stained with bronze; not complete
9	Hunsbury, Northants.	4th cent. BC –1st cent. AD	Δ	38	25	—	A2	SP8
9	Hunsbury, Northants.	4th cent. BC –1st cent. AD	O	71	58	—	F1	SP3 'Bronze dross inside

Table D –*continued*

Table ref.	Provenance	Date	Shape	Size mm O. Dia. or side*	Height	Capacity cm³	Type	Fig. No., Museum Register No., etc.
NMAS†	Finavon, Tayside	1st cent. BC	O	38	23	3·5	B1	HH 397
10	Meare, Somerset	EIA	△	63	38	*c*. 25	A2	
10	Meare, Somerset	EIA	△	—	66		A2	
10	Meare, Somerset	EIA	△	10	63		A2	
11	Maiden Castle	1st cent. BC	△	11	68	—	A1	Fig. 50, 2
11	Maiden Castle	1st cent. BC	O	68	50	—	B1	'Probably a crucible'
11	Maiden Castle	1st cent. AD	O	230	96		—	Very thick, saucer-shaped dish with flat base; bronze inside
11	Maiden Castle	1st cent. AD	O	96	100		B1	
12	Glastonbury, Som.	EIA	△	100	38		A1	Incomplete; (Fig. 50, 1)
12	Glastonbury, Som.	EIA	O	30	12		B3	} Predominates
12	Glastonbury, Som.	EIA	O	30	33		B3	
13	Cullykhan, Grampian	EIA	Lugged	80	100		D1	
14	Crickley Hill, Glos.	EIA	Lugged	40	40	9·0	D1	
15	Coygan Camp, Dyfed.	3rd cent. AD	Lugged	80	75		D1	
16	Gussage All Saints	EIA	△	65	—	—	A1	Many sherds
17	Mucking, Essex	EIA	△	50	50		A2	
18	Caistor by Norwich	Roman	△				A1	Sherds
19	Stibbes, Leics.	Roman	O	70	—		B1	
20	Gestingthorpe, Essex	Roman	O	95	15		G	
21	Chesterfield, Derbys.	Roman	O	40			B1	One inside the other
22	Verulamium (St. Albans)	Roman	O	47			G	
Corbridge Museum	Corbridge, Northumb.	Roman	O	75	75	175	B1	With lip, 100 across
Corbridge Museum	Corbridge, Northumb.	Roman	O	35	45	22	B1	No lip
Corbridge Museum	Corbridge, Northumb.	Roman	O	33	60	22	G	
Corbridge Museum	Corbridge, Northumb.	Roman					B1A	With side hole
23	Caerwent, Glam.	Roman	O	65	27		B1	
24	Huntsham, Heref.	Roman	O	54	110		C1	
24	Huntsham, Heref.	Roman	O	44	83		C2	
25	Sutton Walls, Heref.	1st cent. BC –3rd cent. AD	D	100	50		A3	Fig. 50, 3; trace of bronze
26	Exeter, Devon	Roman	△	69	43		A1	Fig. 50, 4; incomplete
27	Wroxeter, Salop	Roman	O	38	38		B1	Contains slag and silver
28	Wilderspool, Lancs.	Roman	O	71	46		B1	Fig. 50, 6
29	Wilderspool, Lancs.	Roman	O	60	44		B1	Fig. 50, 27
30	Templebrough, Yorks.	Roman	O	53	35		B1	
30	Templebrough, Yorks.	Roman	O	36	28		B1	Incomplete
31	York (Feasegate)	Roman or med.	O	—	—		—	See Table 57 for contents
32	London (St. Helen's Place)	Roman	O	140	71		B1	
33	Lullingstone, Kent	4th cent.	O	94	84		B3	Conical; incomplete
34	Great Casterton, Leics.	Roman	O	81	64		B3	Conical; (Fig. 50, 5)
35	Heronbridge, Chester	1st–2nd cent. AD	O	46	53		B1	Fig. 50, 7; another similar with lip
35	Heronbridge, Chester	1st–2nd cent. AD	O	25	20		B2	Flat bottom
36	Wakefield (?)	Roman	O	53	79		C1	Narrow neck, (Fig. 50, 8)
36	Castor, Northants	Roman	O	—	—		C1	Narrow neck, (Fig. 50, 8)
NMAS	Gask, Tayside	Roman (?)	O	114	76		B1	HR 1070; thick-walled, red deposit inside
NMAS	Broch, near Mousa, Shetland	1st cent. AD	D	46	23	3·5	A3	GA 501

Table D – ***continued***

Table ref.	Provenance	Date	Shape	Size mm O. Dia. or side*	Size mm Height	Capacity cm³	Type	Fig. No., Museum Register No., etc.
NMAS	Foshigarry, N. Uist	1st–5th cent. AD	O	28	15	0·5	B3	GNA 303
37	Dun Beag Broch, Skye	1st–5th cent. AD	O	36	13	0·5	B3	GT 1071 (NMAS); *Proc. Soc. Antiq. Scotl.* 1920–21, **55**, 110
37	Dun Beag Broch, Skye	1st–5th cent. AD	O	33	—	—	B1	GA 1072; bottom missing (NMAS)
38	Traprain Law	2nd cent. AD	Δ	30	28	7·5	A2	—
38	Traprain Law	2nd cent. AD	Δ	48	38	23·0	A2	FE/2 XII, 15, 192 (bronze inside) (NMAS)
39	Godmanchester, Cambs.	Roman (3rd cent.)	O	135	100		F2	
40	Nettleton, Som.	Roman	O	25	50		E	Lidded
41	Fishbourne, Sussex	Roman	O	45 × 55	40		J1	
42	Silchester, Hants.	Roman	O	105	150		G	Pointed bottom
22	Verulamium (St. Albans)	Roman (90 AD)	O	65	35			With lip
22	Verulamium (St. Albans)	Roman (90 AD)	O	48	33		B3	
43	Bac Mhic Connain, N. Uist.	3rd cent. AD	Δ	53	46	A2		
43	Bac Mhic Connain, N. Uist.	3rd cent. AD	Δ	38	33	10	A2	GNB 89
43	Bac Mhic Connain, N. Uist.	3rd cent. AD	Δ	35	23	10	A2	GNB 88
43	Bac Mhic Connain, N. Uist.	3rd cent. AD	Δ	38	38	19	A2	GNB 87 (Bronze outside)
44	Sutton Courtney, Berks.	5th cent. AD	Bag-shaped	'small'		—	—	
45	Garranes, Co. Cork.	5th–6th cent. AD	Δ	Various		2·5–30	A2	Very similar to Meare
45	Garranes, Co. Cork	5th–6th cent. AD	O	Various		20–100	B4	Globular. (Fig. 50, 13); one clay, 100 cc; rest stone, 20–50 cc
46	Ballinderry Crannog	4th–8th cent. AD	Δ	50	58	20	A2	E6 555; fragments of 45
46	Ballinderry Crannog	4th–8th cent. AD	Lugged	—	—	2·0	D1	E6 557C
NMI‡	Lettermacaward, Donegal	5th cent. AD	O			20	B1	
47	Lough Faughan Crannog	EC§	Δ	81	71		A2	(Fig. 50, 9)
48	Irish	EC (?)	O			43	—	
49	Nendrum, Co. Down	6th–13th cent. AD	D	76	50	*c.* 75	A3	1925, 53, incomplete, red-stained outside (Fig. 50, 10)
49	Nendrum, Co. Down	6th–13th cent. AD	Lugged	23 × 15	35	5·0	D1	1925, 53, probably unused (Fig. 50, 11)
49	Nendrum, Co. Down	6th–13th cent. AD	O	88	50		—	Stone, with red deposit
49	Nendrum, Co. Down	6th–13th cent. AD	Lugged			4·0	D1	1925, 53
49	Aghaloghan, Antrim	EC	0–Δ			4·0	B1 A2	1925–53; spherical at bottom with rim (Fig. 50, 14)
49	Aghaloghan, Antrim	EC	Δ			7·0	A2	1910–698 A
49	Aghaloghan, Antrim	EC	O–Δ			4·0	B1/ A2	1910–698 B
49	Aghaloghan, Antrim	EC	O			4·0	B1	1910–698 C
NMI	Corraneary Crannog, Co. Cavan	EC	Lugged			5·0	D2	Knob on top (Fig. 50, 12)
NMI	Corranèary Crannog, Co. Cavan	EC	O + Δ			*c.* 10	B1 and A2	Some have red slag inside
50	Buston Crannog, Strathclyde	SIA¶	Δ	50	38	11	A2	HV 145 (NMAS)

Table D – *continued*

Table ref.	Provenance	Date	Shape	Size mm O. Dia. or side*	Height	Capacity cm³	Type	Fig. No., Museum Register No., etc.
50	Buston Crannog, Strathclyde	SIA	Δ	61	50	—	A2	HV 144, incomplete, two-layered (NMAS)
51	Mote of Mark, Dumfries & Galloway	EC	Δ				A	With marks of tongs inside
52	Birsay, Orkney	Pre-Norse	Lugged				D1	
NMAS	Cinn Trolla Broch, Highland	SIA	Δ				A1	Sherd, GL 75
NMAS	Ousdale, Broch, Highland	SIA	O	64			B6	With base
NMAS	Sanday, Orkney	SIA	Δ				A1	GA 965
NMAS	Sanday, Orkney	SIA	O				B1	
53	Chalk Lane, Northants.	Saxo-med.	O	60	60	60	B6	Pedestal base (Fig. 50, 26)
54	Thetford, Norfolk	Saxon	Bag				F1	
55	Lincoln (Flaxengate)	9th–10th cent.	Bag ?				—	Sherds only
56	Dunadd, Strathclyde	6th–11th cent. AD	O	40	18	9·0	B2	Flat-bottomed
56	Dunadd, Strathclyde	6th–11th cent. AD	O	40	63	31	B1	With handle (Fig. 50, 19)
56	Dunadd, Strathclyde	6th–11th cent. AD	O	89	81	155	B1	Bronze inside (Fig. 50, 18)
56	Dunadd, Strathclyde	6th–11th cent. AD	Lugged	56	48	20·5	D1	Bronze inside (Fig. 50, 18)
56	Dunadd, Strathclyde	6th–11th cent. AD	Lidded	38	18	6·0	E1	Remains of handle on top and bronze slag outside (HPO/178.272)
56	Dunadd, Strathclyde	6th–11th cent. AD	Lugged	40	48	7·5	D1	
57	Dinas Powys, Glam.	7th cent. AD	Lidded	50	38		E1	(Fig. 50, 15)
NMI	Lagore Crannog, Meath	7th–11th cent. AD	Δ	50	63		A2	E14 509, incomplete
NMI	Lagore Crannog, Meath	7th–11th cent. AD	Δ	68	76		A2	E14 372A, incomplete
NMI	Lagore Crannog, Meath	7th–11th cent. AD	O	38	23		B2	E14 415, (Fig. 50, 17)
NMI	Lagore Crannog, Meath	7th–11th cent. AD	Δ	43	38		A2	E14 396, incomplete
NMI	Lagore Crannog, Meath	7th–11th cent. AD	Δ	96	89		A2	E14 369, incomplete
NMI	Lagore Crannog, Meath	7th–11th cent. AD	Δ	56	46	25	A2	E14 486
NMI	Lagore Crannog, Meath	7th–11th cent. AD	Δ	38	46	15	A2	E14 498
NMI	Lagore Crannog, Meath	7th–11th cent. AD	O	58	83	35	B1	E14 428 } All with remains of integral lid
NMI	Lagore Crannog, Meath	7th–11th cent. AD	O	—	—	47	B1	E14 424 } All with remains of integral lid
NMI	Lagore Crannog, Meath	7th–11th cent. AD	O	50	76	30	B1	E14 421 } All with remains of integral lid
NMI	Lagore Crannog, Meath	7th–11th cent. AD	O	50	76	25	B1	E14 433 } All with remains of integral lid
NMI	Lagore Crannog, Meath	7th–11th cent. AD	O = O	43 × 63	76	30	B1	E14 431 } All with remains of integral lid
NMI	Lagore Crannog, Meath	7th–11th cent. AD	Lugged	—	—	*c.* 5·0	D1	E14 458
NMI	Lagore Crannog, Meath	7th–11th cent. AD	O–(3-legged)	61	68	33	B5	(Fig. 50, 16)
58	Lough Gur; Carraig Aille No. 2	10th cent. AD	O	43	40	9	B1	511; with green stains and knob
58	Lough Gur; Carraig Aille No. 2	10th cent. AD	Pinched	35	30	6	D1	509
58	Lough Gur; Carraig Aille No. 2	10th cent. AD	Pinched	30	25	4	D1	487; with green stains

Table D *–continued*

Table ref.	Provenance	Date	Shape	Size mm O. Dia. or side*	Size mm Height	Capacity cm³	Type	Fig. No., Museum Register No., etc.
59	Oxford (Radcliffe Camera)	Late Saxon	Bag-shaped	82	45		F1	Covered with slag
59	Oxford (Cornmarket)	Late Saxon	Bag-shaped	100	70		F1	Fig. 50, 20
59	Oxford (Angel Inn)	Late Saxon	Bag-shaped	53	30		F1	
NMAS	Jarlshof, Shetland	9th–10th cent. AD	O	35	48		B1	S.A. 699; bronze inside (Fig. 50, 21)
60	Wadsley, Yorks.	Med.	O	40	50		G1	J 1928–14 (Fig. 50, 22)
61	R. Nene, Northampton Castle	Med. (?)	O	144	223		H1	Flat-bottomed with lip; similar to modern crucible shown in Fig. 52
62	Charterhouse, Som.	Med.	O	50	76	*c.* 50	G	Similar to Wadsley
62	Charterhouse, Som.	Med.	O	35	45	*c.* 15	G	Similar to Wadsley
63	Cheddar, Som.	Med.	D	45	45		A3	
63	Cheddar, Som.	Med.	O	40	40			
63	Cheddar, Som.	Med.	O	80	?			
63	Cheddar, Som.	Med.	O	30	40		B1	
64	Winchester, Hants.	12th cent.	O	170	90		D1	Bag-shaped
65	Elgin (Nicholson's Garage)	14th cent.	Lugged	60	50			

Notes
*For circular crucibles this is the maximum external diameter; for triangular crucibles the longest side, measured externally; for D-shaped crucibles, the vertical stroke of the D.

† National Museum of Antiquities of Scotland, Edinburgh

‡ National Museum of Ireland, Dublin

§ EC = Early Christian

¶ SIA = Scottish Iron Age (*c.* 500 BC–AD 600)

References

1 S. NEEDHAM: *Proc. Prehist. Soc.*, 1980, **46**, 177–215.
2 M. U. JONES: in 'Aspects of early metallurgy', 120; 1977, London. The Bronze Age crucibles from this site are not yet published.
3 C. MUSSON: Breiddin, 1970–1; private communication.
4 P. T. CRADDOCK: 'Report on material from Rathgall', (exc. B. Raftery); 1981, Brit. Mus. Res. Lab.
5 J. R. C. HAMILTON: 'Excavations at Jarlshof, Shetland'; 1956, Edinburgh, HMSO.
6 G. D. LIVERSAGE: *Proc. R. Ir. Acad.*, 1968, **66**, (C2), 53–233.
7 H. N. SAVORY: *Oxon.*, 1937, **2**, 3.
8 E. H. GODDARD: *Wilts, Archaeol. Nat. Hist. Mag.*, 1913, 38, 109.
9 C. I. FELL: *Archaeol. J.*, 1936, **93**, 57–100.
10 H. ST. G. GRAY and A. BULLEID: 'Meare Lake Village', Vol. 1; 1948, Taunton.
11 R. E. M. WHEELER: 'Maiden Castle', Res. Rep. No. 12; 1943, London, Soc. Antiq.
12 A. BULLEID and H. ST. G. GRAY: 'Glastonbury Lake Village', (2 Vols.); 1911 and 1917, Glastonbury Antiq. Soc.
13 C. GREIG: *Current Archaeol.*, 1972, **32**, 227–31.
14 P. DIXON and R. SAVAGE: personal communication, 1969.
15 G. J. WAINWRIGHT: personal communication, 1965.
16 M. G. SPRATLING, *et. al.*: in 'Proc. of 16th internat. symposium on archaeometry', (eds. E. A. Slater and J. O. Tate); 1976, Edinburgh.
17 M. U. JONES: in 'Aspects of early metallurgy', 117–20; 1977.
18 R. F. TYLECOTE: *Bull. Hist. Metall. Group*, 1969, **3**, (2), 46–7.
19 E. BLANK: 'Ratae Coritanorum'; 1971, London.
20 J. DRAPER: *East Anglian Archaeol.*, 1985, **25**.
21 T. W. COURTNEY and P. BORNE: 'Chesterfield; the recent archaeological discoveries'; 1975, Chesterfield Archaeol. Res. Committee.
22 A. D. MCWHIRR: personal communication.
23 G. W. DUNNING: personal communication, 1970.
24 The material from Huntsham, Heref., was examined for the Director of excavations, N. P. Bridgewater.
25 K. M. KENYON: *Archaeol. J.*, 1953, **110**, 1–87.
26 A. FOX: 'Roman Exeter. Excavations in the war-damaged areas'; 1952, Manchester University Press.
27 J. P. BUSHE-FOX: 'Excavations at Wroxeter, 1914'; 1916, Res. Committee, Soc. Antiq., London, Rep. No. 4.
28 T. MAY: *Iron Coal Trades Rev.*, 1905, **71**, 427.

29 J. HINCHCLIFFE: now in the Warrington Museum.
30 T. MAY: 'The Roman forts at Templeborough, near Rotherham'; 1922.
31 J. DYER and P. WENHAM: *Yorks. Archaeol. J.*, 1958, **39**, 419–26.
32 R. E. M. WHEELER: 'London in Roman times'; 1930, London Mus. Cat. No. 3.
33 G. W. MEATES *et al*: *Archaeol. Cantiana*, 1950, 63, 1–49. (Fig. 6).
34 P. CORDER (ed.): 'The Roman town and villa at Great Casterton, Rutland'; 1951, University of Nottingham.
35 B. R. HARTLEY: *J. CNWAAHS.*, 1954, **41**, 1–14.
36 'Antiquities of Roman Britain', 78, Fig. 39; 1951, London, Trustees of the Brit. Mus.
37 J. G. CALLENDER: *Proc. Soc. Antiq. Scotl.*, 1920–1, **55**, 110.
38 A. O. CURLE and J. E. CREE: *Proc. Soc. Antiq. Scotl.*, 1915–16, **50**, 64–144.
39 H. J. M. GREEN (Director of Excavations): personal communication.
40 W. J. WEDLAKE: 'The excavation of the shrine of Apollo at Nettleton, Wilts. 1956–1971'; 1982, Soc. Antiq. London, Res. Rep. No. XL.
41 B. W. CUNLIFFE: personal communication.
42 J. BAYLEY: AM Lab. Report No. 3407. Finds in Reading Museum.
43 J. G. CALLENDER: *Proc. Soc. Antiq. Scotl.*, 1921–2, **66**, 42–66.
44 E. T. LEEDS: *Archaeologia*, 1922–3, **73**, 147–92.
45 S. P. O'RIORDAIN: *Proc. R. Ir. Acad.* (C), 1941, **47**, 77–150.
46 H. O'N. HENCKEN: *Proc. R. Ir. Acad.* (C), 1941, **47**, 1–76.
47 A. E. P. COLLINS: *Ulster. J. Archaeol.*, (3rd ser.), 1955, **18**, 45–82.
48 R. J. MOSS: *Proc. R. Ir. Acad.* (C), 1924–7, **37**, 175–93.
49 Now in the museum at Belfast.
50 R. MUNRO: *Archaeol. Hist. Collect. Ayr Wigtown*, 1882, **3**, 19.
51 N. SWINDELLS and L. LAING: in 'Aspects of early metallurgy', 121–34; 1977.
52 C. L. CURLE: *Proc. Soc. Antiq., Scotl.*, 1972–4, **105**, 301–7.
53 J. H. WILLIAMS: 'St. Peters St, Northampton; Excavations 1973–76'; 1979, Northampton Development Corp. Archaeol. Monog. No. 2.
54 J. BAYLEY: in A. Rogerson and C. Dallas, 'Excavations in Thetford, 1948–59 and 1974–7'; East Anglian Archaeol., (forthcoming).
55 J. BAYLEY: *PACT*, 1982, **7**, 487–96.
56 D. CHRISTISON *et al.*: *Proc. Soc. Antiq. Scotl.*, 1905, **39**, 259–322.
57 L. ALCOCK: *Archaeology*, 1964, **17**, (2), 104–11.
58 S. P. O'RIORDAIN: *Proc. R. Ir. Acad.* (C), 1948–50, **52**, 39–111.
59 E. M. JOPE: *Oxon.*, 1952–3, 17–18, 77–111.
60 Sheffield City Museum; by courtesy of the Director, H. R. Singleton.
61 In Northampton Museum.
62 In Taunton Museum; probably connected with lead-working.
63 P. A. RAHTZ: 'The Saxon and medieval palaces at Cheddar; excavations 1960–62', 256; 1979, BAR British Ser. 65.
64 B. W. CUNLIFFE: personal communication.
65 S. CRACKNELL: personal communication, 4 Feb. 1977.

TABLE OF ELEMENTS

Element	Symbol	Specific gravity (density gcm^{-3})	Melting point °C
Aluminium	Al	2·70	659·7
Antimony	Sb	6·62	630·5
Arsenic	As	5·73	
Beryllium	Be	1·8	1278 ± 5
Bismuth	Bi	9·75	271·3
Boron	B	3·33	
Cadmium	Cd	8·65	320·9
Calcium	Ca	1·54	
Carbon	C	3·52	
Cerium	Ce	6·79	640
Chlorine	Cl		
Chromium	Cr	6·73	1890
Cobalt	Co	8·71	1495
Copper	Cu	8·95	1083
Fluorine	F		
Germanium	Ge	5·46	958·5
Gold	Au	19·3	1063
Hydrogen	H		
Indium	In	7·28	156·1
Iron	Fe	7·88	1535
Lead	Pb	11·34	327·4
Lithium	Li	0·53	186
Magnesium	Mg	1·74	651
Manganese	Mn	7·42	1260
Mercury	Hg	13·60	−38·9
Molybdenum	Mo	9·01	2620 ± 10
Nickel	Ni	8·9	1455
Niobium	Nb	8·4	1950
Nitrogen	N		
Oxygen	O		
Phosphorus	P	1·83	44·1
Platinum	Pt	21·37	1773·5
Potassium	K	0·87	62·3
Selenium	Se	4·8	217
Silicon	Si	2·42	1420
Silver	Ag	10·53	960·8
Sodium	Na	0·97	97·5
Sulphur	S	2·1	112·8
Tellurium	Te	6·25	452
Thallium	Tl	11·86	303·5
Tin	Sn	7·29	231·9
Titanium	Ti	4·5	1800
Tungsten	W	19·1	3370
Vanadium	V	5·69	1710
Zinc	Zn	7·16	419·5
Zirconium	Zr	6·44	1900

TECHNICAL GLOSSARY

Adit Strictly a way into a mine but usually used for a drainage channel along which access can be made to the working levels of a mine.

Alluvial deposits Mineral deposits formed in or bordering on streams and valleys owing to the weathering of veins intersecting the valleys. Some of them are at quite considerable depth below quaternary gravels.

Annealing The process of softening a metal hardened by cold-working (e.g. hammering). The lowest temperature at which a metal will soften varies with the degree of cold-working, greater amounts of work tending to reduce it.

Assay To test the quality of an ore or metal; in early times by making use of simple chemical reactions which extracted the metal in visible and weighable form. Today calibrated physical instruments are used such as spectrographs.

Austenite A non-magnetic form of iron normally existing only at high temperature (above *c.* 720°C). Carbon can dissolve in it up to about 1·8% at 1 150°C and diffuse readily.

Bloom or bloomery iron Iron that has been produced in a solid condition directly as a result of the reduction (e.g. smelting) of iron ore. Pure iron melts at 1 535°C, but bloomery iron has usually never been heated above *c.* 1 250°C. The carbon content is variable but usually low. High-carbon bloomery irons have properties similar to modern carbon steels.

Blowing house A building set aside for the smelting of tin ore (black tin) with the aid of a furnace and water-wheel driven bellows. It sometimes also contained a stamp mill for breaking up the ore driven from the same wheel.

Brass An alloy of copper and zinc. It used to be made by melting copper in contact with calamine ($ZnCO_3$) under charcoal in a crucible. But in Europe from about 1750 it was made by melting the two metals together.

Cast iron An impure iron, containing more than 1·9% C and other impurity elements such as silicon and phosphorus, which has been formed from the liquid condition. Very brittle and not malleable either hot or cold. Cast iron exists in two forms, white and grey, which describes the appearance of the surface exposed when it is fractured. In grey iron, most of the carbon exists as graphite and the silicon content usually exceeds 1%. In white iron most of the carbon exists as cementite. Its production is favoured by fast cooling rates and low silicon content. Both irons are brittle – grey iron because of the lack of strength and disposition of the graphite, and white because of the extreme hardness and brittleness of cementite.

Catalan hearth Hearth blown by a 'trompe' for making wrought iron by the direct process. Used in Catalonia, Ariège, and some other regions bordering the Mediterranean; often used loosely for a bellows-blown low hearth from which the bloom is extracted through the top.

Chafery A reheating furnace for semi-finished iron blooms in a forge or finery. Can be fueled with coal or coke as in a smith's fire.

Cementite A compound or carbide of iron with the formula Fe_3C. Very hard and brittle, forming one of the constituents of pearlite. It also appears as a separate constituent in the grain boundaries of wrought iron containing about 0·02% C and in irons containing more than 0·89% C. In the latter case it may produce a Widmanstätten structure. It also appears in white cast iron in the pearlite and as a separate constituent. The carbon in cementite is normally referred to as 'combined carbon' to distinguish it from the form of carbon known as graphite.

Cold-working Metals, when hammered at low temperatures, become hardened and stronger. If the temperature of working is increased, a point is reached at which hardening no longer occurs, i.e. the hot-working temperature is reached. The dividing line between hot and cold-working for lead is about room temperature; for pure iron it is about 600°C.

Core box A wooden mould usually in two or more parts in which were made the sand-clay cores intended to be placed in clay or metal moulds to make the hollow parts of castings – such as the hole in a socketed BA axe or the handle of a medieval cauldron.

Crucible A container for melting metal normally made from a refractory clay, often graphitized. Now mainly of silicon carbide.

Cupellation A process used for extracting silver and gold from lead. The principle involves the oxidation of the lead to litharge (PbO) in a shallow, dish-shaped crucible usually made of bone-ash, leaving the precious metals behind as a molten globule. A temperature of 1 000°C is needed. The litharge volatilizes or is skimmed off.

Dendrites A fern- or leaf-like growth formed by a solid metal or constituent growing from the liquid. Many pure metals and alloys solidify in this way, and some constituents of slags, such as magnetite in fayalite.

Elling hearth A hole in the ground in which wood was burnt for the production of potash. The potash was recovered by dissolving the soluble constituents of the ashes in water, which could then be concentrated by evaporation. The potash may be used as a mordant in dyeing and for the production of soap. This term has also been used to describe ore-roasting hearths; it stems from an Anglo-Saxon word *eilding* meaning firing or fuel.

Feeders Metal that is kept liquid during the time that the cast object takes to solidify so as to offset porosity due to solidification shrinkage. Most metals contract on solidification from 2–7%. Much of this may be filled by a properly located and designed feeder which could be discarded after use and remelted. But such a technique was rarely necessary in the thin castings of the Bronze Age.

Ferrite A magnetic form of iron, almost devoid of carbon but capable of containing various amounts of other elements such as phosphorus.

Fettling Repairing the clay lining of a furnace.

Fire-setting Removing rock or ore from a rock face by lighting a fire against it and quenching the rock with water, so as to make the sudden contraction break the brittle rock.

Free-running temperature Pure metals have sharply

defined melting and freezing temperatures; glasses and slags, on the other hand, do not. They soften gradually as the temperature is increased until their viscosity is low enough to permit them to be worked or poured. The temperature at which this may be said to occur is the free-running temperature.

Gangue The unwanted part of the ore that has been removed or cannot be removed by mineral dressing. Normally removed in smelting by fluxing.

Gassed Some molten metals will dissolve gases, which are released upon solidification and which give rise to porosity in the casting. The term 'gassed' or 'gassy' describes the result. It can sometimes be avoided by suitable pre-treatment such as adding a small amount of cuprous oxide to a hydrogen-containing (over-poled) copper.

Gates The narrow channels through which molten metal enters the casting (sometimes referred to as 'jets'). These are situated between the funnel-shaped runner-bush, which receives the metal from the crucible, and the casting itself. The term may combine the gates and the runner-bush all in a 'running and gating' system.

Graphite In cast irons containing more than *c.* 1% silicon and slowly cooled, most of the carbon occurs as graphite which is a form of carbon possessing a soft, plate-like structure.

Grey iron (*see* cast iron)

Hammer scale The scale removed from iron during forging. This consists of metal which has reacted with air and which has thereby been converted mainly into oxides of iron.

Hardness The hardness of metals is usually measured by indentation tests. The hardness is estimated from the size or depth of an indentation made with a loaded ball or pyramidal diamond. Two popular scales of hardness are the diamond pyramid (HV) and the Brinell (HB). Between 0 and 300 these systems are roughly equivalent. For hardnesses above 300 only the diamond pyramid system has been used in this work.

Induced draught Furnace operated by means of a draught created by the chimney effect of hot air rising, due to the difference in density between the ingoing cold air and the heated air.

Ingot A cast block of metal which may be intended for further working (i.e. into a billet) or merely reused on some future occasion.

Investment The result of wrapping, pressing or casting the moulding material around the pattern. The pattern material must be capable of being converted into a liquid so as to remove it from inside the mould. Suitable pattern materials are waxes, plastics or the low melting-point metals (mercury, tin or lead). The best known is the lost-wax or *cire perdue* process.

Litharge The lead oxide α PbO. White to yellow in colour, but often reddish due to cuprous oxide. Formed as a result of cupellation.

Martensite A hard product produced by quenching iron containing carbon from temperatures above 720°C. The hardness depends on the carbon content. In order to produce a structure containing only martensite, the temperature must be above a certain figure, depending on the carbon content, and the rate of quenching must be extremely high. Suitable rates are obtained by quenching into cold water or brine.

Matte A liquid or solid mixture of sulphides, usually FeS and Cu_2S, but other sulphides may dissolve in the mixture. Usually present at some stage in the production of copper from sulphide minerals.

Neumann lines or bands Markings on ferrite which occur as a result of great shock at low temperatures, i.e. below about 500°C. In terrestrial iron they indicate severe cold-hammering; in meteorites they are probably caused by shock on entering the earth's atmosphere. They disappear when the metal is heated above *c.* 600°C.

Niello A mixture of sulphides, usually copper and silver, used as a black decorative inlay on silver and some other metals.

Patina The surface layer resulting from the reaction between a metal and its environment. May contain oxides, sulphates, chlorides and complex compounds of the underlying metals, and can entrap soil etc. in which the metal is corroding. The reaction may go to completion so that what started as a patina becomes the final or only product, as in some Egyptian cats, and tin ingots.

Pearlite One of the constituents of iron containing carbon in excess of about 0·02%; a mixture of ferrite and cementite, usually laminated in form. It is a product of slow cooling but may appear in conjunction with martensite and troostite (*q.v.*) when the rate of quenching is not high enough to form martensite only.

Pig In the casting of iron from the 16th century onwards, the product was often designed to consist of a large mass (the sow) together with the smaller masses branching off it (the pigs). This term has now become more generalized so that it relates to large lumps of a wide range of metals which, like the ingot, are destined for breaking up and remelting.

Puddling furnace A reverberatory furnace used for converting cast iron into wrought iron, using a stirring or 'puddling' action in the later stages when the iron becomes pasty.

Reverberatory furnace A furnace in which the heat is reflected down onto the charge from the roof. The furnace is of the induced draught type with a tall chimney at one end and a fire box at the other. It needs a long-flame fuel such as coal or wood (*not* charcoal or coke) to transfer the heat towards the near-horizontal roof. Normally the fuel is separated from the charge. Used for remelting cast iron, bronze and bell metal. Often known as an 'air furnace' and latterly used for puddling, steelmaking and non-ferrous smelting.

Risers When metal is poured into a mould air has to be displaced. A riser is a vent which lets out the air and allows the metal to enter the mould. When the metal is seen in the riser, pouring is stopped.

Smelting Involves a chemical reaction between the ore and the fuel, or between a heated sulphide ore and the atmosphere. Most smelting processes are carried out above the melting point of the metal concerned, the main exception being iron.

Sorbite A fine dispersion of cementite in ferrite, formed by tempering martensite between 500 and 700°C.

Steel An alloy of iron and carbon in which the carbon content does not exceed about 1·8%. Some modern steels do not contain carbon in more than trace amounts, but are alloys of iron and some other element(s); others contain iron, carbon and other alloying elements.

Stope, stoping A mine level or drift is driven to give access to a vein. A stope is the cavern cut into the vein to extract the ore. It may be done *overhand* i.e. above or *underhand* below the level of the drift.

Stream tin Alluvial tin ore was often deposited in old river beds which are now filled with immense thicknesses of gravel and detritus. In order to get at this ore it was necessary to dig down through the bed with the aid of shafts not unlike well-shafts. This is what is referred to as tin-streaming. The shafts may be quite shallow and merge into trenches, or reach depths of ten or more metres.

Tempering The operation of softening the hard and brittle constituent, martensite, by heating it for a short time at temperatures between 100 and 650°C.

Troostite If the rate of quenching is not high enough to produce martensite, some troostite may be formed. A structure containing troostite is softer than that containing only martensite. This is now known to be a fine form of pearlite and is often referred to as such. However, it is useful to keep this term for the optically unresolvable, nodular-type growths that emanate from prior austenite grain boundaries.

Troy weight A system of weights usually used for precious metals. A troy ounce is 1·1 times the avoirdupois ounce. There used to be a troy pound which contained 12 troy ounces, but this is now only used in the silver market. A troy ounce is divided into 20 pennyweights (dwt) and the pennyweight into 24 grains. The weight of the grain is the same in both avoirdupois and troy weights. The silver content of lead and copper is often expressed for convenience in grams of precious metal per metric tonne (or parts per million – ppm). This avoids the small numbers involved when the precious metal content is expressed as a percentage. 30 g t^{-1} is equal to 0·003%.

White iron (*see* cast iron)

Widmanstätten Structure The structure occurring in steels which have been fairly rapidly cooled from high temperatures (~1 000°C). Ejection of ferrite or cementite takes place along certain crystal planes forming a mesh-like arrangement. The same type of structure occurs in the octahedrite meteorites; Fig. 6 (Chapter 1) shows a typical example of this structure, though on a much coarser scale than occurs with steels.

Wrought iron Iron made either by the direct (bloomery) process or resulting from a conversion process such as puddling or fining. It is usually quite heterogeneous with low but variable carbon content and some slag. This may be reduced by cutting and welding (piling) many times, thus evening out the carbon content and expelling the slag. It is normally low in Mn, Cu, Ni but may contain appreciable amounts of P.

JOURNALS CONSULTED AND ABBREVIATIONS

Am. Antiq.: American Antiquity, Wisconsin, USA
Am. J. Archaeol.: American Journal of Archaeology, Baltimore, USA
Antiquity: Cambridge, UK
Antiq. J.: Antiquaries Journal, Society of Antiquaries, London
Archaeologia: Society of Antiquaries, London
Archaeology: The Archaeological Institute of America, New York
Archaeol. Aeliana: Archaeologia Aeliana, Society of Antiquaries, Newcastle upon Tyne
Archaeol. Atlantica: Archaeologia Atlantica, Germany
Archaeol. Austriaca: Archaeologica Austriaca, Vienna
Archaeol. Cambrensis: Archaeologia Cambrensis, Cardiff
Archaeol. Cantiana: Archaeologia Cantiana, Kent Archaeological Society, Maidstone
Archaeol. Eisenhüttenwes.: Archiv für das Eisenhüttenwesen, Düsseldorf
Archaeol. J.: Archaeological Journal, Royal Archaeological Institute of Great Britain and Ireland, London
Archaeometry: Oxford
Ars Orient.: Ars Orientalis
BA: Bronze Age
Berks. Archaeol. J.: Berkshire Archaeological Journal, Reading
Birm. Archaeol. Soc.: Birmingham Archaeological Society
Bonner Jahrb.: Bonner Jahrbücher, Köln
Brigantian: Journal of the Huddersfield and District Archaeological Society
Britannia: London
Br. Archaeol. Rep.: British Archaeological Reports, Oxford
Br. Mus. Q.: British Museum Quarterly, London
Br. Numismatic J.: British Numismatic Journal, London
Br. Weld. J.: British Welding Journal, Cambridge
Brycheiniog: Journal of the Brecknock Society
Bull. Board Cel. Stud.: Bulletin of the Board of Celtic Studies, Cardiff
Bull. Hist. Metall. Group: Bulletin of the Historical Metallurgy Group, London
Bull. Inst. Archaeol. Univ. Lond.: Bulletin of the Institute of Archaeology, University of London
Bull. Inst. Min. Metall.: Bulletin of the Institution of Mining and Metallurgy, London
Bull. PDMHS: Bulletin of the Peak District Mines Historical Society, Matlock, Derbyshire
Bull. Wealden Iron Res. Group: Bulletin of the Wealden Iron Research Group, Crawley
Can. Mineral: Canadian Mineral Industry, Ottawa

Carmarthen Antiq.: Carmarthen Antiquary
CBA: Council for British Archaeology
Ceredigion: Journal of the Cardiganshire Antiquaries Society
Chem. Ind.: Chemistry and Industry, Chemical Society, London
CIL: Corpus Inscriptorium Latinarum
Conservation: Studies in Conservation, London
Corn. Archaeol.: Cornish Archaeology, Camborne
Current Archaeol.: Current Archaeology, A. and W. Selkirk, London
Derbys. Archaeol. J.: Journal of the Derbyshire Archaeological and Natural History Society
Discovery: London
Dorset Archaeol. J.: Dorset Archaeological Journal, Dorchester
EA Archaeol.: East Anglian Archaeology, Norwich
EA News: Edgar Allen News, Sheffield
Early Medieval Stud.: Early Medieval Studies, Stockholm
EBA: Early Bronze Age
Econ. Hist. Rev.: Economic History Review, London
EIA: Early Iron Age
Engl. Hist. Rev.: English Historical Review, London
Eng. Min. J.: Engineering and Mining Journal, New York
Essex Archaeol. Hist.: Essex Archaeology and History Society
Flints. Hist. Soc.: Flintshire Historical Society
Foundry Trade J.: Foundry Trade Journal, Reigate
Frühmittelalterliche Stud.: Berlin
Geol. Mag.: Geological Magazine, London
Germania: Germania, Berlin
Glastonbury Antiq. Soc.: Glastonbury Antiquarian Society
Gold. Bull.: Gold Bulletin, Marshalltown, S. Africa
Herts. Archaeol.: Hertfordshire Archaeology, Hertford, UK
IA: Iron Age
Illus. London News: Illustrated London News, London
Ind. Archaeol.: Industrial Archaeology, Newton Abbot, Devon
Ind. Archaeol. Rev.: Industrial Archaeology Review, Oxford
Inst. Geol. Sci.: Institute of Geological Science, London
Int. J. Naut. Archaeol.: International Journal of Nautical Archaeology and Underwater Exploration, London
Inventaria Archaeol. GB: Inventaria Archaeologica, London
Ir. Archaeol. Res. Forum: Irish Archaeological Research Forum, Belfast
Iron and Steel: London
Iron Coal Trades Rev.: Iron and Coal Trades Review, Reigate
J. Archaeol. Sci.: Journal of Archaeological Science, London
J. Br. Archaeol. Assoc.: Journal of the British Archaeological Association, London
J. Chem. Soc.: Journal of the Chemical Society, London
J. CNWAAHS: Journal of the Chester and North Wales Architectural, Archaeological and Historic Society, Chester
J. Cork Hist. Archaeol. Soc.: Journal of the Cork Historical and Archaeological Society, Cork
J. Galway Hist. Archaeol. Soc.: Journal of the Galway Historical and Archaeological Society
J. Hist. Metall. Soc.: Journal of the Historical Metallurgical Society, UK
J. Inst. Met.: Journal of the Institute of Metals, London
J. Iron Steel Inst.: Journal of the Iron and Steel Institute, London
J. Kerry Archaeol. Hist. Soc.: Journal of the Kerry Archaeological and Historical Society, Cork, Eire
J. R. Anthropol. Inst.: Journal of the Royal Anthropological Institute of Great Britain and Ireland, London
J. R. Inst. Corn.: Journal of the Royal Institution of Cornwall, Truro, UK
J. Rom Stud.: Journal of Roman Studies, London
J. R. Soc. Antiq. Irel.: Journal of the Royal Society of Antiquaries of Ireland, Dublin
J. S. Afr. Inst. Min. Metall.: Journal of the South African Institution of Mining and Metallurgy, Johannesburg
J. Soc. Chem. Ind.: Journal of the Society of Chemical Industry, London
Man: Royal Anthropological Institute, London
Masca J.: Masca Journal, Philadelphia, USA
Medieval Archaeol.: Medieval Archaeology, London
Mém. Antiq. Nord: Mémoires des Antiquités du Nord
Mem. Geol. Surv. GB: Memoirs of the Geological Survey of Great Britain
Metallurgia: Reigate
Metall. Mater. Sci.: The Metallurgist and Materials Scientist, London
Metall. Mater. Technol.: The Metallurgist and Materials Technologist, London
Metall. Rev.: Metallurgical Reviews, The Metals Society, London
Met. Bull.: Metal Bulletin, London
Mét. Civilis.: Métaux et Civilisation, Nancy, France
Met. Mater.: Metals and Materials, The Metals Society, London
Met. Prog.: Metals Progress, Cleveland, Ohio
Mikrochim. Acta: Mikrochimica Acta, Vienna
Min. J.: Mining Journal, London
MIT: Massachusetts Institute of Technology
Naturalist: London
Nature: Weekly Journal of Science, London
Nawpa Pacha: Institute of Andean Studies, Berkeley, California
N. Munster Antiq. J.: North Munster Antiquarian Journal, Limerick, Eire
Norf. Archaeol.: Norfolk Archaeology, Norwich
Northants. Archaeol.: Northamptonshire Archaeology, Northampton, UK
Numismatic Chron.: Numismatic Chronicle, London
Numismatic Circ.: Numismatic Circular, London
Oxon.: Oxoniensa, Oxford
Philos. Trans.: Philosophical Transactions of the Royal Society, London.
Post Medieval Archaeol.: Post Medieval Archaeology, London
Proc. Cambs. Antiq. Soc.: Proceedings of the Cambridgeshire Antiquarian Society, Cambridge
Proc. Hants. Field Club: Proceedings of the Hampshire Field Club, Gloucester
Proc. Prehist. Soc.: Proceedings of the Prehistoric Society, Cambridge
Proc. R. Ir. Acad.: Proceedings of the Royal Irish Academy, Dublin
Proc. SANHS: Proceedings of the Somerset Archaeological and Natural History Society, Taunton

Proc. Soc. Antiq. Lond.: Proceedings of the Society of Antiquaries of London

Proc. Soc. Antiq. Newcastle: Proceedings of the Society of Antiquaries of Newcastle upon Tyne

Proc. Soc. Antiq. Scotl.: Proceedings of the Society of Antiquaries of Scotland, Edinburgh

Proc. Som. Archaeol. Soc.: Proceedings of the Somerset Archaeological Society, Taunton

Proc. Speleol. Soc.: Proceedings of the University of Bristol Spelaeological Society

Proc. Thoresby Soc.: Proceedings of the Thoresby Society, Leeds

Proc. Univ. Durham Philos. Soc.: Proceedings of the University of Durham Philosophical Society

R. Archaeol. Inst.: Royal Archaeological Institute, London

Rev. Archéol.: Revue Archéologique, Paris

Rev. Hist. Mines Métall.: Revue d'Histoire des Mines et Métallurgie, Geneva

Rev. Métall.: Revue de Métallurgie, Paris

R. Numismatic Soc. Spec. Proc.: Royal Numismatic Society Special Proceedings, London

SAM: Studien zu den Anfängen der Metallurgie, Berlin

Sci. Archaeol.: Science and Archaeology, Stafford, UK

Science: American Association for the Advancement of Science, Washington DC

Sci. Proc. R. Dublin Soc.: Scientific Proceedings of the Royal Dublin Society

Seaby's Coin Medal Bull.: Seaby's Coin and Medal Bulletin, London

Sibrium: Centro di Studi Preistorice ed Archeologici, Varese, Italy

Soc. Antiq. Lond.: Society of Antiquaries, London

Soc. Medieval Archaeol.: Society for Medieval Archaeology, London

Stahl Eisen: Stahl und Eisen, Verein Deutscher Eisenhüttenleute, Dusseldorf

Studi Etruschi: Florence, Italy

Suffolk Inst. Archaeol.: Suffolk Institute of Archaeology

Surrey Archaeol. Collect.: Surrey Archaeological Collections, Guildford

Sussex Archaeol. Collect.: Sussex Archaeological Collections, Lewes

Sussex Notes Queries: Sussex Notes and Queries, Lewes

Tech. Civilis.: Techniques et Civilisations, Paris

Tech. Cult.: Technology and Culture, USA

Tin Uses: Tin and its Uses, Greenford, Middx., UK

Trab. Soc. Portug. Antropol. Etnol.: Trabalhos de Antropologia e Etnologia Soc. Portugal, Lisbon

Trans. AASFC.: Transactions of the Anglesey Antiquarian Society and Field Club, Bangor

Trans. Am. Philos. Soc.: Transactions of the American Philosophical Society, Philadelphia

Trans. BGAS: Transactions of the Bristol and Gloucestershire Archaeological Society, Bristol

Trans. Br. Ceram. Soc.: Transactions of the British Ceramic Society, London

Trans. BWAS: Transactions of the Birmingham and Warwickshire Archaeological Society

Trans. Cardiff Nat. Soc.: Transactions of the Cardiff Naturalists' Society, Cardiff

Trans. CWAAS.: Transactions of the Cumberland and Westmorland Antiquarian and Archaeological Society, Carlisle

Trans. Devon. Assoc.: Transactions of the Devon Association for the Advancement of Science, Literature and Art, Exeter

Trans. DGNHAS: Transactions of the Dumfriesshire and Galloway Natural History and Antiquarian Society, Dumfries

Trans. Fed. Inst. Min. Eng.: Transactions of the Federated Institution of Mining Engineers, Newcastle upon Tyne

Trans. Geol. Soc. Glasgow: Transactions of the Geological Society, Glasgow

Trans. HSLC: Transactions of the Historical Society for Lancashire and Cheshire, Liverpool

Trans. Inst. Min. Eng.: Transactions of the Institution of Mining Engineers, London

Trans. Inst. Min. Metall.: Transactions of the Institution of Mining and Metallurgy, London

Trans. Inst. Weld.: Transactions of the Institute of Welding, London

Trans. LCAS: Transactions of the Lancashire and Cheshire Archaeological Society

Trans. LSSAS: Transactions of the Lichfield and South Staffs. Archaeological Society, Lichfield

Trans. Newcomen Soc.: Transactions of the Newcomen Society, London

Trans. NHSN: Transactions of the Natural History Society of Northumberland, Newcastle

Trans. Proc. Birm. Archaeol. Soc.: Transactions and Proceedings of the Birmingham Archaeological Society

Trans. R. Geol. Soc. Corn.: Transactions of the Royal Geological Society of Cornwall, Camborne

Trans. R. Hist. Soc.: Transactions of the Royal Historical Society, London

Trans. SANHS: Transactions of the Shropshire Archaeological and Natural History Society, Shrewsbury

Trans. SSAHS: Transactions of the South Staffs. Archaeological and Historical Society

Trans. WNFC: Transactions of the Woolhope Naturalists' Field Club, Hereford

Ulster J. Archaeol.: Ulster Journal of Archaeology

VCH: Victoria County History, London

Weld. Met. Fabr.: Welding and Metal Fabrication, London

Wilts. Archaeol. Nat. Hist. Mag.: Wiltshire Archaeological Society and Natural History Magazine, Devizes

Yorks. Archaeol. J.: Yorkshire Archaeological Journal

PRINCIPAL WORKS CONSULTED

1 H. H. COGHLAN: 'Notes on the prehistoric metallurgy of copper and bronze in the Old World', 1951, (revised 1975), Oxford, Pitt Rivers Museum.

2 H. H. COGHLAN: 'Notes on the prehistoric and early iron in the Old World', 1956, (revised 1977), Oxford, Pitt Rivers Museum.

3 H. R. SCHUBERT: 'History of the British iron and steel industry from *c.* 450 BC to AD 1775', 1957, London, Routledge and Kegan Paul.

4 E. STRAKER: 'Wealden iron', 1931, London; (facsimile reprint, 1969, Newton Abbot).

5 J. EVANS: 'The ancient bronze implements, weapons and ornaments of Great Britain and Ireland', 1881, London.

6 G. AGRICOLA: 'De re metallica', (trans. H. C. and

L. H. Hoover), 1950, London (1st edn. Mining Magazine, 1912, London).

7 V. BIRINGUCCIO: 'Pirotechnia', (trans. C. S. Smith and M. T. Gnudi), 1959, New York, Basic Books.

8 L. AITCHISON: 'A history of metals', 2 Vols., 1960, London.

9 J. PERCY: 'Metallurgy; fuel; fireclays; copper; zinc; brass etc.', 1861, London, Murray.

10 J. PERCY: 'Metallurgy; iron and steel', 1864, London, Murray.

11 J. PERCY: 'Metallurgy; lead', 1870, London, Murray.

12 J. PERCY: 'Metallurgy; refractory materials and fuel', 1875, London, Murray.

13 R. J. FORBES: 'Metallurgy in Antiquity', 1950, Leiden; 'Studies in ancient technology', vols. 7–9, 1964, Leiden, Brill.

14 B. NEUMANN: 'Die ältesten Verfahren der Erzeugung technischen Eisens', Freiberger Forschungshefte, 'Kultur und Technic', D.6., 1954, Berlin.

15 L. BECK: 'Geschichte des Eisens', 5 vols., 1884–1903, Braunschweig.

16 O. DAVIES: 'Roman mines in Europe', 1935, Oxford.

17 J. W. GOUGH: 'The mines of Mendip', 1930, Oxford; (reprinted, 1967, Newton Abbot).

18 J. D. KENDALL: 'The iron ores of Great Britain and Ireland', 1893, London.

19 A. FELL: 'The early iron industry of Furness and district', 1908, Ulverston.

20 C. R. DODWELL (trans. and ed.): 'Theophilus; the various arts', 1961, London, Nelson.

21 J. G. HAWTHORNE and C. S. SMITH (trans. and ed.): 'On Divers Arts – the treatise of Theophilus', 1963, Chicago.

Indexes

SUBJECT INDEX

TOPOGRAPHICAL INDEX